Methods in Enzymology

Volume 169

PLATELETS: RECEPTORS, ADHESION, SECRETION

PART A

METHODS IN ENZYMOLOGY

Methods in Enzymology

Volume 169

Platelets: Receptors, Adhesion, Secretion

Part A

EDITED BY

Jacek Hawiger

DEPARTMENT OF MEDICINE
HARVARD MEDICAL SCHOOL
NEW ENGLAND DEACONESS HOSPITAL
BOSTON, MASSACHUSETTS

ACADEMIC PRESS, INC.
Harcourt Brace Jovanovich, Publishers
San Diego New York Berkeley Boston
London Sydney Tokyo Toronto

ACADEMIC PRESS, INC.
San Diego, California 92101

United Kingdom Edition published by
ACADEMIC PRESS LIMITED
24-28 Oval Road, London NW1 7DX

LIBRARY OF CONGRESS CATALOG CARD NUMBER: 54-9110

ISBN 0-12-182070-X (alk. paper)

PRINTED IN THE UNITED STATES OF AMERICA
89 90 91 92 9 8 7 6 5 4 3 2 1

Table of Contents

Section I. Isolation of Platelets

Section II. Platelet Adhesion, Aggregation, and Turnover

Section III. Platelet Secretion

A. Dense Granule Constituents

B. α-Granule Constituents

C. Lysosomal Enzymes

Section IV. Regulation of Platelet Function

Contributors to Volume 169

Article numbers are in parentheses following the names of contributors.
Affiliations listed are current.

HARRY N. ANTONIADES (18), *Department of Nutrition, Harvard School of Public Health, Boston, Massachusetts 02115*

JAVIER BATLLE (21), *Servicio de Hematología y Hemoterapia, Hospital Juan Canalejo, 15006 La Coruña, Spain*

HANS R. BAUMGARTNER (5), *F. Hoffmann-La Roche & Co., Ltd., CH-4002 Basel, Switzerland*

ELIZABETH BELMONTE (32), *Departments of Medicine and Pathology, University of Pennsylvania, Philadelphia, Pennsylvania 19104*

CINDY L. BERMAN (27), *Division of Hematology/Oncology, New England Medical Center, Tufts University, Boston, Massachusetts 02111*

DAVID BESSMAN (13), *Department of Internal Medicine, University of Texas Medical Branch at Galveston, Galveston, Texas 77550*

LAWRENCE F. BRASS (31, 32), *Departments of Medicine and Pathology, University of Pennsylvania, Philadelphia, Pennsylvania 19104*

K. M. BRINKHOUS (12), *Department of Pathology, University of North Carolina, Chapel Hill, North Carolina 27599*

M. JOHAN BROEKMAN (34), *Division of Hematology/Oncology, Department of Medicine, New York Veterans Administration Medical Center, and Cornell University Medical College, New York, New York 10010*

JAMES L. CATALFAMO (4), *Wadsworth Center for Laboratories and Research, New York State Department of Health, Albany, New York 12201*

DESIRE COLLEN (25), *Department of Medical Research, Katholieke Universiteit Leuven, Herestraat, B-3000, Leuven, Belgium*

ROBERT W. COLMAN (24), *Thrombosis Research Center, Temple University School of Medicine, Philadelphia, Pennsylvania 19140*

STUART L. COOPER (8), *Department of Chemical Engineering, University of Wisconsin, Madison, Wisconsin 53706*

CAROL A. DANGELMAIER (16, 17, 29), *Department of Biochemistry, University of Bergen, Bergen, Norway*

THOMAS F. DEUEL (20), *Department of Biological Chemistry, Jewish Hospital at Washington University Medical Center, St. Louis, Missouri 63110*

W. JEAN DODDS (4), *Wadsworth Center for Laboratories and Research, New York State Department of Health, Albany, New York 12201*

AMIRAM ELDOR (7), *Department of Hematology, Hadassah University Hospital, Jerusalem, Israel 91120*

MONY M. FROJMOVIC (11), *Department of Physiology, McGill University, Montreal, Quebec, H3G 1Y6 Canada*

GLENNA L. FRY (6), *Department of Medicine, University of Iowa, Iowa City, Iowa 52242*

ZVI FUKS (7), *Department of Radiation and Clinical Oncology, Hadassah University Hospital, Jerusalem, Israel 91120*

BARBARA C. FURIE (27), *Division of Hematology/Oncology, New England Medical Center, Tufts University, Boston, Massachusetts 02111*

BRUCE FURIE (27), *Division of Hematology/Oncology, New England Medical Center, Tufts University, Boston, Massachusetts 02111*

ADRIAN L. GEAR (11), *Department of Biochemistry, University of Virginia, Charlottesville, Virginia 22908*

HANS J. GEUZE (26), *Department of Hematology, University Hospital, Utrecht, The Netherlands*

Gail L. Griffin (20), *Department of Medicine, Jewish Hospital at Washington University Medical Center, St. Louis, Missouri 63178*

PETER C. HARPEL (23), *Department of Medicine, Division of Hematology/Oncology, New York Hospital, Cornell Medical Center, New York, New York 10021*

RICHARD J. HASLAM (37), *Departments of Pathology and Biochemistry, McMaster University, Hamilton, Ontario, L8N 3Z5 Canada*

JACEK HAWIGER (2, 15), *Division of Experimental Medicine, Harvard Medical School, New England Deaconess Hospital, Boston, Massachusetts 02215*

W. ANDREW HEATON (14), *American Red Cross Blood Services, Tidewater Region, Norfolk, Virginia 23501*

ANTHON DU P. HEYNS (14), *Blood Platelet Research Unit, Department of Haematology, University of the Orange Free State, Bloemfontein 9300, South Africa*

JOHN C. HOAK (6), *Department of Medicine, University of Iowa, Iowa City, Iowa 52242*

HOLM HOLMSEN (16, 17, 29), *Department of Biochemistry, University of Bergen, Bergen, Norway*

JOHN C. HOLT (19), *Rorer Technology, Inc., King of Prussia, Pennsylvania 19406*

JUNG SAN HUANG (20), *Department of Biochemistry, St. Louis University School of Medicine, St. Louis, Missouri 63178*

KIYOSHI ITOH (35), *Kyoto Research Laboratories, Kaken Pharmaceutical Company, Ltd., Kobe University School of Medicine, Kobe 650, Japan*

PETER C. JOHNSON (33), *Department of Plastic Surgery, University of Pittsburgh School of Medicine, Pittsburgh, Pennsylvania 15261*

J. HEINRICH JOIST (14), *Division of Hematology, St. Louis University Medical Center, St. Louis, Missouri 63104*

NORIO KAJIKAWA (35), *Kyoto Research Laboratories, Kaken Pharmaceutical Company, Ltd., Kobe University School of Medicine, Kobe 650, Japan*

JUN-ICHI KAMBAYASHI (36), *The Second Department of Surgery, Osaka University Medical School, Osaka 553, Japan*

USHIO KIKKAWA (35), *Department of Biochemistry, Kobe University School of Medicine, Kobe 650, Japan*

AKIRO KIMURA (20), *Department of Internal Medicine, Research Institute for Nuclear Medicine, Hiroshima University, Hiroshima, Japan*

RAELENE L. KINLOUGH-RATHBONE (1), *Department of Pathology, McMaster University, Hamilton, Ontario, L8N 3Z5 Canada*

LESLIE KUSHNER (9), Department of Surgery, Beth Israel Hospital, Boston, Massachusetts 02215

LAWRENCE L. K. LEUNG (23), *Department of Medicine, Division of Hematology, Stanford University School of Medicine, Stanford, California 94305*

RICHARD F. LEVINE (7), *Veterans Administration Medical Center, Washington, D.C. 20422*

JACK N. LINDON (9), *Department of Surgery, Beth Israel Hospital, Boston, Massachusetts 02215*

MARIA FERNANDA LOPEZ-FERNANDEZ (21), *Servicio de Hematologia, Hospital Juan Canalejo, 15006 La Coruña, Spain*

LINDSEY A. MILES (25), *Department of Immunology, Research Institute of Scripps Clinic, La Jolla, California 92037*

JOHN G. MILTON (11), *Department of Physiology, McGill University, Montreal, Quebec, H3G 1Y6 Canada*

DEANE F. MOSHER (8), *Department of Medicine, University of Wisconsin, Madison, Wisconsin 53706*

RETO MUGGLI (5), *F. Hoffmann-La Roche & Co., Ltd., CH-4002 Basel, Switzerland*

J. FRASER MUSTARD (1), *Department of Pathology, McMaster University, Hamilton, Ontario L8N 3Z5, Canada*

RALPH L. NACHMAN (23), *Department of Medicine, Division of Hematology/Oncology, New York Hospital, Cornell Medical Center, New York, New York 10021*

STEFAN NIEWIAROWSKI (3, 19), *Thrombosis Research Center, Temple University, Philadelphia, Pennsylvania 19140*

YASUTOMI NISHIZUKA (35), *Department of Biochemistry, Kobe University School of Medicine, Kobe 650, Japan*

MARIAN A. PACKHAM (1), *Department of Biochemistry, University of Toronto, Toronto, Ontario, Canada*

PANAYOTIS PANTAZIS (18), *Harvard School of Public Health, Boston, Massachusetts 02115*

KINAM PARK (8), *Department of Chemical Engineering, University of Wisconsin, Madison, Wisconsin 53706*

EDWARD F. PLOW (25), *Department of Immunology, Research Institute of Scripps Clinic, La Jolla, California 92037*

MARJORIE S. READ (12), *Department of Pathology, University of North Carolina, Chapel Hill, North Carolina 27599*

ROBERT D. ROSENBERG (30), *Massachusetts Institute of Technology, Cambridge, Massachusetts 02139*

BOGUSLAW RUCINSKI (3), *Thrombosis Research Center, Temple University, Philadelphia, Pennsylvania 19140*

ZAVERIO M. RUGGERI (21), *Department of Basic and Clinical Research, Division of Experimental Hemastasis, Scripps Clinic and Research Foundation, La Jolla, California 92037*

Kjell S. Sakariassen (5), *F. Hoffmann-La Roche & Co., Ltd., CH-4002 Basel, Switzerland*

MASATO SAKON (36), *The Second Department of Surgery, Osaka University Medical School, Osaka 553, Japan*

EDWIN W. SALZMAN (9, 33), *Department of Surgery, Beth Israel Hospital and, Harvard Medical School, Boston, Massachusetts 02215*

ALVIN H. SCHMAIER (24), *Department of Medicine, Thrombosis Research Center, Temple University School of Medicine, Philadelphia, Pennsylvania 19140*

ROBERT M. SENIOR (20), *Department of Medicine, Jewish Hospital at Washington University Medical Center, St. Louis, Missouri 63178*

SANFORD J. SHATTIL (31), *Department of Medicine, Hematology/Oncology Section, Hospital of the University of Pennsylvania, Philadelphia, Pennsylvania 19104*

JAN J. SIXMA (26), *Department of Hematology, University Hospital, Utrecht, The Netherlands*

HENRY S. SLAYTER (22, 28), *Laboratory of Structural Molecular Biology, Dana-Farber Cancer Institute, and Department of Physiology and Biophysics, Harvard Medical School, Boston, Massachusetts 02115*

JAN-WILLEM SLOT (26), *Department of Hematology, University Hospital, Utrecht, The Netherlands*

SHEILA TIMMONS (2), *Division of Experimental Medicine, Department of Medicine, New England Deaconess Hospital and Harvard Medical School, Boston, Massachusetts 02215*

MARIANNE VANDERWEL (37), *Medicorp, Inc., Montreal, Quebec, H4P 2R2 Canada*

ISRAEL VLODAVSKY (7), *Department of Radiation and Clinical Oncology, Hadassah University Hospital, Jerusalem, Israel 91120*

J. ANTHONY WARE (33), *Department of Medicine, Beth Israel Hospital and, Harvard Medical School, Boston, Massachusetts 02215*

ERIK L. YEO (27), *Division of Hematology/Oncology, New England Medical Center, Tufts University, Boston, Massachusetts 02111*

THEODORE S. ZIMMERMAN (21), *Department of Basic and Clinical Research, Division of Experimental Hemastasis, Scripps Clinic and Research Foundation, La Jolla, California 92037*

MARJORIE B. ZUCKER (10), *Department of Pathology, New York University Medical Center, New York, New York 10016*

Preface

Blood platelets are tiny corpuscles continuously surveying the inner lining of blood vessels, the vascular endothelium. Any break in the continuity of the vessel wall, leading to hemorrhage, or a break in the atherosclerotic plaque is met with an instant response from the platelets, which contact the zone of injury, spread, and aggregate, forming thrombi that seal off the break. The membrane is transformed into batteries of receptors for adenine nucleotides, catecholamines, adhesive molecules, and coagulation proteins. The platelet interior receives a flux of calcium, flexes its contractile apparatus, and becomes a furnace for enzymatic oxidation of arachidonic acid to endoperoxides, which are then transformed to thromboxane A_2, a potent vasoconstrictor and platelet agonist. Platelet storage granules secrete their constituent molecules, adenosine diphosphate, serotonin, adhesive proteins, and coagulation factor V, targeted toward their own receptors and receptors on other platelets. Coagulation enzymes, assembled on the platelet membrane as the prothrombinase complex, generate thrombin. The platelet, hardly a cell, has gained recognition in human cell biology as a cellular element endowed with a membrane that bears the highest density of receptors per surface unit area among the known blood cells. Mitogenic growth factors, stored in and secreted from platelet α granules, stimulate migration and proliferation of vascular smooth muscle cells and fibroblasts producing the extracellular matrix.

An increasing number of biochemists, cell biologists, hematologists, pharmacologists, and pathologists are working with blood platelets as a useful system to study the processes of adhesion, secretion, and their regulation.

This volume (Part A) is devoted to the methods of platelet isolation and to the study of their adhesive and secretory functions. A later volume (Part B) will include methods for the isolation of subcellular organelles and their characterization. A wide spectrum of methods for the analysis of platelet receptors will constitute the major thrust of Part B.

The idea for these volumes, comprising modern methods of platelet analysis, germinated when the late Editor-in-Chief, Sidney P. Colowick, encouraged me to embark on this task. Dr. Colowick's untimely death left us not only without one of the founding fathers of this monumental series but also without a most inspiring friend and advisor. His legacy cannot be measured in words. He was an example of exquisite intellectual elegance in every aspect of his scientific activities. His gentle discourse and unassuming demeanor were matched by his all-encompassing mind and heart. These two volumes on platelets are dedicated to him.

It has been a genuine pleasure for me to interact with the staff of Academic Press. They have been helpful and patient in dealing with the difficulties encountered in the preparation of these volumes. In my unit at the New England Deaconess Hospital, Ms. Sheila Timmons and Ms. Marie Bingyou were enormously resourceful in helping me organize and edit these volumes. Finally, I offer my deepest appreciation to my colleagues, the contributing authors, for their wisdom, expertise, and scholarship.

JACEK HAWIGER

METHODS IN ENZYMOLOGY

VOLUME I. Preparation and Assay of Enzymes
Edited by SIDNEY P. COLOWICK AND NATHAN O. KAPLAN

VOLUME II. Preparation and Assay of Enzymes
Edited by SIDNEY P. COLOWICK AND NATHAN O. KAPLAN

VOLUME III. Preparation and Assay of Substrates
Edited by SIDNEY P. COLOWICK AND NATHAN O. KAPLAN

VOLUME IV. Special Techniques for the Enzymologist
Edited by SIDNEY P. COLOWICK AND NATHAN O. KAPLAN

VOLUME V. Preparation and Assay of Enzymes
Edited by SIDNEY P. COLOWICK AND NATHAN O. KAPLAN

VOLUME VI. Preparation and Assay of Enzymes *(Continued)*
Preparation and Assay of Substrates
Special Techniques
Edited by SIDNEY P. COLOWICK AND NATHAN O. KAPLAN

VOLUME VII. Cumulative Subject Index
Edited by SIDNEY P. COLOWICK AND NATHAN O. KAPLAN

VOLUME VIII. Complex Carbohydrates
Edited by ELIZABETH F. NEUFELD AND VICTOR GINSBURG

VOLUME IX. Carbohydrate Metabolism
Edited by WILLIS A. WOOD

VOLUME X. Oxidation and Phosphorylation
Edited by RONALD W. ESTABROOK AND MAYNARD E. PULLMAN

VOLUME XI. Enzyme Structure
Edited by C. H. W. HIRS

VOLUME XII. Nucleic Acids (Parts A and B)
Edited by LAWRENCE GROSSMAN AND KIVIE MOLDAVE

Volume XIII. Citric Acid Cycle
Edited by J. M. Lowenstein

Volume XIV. Lipids
Edited by J. M. Lowenstein

Volume XV. Steroids and Terpenoids
Edited by Raymond B. Clayton

Volume XVI. Fast Reactions
Edited by Kenneth Kustin

Volume XVII. Metabolism of Amino Acids and Amines (Parts A and B)
Edited by Herbert Tabor and Celia White Tabor

Volume XVIII. Vitamins and Coenzymes (Parts A, B, and C)
Edited by Donald B. McCormick and Lemuel D. Wright

Volume XIX. Proteolytic Enzymes
Edited by Gertrude E. Perlmann and Laszlo Lorand

Volume XX. Nucleic Acids and Protein Synthesis (Part C)
Edited by Kivie Moldave and Lawrence Grossman

Volume XXI. Nucleic Acids (Part D)
Edited by Lawrence Grossman and Kivie Moldave

Volume XXII. Enzyme Purification and Related Techniques
Edited by William B. Jakoby

Volume XXIII. Photosynthesis (Part A)
Edited by Anthony San Pietro

Volume XXIV. Photosynthesis and Nitrogen Fixation (Part B)
Edited by Anthony San Pietro

Volume XXV. Enzyme Structure (Part B)
Edited by C. H. W. Hirs and Serge N. Timasheff

Volume XXVI. Enzyme Structure (Part C)
Edited by C. H. W. Hirs and Serge N. Timasheff

VOLUME XXVII. Enzyme Structure (Part D)
Edited by C. H. W. HIRS AND SERGE N. TIMASHEFF

VOLUME XXVIII. Complex Carbohydrates (Part B)
Edited by VICTOR GINSBURG

VOLUME XXIX. Nucleic Acids and Protein Synthesis (Part E)
Edited by LAWRENCE GROSSMAN AND KIVIE MOLDAVE

VOLUME XXX. Nucleic Acids and Protein Synthesis (Part F)
Edited by KIVIE MOLDAVE AND LAWRENCE GROSSMAN

VOLUME XXXI. Biomembranes (Part A)
Edited by SIDNEY FLEISCHER AND LESTER PACKER

VOLUME XXXII. Biomembranes (Part B)
Edited by SIDNEY FLEISCHER AND LESTER PACKER

VOLUME XXXIII. Cumulative Subject Index Volumes I–XXX
Edited by MARTHA G. DENNIS AND EDWARD A. DENNIS

VOLUME XXXIV. Affinity Techniques (Enzyme Purification: Part B)
Edited by WILLIAM B. JAKOBY AND MEIR WILCHEK

VOLUME XXXV. Lipids (Part B)
Edited by JOHN M. LOWENSTEIN

VOLUME XXXVI. Hormone Action (Part A: Steroid Hormones)
Edited by BERT W. O'MALLEY AND JOEL G. HARDMAN

VOLUME XXXVII. Hormone Action (Part B: Peptide Hormones)
Edited by BERT W. O'MALLEY AND JOEL G. HARDMAN

VOLUME XXXVIII. Hormone Action (Part C: Cyclic Nucleotides)
Edited by JOEL G. HARDMAN AND BERT W. O'MALLEY

VOLUME XXXIX. Hormone Action (Part D: Isolated Cells, Tissues, and Organ Systems)
Edited by JOEL G. HARDMAN AND BERT W. O'MALLEY

VOLUME XL. Hormone Action (Part E: Nuclear Structure and Function)
Edited by BERT W. O'MALLEY AND JOEL G. HARDMAN

Volume XLI. Carbohydrate Metabolism (Part B)
Edited by W. A. Wood

Volume XLII. Carbohydrate Metabolism (Part C)
Edited by W. A. Wood

Volume XLIII. Antibiotics
Edited by John H. Hash

Volume XLIV. Immobilized Enzymes
Edited by Klaus Mosbach

Volume XLV. Proteolytic Enzymes (Part B)
Edited by Laszlo Lorand

Volume XLVI. Affinity Labeling
Edited by William B. Jakoby and Meir Wilchek

Volume XLVII. Enzyme Structure (Part E)
Edited by C. H. W. Hirs and Serge N. Timasheff

Volume XLVIII. Enzyme Structure (Part F)
Edited by C. H. W. Hirs and Serge N. Timasheff

Volume XLIX. Enzyme Structure (Part G)
Edited by C. H. W. Hirs and Serge N. Timasheff

Volume L. Complex Carbohydrates (Part C)
Edited by Victor Ginsburg

Volume LI. Purine and Pyrimidine Nucleotide Metabolism
Edited by Patricia A. Hoffee and Mary Ellen Jones

Volume LII. Biomembranes (Part C: Biological Oxidations)
Edited by Sidney Fleischer and Lester Packer

Volume LIII. Biomembranes (Part D: Biological Oxidations)
Edited by Sidney Fleischer and Lester Packer

Volume LIV. Biomembranes (Part E: Biological Oxidations)
Edited by Sidney Fleischer and Lester Packer

VOLUME LV. Biomembranes (Part F: Bioenergetics)
Edited by SIDNEY FLEISCHER AND LESTER PACKER

VOLUME LVI. Biomembranes (Part G: Bioenergetics)
Edited by SIDNEY FLEISCHER AND LESTER PACKER

VOLUME LVII. Bioluminescence and Chemiluminescence
Edited by MARLENE A. DELUCA

VOLUME LVIII. Cell Culture
Edited by WILLIAM B. JAKOBY AND IRA PASTAN

VOLUME LIX. Nucleic Acids and Protein Synthesis (Part G)
Edited by KIVIE MOLDAVE AND LAWRENCE GROSSMAN

VOLUME LX. Nucleic Acids and Protein Synthesis (Part H)
Edited by KIVIE MOLDAVE AND LAWRENCE GROSSMAN

VOLUME 61. Enzyme Structure (Part H)
Edited by C. H. W. HIRS AND SERGE N. TIMASHEFF

VOLUME 62. Vitamins and Coenzymes (Part D)
Edited by DONALD B. MCCORMICK AND LEMUEL D. WRIGHT

VOLUME 63. Enzyme Kinetics and Mechanism (Part A: Initial Rate and Inhibitor Methods)
Edited by DANIEL L. PURICH

VOLUME 64. Enzyme Kinetics and Mechanism (Part B: Isotopic Probes and Complex Enzyme Systems)
Edited by DANIEL L. PURICH

VOLUME 65. Nucleic Acids (Part I)
Edited by LAWRENCE GROSSMAN AND KIVIE MOLDAVE

VOLUME 66. Vitamins and Coenzymes (Part E)
Edited by DONALD B. MCCORMICK AND LEMUEL D. WRIGHT

VOLUME 67. Vitamins and Coenzymes (Part F)
Edited by DONALD B. MCCORMICK AND LEMUEL D. WRIGHT

VOLUME 81. Biomembranes (Part H: Visual Pigments and Purple Membranes, I)
Edited by LESTER PACKER

VOLUME 82. Structural and Contractile Proteins (Part A: Extracellular Matrix)
Edited by LEON W. CUNNINGHAM AND DIXIE W. FREDERIKSEN

VOLUME 83. Complex Carbohydrates (Part D)
Edited by VICTOR GINSBURG

VOLUME 84. Immunochemical Techniques (Part D: Selected Immunoassays)
Edited by JOHN J. LANGONE AND HELEN VAN VUNAKIS

VOLUME 85. Structural and Contractile Proteins (Part B: The Contractile Apparatus and the Cytoskeleton)
Edited by DIXIE W. FREDERIKSEN AND LEON W. CUNNINGHAM

VOLUME 86. Prostaglandins and Arachidonate Metabolites
Edited by WILLIAM E. M. LANDS AND WILLIAM L. SMITH

VOLUME 87. Enzyme Kinetics and Mechanism (Part C: Intermediates, Stereochemistry, and Rate Studies)
Edited by DANIEL L. PURICH

VOLUME 88. Biomembranes (Part I: Visual Pigments and Purple Membranes, II)
Edited by LESTER PACKER

VOLUME 89. Carbohydrate Metabolism (Part D)
Edited by WILLIS A. WOOD

VOLUME 90. Carbohydrate Metabolism (Part E)
Edited by WILLIS A. WOOD

VOLUME 91. Enzyme Structure (Part I)
Edited by C. H. W. HIRS AND SERGE N. TIMASHEFF

VOLUME 92. Immunochemical Techniques (Part E: Monoclonal Antibodies and General Immunoassay Methods)
Edited by JOHN J. LANGONE AND HELEN VAN VUNAKIS

VOLUME 106. Posttranslational Modifications (Part A)
Edited by FINN WOLD AND KIVIE MOLDAVE

VOLUME 107. Posttranslational Modifications (Part B)
Edited by FINN WOLD AND KIVIE MOLDAVE

VOLUME 108. Immunochemical Techniques (Part G: Separation and Characterization of Lymphoid Cells)
Edited by GIOVANNI DI SABATO, JOHN J. LANGONE, AND HELEN VAN VUNAKIS

VOLUME 109. Hormone Action (Part I: Peptide Hormones)
Edited by LUTZ BIRNBAUMER AND BERT W. O'MALLEY

VOLUME 110. Steroids and Isoprenoids (Part A)
Edited by JOHN H. LAW AND HANS C. RILLING

VOLUME 111. Steroids and Isoprenoids (Part B)
Edited by JOHN H. LAW AND HANS C. RILLING

VOLUME 112. Drug and Enzyme Targeting (Part A)
Edited by KENNETH J. WIDDER AND RALPH GREEN

VOLUME 113. Glutamate, Glutamine, Glutathione, and Related Compounds
Edited by ALTON MEISTER

VOLUME 114. Diffraction Methods for Biological Macromolecules (Part A)
Edited by HAROLD W. WYCKOFF, C. H. W. HIRS, AND SERGE N. TIMASHEFF

VOLUME 115. Diffraction Methods for Biological Macromolecules (Part B)
Edited by HAROLD W. WYCKOFF, C. H. W. HIRS, AND SERGE N. TIMASHEFF

VOLUME 116. Immunochemical Techniques (Part H: Effectors and Mediators of Lymphoid Cell Functions)
Edited by GIOVANNI DI SABATO, JOHN J. LANGONE, AND HELEN VAN VUNAKIS

VOLUME 117. Enzyme Structure (Part J)
Edited by C. H. W. HIRS AND SERGE N. TIMASHEFF

VOLUME 118. Plant Molecular Biology
Edited by ARTHUR WEISSBACH AND HERBERT WEISSBACH

VOLUME 119. Interferons (Part C)
Edited by SIDNEY PESTKA

VOLUME 120. Cumulative Subject Index Volumes 81–94, 96–101

VOLUME 121. Immunochemical Techniques (Part I: Hybridoma Technology and Monoclonal Antibodies)
Edited by JOHN J. LANGONE AND HELEN VAN VUNAKIS

VOLUME 122. Vitamins and Coenzymes (Part G)
Edited by FRANK CHYTIL AND DONALD B. MCCORMICK

VOLUME 123. Vitamins and Coenzymes (Part H)
Edited by FRANK CHYTIL AND DONALD B. MCCORMICK

VOLUME 124. Hormone Action (Part J: Neuroendocrine Peptides)
Edited by P. MICHAEL CONN

VOLUME 125. Biomembranes (Part M: Transport in Bacteria, Mitochondria, and Chloroplasts: General Approaches and Transport Systems)
Edited by SIDNEY FLEISCHER AND BECCA FLEISCHER

VOLUME 126. Biomembranes (Part N: Transport in Bacteria, Mitochondria, and Chloroplasts: Protonmotive Force)
Edited by SIDNEY FLEISCHER AND BECCA FLEISCHER

VOLUME 127. Biomembranes (Part O: Protons and Water: Structure and Translocation)
Edited by LESTER PACKER

VOLUME 128. Plasma Lipoproteins (Part A: Preparation, Structure, and Molecular Biology)
Edited by JERE P. SEGREST AND JOHN J. ALBERS

VOLUME 129. Plasma Lipoproteins (Part B: Characterization, Cell Biology, and Metabolism)
Edited by JOHN J. ALBERS AND JERE P. SEGREST

VOLUME 130. Enzyme Structure (Part K)
Edited by C. H. W. HIRS AND SERGE N. TIMASHEFF

VOLUME 131. Enzyme Structure (Part L)
Edited by C. H. W. HIRS AND SERGE N. TIMASHEFF

VOLUME 132. Immunochemical Techniques (Part J: Phagocytosis and Cell-Mediated Cytotoxicity)
Edited by GIOVANNI DI SABATO AND JOHANNES EVERSE

VOLUME 133. Bioluminescence and Chemiluminescence (Part B)
Edited by MARLENE DELUCA AND WILLIAM D. MCELROY

VOLUME 134. Structural and Contractile Proteins (Part C: The Contractile Apparatus and the Cytoskeleton)
Edited by RICHARD B. VALLEE

VOLUME 135. Immobilized Enzymes and Cells (Part B)
Edited by KLAUS MOSBACH

VOLUME 136. Immobilized Enzymes and Cells (Part C)
Edited by KLAUS MOSBACH

VOLUME 137. Immobilized Enzymes and Cells (Part D)
Edited by KLAUS MOSBACH

VOLUME 138. Complex Carbohydrates (Part E)
Edited by VICTOR GINSBURG

VOLUME 139. Cellular Regulators (Part A: Calcium- and Calmodulin-Binding Proteins)
Edited by ANTHONY R. MEANS AND P. MICHAEL CONN

VOLUME 140. Cumulative Subject Index Volumes 102–119, 121–134

VOLUME 141. Cellular Regulators (Part B: Calcium and Lipids)
Edited by P. MICHAEL CONN AND ANTHONY R. MEANS

VOLUME 142. Metabolism of Aromatic Amino Acids and Amines
Edited by SEYMOUR KAUFMAN

VOLUME 143. Sulfur and Sulfur Amino Acids
Edited by WILLIAM B. JAKOBY AND OWEN GRIFFITH

VOLUME 144. Structural and Contractile Proteins (Part D: Extracellular Matrix)
Edited by LEON W. CUNNINGHAM

VOLUME 145. Structural and Contractile Proteins (Part E: Extracellular Matrix)
Edited by LEON W. CUNNINGHAM

VOLUME 146. Peptide Growth Factors (Part A)
Edited by DAVID BARNES AND DAVID A. SIRBASKU

VOLUME 147. Peptide Growth Factors (Part B)
Edited by DAVID BARNES AND DAVID A. SIRBASKU

VOLUME 148. Plant Cell Membranes
Edited by LESTER PACKER AND ROLAND DOUCE

VOLUME 149. Drug and Enzyme Targeting (Part B)
Edited by RALPH GREEN AND KENNETH J. WIDDER

VOLUME 150. Immunochemical Techniques (Part K: *In Vitro* Models of B and T Cell Functions and Lymphoid Cell Receptors)
Edited by GIOVANNI DI SABATO

VOLUME 151. Molecular Genetics of Mammalian Cells
Edited by MICHAEL M. GOTTESMAN

VOLUME 152. Guide to Molecular Cloning Techniques
Edited by SHELBY L. BERGER AND ALAN R. KIMMEL

VOLUME 153. Recombinant DNA (Part D)
Edited by RAY WU AND LAWRENCE GROSSMAN

VOLUME 154. Recombinant DNA (Part E)
Edited by RAY WU AND LAWRENCE GROSSMAN

VOLUME 155. Recombinant DNA (Part F)
Edited by RAY WU

VOLUME 156. Biomembranes (Part P: ATP-Driven Pumps and Related Transport: The Na,K-Pump)
Edited by SIDNEY FLEISCHER AND BECCA FLEISCHER

VOLUME 157. Biomembranes (Part Q: ATP-Driven Pumps and Related Transport: Calcium, Proton, and Potassium Pumps)
Edited by SIDNEY FLEISCHER AND BECCA FLEISCHER

VOLUME 158. Metalloproteins (Part A)
Edited by JAMES F. RIORDAN AND BERT L. VALLEE

VOLUME 159. Initiation and Termination of Cyclic Nucleotide Action
Edited by JACKIE D. CORBIN AND ROGER A. JOHNSON

VOLUME 160. Biomass (Part A: Cellulose and Hemicellulose)
Edited by WILLIS A. WOOD AND SCOTT T. KELLOGG

VOLUME 161. Biomass (Part B: Lignin, Pectin, and Chitin)
Edited by WILLIS A. WOOD AND SCOTT T. KELLOGG

VOLUME 162. Immunochemical Techniques (Part L: Chemotaxis and Inflammation)
Edited by GIOVANNI DI SABATO

VOLUME 163. Immunochemical Techniques (Part M: Chemotaxis and Inflammation)
Edited by GIOVANNI DI SABATO

VOLUME 164. Ribosomes
Edited by HARRY N. NOLLER, JR. AND KIVIE MOLDAVE

VOLUME 165. Microbial Toxins: Tools for Enzymology
Edited by SIDNEY HARSHMAN

VOLUME 166. Branched-Chain Amino Acids
Edited by ROBERT HARRIS AND JOHN R. SOKATCH

VOLUME 167. Cyanobacteria
Edited by LESTER PACKER AND ALEXANDER N. GLAZER

VOLUME 168. Hormone Action (Part K: Neuroendocrine Peptides) (in preparation)
Edited by P. MICHAEL CONN

VOLUME 169. Platelets: Receptors, Adhesion, Secretion (Part A)
Edited by JACEK HAWIGER

VOLUME 170. Nucleosomes (in preparation)
Edited by PAUL M. WASSARMAN AND ROGER D. KORNBERG

VOLUME 171. Biomembranes (Part R: Transport Theory: Cells and Model Membranes) (in preparation)
Edited by SIDNEY FLEISCHER AND BECCA FLEISCHER

VOLUME 172. Biomembranes (Part S: Transport: Membrane Isolation and Characterization) (in preparation)
Edited by SIDNEY FLEISCHER AND BECCA FLEISCHER

VOLUME 173. Biomembranes (Part T: Cellular and Subcellular Transport: Eukaryotic (Nonepithelial) Cells) (in preparation)
Edited by SIDNEY FLEISCHER AND BECCA FLEISCHER

Section I

Isolation of Platelets

[1] Isolation of Human Platelets from Plasma by Centrifugation and Washing

By J. FRASER MUSTARD, RAELENE L. KINLOUGH-RATHBONE, and MARIAN A. PACKHAM

Blood platelets are readily activated by a number of stimuli to change their shape, extend pseudopods, aggregate, release the contents of their storage granules, form thromboxanes and prostaglandins, and provide sites on their membranes which take part in two steps of the intrinsic coagulation pathway.[1,2] An ideal isolation method should avoid activation of platelets, or any of these consequences of activation, and provide a suspension of platelets that responds to stimulation in the same was as do platelets in the circulation. In addition, it is desirable to have the platelets maintain this responsiveness for several hours to permit *in vitro* experiments.[2-4]

Although it might appear that plasma would be the ideal suspending medium, the anticoagulants that must be used have major effects on the response of platelets. Many preparations of heparin cause some aggregation of the platelets[5,6]; hirudin is expensive and does not block activated coagulation factors other than thrombin, e.g., factor Xa; and EDTA, because of its strong chelation of divalent cations, has a deleterious effect on membrane glycoproteins participating in platelet aggregation. Many studies have been done with citrated platelet-rich plasma. The low concentration of ionized calcium in this medium introduces a major artifact because when human platelets are brought into close contact with each other in such a medium (for example by ADP) the arachidonate pathway is activated, resulting in thromboxane formation and the release of granule

[1] M. A. Packham and J. F. Mustard, *in* "Blood Platelet Function and Medicinal Chemistry" (A. Lasslo, ed.), p. 61. Elsevier, New York, 1984.

[2] R. L. Kinlough-Rathbone, M. A. Packham, and J. F. Mustard, *in* "Methods in Hematology: Measurements of Platelet Function" (L. A. Harker and T. S. Zimmerman, eds.), p. 64. Churchill-Livingstone, Edinburgh, Scotland, 1983.

[3] J. F. Mustard, D. W. Perry, N. G. Ardlie, and M. A. Packham, *Br. J. Haematol.* **22,** 193 (1972).

[4] R. L. Kinlough-Rathbone, J. F. Mustard, M. A. Packham, D. W. Perry, H.-J. Reimers, and J.-P. Cazenave, *Thromb. Haemostasis* **37,** 291 (1977).

[5] C. Eika, *Scand. J. Haematol.* **9,** 480 (1972).

[6] M. B. Zucker, *Thromb. Diath. Haemorrh.* **33,** 63 (1975).

contents.[7-9] Human platelets in a medium with approximately physiological concentrations of ionized calcium do not behave in this way; although they aggregate in response to ADP, the arachidonate pathway is not activated and granule contents are not released. A major advantage of studying platelet reactions in artificial media instead of in citrated plasma is that physiological concentrations of ionized calcium can be used.

Like other cells, platelets maintain their functions best in media that are similar to tissue culture media. Protective protein (albumin is most commonly used) prevents them from adhering to the sides of containers and lessen the chance of activation during isolation and handling.[3,4,10,11] A source of metabolic energy (glucose) is essential,[12,13] and the pH should be controlled at the pH of plasma (7.35).[14,15] Buffer such as Tris [tris(hydroxymethyl)aminomethane] should be avoided[16] because, like other amines,[17] it inhibits some platelet responses and potentiates others.

Another problem is the development of refractoriness of platelets to ADP-induced aggregation[18-20]; when platelets have been exposed to ADP, they aggregate to a lesser extent (or not at all) when more ADP is added with stirring. Since platelets contain ADP in their amine storage granules, release of small amounts of ADP during the isolation of platelets, and during storage, may make the platelets unresponsive to ADP. To prevent this, enzyme systems that remove ADP should be included in platelet-suspending media.[2-4] (It should be pointed out that plasma itself contains

[7] J. F. Mustard, D. W. Perry, R. L. Kinlough-Rathbone, and M. A. Packham, *Am. J. Physiol.* **228,** 1757 (1975).

[8] D. E. Macfarlane, P. N. Walsh, D. C. B. Mills, H. Holmsen, and H. J. Day, *Br. J. Haematol.* **30,** 457 (1975).

[9] B. Lages and H. J. Weiss, *Thromb. Haemostasis* **45,** 173 (1981).

[10] M. A. Packham, G. Evans, M. F. Glynn, and J. F. Mustard, *J. Lab. Clin. Med.* **73,** 686 (1969).

[11] O. Tangen, M. L. Andrae, and B. E. Nilsson, *Scand. J. Haematol.* **11,** 241 (1973).

[12] R. L. Kinlough-Rathbone, M. A. Packham, and J. F. Mustard, *J. Lab. Clin. Med.* **75,** 780 (1970).

[13] R. L. Kinlough-Rathbone, M. A. Packham, and J. F. Mustard, *J. Lab. Clin. Med.* **80,** 247 (1972).

[14] M. B. Zucker, *Thromb. Diath. Haemorrh. Suppl.* **42,** 1 (1970).

[15] P. Han and N. G. Ardlie, *Br. J. Haematol.* **26,** 373 (1974).

[16] M. A. Packham, M. A. Guccione, M. Nina, R. L. Kinlough-Rathbone, and J. F. Mustard, *Thromb. Haemostasis* **51,** 140 (1984).

[17] R. L. Kinlough-Rathbone, M. A. Packham, and J. F. Mustard, *Thromb. Haemostasis* **52,** 75 (1984).

[18] J. R. O'Brien, *Nature (London)* **212,** 1057 (1966).

[19] S. Holme, J. J. Sixma, J. Wester, and H. Holmsen, *Scand. J. Haematol.* **18,** 267 (1977).

[20] T. J. Hallam, P. A. Ruggles, M. C. Scrutton, and R. B. Wallis, *Thromb. Haemostasis* **47,** 278 (1982).

enzymes that dephosphorylate ADP.[21]) The most commonly added enzyme is apyrase (EC 3.6.1.5), prepared from potatoes,[22] although creatine phosphate/creatine phosphokinase (EC 2.7.3.2) has also been used. However, platelets that are refractory to ADP-induced aggregation can be aggregated by other aggregating and release-inducing agents. It is almost impossible to prepare a platelet suspension that is unresponsive to strong agonists such as thrombin.

Methods

Collection of Blood

Because so many drugs affect platelet reactions (particularly aspirin and other nonsteroidal antiinflammatory drugs) donors should be carefully questioned about the drugs they have taken during the previous 2 weeks. The donors should not be stressed and it is preferable that blood be collected in the morning so that the donors are fasting and have not smoked for a number of hours.

Collect blood from a forearm vein with or without venous occlusion. If venous occlusion is used, the sphygmomanometer cuff should not be inflated above 60 mmHg (8 kPa). Care should be taken to avoid prolonged stasis and anoxia. A 19-gauge needle, or larger, should be used. Blood should be mixed immediately with the anticoagulant and frothing should be avoided during the mixing procedure. The first 1–2 ml of blood may be discarded to avoid the effects of traces of thrombin that have been shown to be generated during venepuncture.

Reagents

Plasticware and Glassware. Containers may be polyethylene, polycarbonate, or siliconized glass. Use a silicone preparation that binds firmly to the glass surface so that small micelles of silicone do not contaminate the platelet preparation. A suitable preparation is Surfasil (dichlorooctamethyltetrasiloxane, Pierce Chemical Co., Rockford, IL) prepared as a 10% solution in carbon tetrachloride or tetrahydrofuran. After coating, the glass surfaces should be thoroughly rinsed and dried.

Acid–Citrate–Dextrose (ACD). The acid–citrate–dextrose anticoagulant solution of Aster and Jandl[23] has the advantage of not only chelating

[21] I. Holmsen and H. Holmsen, *Thromb. Diath. Haemorrh.* **26,** 177 (1971).
[22] J. Molnar and L. Lorand, *Arch. Biochem. Biophys.* **93,** 353 (1961).
[23] R. H. Aster and J. H. Jandl, *J. Clin. Invest.* **43,** 843 (1964).

the calcium in the blood and thus preventing the activation of coagulation, but also of lowering the pH of the blood to 6.5; platelets do not aggregate readily at this pH. Prepare this anticoagulant by dissolving 25 g of trisodium citrate dihydrate, 15 g of citric acid monohydrate, and 20 g of dextrose in 1 liter of distilled water. This solution has an osmolarity of 450 mOsm/liter and a pH of about 4.5. One part of this anticoagulant is required for each six parts of blood.

Stock Solutions. Stock solutions for preparing Tyrode's buffer used for the preparation of washed platelets (may be stored 2 weeks at 4°) are as follow:

Stock I: This consists of NaCl (160 g), KCl (4.0 g), $NaHCO_3$ (20 g), and NaH_2PO_4 (1.0 g) made up to 1 liter with distilled water

Stock II: $MgCl_2 \cdot 6H_2O$ (20.33 g) made up to 1 liter to produce a 0.1 *M* solution

Stock III: $CaCl_2 \cdot 6H_2O$ (21.91 g) made up to 1 liter to produce a 0.1 *M* solution

Tyrode's Solution. Tyrode's solution containing albumin and apyrase (washing and resuspending fluid) is made up as follows: Add 750 ml of distilled water to 50 ml of stock I, add 10 ml of stock II (with mixing), and 20 ml of stock III (with mixing). Dissolve 3.5 g Pentex fraction V bovine serum albumin (Miles Laboratories, Kankakee, IL) and 1 g of dextrose in this solution and bring to a total volume of 1 liter with distilled water. Adjust the pH to 7.35 and the osmolarity to 290 mOsm/liter by addition of NaCl if necessary. Human serum albumin may be used in place of bovine albumin. Add apyrase in the concentrations indicated below.

Optional Final Resuspending Fluid. Eagle's medium (Cat. No. 410–1100, Gibco, Grand Island, NY) is supplemented with the following additions: 2.2 g/liter $NaHCO_3$, 0.35% albumin, and apyrase.

Apyrase. Apyrase can be prepared by a modification of the method of Molnar and Lorand[2,22] in a two-stage procedure:

Stage I is conducted at room temperature. Wash and slice potatoes (10 kg) and homogenize in a Waring blender with 1 liter of distilled water. Stir the homogenate for 30 min and then centrifuge it at 900 *g* for 10 min. Filter the supernatant fluid through multiple layers of cheesecloth and measure the volume of the effluent.

Stage II is conducted at 0–4°. Add $CaCl_2$ to the supernatant in sufficient quantity to bring the solution to 0.025 *M* $CaCl_2$, stir for 15 min, allow the mixture to settle for 1 hr, and centrifuge for 20 min at 3500 *g*. Resuspend the precipitate in 1 *M* $CaCl_2$ to a volume that is approximately one-tenth that of the original effluent and stir for 60 min before centrifuging for 20 min at 3500 *g*. Dialyze the supernatant against 0.1 *M* KCl (use

20 liters for material from 10 kg of potatoes) for 24 hr and then centrifuge for 20 min at 3500 *g*. Fractionate the supernatant fluid by adding 3 vol of ice-cold saturated ammonium sulfate solution to each volume of supernatant fluid. Stir for 40 min and then centrifuge for 15 min at 3500 *g*. Dissolve the precipitate in a minimum volume of distilled water and dialyze against several changes of 0.154 *M* (0.9%) NaCl.

Finally, assay the protein by the Folin–Ciocalteau method[24] and dilute to a concentration of 3 mg/ml. Store the enzyme in small aliquots at −20°. The nucleotidase activity of the apyrase preparation can be determined by measuring the rate at which it degrades [^{14}C]ATP to [^{14}C]ADP and [^{14}C]AMP.[2,25] Some commercially available preparations are not suitable because they contain impurities and 5′-nucleotidase activity that hydrolyzes [^{14}C]AMP to [^{14}C]adenosine. Use a concentration of apyrase in the final suspending medium that is capable of converting 0.25 μM ATP to AMP in 120 sec at 37°. Alternatively, choose an apyrase concentration that maintains platelet sensitivity to ADP, but does not have an appreciable inhibitory effect on ADP-induced aggregation (tested in the presence of fibrinogen). If too much apyrase is used, ADP-induced aggregation will be inhibited and if too little is used, the platelets will become refractory to ADP.

Fibrinogen. Fibrinogen can be partially purified by the method of Lawrie *et al.*[26] to remove contaminating factor XIII, plasminogen, and fibronectin.

Heparin. Some preparations of heparin cause platelet aggregation[5,6] and cause platelets to stick to the walls of their container. Batches of heparin should be screened before use to ensure that this does not occur.[2]

Preparation of Suspensions of Washed Human Platelets

The procedure can be used to prepare platelets from 120–150 ml of blood, but can be adapted for larger volumes provided the proportions of washing and resuspending solutions are maintained. The platelets are kept at 37° throughout this procedure.

Collect blood into ACD anticoagulant (6/1, v/v) in 50-ml polycarbonate tubes. Centrifuge at 190 *g* for 15 min at 37° to obtain platelet-rich plasma and transfer the plasma to fresh, 50-ml conical polycarbonate tubes with a siliconized Pasteur pipet or a plastic syringe. (Note: Occasionally a lower *g* force may be required, depending on the characteristics of the

[24] J. M. Clark, Jr., and R. L. Switzer, "Experimental Biochemistry," 2nd Ed. Freeman, San Francisco, California, 1977.

[25] M. A. Packham, N. G. Ardlie, and J. F. Mustard, *Am. J. Physiol.* **217**, 1009 (1969).

[26] J. S. Lawrie, J. Ross, and G. D. Kemp, *Biochem. Soc. Trans.* **7**, 693 (1979).

platelets, plasma, and red cells of individual donors.) Centrifuge the platelet-rich plasma at 2500 *g* for 15 min at 37°, discard as much of the plasma as possible, and gently resuspend the platelets, using a siliconized Pasteur pipet, in 10 ml of Tyrode's solution containing heparin (50 U/ml), albumin (0.35%), and apyrase (use approximately 10 times the amount of apyrase that is required to stabilize platelets in suspending medium), pH 7.35, 37°. It is *essential* that platelets be left in this solution at 37° for at least 15 min. Centrifuge at 1200 *g* for 10 min at 37°, discard the supernatant fluid, and gently resuspend the platelets in Tyrode's solution containing albumin (0.35%) and the same amount of apyrase as used in the first washing solution. If there are a few red cells at the bottom of the tube, avoid resuspending them with the platelets. Centrifuge once more (1200 *g* for 15 min), discard the supernatant fluid, and gently resuspend the platelets in Tyrode's solution containing albumin (0.35%) and apyrase.

Platelets prepared using this method and stored in a water bath at 37° remain responsive to ADP in the presence of fibrinogen (75–200 μg/ml) for 3–4 hr.[4] If even longer periods of storage are to be used, HEPES buffer (5 m*M*, *N*-2-hydroxyethylpiperazine-*N'*-2-ethanesulfonic acid, Sigma) may be added to the suspending medium.[27] Platelets in these media are also responsive to all other aggregating and release-inducing agents such as collagen, thrombin, sodium arachidonate, the divalent cation ionophore A23187, and platelet-activating factor among others. Additional fibrinogen is not required when a release-inducing agent is used since fibrinogen released from the platelet α granules in response to release-inducing agonists is sufficient to support platelet aggregation. In this calcium-containing medium, platelets do not respond consistently to epinephrine,[2] even in the presence of fibrinogen, unless traces of another agonist are present.[28]

Comments

Potential Problems and Pitfalls

Problems are apt to arise when the procedure is shortened or necessary reagents are omitted. Maintaining the temperature at 37° ensures that the platelets are not activated by cooling and rewarming. The times of incubation in the washing procedure must not be shortened below 15 min; they may be longer without deleterious effects. It is essential that a protective protein be present. Albumin is most satisfactory and may be used at

[27] S. Timmons and J. Hawiger, *Thromb. Res.* **12,** 297 (1978).

[28] J. F. Mustard, C. Lalau-Keraly, D. W. Perry, and R. L. Kinlough-Rathbone, *Fed. Proc., Fed. Am. Soc. Exp. Biol.* **44,** 1127 (1985).

concentrations between 0.35% and 4%. The latter concentration is similar to that in plasma, but investigators should be aware of the ability of albumin to bind drugs, fatty acids, etc. For example, the concentration of arachidonic acid required to stimulate platelets in 4% albumin is much greater than that required in 0.35% albumin. If albumin is omitted entirely, some activation occurs when the platelets are merely stirred; prelabeled platelets release 5 to 10% of their [^{14}C]serotonin and lose 2% of their ^{51}Cr upon stirring in an aggregometer cuvette, whereas in the presence of 0.35% albumin only 0.1 to 0.4% of the [^{14}C]serotonin is released and only 0.5% of the ^{51}Cr.[4] In addition, without albumin in the suspending medium, the response to ADP (in the presence of fibrinogen) is variable.

Omission of apyrase or the creatine phosphate/creatine phosphokinase system, either of which removes any ADP that accumulates in the suspending medium, causes the platelets to become progressively more and more refractory to ADP, although they will continue to respond to other aggregating agents.

Omission of Ca^{2+} from the suspending medium introduces the same artifact that occurs in citrated platelet-rich plasma (Table I). When human platelets are brought into close contact in a medium without added calcium, the arachidonate pathway is activated, thromboxane A_2 forms, the platelets release some of their granule contents, and second phase aggregation occurs; aspirin and other nonsteroidal antiinflammatory drugs inhibit this secondary activation by blocking thromboxane A_2 formation. ADP causes only the primary phase of aggregation in a medium containing 1–2 m*M* Ca^{2+}, regardless of the concentration of ADP used. However, if traces of release-inducing stimuli are present, addition of ADP may lead to activation of the arachidonate pathway, even in the presence of 2 m*M* calcium.[3] The omission of Ca^{2+} from platelet-suspending media has led to many results of dubious significance in regard to physiological conditions.

Advantages of the Method

The advantages of this method are that large volumes of blood can be handled readily, the reagents required are inexpensive, the platelet count in the final suspension can be adjusted to any desired number, platelets respond to aggregating and release-inducing agents in a manner similar to platelets in native plasma or plasma anticoagulated with hirudin, and maintain this responsiveness for hours.[4] The constituents of the medium can be varied, or additions made, as required for experimental purposes. Thrombin can be used as an agonist without the problem of the formation

[29] B. Lages, M. C. Scrutton, and H. Holmsen, *J. Lab. Clin. Med.* **85,** 811 (1975).
[30] M. Radomski and S. Moncada, *Thromb. Res.* **30,** 383 (1983).

TABLE I
EFFECT OF Ca^{2+} ON RELEASE OF [^{14}C]SEROTONIN[a]

Ca^{2+} in platelet-suspending medium	ASA (500 μM)	^{14}C in suspending fluid (% of total in platelets)
2 mM	−	0.3
2 mM	+	0.2
0	−	43.0
0	+	0.4

[a] Prelabeled human platelets stimulated with 2.5 μM ADP. Fibrinogen (200 μg/ml) and ADP were added to the platelet suspensions stirred at 1100 rpm in aggregometer cuvettes at 37°. The release of [^{14}C]serotonin from prelabeled platelets was measured after 3 min, as described elsewhere.[4] In the presence of 2 mM Ca^{2+}, only the primary phase of platelet aggregation occurred. When no Ca^{2+} was added to the medium, ADP induced two phases of aggregation accompanied by the release of platelet granule contents. This second phase was blocked by aspirin (ASA), added 5 min before ADP in this experiment.

of large amounts of fibrin that occurs when thrombin is added to platelet-rich plasma. The morphological appearance of the platelets by electron microscopy is similar to that of platelets in plasma.[4] Since the platelet suspensions are kept at 37°, experiments can be readily done at this temperature so that conditions more closely resemble the *in vivo* situation.

Although there are a number of other isolation procedures, most of them are less suitable for large volumes of blood and are more costly. Gel filtration[11,29] suffers from the problem that platelet counts are low because of dilution on the column and proteins of high molecular weight may be eluted with the platelets unless special precautions are taken.[27] Addition of PGI_2 to prevent activation of platelets during isolation appears to be satisfactory,[30] but is prohibitively expensive for routine use in most laboratories. Separation on density gradients of albumin,[31] Stractan,[32] or sodium metrizoate[33] is not suitable for large-scale preparations.

The method of isolation described here has proved to be most satisfactory for many types of *in vitro* experiments involving aggregating agents, inhibitors, and various conditions that affect platelet responses. A major

[31] P. N. Walsh, D. C. B. Mills, and J. G. White, *Br. J. Haematol.* **36,** 281 (1977).

[32] L. Corash, B. Shafer, and M. Perlow, *Blood* **52,** 726 (1978).

[33] P. Ganguly, and W. J. Sonnichsen, *J. Clin. Pathol.* **26,** 635 (1973).

advantage is that the response of platelets to ADP (in the presence of added fibrinogen) remains similar to that in native plasma. Aggregation in response to ADP, without thromboxane production or release of granule contents, is the characteristic of a good preparation.

[2] Isolation of Human Platelets by Albumin Gradient and Gel Filtration

By SHEILA TIMMONS and JACEK HAWIGER

Human blood platelets interact with several plasma proteins that participate in the formation of a hemostatic plug.[1] To study the interactions of plasma proteins with their receptors on the platelet membrane a preparation of platelets free of plasma proteins is needed in order to assure that they will not interfere with the binding of labeled ligands. A number of methods provide platelets that have been used for testing the effect of plasma proteins and general requirements have been formulated.[2] The methods which employ repeated washing and centrifugation work best when a relatively large volume of blood is used.[3] Isolation of platelets from small samples of blood, i.e., 20–40 ml, by repeated washing and centrifugation usually ends with a poor yield of platelets. In some instances washed platelets are treated with formalin for agglutination studies.[4] Such treated platelets cannot be used for studying processes requiring metabolically intact platelets, e.g., signal transduction generated by ADP, epinephrine, and arachidonic acid.

The second group of methods of separation of platelets from plasma proteins includes gel filtration and/or centrifugation over a cushion or gradient made of albumin, Ficoll, or stractan.[5-12] The gel filtration method

[1] J. Hawiger, *in* "Hemostasis and Thrombosis: Basic Principles and Clinical Practice" (R. Colman, J. Hirsh, V. Marder, and E. Salzman, eds.), 2nd Ed., p. 162. Lippincott, Philadelphia, Pennsylvania, 1987.

[2] H. J. Day, H. Holmsen, and M. B. Zucker, *Thromb. Diath. Haemorrh.* **33,** 648 (1975).

[3] J. F. Mustard, D. W. Perry, N. H. Ardlie, and M. A. Packham, *Br. J. Haematol.* **22,** 193 (1972).

[4] D. E. Macfarlane, J. Stibble, E. P. Kirby, and M.B. Zucker, *Thromb. Diath. Haemorrh.* **34,** 306 (1975).

[5] S. Timmons and J. Hawiger, *Thromb. Res.* **12,** 298 (1978).

introduced by Tangen and colleagues[6] posed some difficulties due to inadvertent activation of platelets, dilution of the platelet suspension, and the use of Tris buffer, which affects platelet function. Subsequent modification of this method improved the quality of the platelet preparation.[9] However, gel-filtered platelets retain a residual quantity of large plasma proteins, such as von Willebrand factor (vWF) and fibrinogen, which may influence the study of platelet receptors for these adhesive macromolecules.[5] A similar problem concerns platelet preparation on an albumin cushion or gradient, originally developed by Walsh.[7] We have adapted these methods of platelet separation from plasma proteins by gel filtration and albumin gradient centrifugation to produce stable, functionally unimpaired platelets which are relatively free of absorbed plasma proteins, including the largest ones.[5] The combined albumin gradient–gel filtration method described herein yields human platelets which are separated from plasma proteins, including vWF and fibrinogen, as judged by a negative response to ristocetin and ADP, respectively. The stability of the obtained platelet preparation has been enhanced by our application of HEPES buffer[5,13] which, in contrast to Tyrode's buffer, is not affected by changes in the $p\mathrm{CO_2}$ of the platelet suspension. Subsequent to our introduction of HEPES buffer to platelet isolation,[5] other workers have since employed HEPES buffer in their experiments with platelets.[14,15] The use of Tris buffer in platelet work is not recommended due to its detrimental effect on calcium fluxes.[16]

Materials and Reagents

Preparation of 50% Albumin. Deionized water (200 ml) is poured in a 1-liter Erlenmeyer flask, followed by 100 g of bovine albumin (fraction V, reagent grade, Miles Laboratories, Naperville, IL). The mixture should

[6] O. Tangen, H. J. Berman, and P. Marfey, *Thromb. Diath. Haemorrh.* **25,** 268 (1971).
[7] P. M. Walsh, *Br. J. Haematol.* **22,** 205 (1972).
[8] R. A. Hutton, M. A. Howard, D. Deykin, and R. M. Hardisty, *Thromb. Diath. Haemorrh.* **31,** 119 (1974).
[9] B. Lages, M. C. Scrutton, and H. Holmsen, *J. Lab. Clin. Med.* **85,** 811 (1975).
[10] D. G. Nichols and J. R. Hampton, *Thromb. Diath. Haemorrh.* **27,** 425 (1972).
[11] L. Corash, B. Shafer, and M. Perlow, *Blood* **52,** 726 (1978).
[12] P. Ganguly and W. J. Sonnichsen, *J. Clin. Pathol.* **26,** 635 (1973).
[13] C. Shipman, Jr., *Proc. Soc. Exp. Med. Biol.* **130,** 305 (1969).
[14] E. I. Peerschke, M. B. Zuker, R. A. Grant, J. J. Egan, and M. M. Johnson, *Blood* **55,** 841 (1980).
[15] Z. M. Ruggeri, L. De Marco, L. Gatti, R. Bader, and R. R. Montgomery, *J. Clin. Invest.* **72,** 1 (1983).
[16] M. A. Packham, M. A. Gucionne, M. Nina, R. L. Kinlough-Rathbone, and J. F. Mustard, *Thromb. Haemostis* **51,** 140 (1984).

remain unperturbed at 4° for 24–48 hr; do not mix until all albumin is dissolved. Since the pH of the dissolved albumin should be between 6.5 and 7.0, one should purchase reagent with specified pH of about 7 in solution. Aliquots (7 ml) are stored at −40° until needed. Then the 50% albumin is thawed and diluted with either HEPES buffer or Ca^{2+}-free Tyrode's buffer (see below) to prepare 25, 17, 12, and 10% solutions. If the platelets are used for studies to examine the effects of ADP, apyrase (0.1 unit/ml) can be added to platelet-rich plasma (PRP) and the albumin solutions employed for preparation of the gradient.[17] We found that addition of apyrase (EC 3.6.1.5) was helpful in assuring the sustained response of platelets to ADP when the common receptor mechanism for binding of fibrinogen and vWF to human platelets was studied.[17]

Purification of Apyrase. Commercially available apyrase (grade I) from Sigma is further purified by the method of Traverso-Cori *et al.*[18] Briefly, 6000 units of apyrase is dissolved in 20 ml of 0.05 *M* potassium succinate in 0.3 *M* KCl buffer, pH 4.0, centrifuged at 10,000 *g* for 10 min, and desalted by passing through a Sephadex G-25 column (50 × 4 cm) that is equilibrated with the potassium succinate buffer, pH 4.0. The peak fractions are pooled, concentrated to 10 ml using an Amicon ultrafiltration unit with a PM30 filter, and then passed through a Sephadex G-150 column (80 × 2.5 cm) equilibrated with 0.05 *M* potassium succinate buffer in 0.3 *M* KCl. The fractions are assayed for protein at 280 nm and for ATPase activity. The unit of enzyme activity is that amount of enzyme that releases 1 μmol P_i/min.[19] The fractions containing the highest ATPase activity per milligram of protein are pooled, dialyzed against phosphate-buffered saline, and concentrated to 10 ml on an Amicon ultrafiltration unit. Note that Sigma currently has available apyrase (grade VIII) of similar specific activity (300–400 units/mg of protein) which may be used.

Preparation of Gel Columns. Sepharose 2B (Pharmacia Fine Chemicals, Piscataway, NJ) or BioGel A-150, 100–200 mesh (Bio-Rad Laboratories, Richmond, CA) is prepared by acetone washing (3–4 vol) followed by thorough rinsing with 5–6 vol of 0.9% NaCl. Sepharose 2B prepared in this way may be refrigerated and used within 2 weeks. On the day of an experiment the Sepharose 2B gel suspended in 0.9% NaCl is warmed to room temperature, if necessary deaerated in a vacuum flask, and poured into 2 × 8 cm columns made of glass syringes. The inner bottom of the

[17] S. Timmons, M. Kloczewiak, and J. Hawiger, *Proc. Natl. Acad. Sci. U.S.A.* **81,** 4935 (1984).

[18] A. Traverso-Cori, S. Traverso, and H. Reyes, *Arch. Biochem. Biophys.* **137,** 133 (1970).

[19] A. Traverso-Cori, H. Chaimovich, and O. Cori, *Arch. Biochem. Biophys.* **109,**173 (1965).

syringe contains a disk of nylon microfilament with mesh opening of 52 μm (Small Parts, Inc., Miami, FL) which is held in place by a plastic ring. The gel should be poured in such a way that air bubbles are not formed. The flow is regulated by a Teflon two-way stopcock to which polyethylene tubing is attached. All parts of the column are siliconized (Sigmacote, Sigma, St. Louis, MO), rinsed, and dried before first use. Before applying the platelet suspension, the gel columns are equilibrated with 100 ml of buffer as specified below. If air bubbles appear, the gel columns are repacked and then washed again.

Buffers. Two buffers for separation of platelets from plasma proteins can be used: HEPES (N-2-hydroxyethyl-1-piperazine-N′-2-ethanesulfonic acid)[5] or Ca^{2+}-free Tyrode.[3] HEPES buffer contains a final concentration of 137 m*M* NaCl, 2.7 m*M* KCl, 1 m*M* $MgCl_2$, 5.5 m*M* glucose, 0.35% albumin, 3 m*M* NaH_2PO_4, and 3.5 m*M* HEPES (Ultrol Grade, Calbiochem, La Jolla, CA). Tyrode's buffer contains a final concentration of 137 m*M* NaCl, 2.7 m*M* KCl, 12 m*M* $NaHCO_3$, 1 m*M* $MgCl_2$, 5.5 m*M* Glucose, and 0.35% bovine albumin. The salts are prepared as 50× concentrated stock solutions. On the day of use the concentrated stock solutions are diluted, glucose and bovine albumin are dissolved into the buffer solution, and the pH adjusted to 7.35 with 1 *N* NaOH or 1 *N* HCl. The same buffer is used to dilute the albumin for gradients, to equilibrate the columns, and to elute the platelets therefrom.

Gel Filtration Method. Polypropylene plasticware or siliconized glassware should be used throughout the procedure. Blood from healthy volunteers, who have not ingested any aspirin for 10 days and who have not taken any other medication for at least 2 days, is collected into 0.12 *M* citrate buffer, pH 6.0 (0.01 *M* citric acid in 0.11 *M* sodium citrate). Nine parts blood is drawn into a syringe containing one part citrate buffer and centrifuged at 160 *g* for 15 min to obtain platelet-rich plasma (PRP). After collecting PRP, the remaining portion of blood is centrifuged at 3000 *g* for 10 min to obtain platelet-free plasma (PFP). The columns containing Sepharose 2B or BioGel A-150 are equilibrated with either HEPES buffer or Ca^{2+}-free Tyrode's buffer and are adjusted to have a flow rate of 1 ml/min. When the equilibrating buffer has just entered the gel, 2.5 ml PRP is applied directly to the top of the gel, avoiding disturbance of the interface. The platelets are eluted as indicated below. Because platelet suspensions prepared by single-step gel filtration contain plasma proteins, such as vWF and fibrinogen, modifications were introduced for a two-step procedure combining albumin cushion/gradient centrifugation and gel filtration.[5]

Combined Albumin Cushion/Gradient and Gel Filtration Method. The procedures are outlined diagrammatically in Fig. 1.

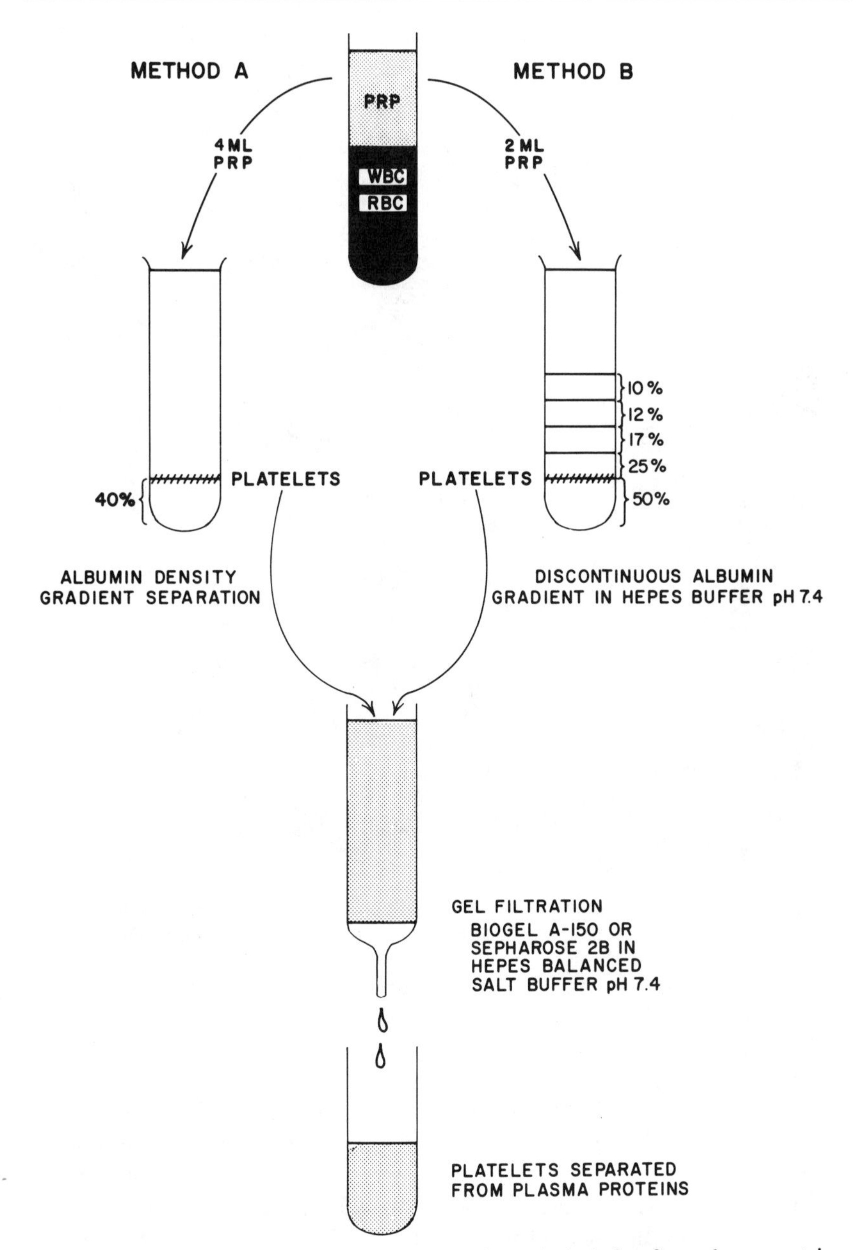

FIG. 1. Diagram outlining methods A and B of platelet isolation from plasma proteins. (Thromb. Res., Vol. 12, Timmons and Hawiger, "Separation of Platelets from Plasma Proteins Including Factor VIII_{vWF} by a Combined Albumin Gradient-Gel Filtration Method Using Hepes Buffer," 1978, Pergamon Journals, Ltd.)

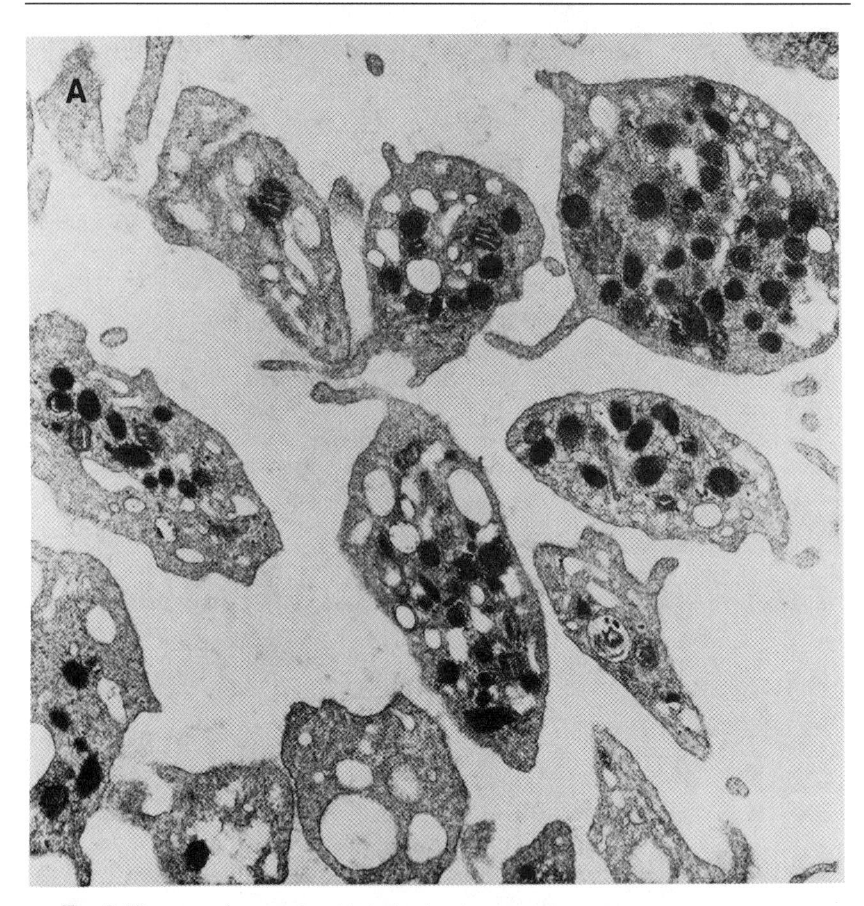

FIG. 2. Electron micrographs of (A) platelets in plasma and (B) platelets separated by the stepwise albumin gradient followed by Sepharose 2B gel filtration. ×32,400. (Thromb. Res., Vol. 12, Timmons and Hawiger, "Separation of Platelets from Plasma Proteins Including Factor VIII_{vWF} by a Combined Albumin Gradient-Gel Filtration Method Using Hepes Buffer," Pergamon Journals, Ltd.)

Method A: PRP (4 ml) is applied to a 50% albumin cushion[7] and centrifuged at 1200 *g* for 15 min. The plasma is removed and, using another pipet, the platelet fraction resting on the top of the albumin cushion is collected, diluted in buffer to 2.5 ml, and applied to either a Sepharose 2B or BioGel A-150 column. Platelets are eluted with the same buffer which was used to equilibrate the column. The opaque fractions in the effluent containing platelets are collected, their absorbance determined at 520 nm, and only the fractions constituting the peak of the elution

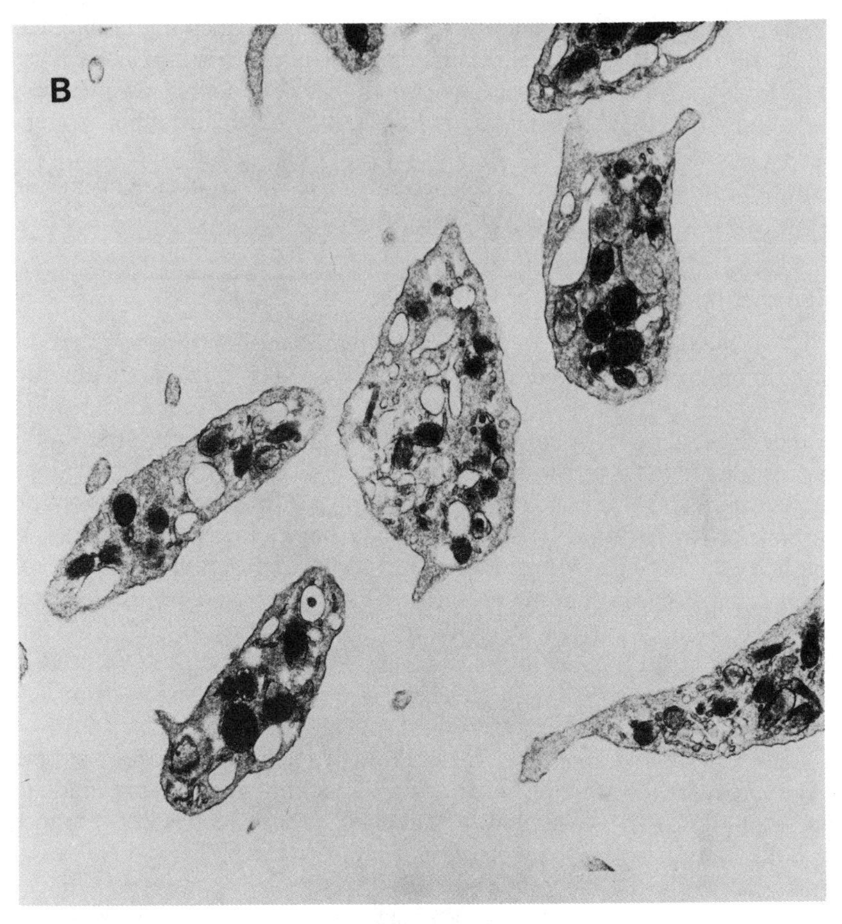

FIG. 2. *(continued)*

profile are pooled for platelet studies. The after-peak fractions of platelets, which are contaminated with vWF, should be discarded.

Method B: PRP (2 ml) is applied to a step-wise discontinuous albumin gradient (50, 25, 17, 12, 10%) and centrifuged at 1200 *g* for 15 min. The supernatant is removed with a Pasteur pipet, being careful to collect from the top of the fluid, and with another siliconized Pasteur pipet the platelet fraction is collected. The platelet fractions from three gradients are diluted in buffer to 2.5 ml, applied to either a Sepharose 2B or BioGel A-150 column, and platelets are eluted as described under method A.

Platelet Counts. These are performed on a Coulter electronic particle counter (Coulter Electronics, Hialeah, FL).

Aggregation Studies. Platelet aggregation is performed by the method of Born[20] in a Payton dual channel aggregometer (Payton Associates, Buffalo, NY). Aggregating reagents used are ADP (Boehringer Mannheim, Indianapolis, IN), ristocetin (Sigma, St. Louis, MO), thrombin (Parke-Davis, Detroit, MI), and low-solubility fibrinogen.[17] Aggregation units are expressed by slope values which represent the change along a tangent line to the steepest increase in light transmission.[21]

Comments

Two modifications of the combined albumin cushion/gradient and gel filtration procedure were compared: method A, which employs an albumin cushion and gel filtration,[8] and method B, which is our modification employing a stepwise discontinuous albumin gradient and gel filtration for separation of platelets from plasma proteins,[5] and provided a platelet preparation which responded well to thrombin. However, the aggregation of platelets in response to ADP is usually higher in method A than in method B. The platelets separated by method A aggregate after addition of ristocetin, indicating that the vWF is still loosely associated with the platelet preparation. In method B, which represents the recommended modification with HEPES buffer, aggregation induced by ristocetin is reduced to a negligible level, indicating that platelets have been separated from adsorbed plasma proteins such as vWF.

Electron micrographs[22,23] (Fig. 2) show platelets in autologous plasma (A) compared with platelets separated by discontinuous albumin gradient and gel filtration (B), documenting that platelets have a normal appearance of both granules and plasma membrane and that the preparation is free of other cells.

Although BioGel A-150 has a higher exclusion limit in comparison with Sepharose 2B the elution profiles from both gels are the same in terms of platelet count, absorbance at 520 nm, and protein content determined by absorbance at 280 nm (Fig. 3).

When the response of platelets to aggregating agents is compared in two buffer systems, the HEPES balanced salt solution introduced by us for platelet work[5] and the conventional Tyrode's buffer, the aggregation response of platelets to ADP and epinephrine is similar during the first hour.

[20] G. V. R. Born, *Nature (London)* **194,** 927 (1962).

[21] C. S. P. Jenkins, D. Meyer, M. D. Dreyfus, and M. J. Larrieu, *Br. J. Haematol.* **28,** 561 (1974).

[22] G. Millonig, *J. Appl. Physiol.* **32,** 1637 (1961).

[23] E. S. Reynolds, *J. Cell Biol.* **17,** 208 (1963).

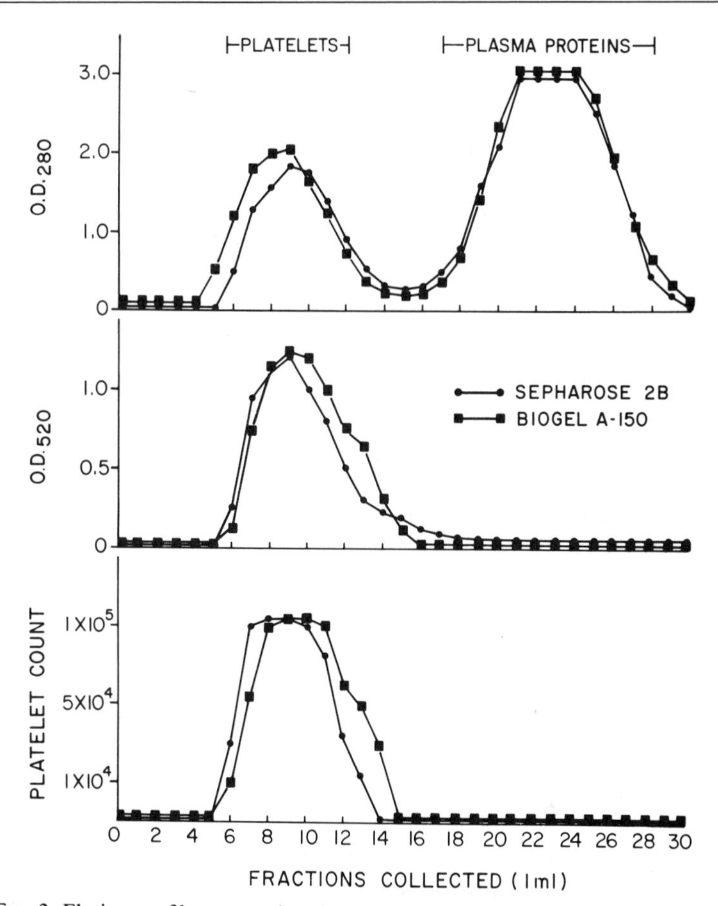

FIG. 3. Elution profiles comparing the patterns obtained when platelets are gel filtered through Sepharose 2B (●) or BioGel A-150 (■).(Thromb. Res., Vol. 12, Timmons and Hawiger, "Separation of Platelets from Plasma Proteins Including Factor VIII_{vWF} by a Combined Albumin Gradient-Gel Filtration Method Using Hepes Buffer," 1978, Pergamon Journals, Ltd.)

However, when aggregation is measured over a longer period of time (2 hr), the platelets separated and suspended in HEPES buffer maintain their responsiveness longer than the platelets suspended in Tyrode's buffer (Table I). It is noteworthy that the pH of the platelet suspension in HEPES buffer remains constant whereas the pH of the platelet suspension in Tyrode's buffer changes significantly upward due to the escape of CO_2.

Although the one-step gel filtration method is quick and simple, under ordinary conditions it does not separate platelets from the vWF protein and trace concentrations of fibrinogen. Changing the matrix to BioGel

TABLE I
EFFECT OF CONDITIONS AND TIME ON *in Vitro* RESPONSIVENESS TO AGGREGATING AGENTS OF PLATELETS SEPARATED BY METHOD B

Condition	Time (hr)			
	0	1	2	3
Tyrode's buffer				
pH	7.4	7.7	8.0	8.0
Aggregation[a] with ADP + Fbg[b]	10	8	0	0
Aggregation[a] with epinephrine[c]	4	2	0	0
Hepes buffer				
pH	7.4	7.4	7.4	7.4
Aggregation[a] with ADP + Fbg[b]	11.5	10	6.5	0
Aggregation[b] with epinephrine[c]	4	4	4	0

[a] Aggregation units expressed as slope value obtained by drawing a tangent line to the steepest rise in the curve and calculating the change per 1 min.

[b] ADP added to have a final concentration of 5 μM and fibrinogen final concentration of 1.5 mg/ml.

[c] Epinephrine added to have a final concentration of 50 μM.

(From Thromb. Res., Vol. 12, Timmons and Howiger, "Separation of Platelets from Plasma Proteins Including Factor VIII_{VWF} by a Combined Albumin Gradient-Gel Filtration Method Using Hepes Buffer," 1978, Pergamon Journals, Ltd.)

A-150, which has a higher exclusion limit than currently used Sepharose 2B, does not improve separation. Platelets separated in this way still aggregate with ristocetin without addition of plasma. In the two-step procedure tested (method A) the platelet-rich plasma is first applied to an albumin cushion and then to a gel column.[8] This has the advantage that platelets can be concentrated by reducing the volume of the resuspending buffer, but it does not separate platelets from plasma vWF as documented by a positive aggregation response to ristocetin. Finally, the use of a stepwise discontinuous albumin gradient prior to gel filtration results in a simple two-step method for separation of platelets from adsorbed plasma proteins including vWF. Platelets obtained through this procedure do not respond to ristocetin unless plasma as a source of vWF is added. Another modification of this procedure involves washing of platelets by centrifugation and resuspending the pellets in buffer before applying to a Sepharose 2B column.[24]

[24] G. A. Marguerie, T. S. Edgington, and E. F. Plow, *J. Biol. Chem.* **255,** 154 (1980).

The use of HEPES buffer, instead of Tyrode's buffer, provides conditions for a preparation of platelets that is more stable and reactive for a substantially longer period of time. The zwitterionic buffer (HEPES) has been developed for biologic systems[25] and has proved satisfactory as a tissue culture buffer.[13] It has a pK of 7.31, allowing the pH to remain constant for several hours. In contrast, commonly used Tyrode's buffer is prone to variation in CO_2 content of the platelet suspension and necessitates a variety of maneuvers to maintain the same level of CO_2.[26] By the use of HEPES buffer the need for control of CO_2 is obviated.

Platelets prepared by method B do not become reactive to ristocetin in the absence of plasma for at least 3 hr, indicating that under these conditions they are not "leaking" vWF protein which has been found to be associated with subcellular membranes and granules.[27] Thus, the system described here as method B is useful for the quantitative study of the interaction of plasma proteins with blood platelets requiring careful control of the environment of the intact, metabolically active platelets.[17,28] A similar procedure of platelet separation has been employed for studies of Na^+/H^+ exchange by using Na^+-free buffer.[29]

[25] N. E. Good, G. D. Winget, W. Winter, T. N. Connolly, S. Izawa, and N. M. Singh, *Biochemistry* **5,** 467 (1966).

[26] A. B. Rogers, *Proc. Soc. Exp. Biol. Med.* **139,** 1100 (1972).

[27] R. L. Nachman and E. A. Jaffe, *J. Exp. Med.* **141,** 1101 (1975).

[28] T. Fujimoto, S. Ohara, and J. Hawiger, *J. Clin. Invest.* **69,** 1212 (1982).

[29] T. M. Connolly and L. E. Limbird, *J. Biol. Chem.* **258,** 3907 (1983).

[3] Isolation and Characterization of Porcine Platelets

By BOGUSLAW RUCINSKI and STEFAN NIEWIAROWSKI

Introduction

The use of porcine platelets is of significance for research in the field of cell biology, thrombosis, and hemostasis. The pig is an important animal model for research of vascular diseases, atherosclerosis, diabetes, and congenital bleeding disorders, such as von Willebrand disease.[1-3] Porcine blood is readily available and the large-scale preparation of porcine platelets is simple and reproducible. A number of physiological and biochemical studies involving porcine platelets have been performed; these platelets are well characterized. Some functional properties of porcine and human platelets are similar, but some biochemical characteristics are different. The adequate characterization of porcine platelets and the availability of purified specific porcine platelet proteins, and their immunoassays, are of importance for investigators working on the role of platelet abnormalities in these diseases.

A number of investigators have developed methods to isolate platelets from porcine blood. Most of the methods are based on the modification of the procedure of Ardlie *et al.*[7] consisting of the collection of blood into acid–citrate–dextrose (ACD), differential centrifugation, and washing in a slightly acid milieu. Some investigators have also applied gel filtration[8] or differential centrifugation and washing at slightly alkaline conditions and in the presence of apyrase (EC 3.6.1.5),[9] i.e., under conditions originally

[1] V. Fuster, E. J. W. Bowie, J. C. Lewis, D. N. Fass, and C. A. Owen, *J. Clin. Invest.* **61,** 722 (1978).
[2] T. R. Griggs, R. P. Reddick, D. Sultzer, and F. M. Brinkhous, *Am. J. Pathol.* **102,** 137 (1981).
[3] W. J. Dodds, *Fed. Proc., Fed. Am. Soc. Exp. Biol.* **41,** 247 (1982).
[4] S. Niewiarowski, E. Regoeczi, G. J. Stewart, A. F. Senyi, and J. F. Mustard, *J. Clin. Invest.* **51,** 685 (1972).
[5] L. Salganicoff and M. H. Fukami, *Arch. Biochem. Biophys.* **153,** 726 (1972).
[6] B. Rucinski, A. Poggi, P. James, J. C. Holt, and S. Niewiarowski, *Blood* **61,** 1072 (1983).
[7] N. G. Ardlie, M. A. Packham, and J. F. Mustard, *Br. J. Haematol.* **19,** 7 (1970).
[8] K. M. Meyers, H. Holmsen, and C. L. Seachord, *Am. J. Physiol.* **243,** R454 (1982).
[9] R. L. Kinlough-Rathbone, A. Chahil, M. A. Packham, and J. F. Mustard, *Lab. Invest.* **32,** 352 (1975).

described for human platelets.[10] Here we describe a method for the large-scale isolation of porcine platelets which is routinely used in our laboratory. This method represents a modification of the procedures originally developed in Mustard's laboratory.[7,11]

Reagents and Buffers

ACD solution: Trisodium citrate, 25 g; citric acid, 14 g; glucose, 20 g dissolved in distilled water up to 1000 ml

Calcium-free Tyrode's buffer (washing medium A): NaCl, 9.5 g; KCl, 0.2 g; $NaHCO_3$, 1 g; Na_2HPO_4, 50 mg; $MgCl_2 \cdot 6H_2O$, 406 g; glucose, 1 g; albumin (free of Ca^{2+} and fatty acids), 3.5 g, and deionized water added to 1000 ml. The osmolarity of the washing fluid used for porcine platelets amounts to 340 and the pH should be adjusted to 6.3 (Table I)

Sodium chloride–citrate (washing medium B): NaCl, 9 g; sodium citrate, 3.8 g, and distilled water added up to 1000 ml; pH should be adjusted to 6.5

Tyrode's albumin buffer (suspending medium C): NaCl, 9 g; KCl, 0.2 g; $NaHCO_3$, 1 g; Na_2HPO_4, 50 mg; $MgCl \cdot 6H_2O$, 203 mg; $CaCl_2 \cdot 6H_2O$, 438 mg; glucose, 1 g; albumin (free of Ca^{2+} and fatty acids), 3.5 g, and deionized water added up to 1000 ml. The pH should be adjusted to 7.35 and osmolarity to 340. The pH can be stabilized with 2.5 m*M* HEPES buffer[12]

Fractionation buffer (medium D): Medium for suspension of isolated platelets for subcellular fractionation consists of 100 m*M* KCL, 50 m*M* Tris chloride, 5 m*M* $MgCl_2$ and 1 m*M* EDTA, pH 6.5

Procedure

Porcine blood is collected at a slaughterhouse into ACD (one part ACD added to six parts blood) with constant mixing to avoid formation of clots. In some procedures, ACD is supplemented with heparin (about 2 U/ml), but we have found that this is not necessary. Blood is sedimented in 20-liter containers for 2–3 hr at 1 gravitational force (*g*); sedimentation and centrifugation steps are carried out at room temperature. The supernatant fluid (representing about 25–35% of the total volume) is siphoned off and centrifuged in 500-ml polypropylene plastic containers at about 200 *g* for 15 min. The sediment contains predominantly red cells and leukocytes.

[10] J. F. Mustard, D. W. Perry, N. G. Ardlie, and M. A. Packham, *Br. J. Haematol.* **22,** 193 (1972).

[11] R. L. Kinlough-Rathbone, A. Chahil, and J. F. Mustard, *Am. J. Physiol.* **226,** 235 (1974).

[12] S. Timmons and J. Hawiger, *Thromb. Res.* **12,** 297 (1978).

TABLE I
COMPOSITION OF WASHING AND RESUSPENDING FLUIDS FOR ISOLATION OF PORCINE PLATELETS

Components	Washing: Medium A	Washing: Medium B	Suspending medium C
NaCl	9.5 g	9.0 g	9.0 g
KCl	0.2 g	—	0.2 g
$NaHCO_3$	1.0 g	—	1.0 g
Na_2HPO_4	0.05 g	—	0.05 g
$MgCl_2 \cdot 6H_2O$	0.4 g	—	0.2 g
$CaCl_2 \cdot H_2O$	—	—	0.438 g
Sodium citrate	—	3.8 g	—
Glucose	1.0 g	—	—
Albumin	3.5 g	—	3.5 g
pH (final)	6.3	6.5	7.35
Volume (final)	1000 ml	1000 ml	1000 ml

The supernatant (containing PRP; pH 6.3–6.5) is transferred again to 500-ml plastic containers and centrifuged for about 15 min at 1500 *g*. Platelet-poor plasma is siphoned off and discarded. A thin layer of plasma covering the sedimented platelets is also carefully removed. Approximately 2.5–3.0 ml of packed platelets is gently suspended into 40 ml washing fluid A or B and transferred to 50-ml plastic conical tubes. The suspension is centrifuged for 15 min at 1500 *g*. The layer of red cells should be left in the bottom of the tube. The platelets are resuspended in washing fluid A for subcellular fractionation or for functional studies. Washing fluid B is used when isolating platelets as a starting material for isolation and purification of platelet-specific proteins. The platelet suspension is centrifuged at 1500 *g* for 15 min. This step should be repeated twice.

The next step depends on the purpose of the experiment. Tyrode's albumin suspending fluid C is used if the platelet suspension is to be used for aggregation or other functional studies. The optimal platelet count for aggregation study is $3-5 \times 10^8$/ml. If the platelets are isolated for subcellular fractionation, they should be resuspended in an ice-cold medium D at a ratio of 10–15 g wet weight of platelets per 16–20 ml of medium.[13] If platelets are isolated as a starting material for purification of platelet factor 4, platelet basic protein, and platelet-derived growth factor (PDGF), they

[13] L. Salganicoff, P. A. Hebda, J. Yandrasitz, and M. H. Fukami, *Biochim. Biophys. Acta* **385,** 394 (1975).

should be resuspended in medium B or in distilled water (5 ml of water/ml of platelets) and disrupted by freezing and thawing three times. The final platelet count should be about $5 \times 10^9 - 10^{10}$/ml.

Comparison of Human and Porcine Platelets

The optimal conditions for separation of human and porcine platelets differ. Differential centrifugation followed by resuspension in Ca^{2+}, Mg^{2+}-Tyrode's buffer with apyrase at pH 7.35[10] appears to be the most suitable method for large-scale separation of fresh human platelets. In our hands, usage of slightly acidic washing fluid A or B gave better results with porcine platelets. It is noteworthy that porcine platelets require high concentrations of salt (osmolarity 340) for their stability as compared to human platelets (osmolarity 290).[4,10,11] Some functional properties of human and porcine platelets are similar. Suspensions of porcine or human platelets can be aggregated by similar ranges of concentrations of ADP in the presence of fibrinogen[14] and by collagen or thrombin.[15] Porcine platelets, in contrast to human platelets, do not aggregate with epinephrine.[15] Sensitivity of human and porcine platelets to thrombin is also similar, as evaluated by measuring release of radiolabeled serotonin and loss of cytoplasmic constituents.[9] However, porcine platelets are not aggregated by sodium arachidonate. Since they respond with shape change, they possess prostaglandin/thromboxane receptors but these have little activity compared to human platelets.[16] Cierniewski *et al.*[17] demonstrated that ^{125}I-labeled fibrinogen binds specifically to ADP-stimulated porcine platelets. The values for the number of fibrinogens binding sites (12,400–25,000/platelet) and for association constant (5×10^8 M^{-1}) are similar to those described for human platelets.[18-20] Grinstein and Furuya identified two glycoproteins (M_r 110,000 and 125,000) on the surface of porcine platelets.[21] They suggested that these glycoproteins are equivalent to glycoprotein IIb and IIIa of the human platelet and may therefore be involved in fibrinogen binding. As

[14] S. Niewiarowski, E. Regeczi, and J. F. Mustard, *Ann. N.Y. Acad. Sci.* **201,** 72 (1972).

[15] D. P. Thomas, S. Niewiarowski, and V. J. Ream, *J. Lab. Clin. Med.* **75,** 607 (1970).

[16] K. M. Meyers, J. B. Katz, R. M. Clemmons, J. B. Smith, and H. Holmsen, *Thromb. Res.* **20,** 13 (1980).

[17] C. S. Cierniewski, M. A. Kowalska, T. Krajewski, and A. Janiak, *Biochim. Biophys. Acta* **714,** 543 (1982).

[18] G. A. Marguerie, T. S. Edgington, and E. F. Plow, *J. Biol. Chem.* **255,** 154 (1980).

[19] J. Hawiger, S. Parkinson, and S. Timmons, *Nature (London)* **283,** 154 (1980).

[20] E. Kornecki, S. Niewiarowski, T. A. Morinelli, and M. Kloczewiak, *J. Biol. Chem.* **256,** 5696 (1981).

[21] S. Grinstein and W. Furuya, *Comp. Biochem. Physiol. B* **78,** 657 (1984).

determined by the immunoblotting technique,[22] anti-human GPIIIa antibody recognizes porcine GPIIIa (Rucinski *et al.*, unpublished observations).

We have isolated and characterized two heparin-binding proteins secreted by thrombin-stimulated porcine platelets; porcine platelet factor 4 and porcine basic protein.[6] The homology of these proteins with human platelet factor 4 and a group of human proteins binding to heparin with low affinity (platelet basic protein, low-affinity platelet factor 4, β-thromboglobulin) was established.[6] Measurements by radioimmunoassay on eight samples of porcine platelet-rich plasma indicate that platelet factor 4 and platelet basic protein amount to 10.6 μg (±5.1 SD)/10^9 platelets and 1.1 μg (±0.59 SD)/10^9 platelets, respectively. In human platelets, the figures for homologous proteins are 12.4 and 24.2 μg, respectively.[6] Porcine PDGF has been purified by our group,[23] and by Stroobant and Waterfield.[24] This protein is strikingly homologous with human PDGF.

Some differences between human and pig platelets should be emphasized. Histamine is a major constituent of dense granules of pig platelets, whereas serotonin is the principal amine of human platelets and of platelets from most species.[25] Fukami *et al.*[26] demonstrated that storage of histamine and serotonin in porcine platelets occurs through independent mechanisms. In contrast to human platelets, which have relatively high levels of calcium in their dense granules, porcine platelets contain high amounts of magnesium and do not secrete calcium when stimulated by thrombin.[27] The total contents of ADP and ATP, and of the secretable fraction of these nucleotides in porcine and in human platelets are similar.[8]

Acknowledgments

The authors wish to acknowledge support from NIH (Grants 14217, HL15226, and HL36579) for their research.

[22] H. Towbin, T. Staehelin, and J. Gordon, *Proc. Natl. Acad. Sci. U.S.A.* **76,** 4350 (1979).

[23] A. Poggi, B. Rucinski, P. James, J. C. Holt, and S. Niewiarowski, *Exp. Cell Res.* **150,** 436 (1984).

[24] P. Stroobant and M. D. Waterfield, *EMBO J.* **1,** 2963 (1984).

[25] M. DaPrada, J. G. Richards, and K. Kettler, *in* "Platelets in Biology and Pathology" (J. L. Gordon, ed.), Vol. 2, p. 101. Elsevier/North-Holland, Amsterdam, The Netherlands, 1981.

[26] M. H. Fukami, H. Holmsen, and K. Ugurbill, *Biochem. Pharmacol.* **33,** 3869 (1984).

[27] K. Ugurbill and H. Holmsen, *in* "Platelets in Biology and Pathology" (J. L. Gordon, ed.), Vol. 2, p. 147. Elsevier/North-Holland, Amsterdam, The Netherlands, 1981.

[4] Isolation of Platelets from Laboratory Animals

By JAMES L. CATALFAMO and W. JEAN DODDS

Known procedures for the separation of human platelets from peripheral blood are frequently adapted with variable success to the isolation of platelets from laboratory animal species.[1] Such procedures adapted to animal platelets are described as modifications of established human methods without specifying the details applied. Existing animal models of thrombosis and hemorrhagic diseases permit comparative studies of structural and functional abnormalities of platelets.[2] In basset hounds, for example, a hereditary signal transduction disorder associated with elevated intraplatelet cyclic AMP and abnormal cAMP phosphodiesterase activity has been identified.[3,4] This disorder provides an unique opportunity to study cAMP's putative role as an antagonist second messenger. Platelets from rats,[5] mice,[6] and pigs[7] exhibit platelet storage pool disease similar to that of humans. Bone marrow transplantation has successfully corrected the prolonged bleeding times observed in mice with the defect.[8] Defects involving decreased numbers of platelet dense granules (Chediak–Higashi syndrome) have been characterized in cats,[9] cattle,[10] mink,[11] and foxes.[12]

[1] H. J. Day, H. Holmsen, and M. B. Zucker, *Thromb. Diath. Haemorrh.* **33,** 648 (1975).
[2] W. J. Dodds, *ILAR News* **30,** R3 (1988).
[3] J. L. Catalfamo, S. L. Raymond, J. G. White, and W. J. Dodds, *Blood* **67,** 1568 (1986).
[4] M. K. Boudreaux, W. J. Dodds, D. O. Slauson, and J. L. Catalfamo, *Biochem. Biophys. Res. Commun.* **140,** 595 (1986).
[5] S. L. Raymond and W. J. Dodds, *Thromb. Diath. Haemorrh.* **33,** 361 (1975).
[6] M. Reddington, E. K. Novak, E. Hurley, C. Medda, M. P. McGarry, and R. T. Swank, Blood **69,** 1300 (1987).
[7] T. M. Daniels, D. N. Fass, J. G. White, and E. J. W. Bowie, *Blood* **67,** 1043 (1986).
[8] M. P. McGarry, E. K. Novak, and R. T. Swank, *Exp. Hematol.* **14,** 261 (1986).
[9] K. M. Meyers, C. L. Seachord, H. Holmsen, and D. J. Prieur, *Am. J. Hematol.* **11,** 241 (1981).
[10] K. M. Meyers, H. Holmsen, C. L. Seachord, G. E. Hopkins, R. E. Borchord, and G. H. Padgett, *Am. J. Physiol.* **237,** R239 (1979).
[11] K. M. Meyers, H. Holmsen, C. L. Seachord, G. Hopkins, and J. Gorham, *Am. J. Hematol.* **7,** 137 (1979).
[12] N. Nes, B. Lium, M. Braend, and O. Sjaastad, *Veterinaertidsskrift* **89,** 313 (1983).

Morphological and biochemical features which are inherent to animal platelets are often exploited to answer unsolved questions regarding human platelets. In contrast to human cells, bovine platelets do not have a well-defined open canalicular system.[13] Studies on the secretory pathway of bovine platelets have recently demonstrated the importance of channels for granule secretion and suggest that new channels rapidly form in response to stimuli.[14] Platelets from a majority of mongrel dogs (>70%) do not aggregate when stirred with arachidonate. These platelets have been used as a source of thromboxane A_2 (TXA_2) to study the response of human platelets to thromboxane.[15] The metabolism of phosphatidylinositol (PI) in thrombin-stimulated horse platelets is different from that described for human platelets.[16] PI is rapidly and completely converted via diacylglycerol to phosphatidic acid, which apparently initiates the activation of phospholipase A_2 and arachidonic acid release.

The lack of risk for transmission of viral diseases such as non-A, non-B hepatitis and AIDS, standardized environmental, health, and genetic factors in laboratory-bred animals, cost, and relative ease of adaptation to various experimental protocols are all factors which influence decisions to use small animals in platelet research. Small animals are ideally suited for studies designed to evaluate *in vitro* and *in vivo* effects of pharmacologic agents on platelet function, development of blood compatible materials for prostheses and artificial organs, and the study of experimentally induced or naturally occurring models of thrombosis and hemorrhagic disease.

While the variation in platelet number, size, and density among various small laboratory animals can be considerable,[17] these differences, especially in relation to platelet density, do not preclude using similar conditions for the isolation of platelets from the peripheral blood of the majority of species. Functionally intact, unactivated platelets can be obtained in high yields from dogs, cats, rabbits, guinea pigs, rats, hamsters, and mice. Red cell contamination of the platelet-rich plasma (PRP) under these conditions is negligible. The method described herein for platelet isolation from small laboratory animals can also be used to isolate human, monkey, horse, cow, sheep, goat, and pig platelets.

[13] D. Zucker-Franklin, K. A. Benson, and K. M. Meyers, *Blood* **65,** 241 (1985).
[14] J. G. White, *Blood* **69,** 878 (1987).
[15] G. J. Johnson, L. A. Leis, and G. S. Francis, *Circulation* **73,** 847 (1986).
[16] E. G. Lapetina, M. M. Billah, and P. Cuatrecasas, *J. Biol. Chem.* **256,** 5037 (1981).
[17] W. J. Dodds, *in* "Platelets: A Multidisciplinary Approach" (G. deGaetano and S. Garattini, eds.), p. 45. Raven, New York, 1978.

Procedures

Successful platelet isolation depends on the quality of blood sample collected, which will be influenced by the collection technique, the general health of the animal, and uniformity of the subgroup or strain used. The techniques described herein reflect our experience with each species as indicated below. Whenever possible, the animal should be fasted overnight and should not have received medications for at least 2 weeks prior to blood collection.

The Choice of Anticoagulant

Trisodium citrate [3.8% (w/v)] is the anticoagulant or choice. Sodium heparin (porcine intestinal mucosa, Sigma Chemical Co., St. Louis, MO) made to 100 units/ml in 0.15 *M* NaCl, or 54 m*M* EDTA in 0.15 *M* NaCl, pH 7.4, can also be used if required by the nature of the experiment. However, both heparin and EDTA may have deleterious effects on platelets. All three anticoagulants should be sterile filtered (0.20-μm disposable filter unit, Nalge Corp., Rochester, NY) prior to use and each is used at a ratio of one part anticoagulant to nine parts whole blood. As the quality of commercially available heparin varies and some preparations reportedly induce spontaneous platelet aggregation, caution is advised. Heparins which contain preservatives should be avoided. Trisodium citrate (0.32% final concentration) is sometimes used for preparation of animal PRP; however, platelet activation is likely to occur at final citrate concentrations below 0.38%. In situations where the animal's hematocrit value is unusually high, it might be advantageous to use citrate at a lower concentration adjusted for increased hematocrit.[18]

Blood Collection

Dog, cat, and rabbit peripheral blood can readily be obtained in adequate amounts from these animals without the use of an anesthetic agent and with minimal discomfort or stress, the latter situation being important for obtaining samples of high quality and yield. A second person, preferably an experienced animal handler, is necessary to gently restrain the animal during the procedure and to facilitate collection. Specific techniques apply for each species and are presented in detail.

Dogs. The dog is placed in a comfortable position (preferably on or over the handler's lap on a counter or the floor) which will allow easy access to the cephalic vein in the foreleg region between the dog's carpus

[18] R. M. Hardisty, *Br. J. Haematol.* **19**, 307 (1970).

and elbow. The vein is occluded manually or with a light tourniquet, palpated, and then *cleanly* punctured using a 21-gauge hypodermic needle attached to a disposable 1-ml plastic syringe. After the initial 1.0 ml of blood is collected, a second syringe containing the appropriate $\frac{1}{10}$ volume of anticoagulant is attached to the needle and the requisite blood volume is slowly and steadily withdrawn. Alternative collection sites include the saphenous veins of the hidlegs or the jugular veins, although the latter are usually reserved for larger volume collections. Samples ranging in volume from 10 to 60 ml can be safely and easily collected from mid- to large-sized dogs at 3-week intervals without inducing hematopoietic or other physiological stress alterations in blood volume or composition.

Cats. Cats should be minimally restrained for blood collection. Fractious animals may require short-acting anesthesia with such agents as ketamine–HCl or the thiobarbituates. A veterinarian should be consulted about the best type and dosage of anesthetic to be used. These drugs also may affect platelet function. Small blood samples of less than 5 ml can be obtained from the cephalic vein as described for the dog. Samples of 5- to 20-ml volume are best obtained by restraining the cat's forelegs firmly within a towel or gloves to avoid being scratched and then gently extending the animal's neck over the edge of a table or counter. The jugular vein is occluded manually, palpated, and a clean puncture is made using a 21-gauge hypodermic needle; the sample is withdrawn by the two-syringe technique as previously described.

Rabbits. The rabbit is placed in a shallow tray and its back is draped with a small towel to reduce its inclination to hop out. The animal is held in place from behind by an assistant who also covers the animal's eyes with his/her hands. Plastic restraining boxes may be used; however, the rabbit still has considerable mobility and if the box is not precisely adjusted serious injury to the animal's back may occur if it jumps. The central ear artery is dilated by rubbing a xylene-soaked cotton applicator along its length. The artery is cleanly punctured using a 22-gauge hypodermic needle attached to a 1-ml plastic syringe. After collecting 1 ml of blood a second syringe containing the $\frac{1}{10}$ volume of anticoagulant is used. Up to 30 ml of blood can be withdrawn without difficulty from a 1.5- to 3-kg rabbit. Larger rabbits can have 40–60 ml withdrawn every 3 weeks. The arterial puncture wound must be held off firmly with a gauze sponge until all blood flow stops. The xylene must then be removed to avoid irritation and damage to the vessel by washing the area with a 70% ethanol-soaked applicator followed immediately by distilled H_2O. Mineral oil is then applied to lubricate the area. Ears treated in this manner will heal rapidly and can be sampled repeatedly during experimental protocols without thrombosis or discomfort to the animal.

Guinea Pigs, Rats, Hamsters, and Mice. For these animals anesthesia is necessary prior to blood collection. Standard sodium pentobarbital solution (Fort Dodge Laboratories, Fort Dodge, IA, or an equivalent source) supplied at a concentration of 64.8 mg/ml is diluted 1:1 with saline to provide a working concentration of 32.4 mg/ml, thereby enhancing the margin of safety during use of small volumes. The anesthetic is administered intraperitoneally, with the animal held upright over a counter, using a 1-ml tuberculin syringe and a 23- or 24-gauge hypodermic needle. The injection should be made lateral to the umbilicus to avoid major organs. The dosage required to induce a deep plane of surgical anesthesia from which the animal will not recover varies with the species.[19] Guinea pigs will require approximately 10 mg sodium pentobarbital/100 g body weight, while rats, hamsters, and mice require approximately 15, 30, and 90 mg/100 g, respectively.

Once the animal has reached the surgical plane of anesthesia (fails to respond to a toe pinch), a midline abdominal incision is made. The organs are moved to one side to facilitate location of the inferior vena cava, which is punctured using a 22-gauge hypodermic needle. The sample is withdrawn into a plastic syringe containing $\frac{1}{10}$ volume of anticoagulant. Ten milliliters of blood is easily obtained from an adult guinea pig as well as from a large, adult rat (350 g) by this technique. Two to three milliliters of blood can be collected from an adult hamster but only 1–2 ml can realistically be obtained with good blood flow from large adult mice. Blood from several of these very small animals is generally pooled to obtain adequate amounts of PRP for study. The amount of blood to be drawn must be accurately estimated beforehand to avoid overanticoagulation of samples. The animal's heart may be gently massaged during collection to obtain maximal volumes. If necessary, a lethal dose of sodium pentobarbital is given to the animal via the heart or chest following the procedure.

We do not recommend cardiac puncture as the method for collecting blood from anesthetized laboratory animals for hemostatic studies, because both platelets and clotting factors are activated by this technique.

Platelet Preparation

The techniques required to isolate platelets from the peripheral blood of all seven species are similar. Only blood samples obtained using a two-syringe technique and with minimal tissue trauma are suitable for platelet preparation. Samples which exhibit signs of red cell hemolysis or

[19] W. V. Lumb and E. W. Jones, "Veterinary Anesthesia." Lea & Febiger, Philadelphia, Philadelphia, 1984.

fibrin formation cannot be used. The freshly obtained anticoagulated blood is gently mixed and transferred to 10- to 15-ml round-bottom polycarbonate centrifuge tubes (Nalge Corp., Rochester, NY). The column of blood should preferably come to within about 4 mm of the top of the tube. Partially filled or an unevenly filled series of tubes will be exposed to centrifugal forces greater than optimal and excessive platelet loss with poor yields will likely occur. The tubes should be balanced and then spun at room temperature and 1800 rpm (650 *g*, radius from center of column of blood to axis of rotation) for three successive 1.5- to 2.0-min intervals in a model UV International (Needham Hts., MA) or equivalent centrifuge equipped with a swinging bucket rotor. Between each spin, the centrifuge is allowed to coast to a stop without braking. The supernatant platelet-rich layer of each sample is collected and transferred using a siliconized or plastic Pasteur pipet to another polycarbonate test tube and capped with Parafilm. The column of remaining packed cells is left undisturbed and is then recentrifuged. Occasionally, a fourth centrifugation may be required to obtain the optimal yield of PRP (85% and about one-third to one-half the starting volume of whole blood). It is important to monitor the distribution of platelets during the first two centrifugations. If they appear to be concentrated close to the packed red cell interface, then the centrifugal force can be adjusted downward to approach 520 *g*, at average radius.

The number of platelets in the PRP is enumerated by either phase-contrast microscopy or electronic particle counting using a Coulter model ZBI (Coulter Electronics, Inc., Hialeah, FL) or equivalent instrument with adjustable upper and lower threshold limits; the optimal threshold settings vary for each species and must be established prior to use of electronic counting techniques. The final platelet number in PRP is standardized to 600,000/μl for rats, mice, and hamsters and to 300,000/μl for the other species using fresh, autologous platelet-poor plasma (PPP). PPP is obtained by spinning the remaining final packed blood cell fraction for 20–30 min at room temperature and 1400–1800 *g* at average radius. If the platelets have not been activated, a noticeable swirling should be observed when the platelet suspension is gently agitated or mixed. Platelets which fail to "swirl" respond weakly to various agonist stimuli. The PRP should be kept at room temperature in a tightly capped tube, and should be utilized within 3 hr of blood collection to assure maximal reactivity. It is generally advisable to allow the platelets to reequilibrate in their PPP from effects of centrifugation for a 30-min "rest" period before use.

Separation of Platelets from Plasma Proteins

Platelet-rich plasma from all seven species can be used to obtain gel-filtered platelets (GFP) essentially as described by Lages *et al.*[20] The platelets (3–5 ml, at 7×10^8 platelets/ml) are separated on a 1.5×30 cm column packed with acetone-washed Sepharose 4B (Pharmacia Fine Chemicals, Piscataway, NJ), equilibrated with Tangen-HEPES buffer[21] (147 mmol/liter of NaCl, 5 mmol/liter KCl, 0.05 mmol/liter $CaCl_2$, 0.1 mmol/liter $MgCl_2$, 5 mmol/liter HEPES, 5.5 mmol/liter glucose, pH 7.4) containing 0.35% bovine albumin (Pentex fraction V, Miles Scientific, Naperville, IL). The column must be siliconized and the nylon mesh support replaced with one having mesh openings of 44 μm (Small Parts, Inc., Miami, FL). Following filtration, the GFP preparation is made 1 m*M* in $CaCl_2$. Under these conditions >85% platelet recovery can be expected from PRP. Human or species-specific fibrinogen (0.1% final concentration) is added to the GFP for most aggregation studies.

Platelets from small laboratory animals can be also separated from nonadsorbed plasma proteins using citrated PRP and the stractan-gradient isolation technique of Corash *et al.*[22] with only minor modification (3.8% sodium citrate as anticoagulant and BSG–citrate adjusted to pH 7.2). Other platelet isolation techniques[23,24] that use albumin-density gradient or citrate–saline–EDTA buffers yield variable results, and are not suited for preparing platelets from small animals. Particularly problematical is the varying pH optimum for washing and maintaining different animal platelets in nonactivated state. The reason for the differences is not clear but may relate to known species variation in pH-dependent erythrocyte fragility[25] and consequent release of ADP during the washing procedure.

Comments

The techniques described here permit the rapid isolation of sufficient numbers of platelets using relatively small blood samples. Such platelets are usually functionally intact and respond to low doses of all major platelet agonists. Unless there is an unquestionable need to study platelets from mice and hamsters their sacrifice is unnecessary and obtaining plate-

[20] B. Lages, M. C. Scrutton, and H. Holmsen, *J. Lab. Clin. Med.* **85,** 811 (1975).

[21] R. C. Carroll and J. M. Gerrard, *Blood* **59,** 466 (1982).

[22] L. Corash, B. Shafer, and M. Perlow, *Blood* **52,** 726 (1978).

[23] P. N. Walsh, *Br. J. Haematol.* **22,** 205 (1972).

[24] N. G. Ardlie, M. A. Packham, and J. F. Mustard, *Br. J. Haematol.* **19,** 7 (1970).

[25] U. Giger, J. W. Harvey, R. A. Yamaguchi, P. K. McNulty, A Chiapella, and E. Beutler, *Blood* **65,** 345 (1985).

lets from these animals is not recommended due to inherent problems with their activation. The use of an anesthetic agent may in some cases be undesirable for the studying of specific platelet function, e.g., serotonin transport. The dose of sodium pentobarbital can be decreased to the minimal amount needed to achieve a suitable plane of anesthesia or an alternate anesthetic can be chosen. The doses indicated above are meant only as guidelines.

Acknowledgments

The authors wish to acknowledge support from NIH Grant HL09902.

Section II

Platelet Adhesion, Aggregation, and Turnover

[5] Measurements of Platelet Interaction with Components of the Vessel Wall in Flowing Blood

By KJELL S. SAKARIASSEN, RETO MUGGLI, and HANS R. BAUMGARTNER

Numerous approaches have been developed in order to study platelet–surface interactions. The interactions have been studied in a variety of systems and evaluated by using radiolabeled platelets,[1] by measuring the decline in platelet count following exposure to test surfaces,[2-4] and by turbidometry.[5] The information obtained with these methods is in most instances limited by the fact that they do not differentiate between platelet–surface and platelet–platelet interactions.[6] Platelet interactions with the vascular subendothelium or other collagenous substrates are rapidly followed by platelet–platelet cohesion. Unambiguous differentiation of these surface-induced interactions is achieved only by morphologic inspection.[6] However, indirect methods such as radiolabeling may be sufficient if nonthrombogenic surfaces are used and/or if platelet aggregation is inhibited.[7,8]

This chapter describes methods which differentiate and quantify the various platelet–surface interactions observed under *in vitro* conditions which mimic the sequence of events leading to thrombosis in various parts of the circulation *in vivo.* The chapter describes perfusion chambers and systems with laminar (nonturbulent) blood flow and the preparation of vascular and artificial surfaces as well as blood perfusates for the study of platelet–surface interactions. By using perfusion systems the importance

[1] J.-P. Cazenave, M. A. Packham, and J. F. Mustard, *J. Lab. Clin. Med.* **82,** 978 (1973).

[2] A. J. Hellem, *Scand. J. Clin. Lab. Invest.* **12** (Suppl. 51), 1 (1960).

[3] E. W. Salzman, *J. Lab. Clin. Med.* **62,** 724 (1963).

[4] Y. J. Legrand, F. Fauvel, G. Kartalis, J. L. Wautier, and J. P. Caen, *J. Lab. Clin. Med.* **94,** 438 (1979).

[5] T. H. Spaet and I. Lejnieks, *Proc. Soc. Exp. Biol. Med.* **132,** 1038 (1969).

[6] V. T. Turitto and H. R. Baumgartner, *in* "Hemostasis and Thrombosis: Basic Principles and Clinical Practice" (R. W. Colman, J. Hirsh, V. J. Marder, and E. W. Salzman, ed.), p. 364. Lippincott, Philadelphia, Pennsylvania, 1982.

[7] K. S. Sakariassen, P. A. Bolhuis, and J. J. Sixma, *Nature (London)* **279,** 636 (1979).

[8] P. A. Bolhuis, K. S. Sakariassen, H. J. Sander, B. N. Bouma, and J. J. Sixma, *J. Lab. Clin. Med.* **97,** 568 (1981).

of rheologic parameters and of various vessel wall components for platelet–surface interactions was elucidated.[9-12] Results obtained with blood from patients served as a framework for characterization of some bleeding disorders.[13-20]

Definitions

The division of platelet–surface interactions into adhesion and thrombus formation is based on morphologic criteria derived from semithin cross sections of expoxy-embedded vessels from *in vivo* and *in vitro* experiments.[10,21] The sequence of events is illustrated in Fig. 1. The initial interaction between a platelet and a surface occurs with a minute part of the platelet membrane. Initially, such an attachment may be reversible. However, with increasing parts of the membrane involved, attachment becomes progressively irreversible. Platelets are classified as *contact platelets* (Fig. 1A) as long as the platelet diameter that is perpendicular to the surface is larger than the membrane fraction attached to the surface. Subsequently, the platelets continue to spread out on the surface. They are classified as *spread platelets* (Fig. 1B) as soon as the membrane fraction attached to the surface has become larger than the platelet diameter perpendicular to the surface. Other platelets interact with the spread platelets and with each other, producing a *mural thrombus* (Fig. 1C). In citrated blood, thrombus formation and growth occur steadily during the first 5 min. Subsequently, the thrombi disappear gradually, leaving a carpet of adherent platelets which cover 100% of the surface (Fig. 2).

[9] H. R. Baumgartner and C. Haudenschild, *Ann. N.Y. Acad. Sci.* **201,** 22 (1976).

[10] H. R. Baumgartner, *Microvasc. Res.* **5,** 167 (1973).

[11] V. T. Turitto and H. R. Baumgartner, *Microvasc. Res.* **9,** 335 (1975).

[12] H. R. Baumgartner, R. Muggli, T. B. Tschopp, and V. T. Turitto. *Thromb. Haemostasis* **35,** 124 (1976).

[13] T. B. Tschopp, H. J. Weiss, and H. R. Baumgartner, *J. Lab. Clin. Med.* **83,** 296 (1974).

[14] H. J. Weiss, T. B. Tschopp, H. R. Baumgartner, I. I. Sussman, M. M. Johnson, and J. J. Egan, *Am. J. Med.* **57,** 920 (1974).

[15] H. J. Weiss, T. B. Tschopp, and H. R. Baumgartner, *N. Engl. J. Med.* **293,** 619 (1975).

[16] T. B. Tschopp, H. J. Weiss, and H. R. Baumgartner, *Experientia* **31,** 113 (1975).

[17] H. J. Weiss, D. Meyer, R. Rabinowitz, G. Pietu, J.-P. Girma, W. J. Vicic, and J. Rogers, *N. Engl. J. Med.* **306,** 326 (1982).

[18] K. S. Sakariassen, M. Cattaneo, A. van der Berg, Z. M. Ruggeri, P. M. Mannucci, and J. J. Sixma, *Blood* **64,** 229 (1984).

[19] K. S. Sakariassen, H. K. Nieuwenhuis, and J. J. Sixma, *Br. J. Haematol.* **59,** 459 (1985).

[20] H. K. Nieuwenhuis, K. S. Sakariassen, W. P. M. Houdijk, P. F. E. M. Nievelstein, and J. J. Sixma, *Blood* **68,** 692 (1986).

[21] H. R. Baumgartner, and R. Muggli, *in* "Platelets in Biology and Pathology" (J. L. Gordon, ed.), p. 23. North-Holland, Amsterdam, The Netherlands, 1976.

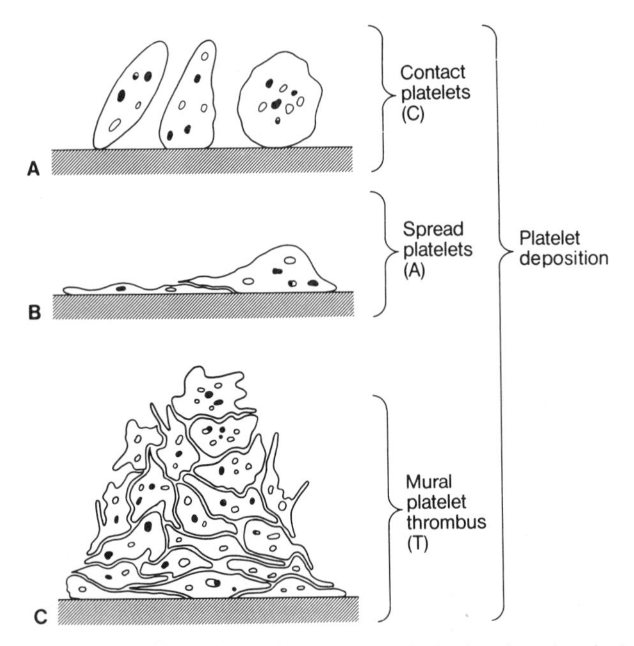

FIG. 1. The sequence of events leading to a mural platelet thrombus includes initial attachment and early phase of spreading (A), spreading and release (B), and platelet–platelet cohesion (aggregation) on spread platelets (C). (A) Three types of contact platelets. The left and the right platelets make contact with the surface with a small portion of their membrane. The middle platelet is attached to the surface with a larger portion of its membrane than the other two platelets. (B) Platelet spreading. The left platelet is completely spread out on the surface and interacts intimately with the surface. Degranulation has occurred. The right platelet is partly spread out on the surface. Spread platelets usually overlap each other partly. (C) Mural platelet thrombus. Platelets at the base of the thrombus are degranulated. Less degranulation occurs in the middle and the more peripheral parts of the thrombus.

Platelet Adhesion

Platelets directly associated with the surface are defined as adherent platelets and comprise both contact *(C)* and spread *(A)* platelets (Fig. 1). Platelet adhesion is quantified as the percentage of the total surface covered with adherent platelets [surface coverage with $C + A + T$ (mural platelet thrombi)], or as the number of adherent platelets per surface unit area. The former quantitation is direct and relies on the morphometry of the semithin sections.[9,12,21] The latter quantitation makes use of radiolabeled platelets and experimental conditions which do not trigger platelet–platelet interaction.[7] Platelet adhesion can also be quantified by means of

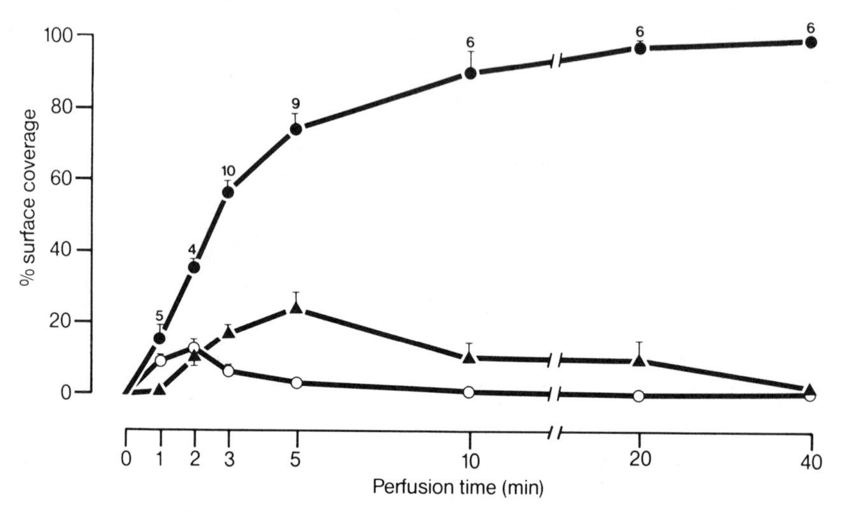

FIG. 2. Time course of platelet adhesion and thrombus formation. Citrated rabbit blood was perfused for the times indicated through the small annular chamber (Table I) with exposed rabbit aorta subendothelium at 40 ml/min (2600 sec^{-1} shear rate). Percentage surface coverage with contact platelets (○), spread platelets (●), and thrombi more than 5 μm in height (▲) are shown (mean ± SE, $n = 4-10$).

automated microdensitometry,[22] and by microscopy of *en face* preparations.[12,23] Differentiation between contact and spread platelets is only achieved by cross-sectional morphometry.[12]

Platelet Deposition

Platelet deposition is defined as the number of platelets deposited per unit surface area as measured by radiolabeled platelets and includes all types of platelet–surface and platelet–platelet interactions.[23] On collagenous surfaces platelet deposition correlates well with the platelet thrombus volume per unit surface area[24]; in the absence of platelet aggregation it serves as a measure for platelet adhesion.[7]

[22] R. Muggli, H.R. Baumgartner, T. B. Tschopp, and H. Keller, *J. Lab. Clin. Med.* **95,** 195 (1980).

[23] K. S. Sakariassen, P. A. M. M. Aarts, P. G. de Groot, W. P. M. Houdijk, and J. J. Sixma, *J. Lab. Clin. Med.* **102,** 522 (1983).

[24] T. B. Tschopp, *Thromb. Res.* **11,** 619 (1977).

Platelet Thrombus Formation, Volume, Height, and Growth

The percentage of the total surface covered with mural thrombi (Fig. 1) more than 5 μm in height is taken as a parameter for *thrombus formation.*[15,21] This term is by definition arbitrary since it includes all thrombi higher than 5 μm regardless of their absolute height. It is a reflection of the frequency of thrombus growth above 5 μm rather than a measure for thrombus volume and thus complementary to the latter one.

The average *thrombus volume* per unit surface is a precise measure of the total platelet mass (excluding a platelet monolayer) attached to a surface but gives no information regarding the shape of the thrombi. Further information is gained by determining *thrombus height* (distance between a thrombus peak and the surface). The most complete information is obtained by combining measurements of platelet adhesion and the distribution of thrombus volumes and heights, respectively.[25] To determine *thrombus growth* several time points are required. Average thrombus growth per minute can then be expressed either in height or in volume per surface area per minute.[25,26]

Important Parameters Affecting Platelet–Surface Interactions

Physical and chemical factors have profound effects on platelet adhesion and growth of mural thrombi. Therefore, they are described briefly below.

Physical Factors

Temperature, Platelet Count, and Hematocrit. Lowering the temperature below 37° decreases platelet adhesion and thrombus formation.[27] Increasing the platelet count and increasing hematocrit enhance the rate of platelet adhesion and thrombus formation.[21] Platelet adhesion increases about 60-fold when the hematocrit is raised from 0 to 40%.[11]

Shear Rate. Platelet–surface interactions are strongly influenced by wall shear rates.[28,29] The shear rate (sec^{-1}) is a parameter used to characterize laminar flow. In tubes, laminar blood flow has a parabolic flow velocity profile with increasing flow velocity of the adjacent streaming layers reach-

[25] H. R. Baumgartner, V. T. Turitto, and H. J. Weiss, *J. Lab. Clin. Med.* **95,** 208 (1980).
[26] H. R. Baumgartner and K. S. Sakariassen, *Ann. N.Y. Acad. Sci.* **454,** 162 (1985).
[27] V. T. Turitto and H. R. Baumgartner, *Haemostasis* **3,** 224 (1974).
[28] V. T. Turitto and H. R. Baumgartner, *Microvasc. Res.* **17,** 38 (1979).
[29] V. T. Turitto, H. J. Weiss, and H. R. Baumgartner, *Microvasc. Res.* **19,** 352 (1980).

ing a maximum at the luminal axis.[30] The rate of radial velocity increase in adjacent layers is defined as the shear rate which is highest at the vessel wall: wall shear rate. At a given average flow rate, the tube radius determines the shear rate. In the human circulation, the wall shear rates vary from 30 to 40 sec^{-1} in the largest veins to 5000 sec^{-1} in the microcirculation.[6]

Experimental findings and theoretical considerations have indicated that the radial platelet transport toward the vessel wall increases with increasing shear rate. Oscillations of erythrocytes in flowing blood have been credited for this effect.[11] This physical effect is substantiated by the close correlation observed between the rate of platelet adhesion and shear rate up to shear rates of 650 sec^{-1}. At shear rates above 650 sec^{-1} the rate of human platelet adhesion to rabbit aorta subendothelium levels off asymptotically.[28] Adhesion is predominantly transport controlled at shear rates below 650 sec^{-1} and becomes progressively more controlled by platelet–surface reaction as shear rates increase above 650 sec^{-1}.[16] Thrombus formation and growth are highly dynamic events.[26] Initial formation and growth of thrombi are also highly shear rate dependent and increase dramatically at shear rates greater than 650 sec^{-1}.[26,29]

Axial Dependence. Platelet deposition decreases along the axis of the flowing blood.[26] The rapid incorporation of platelets into upstream thrombi depletes the boundary layer (the blood layer streaming adjacent to the surface) of platelets, resulting in decreased platelet adhesion and thrombus formation on the downstream areas of the surface. Effects opposing this axial decrement include high shear rates which tend to reestablish the original platelet concentration in the boundary layer[31] and translocation of platelets from upstream thrombi to downstream areas of the surface.[12]

Conclusions. Physical factors have multiple effects and strongly affect the results. Hence, experimental conditions must be carefully chosen and controlled. The generation of time curves is recommended. To avoid secondary effects we advise the use of short perfusion times (<5 min at shear rates >650 sec^{-1}) and the evaluation of the upstream part of the exposed surface.

[30] H. L. Goldsmith, *in* "Progress in Hemostasis and Thrombosis" (T. H. Spaet, ed.), p. 97, Grune & Stratton, New York, 1972.

[31] K. S. Sakariassen, E. Fressinaud, J. P. Girma, H. R. Baumgartner, and D. Meyer, *Blood* **67**, 1515 (1986).

Chemical Factors

Reactive sites of the subendothelial surface[9,31–35] and of the platelet membrane[14,16,20] are involved in platelet adhesion. Divalent cations and von Willebrand factor are important for normal platelet adhesion and thrombus growth.[13,35–38] In addition, prostaglandin I_2 and possibly other humoral factors released from vessel segments may inhibit and/or enhance these interactions.[39] Components stored in specific compartments of the platelet and discharged during platelet–surface interactions promote thrombus formation and growth and also affect platelet adhesion. Components released from upstream deposited platelets may activate platelets in the bloodstream and enhance their capacity to interact with the surface and/or with other platelets further downstream. Accelerated thrombus formation and growth, as observed at shear rates above 650 sec^{-1}, have been attributed partly to liberation of adenosine diphosphate from erythrocytes in addition to their physical effect of increasing radial platelet transport.[40,41]

Perfusion Chambers

Annular perfusion chambers[10,42] and parallel-plate perfusion chambers[22,23,43] are used to investigate platelet interactions with the subendothelium of vessel segments, with cultured vascular cells, and with their extracellular matrix components. It has been experimentally established that blood flow is laminar in these chambers.[23,44,45] The devices allow

[32] H. R. Baumgartner, *Thromb. Haemostasis* **37,** 1 (1977).
[33] P. Birembaut, Y. J. Legrand, J. Bariety, R. Breton, F. Fauvel, M. F. Belair, G. Pignaud, and J. P. Caen, *J. Histochem. Cytochem.* **30,** 75 (1982).
[34] H. V. Stel, K. S. Sakariassen, P. G. de Groot, J. A. van Mourik, and J. J. Sixma, *Blood* **65,** 85 (1985).
[35] W. P. M. Houdijk, K. S. Sakariassen, P. F. E. M. Nievelstein, and J. J. Sixma, *J. Clin. Invest.* **75,** 531 (1985).
[36] H. J. Weiss, V. T. Turitto, and H. R. Baumgartner, *J. Lab. Clin. Med.* **92,** 750 (1978).
[37] K. S. Sakariassen, M. Ottenhof-Rovers, and J. J. Sixma, *Blood* **63,** 996 (1984).
[38] V. T. Turitto, H. J. Weiss, and H. R. Baumgartner, *J. Clin. Invest.* **74,** 1730 (1984).
[39] H. R. Baumgartner, R. Muggli, and T. B. Tschopp, *in* "Cellular Response Mechanisms and Their Biological Significance" (A. Rotman, F.A. Meyer, C. Gitler, and A. Silberberg, eds.), p. 17. Wiley, London, 1980.
[40] G. V. R. Born and A. Wehmeier, *Nature (London)* **282,** 212 (1979).
[41] V. T. Turitto and H. J. Weiss, *Science* **207,** 541 (1980).
[42] V. T. Turitto and H. R. Baumgartner, *Thromb. Diath. Haemorrh. Suppl.* **60,** 17 (1974).
[43] E. F. Grabowski, K. K. Herther, and P. Didisheim, *J. Lab. Clin. Med.* **88,** 368 (1976).
[44] V. T. Turitto, *Chem. Eng. Sci.* **30,** 503 (1975).
[45] P. A. M. M. Aarts, J. A. T. M. van Broek, G. D. C. Kuiken, J. J. Sixma, and R. M. Heethaar, *J. Biomech.* **17,** 61 (1984).

studies at shear rates throughout the physiological range, and measurements of platelet–surface interactions with a variety of techniques, including quantitative differentiation of platelet adhesion and thrombus formation and growth.

Annular Perfusion Chambers

The annular perfusion chambers[10] are made of Plexiglas and consist of a central rod (length = 75 mm and ø = 3.5 mm) and an outer cylinder (Fig. 3A). Prior to perfusion, an everted artery segment is mounted on the middle of the rod. The rod and outer cylinder are then assembled (Fig. 3B). The inlet and outlet portals of the chamber are circular ($ø_{inner}$ = 3.0 mm). Blood flows from the annulus to the circular outlet portal through four cylindrical holes (ø = 1.6 mm) placed symmetrically around the base of the rod. By varying the flow rate a limited range of shear rates can be achieved. For larger variations, the distance between the inner wall of the outer cylinder and the subendothelial surface, the *effective annular width,* is varied.[46] Chamber dimensions, the corresponding values for $f(k)$ as a function of the physical parameters of the system, and the shear rates at 20 ml/min flow rate are summarized in Table I. The effective annular width should be larger than 0.3 mm, because variations in vessel wall thickness may affect the local shear rate significantly, particularly at high shear rate.

The wall shear rates of annular chambers have been experimentally determined[45] by means of laser-Doppler velocimetry,[47] and were according to theoretical predictions given by the formula[48]:

$$\gamma = f(k)(Q_{av}/r_e^3)$$

where γ is the vessel wall shear rate (sec^{-1}); Q_{av}, the average annular flow rate (ml/sec); r_e, the effective annular width (cm); and $f(k)$, a function of the physical parameters of the system (Table I). Thus, any wall shear rate can be achieved by selecting a chamber with the appropriate effective annular width and calculating the required flow rate by using the above formula (Table I).

Parallel-Plate Perfusion Chambers

The parallel-plate perfusion chambers designed by Sakariassen *et al.*[23] are made of Plexiglas. Their flow slits have a rectangular cross section. A

[46] V. T. Turitto and H. R. Baumgartner, *Thromb. Diath. Haemorrh. Suppl.* **60,** 17 (1974).
[47] L. E. Drain, "The Laser-Doppler Technique." Wiley, New York, 1980.
[48] R. B. Bird, W. F. Stuart, and E. M. Lighthill, "Transport Phenomena." Wiley, New York, 1979.

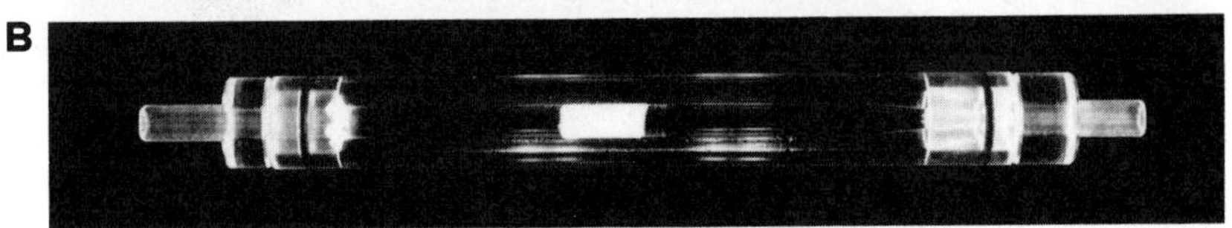

FIG. 3. Photograph of annular perfusion chamber, original model (Table I). (A) Disassembled perfusion chamber. Central rod with mounted artery segment and outer cylinder. (B) Assembled perfusion chamber with artery segment mounted on the rod.

depression in the round knob holds the cover slip (18 × 18 × 0.16 mm Menzelgläser, Braunschweig, FRG and 18 × 22 × 0.18 mm Thermanox, Miles Laboratories, Inc., Naperville, IL) coated with vessel wall components (Fig. 4). The dimensions of the flow slit are 143 mm (length) × 10.0 mm (width) × 0.6–2.0 mm (height). Variations of the slit height and/or flow rate allow production of various shear rates. The inlet and outlet portals are circular, but taper gradually off to the rectangular dimensions. The round knob with the cover slip is positioned 91 mm downstream from the circular flow inlet portal.

The shear rates of the parallel-plate chambers have been experimentally determined,[23] and were in accordance to theoretical predictions given by the formula[49]:

$$\gamma_{cs} = f(k)(6Q/ab^2)$$

where γ_{cs} is the shear rate at the centerline of the cover slip (sec^{-1}); $Q =$ average flow rate (ml/sec); $a =$ slit width (cm), $b =$ slit height (cm), and $f(k) =$ a function of the physical parameters of the system (for shear rates $> 100\ sec^{-1}$ $f(k) \sim 1.03$). By this formula, any shear rate for parallel-plate chambers can be derived. Slit heights smaller than 0.6 mm may introduce blood flow disturbances, due to the transition from circular to rectangular geometry.

[49] H. Richter, "Röhrhydraulik." Springer-Verlag, Berlin, Federal Republic of Germany, 1971.

TABLE I
DIMENSIONS WHICH DETERMINE WALL SHEAR RATE IN ANNULAR CHAMBERS

Perfusion chamber	Annular width (mm)	Artery	Average artery wall thickness (mm)	Average effective annular width (mm)	$f(k)$	Wall shear rate at 20 ml/min (sec^{-1})
I (large)[a]	7.75	Rabbit aorta	0.10	7.65	2.020	2
II (original)[a]	1.30	Rabbit aorta	0.10	1.20	0.516	100
	1.30	Human renal	0.25	1.05	0.430	120
	1.30	Human umbilical	0.15	1.15	0.486	110
III[b]	0.75	Human umbilical	0.15	0.60	0.274	420
IV (small)[a]	0.45	Rabbit aorta	0.10	0.35	0.170	1300

[a] Original designations.[27]
[b] "Small" chamber for umbilical arteries.[18]

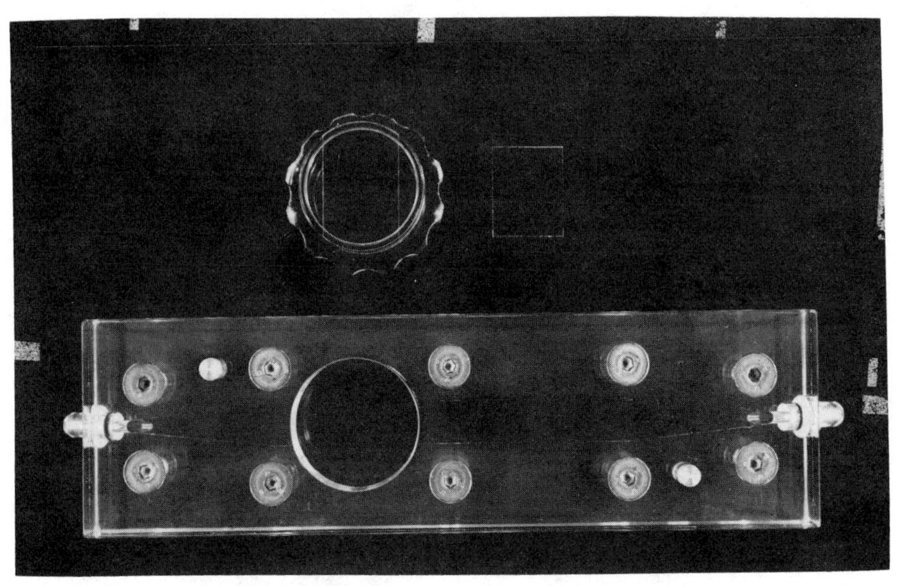

FIG. 4. Photograph of parallel-plate perfusion chamber. The knob and the cover slip which fits the depression of the knob are disassembled.

Test for Blood Flow Laminarity

Experimental determination of the blood flow velocity profile is required to establish whether the blood flow is laminar or turbulent. Only if laminarity is established can the shear rate be derived.[23,49]

Preparation of Test Surfaces

Subendothelium of Human Blood Vessels

The subendothelium is defined as the vessel wall compartment localized between the internal elastic lamina and the endothelium of normal vessels. Human arteries from umbilical cords and from adult postmortem renal arteries as well as vessels from surgical specimens can be used in the annular perfusion chambers.[50,51] Their luminal diameter must match the diameter of the perfusion chamber rod. We recommend the use of umbili-

[50] T. B. Tschopp, H. R. Baumgartner, K. Silberbauer, and H. Sinzinger, *Haemostasis* **8,** 19 (1979).

[51] K. S. Sakariassen, J. D. Banga, P. G. de Groot, and J. J. Sixma, *Thromb. Haemostas.* **52,** 60 (1984).

cal arteries because of their easy access, their morphological homogeneity, and their comparatively thin vessel wall, which allow studies at all physiological shear rates.

The umbilical arteries are flushed immediately after delivery with 0.2 *M* Tris–HCl buffer, pH 7.4, containing 200 U penicillin G/ml and 0.2 mg streptomycine/ml. The two arteries are subsequently dissected free from the surrounding jellylike tissue and removed from the cord. Irregularities of the adventitia which are still present are carefully torn off with a pair of tweezers in order to produce uniform wall thickness of approximately 0.15 mm (Table I). The artery preparations (usually about 60 cm long) are stored at 4° in Tris–HCl buffer containing antibiotics, and used in perfusion experiments 1 to 3 days after delivery. Before an artery is mounted on the rod, it is washed three times at 22° with Tris–HCl buffer, pH 7.4, and everted on a bud probe (see rabbit aorta for detailed description). Subsequently the luminal surface is exposed to air for 5 min, washed three times with Tris–HCl buffer again, and cut into segments of approximately 1.0 cm length. Each segment is then mounted on the middle of a rod. Both edges are carefully trimmed with a razor blade under a magnifying glass to remove tissue which might protrude into the annulus during the perfusion. Very fresh segments are in some cases difficult to mount on the rod. The mechanical stress during the everting procedure and the exposure to air followed by washing removes the endothelium completely and produces a smooth subendothelial surface.

Rabbit Aorta Subendothelium

Deendothelialization is performed *in situ* in anesthetized rabbits by means of the balloon catheter technique.[10,52,53] Subsequent studies have shown that *in situ* perfusion with cold buffer followed by dissection and eversion of the aorta is sufficient for removing the endothelium.

A detailed description for the preparation of rabbit aorta denuded of endothelium by balloon catheter injury was given recently.[54] Deendothelialized rabbit aorta without ballooning is produced as follows: Anesthetized rabbits, weighing 3.2–4.0 kg, are perfused with approximately 1 liter of phosphate-buffered saline (PBS: 58 m*M* Na_2HPO_4, 1 m*M* NaH_2PO_4, 75 m*M* NaCl), pH 7.4, from the thoracic aorta and exsanguinated through a catheter placed in a carotid artery at the same time. Initially the perfusate

[52] H. R. Baumgartner, *Z. Gesamte Exp. Med.* **137,** 227 (1963).

[53] H. R. Baumgartner, M. B. Stemerman, and T. H. Spaet, *Experientia* **27,** 283 (1971).

[54] V. T. Turitto and H. R. Baumgartner, *in* "Measurements of Platelet Function" (L. A. Harker and T. S. Zimmerman, ed.), p. 46. Churchill-Livingstone, Edinburgh, Scotland, 1983.

is kept at 37° (200 ml); the heating coil is then transferred into ice water without interruption of the perfusion. When the outflow of blood from the carotid artery has ceased, the vena cava is cannulated for the outflow of the perfusate. Under constant perfusion with cold PBS the abdominal aorta distal of the renal arteries is carefully dissected, removed from the animal, and immersed in cold PBS. The isolated abdominal aorta which should be totally free of fat tissue and branches is then mounted on a rod of approximately 2 mm in diameter. One end of the vessel is then immobilized by ligation and the vessel everted over itself so that the intimal surface is now facing outward. The everted aorta is cut into segments of 7 to 20 mm length, placed in 0.2 *M* Tris buffer, pH 7.4, and kept at 4° for a period of up to 4 weeks. The buffer is exchanged weekly. The duration of storage after the preparation of the vessels is important. Very fresh aortas, used within 0.5 hr after exsanguination of the rabbit, show inhibitory activity with respect to platelet adhesion compared to vessels stored 1 to 4 weeks.[39] Usually segments prepared in this way and stored for 8–28 days are referred to as rabbit aorta subendothelium.

Cells of the Vessel Wall

Cells grown in culture on plastic cover slips which fit the depression of the knob (Fig. 4) can be used as test surfaces in the parallel-plate perfusion chambers.[23,51] Plastic cover slips (Thermanox as above) resistant to organic solvents used for fixation and embedding are a better substratum for cell attachment than glass cover slips. So far limited experience has been gained with confluent layers of endothelial cells, vascular smooth muscle cells, and fibroblasts (unpublished results).

Extracellular Matrix of Cultured Vessel Wall Cells

The extracellular matrix produced by vascular endothelial cells and smooth muscle cells can also be used as a test surface in the parallel-plate perfusion chambers. The cells are cultured on plastic cover slips and, after having reached confluency, they are removed by exposure to 0.5% Triton in distilled water (v/v) at 22° and gentle shaking for 30 min. The surfaces are subsequently washed six times with PBS, pH 7.4, and gentle shaking.[55] The matrices are stored in the same buffer at 22° and used in perfusion experiments within 2 hr. Some residual cell membrane fragments may remain attached to the extracellular matrix.

[55] D. Gospodarowicz, D. Delgano, and I. Vlodavsky, *Proc. Natl. Acad. Sci. U.S.A.* **77**, 4094 (1980).

Collagens

For perfusion experiments with fibrillar collagen, the collagen is coated onto glass or plastic cover slips which are exposed to blood in parallel-plate perfusion chambers.[22,23] The cover slips are washed in chromic acid (glass) and ethanol (plastic) at 22° for 2 hr, rinsed with deionized water four times at the same temperature, and dried at 70° for 2 hr. A fibrillar collagen suspension, of 1.0 to 1.2 ml (1 mg/ml), is sprayed simultaneously and in multiple runs onto 10 glass and plastic cover slips, respectively, by means of a retouching air brush (Badger, model No. 100 Il, Badger Air-Brush Co., Franklin Park, IL) at a nitrogen operating pressure of 1 atm.[22,23] The finely dispersed collagen droplets dry at 22° between individual runs. The collagen concentration achieved in this way amounts to 25 $\mu g/cm^2$ on the average.[23] The collagen coated cover slips are stored at 22° for about 16 hr before they are used in perfusion experiments.

Binding of purified proteins, e.g., von Willebrand factor and fibronectin, to the collagen meshwork is achieved by incubation of the collagen-coated surface with proteins in 0.2 *M* modified Tyrode's buffer (130 m*M* NaCl, 2 m*M* KCl, 12 m*M* $NaHCO_3$, 2.5 m*M* $CaCl_2$, 0.9 m*M* $MgCl_3$, pH 7.4).[31,35] An amount of 0.5 ml protein solution is applied to the collagen-coated cover slip. Following incubation at 22° without agitation for 20 min the protein solution is removed, and the surface is washed with the same buffer in order to remove nonspecifically bound proteins. Air drying of the protein-treated collagen surfaces must be avoided. They are then used immediately in perfusion experiments.

Preparation of Perfusates

Various perfusates are used, including nonanticoagulated blood, anticoagulated blood, and suspensions of washed platelets with washed erythrocytes in defined buffers. When the perfusate is recirculated, at least 20 ml of perfusion fluid is necessary. The choice of perfusate depends on the aim of the investigation.

Nonanticoagulated Blood

The perfusion of nonanticoagulated ("native") blood from normal individuals,[39,56] patients with bleeding disorders,[36,38] patients with coagulation defects,[57] as well as drug-treated human volunteers,[58] and experimen-

[56] H. R. Baumgartner, *Schweiz. Med. Wochenschr.* **106,** 1367 (1976).

[57] H. J. Weiss, V. T. Turitto, W. J. Vicic, and H. R. Baumgartner, *Blood* **63,** 1004 (1984).

[58] H. J. Weiss, V. T. Turitto, W. J. Vicic, and H. R. Baumgartner, *Thromb. Haemostasis* **45,** 136 (1981).

tal animals,[26,59,60] yielded new information regarding the role of coagulation factors and thrombin generation for the growth and stability of platelet thrombi at various shear rates. Nonanticoagulated blood is directly drawn from a blood vessel through a chamber at a preselected flow rate by a pump placed distally of the chamber (Fig. 5).[61]

Citrated Blood

Citrated blood from normal individuals and patients with bleeding disorders was used in a number of comparative studies in order to unravel pathophysiological mechanisms.[13-20] Addition to normal citrated blood of polyclonal or monoclonal antibodies directed against various plasma proteins[35,61-63] and against platelet membrane glycoproteins[64-66] was used as a tool to identify and clarify the role of components involved in platelet–surface interactions.

Blood is drawn from human volunteers into polystyrene syringes containing $\frac{1}{10}$ volume of 108 m*M* trisodium citrate by means of a Butterfly-19 infusion set. Hematocrit and platelet counts are performed in undiluted EDTA blood. The plasma concentration of citrate is adjusted to 20 m*M* by adding an appropriate volume (V_c) of 108 m*M* trisodium citrate depending on the EDTA hematocrit (% Hct) and the volume of the blood sample (V) according to the formula:

$$V_c = V(9.2 - 0.18\% \text{ Hct})/88$$

The plasma citrate concentration is critical for the rate of platelet adhesion and thrombus formation.[12,51,56,59] At concentrations which inhibit coagulation (15–20 m*M*) platelet thrombus growth is also strongly inhibited.[12,26,56] Platelet thrombus formation on the subendothelium is abolished at citrate concentrations >28 m*M*[59] Its effect on adhesion is shear rate dependent.[39]

[59] H. R. Baumgartner, *Haemostasis* **8,** 340 (1979).

[60] H. R. Baumgartner, R. Muggli, and T. B. Tschopp, *in* "Advances in Pharmacology and Therapeutics II" (H. Yoshida, Y. Hagihara, and S. Ebashi, eds.), p. 21. Pergamon, Oxford, England, 1982.

[61] H. R. Baumgartner, T. B. Tschopp, and D. Meyer, *Br. J. Haematol.* **44,** 208 (1980).

[62] H. V. Stel, K. S. Sakariassen, B. J. Scholte, T. H. van der Kwast, P. G. de Groot, J. J. Sixma, and J. A. van Mourik. *Blood* **63,** 1408 (1984).

[63] D. Meyer, H. R. Baumgartner, and T. S. Edgington, *Br. J. Haematol.* **57,** 609 (1984).

[64] J. P. Caen, H. Michel, G. Tobelem, E. Bodevin, and S. Levy-Toledano, *Experientia* **33,** 91 (1977).

[65] C. Ruan, G. Tobelem, A. McMichael, L. Drouet, Y. Legrand, L. Degos, N. Kieffer, H. Lee, and J. P. Caen, *Br. J. Haematol.* **49,** 511 (1981).

[66] K. S. Sakariassen, P. F. Nievelstein, B. S. Coller, and J. J. Sixma, *Br. J. Haematol.* **63,** 681 (1986).

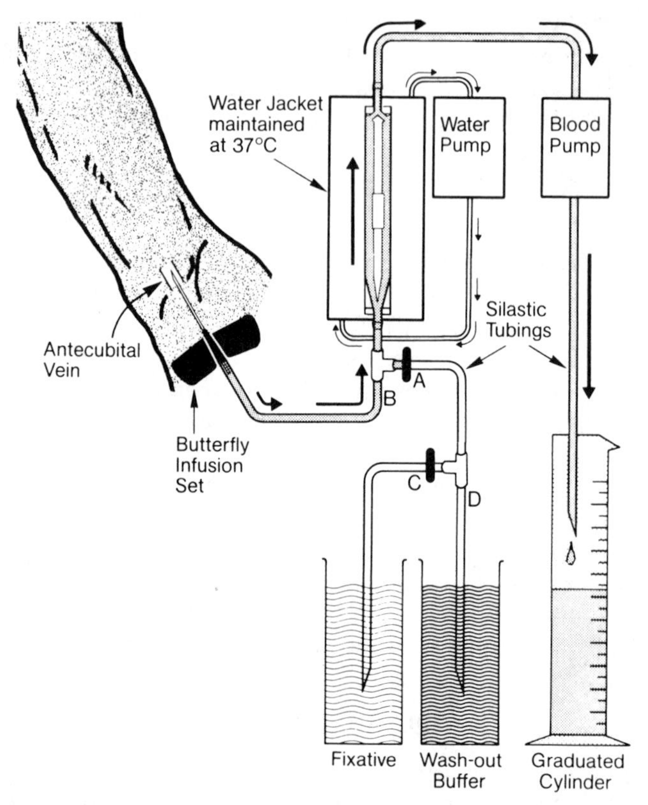

FIG. 5. Schematical illustration of perfusion system for nonanticoagulated blood. The perfusion chamber is maintained at 37° by water jacket. Steps of the perfusion procedure include the following: (1) Set up the system with connections open at A and D, closed at B and C. (2) Start buffer perfusion to control the flow rate and to fill the system with buffer. Continue perfusion until all air bubbles are removed. (3) Close A, open B. (4) Puncture vein, collect "prevalue" blood, connect infusion set at B, and start pump (= status shown in figure). (5) Control flow rate by checking blood volume in graduated cylinder. (6) To terminate blood perfusion: open A, close B. (7) To terminate buffer wash-out: open C, close D. (8) To terminate fixative perfusion: stop pump.

Reconstituted Blood

Blood anticoagulated with trisodium citrate (20 m*M* in plasma, see above) is used for preparing reconstituted blood. By separation of platelets, erythrocytes, and plasma from citrated blood, and by substituting plasma with defined buffers containing various purified plasma proteins, an unlimited number of perfusates can be produced.[7,18,35,37,67-69] Such perfusates are useful to elucidate the role of plasma ions and proteins in platelet adhesion and thrombus formation. Perfusates with blood components from two individuals can be mixed when their blood groups match, and when other immunological cross reactions are absent.[19] Platelets are washed in buffer A (290 mOsm) by centrifugation in order to remove autologous plasma components.[7]

Buffer A

4 m*M* KCl
107 m*M* NaCl
20 m*M* $NaHCO_3$
2 m*M* Na_2SO_4
20 m*M* citric acid
20 m*M* trisodium citrate
5 m*M* D-glucose

The separation of platelets from plasma is outlined in Table II. To facilitate the resuspension of platelets the pH of the various platelet suspensions is kept between 6.1 and 6.2, since aggregation caused by centrifugation occurs at a pH above 6.3.[70]

Radiolabeling of platelets is performed after the first resuspension in buffer A at pH 6.0 (Table II). Either $^{51}CrO_4$ or ^{111}In-oxine complex is used.[7,23] About 1.0×10^{12} platelets/liter are incubated with 1.0 μCi $^{51}CrO_4$/ml at 22° and gentle agitation. After incubation of 20 min, the platelets are washed three times by centrifugation and resuspension as outlined in Table II. They are finally resuspended in plasma or in citrated human albumin solution (C-HAS). Triple centrifugation is needed for removal of $^{51}CrO_4$, which is not taken up by the platelets. The labeling efficiency amounts to less than 10% and the intraplatelet $^{51}CrO_4$ after the third centrifugation is larger than 95%. Labeling of platelets with ^{111}In-oxine is also performed in buffer A at pH 6.0; with 1.5×10^{12} platelets/liter and 10.0 μCi ^{111}In-oxine/ml incubated at 37° and gentle agitation for 15 min. The labeling efficiency is enhanced by the addition of 45 μl oxine (6.5

[67] J. J. Sixma, K. S. Sakariassen, N. H. Beeser-Visser, M. Ottenhof-Rovers, and P. A. Bolhuis, *Blood* **63,** 128 (1984).

[68] K. S. Sakariassen, P. A. Bolhuis, M. Blombäck, L. Thorell, B. Blombäck, and J. J. Sixma, *Thromb. Haemostasis* **52,** 144 (1984).

[69] J. J. Sixma, K. S. Sakariassen, H. V. Stel, W. P. M. Houdijk, D. W. in der Maur, R. J. Hamer, P. G. de Groot, and J. A. van Mourik, *J. Clin. Invest.* **74,** 736 (1984).

[70] R. H. Aster and J. H. Jandel, *J. Clin. Invest.* **43,** 843 (1964).

TABLE II
PLATELET WASHING

Preparation of platelet-rich plasma

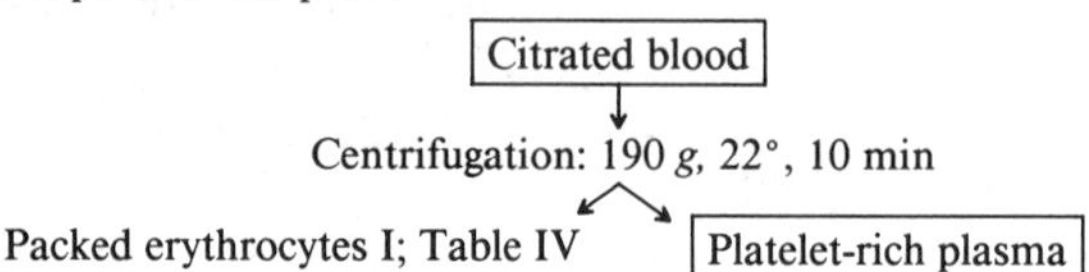

Preparation of platelet suspension in buffer A; platelet labeling and/or aspirin treatment

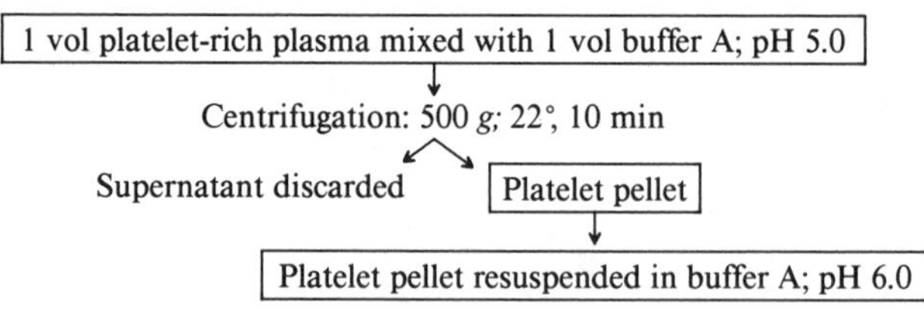

This platelet suspension is used for labeling of platelets with $^{51}CrO_4$ or with ^{11}In-oxine and/or for aspirin treatment

Platelet washing

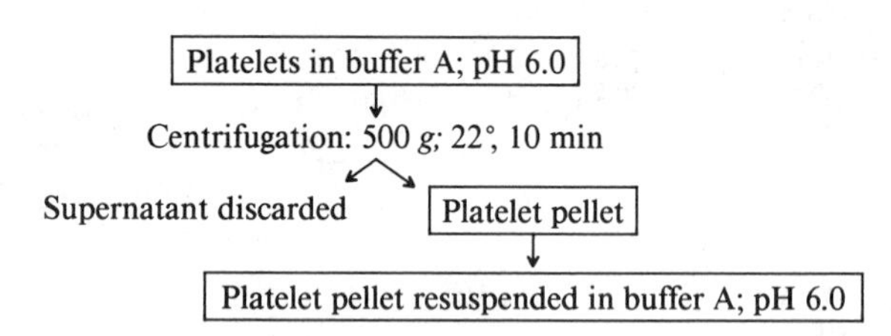

This procedure is twice repeated. The platelets are finally resuspended in HAS, C-HAS, or citrated plasma to give the desired platelet count

mg/liter 95% ethanol) to 10 ml platelet suspension,[71] and amounts to more than 90%. One wash is enough to remove ^{111}In-oxine not taken up by the platelets. In experiments with aspirin-treated platelets, the treatment is performed simultaneously with the labeling.[23] Aspirin (10 μM) is incubated with 1.5×10^{12} platelets/liter for 15 min; this is enough to inhibit platelet cyclooxygenase fully, as indicated by the absence of malondialdehyde production following stimulation with trypsin.[72]

[71] A. du P. Heyns, P. N. Badenhorst, H. Pieters, M. G. Lötter, P. C. Minnaar, L. J. Duyvené de Wit, O. R. van Reenen, and E. P. Retief, *Thromb. Haemostasis* **42,** 1973 (1980).

[72] J. B. Smith, C.M. Ingerman, and M. J. Silver, *J. Lab. Clin. Med.* **88,** 167 (1976).

Table III shows an example of an experiment in which plasma and platelets of two individuals are exchanged. This approach is useful to differentiate between platelet and plasmatic defects.[19]

Substitution of citrated plasma with a defined buffer solution can be performed with citrated human albumin solution (C-HAS) having an osmolarity of 290 mOsm.[7,37]

C-HAS

4 m*M* KCl
107 m*M* NaCl
20 m*M* $NaHCO_3$
2 m*M* Na_2SO_4
2.5 m*M* $CaCl_2$
5 m*M* glucose
20 m*M* trisodium citrate
4% (w/v) lyophilized human albumin (A-9511, Sigma, St. Louis, MO)

The C-HAS is dialyzed overnight against albumin-free C-HAS at 4°, centrifuged (3000 *g* at 4° for 20 min), and filtered in order to remove impurities in the albumin fraction. The pH is adjusted with 0.1 *N* HCl to 7.4 prior to use.

C-HAS, washed platelets (Table II), and washed erythrocytes (Table IV) are combined to prepare citrated plasma-free reconstituted blood. The hematocrit of the packed erythrocytes III is on the average 96%. Usually platelet pellet and packed erythrocytes III are resuspended in C-HAS to give a final hematocrit of 40% and a platelet count of 1.5×10^{11}/liter.

Platelet adhesion to artery subendothelium in citrated plasma-free reconstituted blood is impaired, and similar to that observed in citrated blood from patients with von Willebrand's disease, subtypes IIA, IIB, and III. Addition of 5 μg purified von Willebrand factor/ml C-HAS normalizes the platelet adhesion.[18,37] Normal platelet adhesion to collagen fibrils requires addition of 5 μg purified von Willebrand factor and 400 μg fibronectin/ml C-HAS in plasma-free reconstituted blood.[35] Physiological amounts of divalent cations are achieved when HAS without citrate is used.[18,37] Some albumin batches are, however, contaminated with thrombin which triggers clotting in the absence of citrate.

Perfusion Procedures

The perfusion procedures with the annular chambers and the parallel-plate chambers are essentially similar.

TABLE III
DIFFERENTIATION BETWEEN A PLATELET AND A PLASMATIC DEFECT[a]

Donor	Washed platelets	Plasma	Erythrocytes	Perfusate[b]
Individual A	A	A	A	I
Individual B	B	B	B	II
Individuals A and B	A	B	B	III
Individuals A and B	B	A	A	IV

[a] Perfusion experiments with reconstituted blood: Protocol of a cross-over study with plasma and platelets from a healthy donor and a patient.

[b] Perfusates I and II serve as controls.

TABLE IV
ERYTHROCYTE WASHING

Preparation of packed erythrocytes

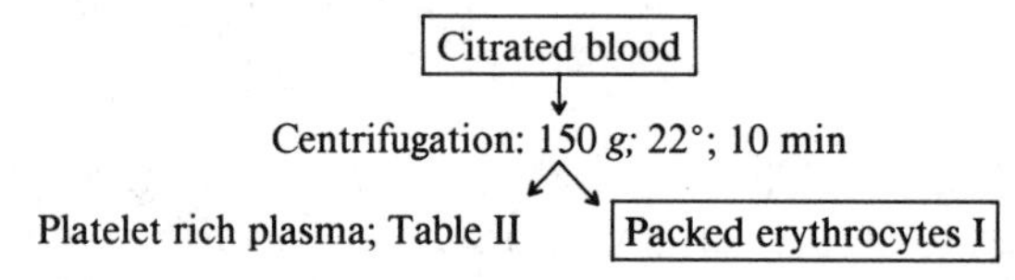

Erythrocyte washing

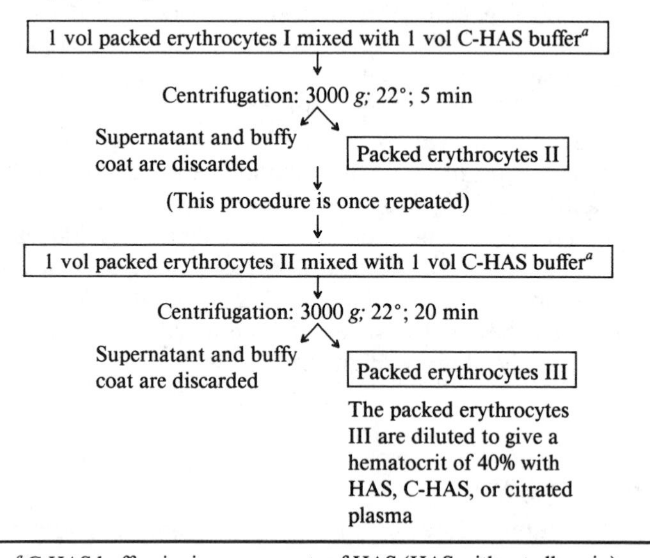

[a] C-HAS buffer, ionic components of HAS (HAS without albumin).

Preparations of Perfusion Systems

All perfusion systems consist of a perfusion chamber (annular or parallel plate), an occlusive roller pump (model 1/10, Perpex Jubile, Werner Meyer AG, Luzern, Switzerland) and silastic tubings ($ø_{inner} = 3.0$ mm, Dow Corning Corp., Midland, MI).

Nonanticoagulated blood is collected into a calibrated cylinder following the passage through the chamber (Fig. 5).[54,56] The roller pump, which is placed after the perfusion chamber, draws blood directly from an antecubital vein into the chamber at the preselected flow rate. The perfusion chamber is kept at 37° by a water jacket. The single steps of the perfusion procedure are outlined in the legend to Fig. 5.

Citrated and reconstituted blood perfusates are recirculated.[10] The perfusate is pumped from a polyethylene container through the perfusion chamber and back into the container. Figure 6 shows and describes the steps involved. The occlusive roller pump produces pulsatile blood flow. The amplitude and the frequency of the pulsations, which both are dependent on the actual flow rate and the type of pump used, affect the platelet–surface interactions.[73] Steady blood flow in the perfusion chamber can be produced by including a funnel between the pump and the perfusion chamber.[73] However, by using the type of pump described above, the effect of the blood flow pulsations is minimized.[73] We recommend the use of this pump and not the funnel system, because of the ease and simplicity of the manual operations during the perfusion procedure.

Prewarming and Preperfusion with Buffer

The perfusion systems with exposed test surfaces are preperfused with prewarmed (37°) PBS, pH 7.4, at the desired shear rate and left filled with PBS at 37° to avoid any air–blood interphase (Figs. 5 and 6). Perfusates used for recirculation are preincubated at 37° for 5 min.

Blood Perfusion. See Figs. 5 and 6. Nonanticoagulated blood from human volunteers is drawn directly from an antecubital vein through the perfusion chamber at 37° by an occlusive roller pump.[54,56] The venipuncture is performed with the Butterfly infusion set No. 19 (or 16) with the arm compressed at 40 mmHg initially and 20 mmHg during perfusion. All perfusates are perfused at preselected flow rates and preselected periods of time. The start of the perfusion time is defined as the moment when the pump starts to draw blood from an antecubital vein or from the blood container.

[73] K. S. Sakariassen, P. A. Bolhuis, and J. J. Sixma, *Thromb. Res.* **19,** 547 (1980).

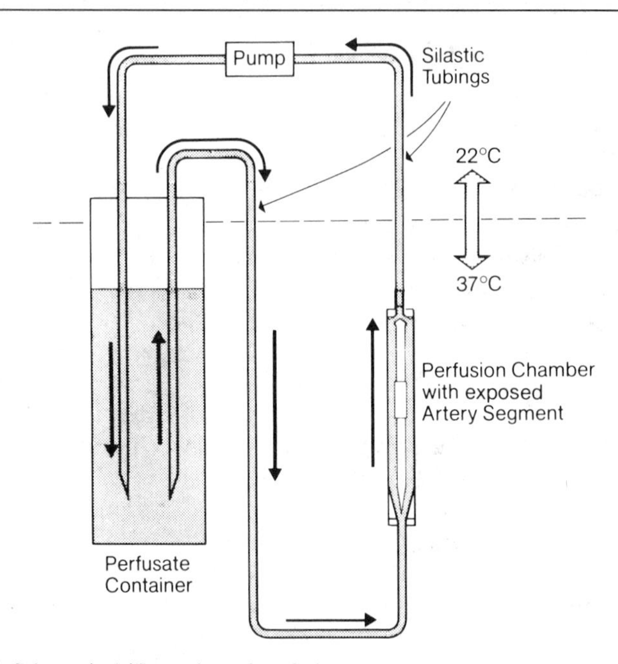

FIG. 6. Schematical illustration of perfusion system for experiments with citrated and reconstituted blood. Perfusion chamber, blood container, and most of the silastic tubings are immersed into a water bath at 37°. Steps of the perfusion procedure include the following: (1) Set up of the system with inflow silastic tubings immersed in a reservoir containing PBS, pH 7.4, at 37° (not shown). (2) Start PBS perfusion to fill the system and to control the flow rate. Remove all air bubbles. (3) Immerse inflow silastic in perfusate container and start pump. (4) Discard outflow of PBS and PBS–perfusate mixture. (5) Immerse outflow silastic in perfusate container to allow recirculation of the perfusate (= status shown in figure). (6) To terminate the perfusion: stop pump, disassemble rod from chamber, immerse it briefly in rinsing buffer and subsequently in fixative.

Control of Blood Flow and Shear Rate

The speed of the roller pump needed to produce preselected flow rate(s) is checked for each experiment, because of variations from day to day between pump speed and flow rate. Blood pressure and blood flow velocity can be measured continuously with transducers at the chamber entrance and exit. The amount of nonanticoagulated blood collected in the graduated cylinder also gives information about the average shear rate during the perfusion. These measurements are crucial if chamber occlusion occurs.[26,74] However, they are not needed when only transient mural thrombi are formed such as with citrated perfusates.

[74] H. R. Baumgartner, H. Kuhn, and T. B. Tschopp, *in* "Atherosclerosis VI" (G. Schettler, A. M. Gotto, G. Middelhoff, A. J. R. Habenicht, and K. R. Jurutka, eds.), p. 649. Springer-Verlag, Berlin, Federal Republic of Germany, 1983.

Termination of Perfusion

Perfusion with nonanticoagulated blood is terminated by switching from blood to buffer (Fig. 5). The Butterfly infusion set is then disconnected from the perfusion system. Subsequently, additional blood can be collected before the needle is removed from the vein. For wash-out and initial fixation, equal volumes are used (usually 2–4 ml); thus, wash-out time depends on the flow rate and may vary from a few seconds to minutes. After initial fixation in the perfusion system, the chamber is disassembled and the rod or cover slip immersed in fresh fixative at 4°.

Perfusion with recirculated perfusates is usually terminated by stopping the pump (Fig. 6). However, a defined wash-out and fixation procedure as described above for nonanticoagulated blood may also be employed. Without wash-out, the chamber is disassembled as quickly as possible, rod or cover slip briefly rinsed in PBS, and then immersed in fresh fixative at 4°.

Fixation

For perfusion and immersion the same fixative is used, namely 2.5% glutaraldehyde in 0.1 *M* cacodylate buffer. Artery segments are removed from the rod after a 20- to 30-min fixation by a longitudinal incision with a razor blade and immersed in fresh fixative at 4° for another 60 min. Cover slips are processed similarly. Artery segments and cover slips may be stored in 0.1 *M* cacodylate buffer containing 7% sucrose and traces of fixative until they are further processed.

Evaluation of Platelet–Surface Interactions in the Absence of Fibrin Formation

A variety of morphometric and other methods have been developed to measure platelet–surface interactions produced with the described devices.[7,9,21–23,39,73] Their resolution and informative value vary considerably. The most informative method, and the method which always discriminates between platelet–surface and platelet–platelet interactions, is the morphometric examination of 0.8- to 1.0-μm thick sections.

Evaluation Using Radiolabeled Platelets

The radioactivity measured does not discriminate between platelet–surface interactions (adhesion) and platelet–platelet cohesion (aggregation). However, platelet adhesion in the absence of aggregation can be measured by using the following protocol[7,8]:

^{51}Cr- or ^{111}In-radiolabeled washed platelets at 1.2×10^{11}/liter (Table II)
Aspirin-treated platelets (Table II)
40 vol% erythrocytes in citrated HAS or citrated plasma
Subendothelium of human umbilical or renal arteries
Perfusions for 1 to 5 min at 800 sec^{-1} shear rate

Adhesion of washed radiolabeled platelets to collagen surfaces is slightly decreased as compared to platelet adhesion in the parent citrated blood. Small numbers of adherent platelets are more precisely determined with ^{111}In-labeled platelets than with ^{51}Cr-labeled platelets, because ^{111}In is a much stronger γ emitter than ^{51}Cr.

Evaluation Using en Face Preparations

En face morphometry of fixed test surfaces attached to glass cover slips is performed after staining with 0.01% fuchsin.[22] We found plastic cover slips not suitable. The optical properties of the available cover slips are apparently not sufficient for microscopy at high magnification.

Staining Solution

0.01% (w/v) fuchsin (C.I. No. 42510, Fluka AG, Buchs SG, Switzerland) in PBS, pH 7.4; store at 4° for 7 days
Filtration of staining solution through Millipore filter (0.22 μm, MF Millipore, Millipore Corp., Bedford, MA)
Extinction of staining solution at 445 μm is adjusted to 0.800 by addition of PBS

The staining is performed at 22° for 60 min in Costar wells (Costar, type 3506, Cambridge, MA) under gentle shaking. The cover slips are subsequently washed three times with PBS for 30 min briefly immersed in tap water and dried overnight at 22°. The cover slips are finally mounted on slides and examined by *en face* microscopy.[21,23]

Morphometry of en Face Preparations. Measurement of surface coverage on cover slips is carried out with a Zeiss microscope fitted with a Planapo 100/1.3 oil immersion objective and Kpl $\times$ 8 oculars and at Optovar intermediate magnification $\times$ 1.25. A green absorption filter (520 nm) is used to enhance the contrast of the red, fuchsin-stained platelets. The surface is evaluated with a 10×10 μm square grid eyepiece micrometer at a total magnification of $\times$ 1000. Under these conditions one square corresponds to a 10×10 μm square on the specimen. The number of grid line intersections falling on platelets and thrombi is counted. The specimen is systematically shifted perpendicular to the direction of flow and the count is repeated. The fraction of counts relative to the total number or intersections gives the percentage surface coverage.[23] Alternatively, the surface can be evaluated by counting the number of platelets per unit area.[75] However, this is only possible under conditions where single platelets are present. Fused platelets in small and large thrombi preclude single platelet counting.

En face evaluation may underestimate surface coverage as the contrast at the periphery of platelets and thrombi may not be sufficient to detect thinly spread parts of platelets. On the other hand, surface coverage is overestimated on surfaces which induce the formation of large thrombi which tend to bend over in the direction of flow.

Automated Microdensitometry. Automated microdensitometry measures the two-dimensional height distribution of platelet deposition on *en face* preparations.[22] It is rapid, but gives no information on the particular state of activation of the platelets (contact, spread, degranulated). The platelets have to be made optically dense (staining) relative to the test surface and the support which has to be flat.

Reproducible staining is mandatory for a quantitative comparison between different experiments. Analysis of the fuchsin-stained preparations is done with a Quantimet 720 image analyzer fitted to a Zeiss universal microscope and interfaced to a PDP 11/10 computer. All measurements are carried out with a Zeiss objective Plan 16/0.35 at Optovar intermediate magnification $\times 2$ and a green interference filter, 550 nm. Under these conditions the total linear magnification is $\times 670$, and a specimen area of 0.42×0.33 mm is resolved into 500,000 picture points. Shading is automatically corrected. Within 10 sec the light transmission relative to the background of all picture points is digitized by the detector module and classified linearly in intensity by the computer in the form of a frequency histogram of 63 intensity classes, gray level numbers (Fig. 7). Background transmission (100%) and 0% transmission correspond to gray level numbers 59 and 5, respectively. The cumulative frequencies up to gray levels 55 and 33 are selected and approximate the percentage of the surface covered with platelets ($\%C + \%A + \%T$; adhesion) and the percentage of the surface covered with thrombi more than 5 μm in height ($\%T$), respectively. The choice of these parameters is based on semiquantitative calibration of Quantimet intensity classifications with conventional morphometry of semithin sections (see below).

Microdensitometry of surfaces exposed to blood gives only the height distribution of the fuchsin-stained material. If one wishes to apply this method to the study of platelets, it is necessary to exclude the presence of other blood-borne and fuchsin-stainable components (e.g., red cells, fi-

[75] E. F. Leonard and L. I. Friedman, *Chem. Eng. Prog. Symp. Ser.* **66,** 59 (1970).

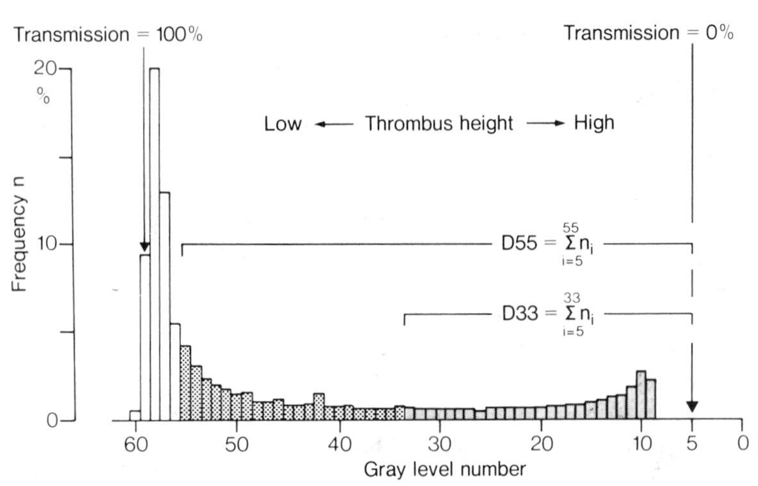

FIG. 7. Typical frequency histogram after automated microdensitometry of a collagen-coated glass slide exposed to rabbit citrated blood at a shear rate of 1000 sec^{-1} for 10 min and stained with fuchsin. A 0.42 by 0.33 mm area was resolved into 500,000 picture points with an image analyzer, and the light transmission is linearly classified as gray level numbers.[22] D55 and D33 are parameters for adhesion and thrombi, respectively.

brin). In addition, quantitative microdensitometry presumes that the extinction coefficient is the same in all platelets, irrespective of their particular state of activation (contact, spread, degranulated).

Evaluation Using Semithin Sections

Evaluation of semithin sections requires dehydration, embedding, cutting, and staining of fixed (see above) artery segments and cover slip test surfaces, respectively. These procedures are described below.

Processing of Artery Segments: Postfixation, Dehydration, and Epon Embedding

Fixed artery segments stored in 7% sucrose/0.1 *M* cacodylate buffer are processed as follows:

Distilled water; 3–5 min
2.0% OsO_4 (w/v) in 0.1 *M* cacodylate buffer, pH 7.4, 4°; agitation by rotation for 1 hr
2 × distilled water; 3–5 min
0.1% (w/v) uranyl acetate in distilled water, pH 4.6; rotation for 30 min (not required for light microscopy)
Distilled water; 3–5 min

Dehydration with graded ethanol concentrations (2 × 75%, 2 × 85%, 3 × 95%, and 4 × 100%); 3–5 min each
3 × propylene oxide; 3–5 min each
1 vol Epon/1 vol propylene oxide; rotation for $1\frac{1}{2}$ hr

Epon composition:
$Epon_{812}$: 48 ml
DDSA (dodecenylsuccinic anhydride): 24 ml
NMA (nadic methyl anhydride): 28 ml
DMP (dimethylaminomethylphenol): 2 ml

The segments are removed from the glass bottles and placed into Epon
The segments are cut *perpendicular* to the flow axis into pieces of approximately 3 mm length
Each piece is placed into the embedding mold (H. Waldner, Riedgrabenweg 28, CH-8050 Zürich, Switzerland) in an oriented manner and fully covered with Epon. If the piece is too large to fit the mold, it is cut once or twice parallel to the flow axis and the pieces put on top of each other
Polymerization, 50°; 1 hr
Polymerization, 63°; 16 hr

Cutting. Semithin sections, approximately 0.8–1.0 μm in thickness, are cut perpendicular to the flow axis with glass knives on an ultramicrotome.

Staining. The cross sections are stained with toluidine blue and basic fuchsin.

Staining Solutions

10 mg toluidine blue (89640, Fluka SG, Buchs SG, Switzerland)/100 ml 2% (w/v) dinatriumtetraborate (Merck, Darmstadt, FRG) in distilled water
10 mg basic fuchsin (54051, Fluka AG, Buchs SG, Switzerland)/100 ml distilled water
Both solutions are heated at 80° until the stains are dissolved, and subsequently filtrated

The specimens are stained at 70° on a hot plate with toluidine blue for 2 min, rinsed with distilled water, dried, and then stained with basic fuchsin for 15 sec. Finally, they are rinsed with distilled water and dried at 70° for 1 hr before they are mounted on microscope slides. It is essential for the quality of the staining that the section is stained with toluidine blue first.

Processing of Test Surfaces on Cover Slips. The test surfaces on the cover slips (glass or plastic) are processed similarly to the artery segments, but with some modifications of the Epon embedding in order to facilitate the removal of cover slips. After immersion in Epon/propylene oxide (1 : 1) pure Epon is applied onto the test surface, and the back of the cover slip wiped clean with propyleneoxide. The following day the plastic cover slip is peeled off the polymerized Epon following shock-freezing the 70° warm specimen on a lump of dry ice. Glass cover slips are dissolved with 38% hydrofluoric acid (Fluka AG, Buchs SG, Switzerland) at 22° for 30 min. The Epon-embedded test surface is subsequently washed three times with distilled water for 10 min, dried at 70° for 2 hr, trimmed, and reembedded in an oriented manner into Epon. Cross sections for light microscopy are cut at well-defined axial positions, similar to that described for the vessel segments. The sections are stained for 45 sec with toluidine blue and for 20 sec with basic fuchsin.

Morphometry of Semithin Sections. Morphometry of 0.8-μm thick cross sections by light microscopy ($\times 1000$) was described in detail by Baumgartner and Muggli.[21] The evaluation is carried out with an eye-piece micrometer in the occular (Fig. 8). The test surface is scored at 10-μm intervals, then shifted systematically along the eye-piece micrometer, and the scoring is repeated (~ 1000 intervals for each perimeter). Scoring is according to the following classification: surface devoid of platelets, naked *(N)*; contact platelet *(C)*; spread platelets and platelet aggregates less than 5 μm in height *(A)*; and thrombi more than 5 μm in height *(T)*.

The percentage of counts of each class of platelet interactions relative to the total number of scorings gives $\%C$, $\%A$, and $\%T$, the fraction of surface covered with contact platelets, platelet accumulations of less than 5 μm and more than 5 μm in height, respectively. Surface coverage with spread platelets ($\%S$) and platelet adhesion are secondary parameters and are the sum of $\%A + \%T$ and $\%C + \%A + \%T = \%C + \%S$, respectively. $\%T$ is a measurement of thrombus formation (Fig. 8).

Thrombus dimensions are assessed by measuring height and volume of all thrombi on the surface by means of image analysis ($\times 310$).[25] The section is projected from a light microscope onto a magnetic tablet (manual optical picture analysis system, Kontron AG, Zürich, Switzerland). Thrombus perimeter and maximum height are traced with the electromagnetic pen. All data are transferred to a microprocessor which computes average thrombus cross-sectional area per unit length of surface evaluated ($\mu m^2/\mu m$), which is equivalent to average thrombus volume per unit surface area ($\mu m^3/\mu m^2$). Refined analysis of thrombus dimensions is obtained by classifying heights and volumes.[25]

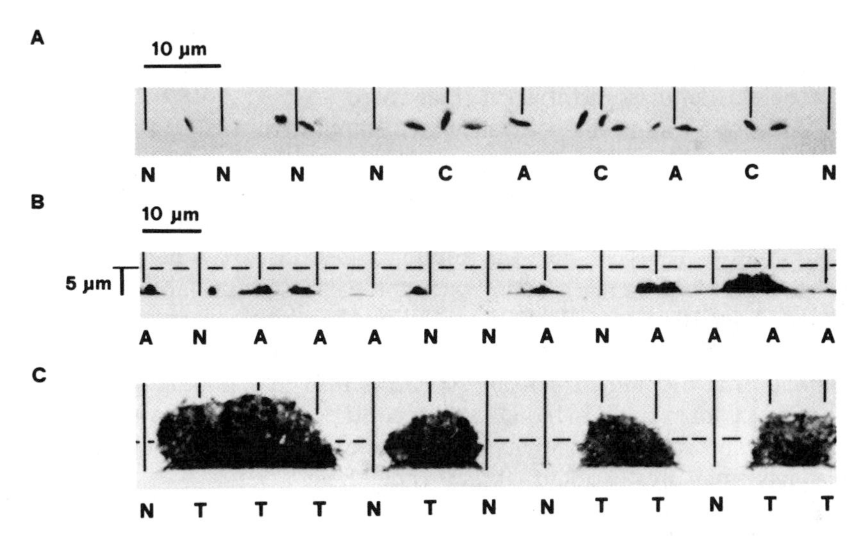

FIG. 8. Morphometric evaluation of platelet–surface interactions from semithin sections. Platelet–surface interactions are evaluated at 10-μm intervals and scored as naked *(N)* contact platelets *(C)*, spread platelets and platelet thrombi <5 μm in height *(A)*, and platelet thrombi >5 μm in height *(T)*. Platelet adhesion is defined as percentage surface coverage with platelets ($\%C + \%A + \%T$). Light micrographs: (A) Platelet adhesion to an Epon surface. Citrated rabbit blood, 800 sec^{-1} shear rate for 5 min, × 1750. (B) Platelet–surface interactions with human fibrilar collagen type III, 5 μg/cm^2. Citrated human blood, 1300 sec^{-1} shear rate for 5 min, × 1300. (C) Platelet–surface interactions with human fibrilar collagen type III, 10 μg/cm^2. Citrated human blood, 1300 sec^{-1} shear rate for 5 min, × 1300.

Theoretical considerations predict that values obtained by morphometry of cross sections are overestimated.[76]

The degree of overestimation is most pronounced (1) when the diameter of the object to be measured is smaller or about equal to the section thickness and (2) at a surface coverage of about 50%. For the thickness of a semithin section ($t = 0.8-1.0$ μm) and an average platelet diameter of $d = 2$ μm instead of a true 50% surface coverage with single platelets a value of about 70% was predicted.[11]

Protein Determination

The mass of platelets deposited per unit surface area is determined with any protein assay of sensitivity in the microgram range, provided that other proteinaceous blood elements (e.g., red cells, fibrin) are absent and the

[76] E. R. Weibel, *Int. Rev. Cytol.* **26,** 235 (1969).

amount of protein coated negligible.[22] The method is rapid, simple, and can be put into practice with standard laboratory equipment, but information on platelet mass distribution is sacrificed.

Disks of 5 mm in diameter are punched out from washed but unfixed collagen-coated plastic slides exposed to citrated blood or reconstituted blood. The proteins on the disk are solubilized with 0.25 ml of 1 *N* NaOH at 37° for 45 min and assayed by a modified Lowry method.[77] To the solubilized proteins 2.5 ml of a solution consisting of 98 parts Na_2CO_3 (2%), 1 part $CuSO_4$ (1%), and 1 part potassium sodium tartrate (1%) is added. After 10 min at 22°, 0.25 ml Folin–Ciocalteau reagent (E. Merck, Darmstadt, FRG) freshly diluted with two parts of water is added. The extinction at 691 nm in a 1-cm cuvette is read in a photometer. Disks coated but not exposed to blood serve as controls. Bovine serum albumin is used as standard. The sensitivity of the method is such that a monolayer of platelets on an area of about 10 mm^2 is detected.

Comparison of Evaluation Methods

The evaluation methods described above and the primary parameters which they measure are summarized in Table V. The approaches were validated by combining two evaluation methods in the same experiment.

Adhesion. In the absence of platelet–platelet cohesion a significant correlation exists between percentage surface coverage with platelets as measured morphometrically and the number of deposited radiolabeled platelets per surface area (Fig. 9, open symbols). However, the slope and the significance of the correlation depend on the surface exposed. The correlation is much weaker when mural thrombi are present (Fig. 9, closed symbols). A much closer correlation exists between the number of radiolabeled platelets per surface area and the average platelet thrombus volume per surface area measured morphometrically (Fig. 9B).[24]

Therefore, it appears imperative to demonstrate absence of platelet–platelet cohesion when radiolabeled platelets are used to determine platelet adhesion. The question whether total inhibition of platelet–platelet cohesion is possible without affecting platelet adhesion has not yet been answered satisfactorily.

The number of human platelets which are needed to fully cover the subendothelial surface of human renal arteries can be derived from the correlation found by Bolhuis *et al.*[8] (Fig. 9A) and amounts to 9.9×10^4 platelets/mm^2. Thus each platelet would occupy an average area of

[77] O. H. Lowry, H. J. Rosenbrough, A. L. Farr, and R. J. Randell, *J. Biol. Chem.* **193,** 265 (1951).

TABLE V
METHODS FOR DETERMINATION OF PLATELET–SURFACE INTERACTIONS

Evaluation method	Primary parameters						
	%C[a]	%A[a]	%T[a]	Percentage surface coverage	Platelet deposition (number/cm^2)	Thrombus height (μm)	Thrombus area ($\mu m^2/\mu m$)
Radiolabeled platelets					+[b]		
Protein assay					+		
Microdensitometry			+	+			
En face morphometry				+			
Section morphometry	+	+	+	%$(C + A + T)$		+	+

[a] See morphometry of semithin sections.

[b] Platelet adhesion to human subendothelium can also be measured with radiolabeled platelets provided platelet aggregation is absent (see Evaluation Using Radiolabeled Platelets).

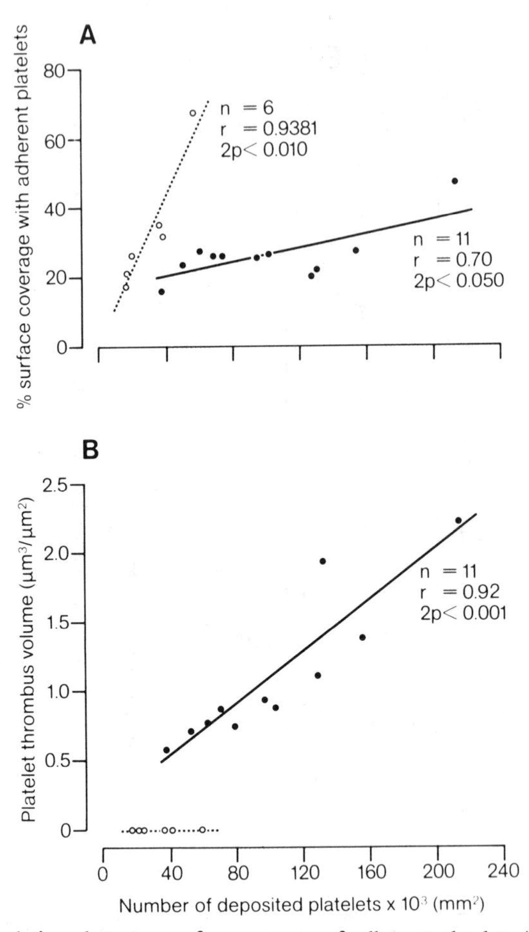

FIG. 9. Correlations between surface coverage of adherent platelets (section morphometry) and number of deposited platelets (^{51}Cr-labeled platelets) (A) and of platelet thrombus volume (section morphometry) and number of deposited platelets (^{51}Cr-labeled platelets) (B). Subendothelium of human renal arteries was exposed in an annular chamber to reconstituted human blood (citrated plasma, 1.2×10^{11} platelets/liter plasma, 40% hematocrit). Data from Bolhuis *et al.*[8] Cover slips coated with equine collagen (Horm, 30 $\mu g/cm^2$) were exposed to reconstituted human blood (citrated plasma, 2×10^{11} platelets/liter plasma, 40% hematocrit).

10.1 μm^2. Assuming an average diameter of 3.2 μm for a nonadherent platelet,[78] the projected circular surface area would correspond to 8.0 μm^2. Platelets forming a monolayer on subendothelium are spread out and

[78] M. M. Frojmovic, J. G. Milton, J. P. Caen, and G. Tobelem, *J. Lab. Clin. Med.* **91,** 109 (1978).

usually overlap each other partially; the figure of 10.1 μm^2 per average human platelet spread on subendothelium thus appears reasonable.

Mural Platelet Thrombi. On surfaces which induce the formation of platelet thrombi such as rabbit or human subendothelium and collagen-coated surfaces the number of deposited platelets correlates with the average thrombus volume per surface area (Fig. 9B).[24] If the protein content of the exposed thrombogenic surface is negligible, a correlation between deposited protein and number of radiolabeled platelets associated with the surface is found.[22]

The number of platelets per unit mural thrombus volume can be derived from correlations such as the one shown in Fig. 9B and amounts to approximately 0.09 human platelets/μm^3. This figure corresponds to an average platelet volume of 11.5 μm^3, a value which is somewhat larger than the 7.4 μm^3 reported for unstimulated human platelets in citrated platelet rich plasma.[19,79] The average values for rabbit platelets are 0.122 platelets/μm^3, corresponding to an average platelet volume of 8.2 μm^3, which again is somewhat larger than the volume reported for rabbit platelets (5.8 μm^3).[24,80] These apparent discrepancies may be due to (1) plasma volume between aggregated platelets as seen by electron microscopy[21] and (2) a larger volume of activated platelets.[81]

Evaluation of Platelet–Surface Interactions in the Presence of Fibrin Formation

Fibrin is deposited on subendothelium in addition to platelets after perfusion of nonanticoagulated blood. There is an inverse and shear rate dependent relationship between platelet and fibrin deposition.[25,26]

Morphometry of Semithin Sections

Coverage of the subendothelial surface with fibrin is measured using the same morphometric principles as for platelets. However, multiple layers of fibrin and fibrin within platelet thrombi as they are observed after perfusion times longer than 5 min are very difficult to quantify morphometrically. An excellent correlation between percentage surface coverage with

[79] L. Corash, *in* "Current Topics in Hematology" (S. Piomelli and S. Yachnin, eds.), p. 99. Liss, New York, 1983.

[80] G. V. R. Born, *J. Physiol. (London)* **209,** 487 (1970).

[81] B. S. Bull and M. B. Zucker, *Proc. Soc. Exp. Biol. Med.* **120,** 296 (1965).

fibrin (%F) and the fibrin determined immunologically by measuring fragment E in plasmin digests of the exposed surface is only found after short perfusion times.[82]

Monitoring of Chamber Occlusion

Particularly at high shear rates progressive chamber occlusion occurs within 10–20 min.[26,74]

Thrombotic masses initially grow on the exposed vessel segment but subsequently also further downstream and occlude the exit holes of the chamber. Pressure transducers (PDCR, Comat, Leicestershire, England) are attached to the chamber entrance and exit to monitor pressure changes (Fig. 5). The pressure difference between chamber entrance and exit is a sensitive measure of chamber occlusion. The drop in blood flow can also be monitored using a flow meter. Blood flow reduction is, however, considerably less sensitive than the increase in pressure difference.

Concluding Remarks

The techniques described in this chapter allow quantitation of platelet–surface interactions with a variety of vascular and artificial surfaces following exposure to numerous blood perfusates under conditions of controlled blood flow.

Combination of two quantitative techniques may improve the yield of information, e.g., half of the test surface is used for radioactive counting and the other half for morphometric evaluation. Panels of test samples are rapidly screened by means of radiolabeled platelets. Refined analysis of platelet–surface interactions is achieved only by means of cross-sectional morphometry. It is, however, important to keep in mind that washed platelets interact less extensively with the surface than nonwashed platelets. This may mask small differences in surface-mediated platelet interactions in comparative studies. The present perfusion systems are not convenient for routine purposes, because some steps are unusually time consuming and require highly trained personnel. However, efforts are being made to computerize the evaluation procedure.

[82] W. Inauen, H. R. Baumgartner, A. Haeberli, and P. W. Straub, *Thromb. Haemostasis* **57**, 306 (1987).

[6] Measurement of Platelet Interaction with Endothelial Monolayers

By GLENNA L. FRY and JOHN C. HOAK

Under physiologic conditions, the vascular endothelium remains nonthrombogenic. Several factors synthesized in the endothelium [prostacyclin (PGI_2), endothelial-derived relaxing factor (EDRF), heparin-like glycosaminoglycans, and thrombomodulin] may contribute to nonthrombogenic properties of the vascular endothelium. To study the interaction of platelets and the endothelium we have developed methods utilizing primary cultures of endothelium from the human umbilical vein, umbilical artery, and bovine aorta. The interaction of platelets with these confluent monolayers can be measured by using a modification of the *in vitro* model system reported by Winkelhake and Nicolson.[1] We have measured both adherence of activated platelets to endothelial monolayers and the effect of released materials from each cell type on the other.

Materials. Powdered medium 199 (with Earle's salts, with L-glutamine, without $NaHCO_3$), Hanks' balanced salt solution (10×), Eagle's basal medium vitamin solution (100×), Eagle's basal medium amino acids solution (100×), calcium- and magnesium-free Dulbecco's phosphate-buffered saline (PBS), and trypsin were purchased from Gibco Laboratories, Grand Island, New York. Trizma base (Tris), *N*-2-hydroxyethylpiperazine-*N*′-2-ethanesulfonic acid (HEPES), hirudin, ethylenediaminetetraacetic acid (EDTA), acetylsalicylic acid (ASA), collagen, and adenosine diphosphate (ADP) were obtained from Sigma Chemical Co., St. Louis, Missouri. Sodium [^{51}Cr] chromate was purchased from E. R. Squibb and Sons, Princeton, New Jersey, and New England Nuclear, Boston, Massachusetts. Heparin was purchased from Abbott Diagnostics, North Chicago, Illinois. Topical bovine thrombin was obtained from Parke, Davis and Company, Detroit, Michigan. Fetal bovine serum (FBS) was obtained from Armour Pharmaceutical Co., Tarrytown, New York. Pentex bovine serum albumin (BSA), fraction V, fatty acid poor, and human fibrinogen were obtained from Miles Laboratories, Inc., Kankakee, Illinois. ASA was also obtained from Mallinckrodt, Inc., St. Louis, Missouri. All other chemicals were of

[1] J. L. Winkelhake and G. L. Nicolson, *J. Natl. Cancer Inst.* **56,** 285 (1976).

reagent grade. Tissue culture plastic supplies were obtained from Falcon Plastics, Division of Bioquest, Oxnard, California, or from Flow Laboratories, Inc., McLean, Virginia. Dansylarginine *N*-(3-ethyl-1, 5-pentanediyl)amide (DAPA) was kindly provided by K. G. Mann, University of Vermont, Burlington, Vermont.

Solutions

Tyrode's solution: 137 m*M* NaCl, 2.7 m*M* KCl, 11.9 m*M* $NaHCO_3$, 0.36 m*M* NaH_2PO_4, 2.0 m*M* $CaCl_2$, 1.0 m*M* $MgCl_2$, 5.5 m*M* dextrose, 0.35% bovine serum albumin

Acid-citrate-dextrose (ACD): 0.065 *M* citric acid, 0.085 *M* trisodium citrate, 0.11 *M* dextrose

Modified Hanks' balanced salt solution (MHBSS): Without $NaHCO_3$, with 15 m*M* HEPES, pH 7.4

Incubation medium (IM): 8.6 g/liter bovine serum albumin, 140.3 m*M* NaCl, 5.8 m*M* KCl, 2.7 m*M* $CaCl_2$, 16.3 m*M* Tris, pH 7.4

ASA solution: ASA was dissolved in either IM or Tyrode's solution

Modified medium 199 (MM-199): 9.87 g/liter medium 199 powder, 10 ml/liter BME vitamin solution (100×), 10 ml/liter BME amino acids solution (100×), 1 g/liter dextrose, 0.1 g/liter neomycin, 2.2 g/liter $NaHCO_3$, pH 7.2-7.3

MM-199 with 20% FBS: Eight parts MM-199 mixed with two parts FBS

Isolation and Culture of Cells. Primary cultures of human umbilical vein endothelial cells and human umbilical artery endothelial cells were prepared according to a slight modification of the method of Jaffe *et al.*[2] Primary cultures of endothelium from bovine aortas were prepared according to the method of Goldsmith *et al.*[3] Cells were suspended in MM-199 with 20% FBS, seeded in 35 × 10 mm tissue culture dishes or Linbro 12-well multiwell plates, and incubated in a 5% CO_2 atmosphere at 37°. After 24 hr, the monolayers were washed with fresh MM-199 with 20% FBS and relayered with the same medium. Confluent monolayers of human endothelium were obtained by adding 2 ml of cell suspension (500,000 cells/ml) to 35-mm dishes and 1 ml to each well of the 12-well plates. Monolayers were used for studies 3 or 4 days after seeding. Other

[2] E. A. Jaffe, R. L. Nachman, C. G. Becker, and C. R. Minick, *J. Clin. Invest.* **52,** 2745 (1973).

[3] J. C.Goldsmith, C. T. Jafvert, P. Lollar, W. G. Owen, and J. C. Hoak, *Lab. Invest.* **45,** 191 (1981).

cell types used for comparative studies were obtained by tissue explants, or enzyme treatment (trypsin or collagenase), with subsequent subculture using 0.25% trypsin-0.05% EDTA.[4,5]

Preparation of Washed Platelet Suspensions.[4] The isolation and washing of platelets was a modification of the procedure described by Mustard *et al.*[6] Venous blood was drawn from healthy, nonfasting donors who had not taken aspirin or other antiplatelet drugs for 10 days, and in the case of women, those who were not using oral contraceptives. Five parts of blood were mixed with one part of ACD anticoagulant. One hundred milliliters of blood, mixed with 20 ml of ACD, was centrifuged in 40-ml siliconized conical centrifuge tubes at 325 *g* for 15 min. The resulting platelet-rich plasma (PRP) was separated and centrifuged at 1000 *g* for 15 min. The platelet pellet was resuspended in 10 ml of Tyrode's solution and 0.4 ml ACD. If aspirin-treated platelets were to be used, the Tyrode's solution contained the aspirin and the platelet number did not exceed 1.5×10^9 platelets/ml. After the addition of 250 units heparin (and 100 μCi $Na_2{}^{51}CrO_4$ for ^{51}Cr-labeled platelets), the platelet suspension was incubated at 37° for 20 min. After centrifugation at 650 *g* for 10 min the pellet was resuspended in 10 ml of Tyrode's solution and 0.4 ml of ACD. After centrifugation at 650 *g* for 10 min the platelet pellet was resuspended in a second wash solution which consisted of 10 ml of Tyrode's solution and 0.2 ml of ACD. A final centrifugation at 650 *g* yielded a pellet which was resuspended in 9 ml of Tyrode's solution. One-half milliliter of this platelet suspension contained $2-3 \times 10^8$ platelets and 50,000-70,000 counts per minute (cpm) of ^{51}Cr radioactivity if labeled. Platelet suspensions were free of aggregates, as determined by phase contrast microscopy. Suspensions were stored at 37° and used within 4 hr after the blood was drawn. When ADP or collagen was to be used with a plasma system, the platelet pellet was resuspended in platelet-poor plasma from the same donor. Nine parts of blood were mixed with one part of 3.8% sodium citrate in siliconized centrifuge tubes and centrifuged at 1000 *g* for 15 min. The supernatant plasma was removed. After the second wash, the pellet was resuspended in the citrated platelet-poor plasma (PPP) and stored at 37° until used.

[4] R. L. Czervionke, J. C. Hoak, and G. L. Fry, *J. Clin. Invest.* **62,** 847 (1978).

[5] G. L. Fry, R. L. Czervionke, J. C. Hoak, J. B. Smith, and D. L. Haycraft, *Blood* **55,** 271 (1980).

[6] J. F. Mustard, D. W. Perry, N. G. Ardlie, and M. A. Packham, *Br. J. Haematol.* **22,** 193 (1972).

Measurement of Platelet Adherence

Adherence of Preformed Platelet Aggregates to Cell Monolayers.[4] Incubation mixture (IM; 0.8–0.9 ml) and 0.5 ml of ^{51}Cr-labeled platelets in Tyrode's solution were stirred in a Payton aggregation module at 1000 rpm and 37° for 2–3 min; 0.1 ml of IM (control) or 0.1 ml of 1.0 U/ml bovine thrombin in IM was added to the platelet suspension and aggregation was monitored on the strip chart recorder. As soon as half-maximal aggregation was indicated (or equivalent time elapsed for IM controls), the platelet suspension was poured onto the cell monolayer, which had previously been washed twice with 1.5 ml of MHBSS. The platelet suspension was poured at half-maximal aggregation because the amount of time required for pouring from the cuvettes was equivalent to the time needed to reach maximal aggregation. In some studies, the excess thrombin was neutralized by adding hirudin (0.1 ml of 4 U/ml) or DAPA (0.1 ml of 20 μM) to the aggregometer cuvette just before pouring the suspension onto the monolayer. The 35 × 10 mm tissue culture dish, containing the cell monolayer and platelet suspension, was placed in a 37° incubator on a rocker platform with a 30° tilt (Bellco Glass, Inc., Vineland, NJ) and rocked continuously at a rate of 10 up and down cycles per minute. After a 30-min incubation, the suspension containing unattached platelets was transferred to a plastic centrifuge tube. After the addition of 1.5 ml of MHBSS, the dish was returned to the rocker for 5 min. This 1.5 ml of MHBSS and an additional 1 ml of MHBSS wash (no rocking) were added to the centrifuge tube containing the unattached platelets. After centrifugation at 1000 *g* for 10 min, ^{51}Cr radioactivity in the tube was counted with a Beckman DP 5000 gamma counter (Beckman Instruments, Palo Alto, CA). The amount of radioactivity remaining with the cell monolayer was determined by solubilizing the cell monolayer and attached platelets with 1.5 ml of 19 mM Na_2CO_3 in 0.1 N NaOH. After a 30-min incubation at 37°, the dish was rinsed with the solution before removing it to a plastic centrifuge tube. One milliliter of the 19 mM Na_2CO_3 in 0.1 N NaOH was added to the dish again, rinsing the dish and adding it to the centrifuge tube. ^{51}Cr Radioactivity was determined. Results are calculated as follows:

$$\text{Percentage adherence} = \{\text{monolayer (cpm)}/[\text{supernatant (cpm)} + \text{monolayer (cpm)}]\}(100)$$

Adherence of Platelets to Monolayers Previously Treated with Thrombin.[4] The cell monolayer was washed twice with 1 ml of MHBSS; 0.6 ml of IM (control) or 1.0 U/ml bovine thrombin in IM was placed on the monolayer and the 12-well multiwell tissue culture plate was incubated without rocking for 5 min at 37°. The preincubation solution was left on the monolayer and 0.3 ml of ^{51}Cr-labeled platelets in Tyrode's solution was immediately added. The multiwell plate was rocked at 37° for 30 min and handled as described above. Volumes of MHBSS added for washes were 0.9 ml for the first wash and 0.6 ml for the second.

Adherence of Platelets to Monolayers Treated with Aspirin.[4] The monolayer was washed twice with MHBSS (1.5 ml for 35-mm dishes and 0.9 ml/well for multiwell plates). IM (control) or ASA in IM was placed on the monolayer (1.0 ml for dishes and 0.6 ml for plates), and the container was incubated with rocking for 30 min at 37°. The ASA or control solution was removed by aspiration, and the dish was washed twice with MHBSS. The appropriate volume of IM (control) or thrombin in IM (for the container being used) was placed on the monolayer followed immediately by untreated or aspirin-treated ^{51}Cr-labeled platelets in Tyrode's solution. For the adherence of performed platelet aggregates to ASA-treated monolayers, the untreated or aspirin-treated ^{51}Cr-labeled platelet suspension was stirred with IM or thrombin in IM, as described previously and the aggregating suspension poured onto the monolayer. After rocking for 30 min at 37°, the dish or plate was handled as described above.

Adherence of ADP or Collagen-Induced Preformed Platelet Aggregates. Platelets were collected, labeled, and washed as described above, except the Tyrode's solution contained 0.1 mg/ml apyrase (EC 3.6.1.5) (0.33–0.34 ATPase U/ml). The platelets were resuspended in Tyrode's solution without $CaCl_2$ after the second wash. Suspensions to be aggregated with ADP also contained 0.05–0.1% human fibrinogen. The platelet suspension (0.5 ml) was mixed with 0.5 ml of Tyrode's solution, without calcium, but with apyrase. Either 50 μl of a 2% collagen suspension or 50 μl of a 1 m*M* ADP solution was added to the platelet suspension in the aggregometer cuvette. At half-maximal aggregation, the suspension was poured onto a cell monolayer to which had been added 0.5 ml of IM with 5.4 m*M* $CaCl_2$ and 1.38% BSA. The incubation, washing, and calculation of results was performed, as previously described. Studies with collagen and ADP may be done with ^{51}Cr-labeled platelets resuspended in platelet-poor plasma containing sodium citrate. However, the endothelial monolayer requires calcium and magnesium to remain intact. In order to prevent interference by thrombin generation, 0.5 ml of ^{51}Cr-labeled platelets suspended in plasma is mixed with IM containing $MgCl_2$ but not $CaCl_2$. Another alternative method would be the addition of a thrombin inhibitor (hirudin or DAPA) with the IM.

Contribution of Platelets to Endothelial Monolayer Prostacyclin Production.[7] Confluent endothelial monolayers in 35 × 10 mm culture dishes were washed twice with MHBSS. Some dishes were pretreated before use with 1 ml of IM or 100 μM ASA in IM for 30 min at 37°. This preincubation solution was removed and the monolayer washed twice with MHBSS. Tyrode's solution (0.5 ml) with or without $2-3 \times 10^8$ platelets (unlabeled platelets prepared as described) was mixed with 0.9 ml of IM in an aggregometer cuvette. After a 2-min preincubation at 37°, with stirring at 1000 rpm, 0.1 ml of IM with or without 1.0 U/ml thrombin was added. At half-maximal aggregation, or the equivalent amount of time (buffer controls), the suspension or buffer control was poured onto the monolayer. In additional studies, the platelets were initially diluted with 0.8 ml of IM. Before the platelets were transferred to the monolayer, the excess thrombin was neutralized at half-maximal aggregation by quickly adding 0.1 ml of IM with 20 μM DAPA or 4 U/ml hirudin. After a 30-min incubation at 37° with rocking, the solution over the monolayer was centrifuged for 2 min at 15,000 *g*. Endothelial prostacyclin release into the supernatant was determined by radioimmunoassay for its stable metabolite 6-ketoprostaglandin (PG) $F_{1\alpha}$ (alpha).[7,8]

General Comments. The method presented herein offers a system to measure the response of endothelial cells to quiescent or activated platelets. Conversely, the products of endothelial cells influence the function of platelets. Although the main limitation of the method is the lack of laminar flow conditions, it is possible to expand this technique to a flow system as described elsewhere.[9]

[7] R. L. Czervionke, J. C. Hoak, D. L. Haycraft, G. L. Fry, and K. G. Mann, *Trans. Assoc. Am. Physicians* **95,** 272 (1982).

[8] R. L. Czervionke, J. B. Smith, J. C. Hoak, G. L. Fry, and D. L. Haycraft, *Thromb. Res.* **14,** 781 (1979).

[9] T. J. Parsons, D. L. Haycraft, J. C. Hoak, and H. Sage, *Thromb. Res.* **43,** 435 (1986).

[7] Measurement of Platelet and Megakaryocyte Interaction with the Subendothelial Extracellular Matrix

By AMIRAM ELDOR, ZVI FUKS, RICHARD F. LEVINE, and ISRAEL VLODAVSKY

Endothelial cells which line the inner surface of blood vessels rest on the subendothelium—a highly thrombogenic basement membranelike extracellular matrix (ECM).[1] One of the main functions of the vascular

METHODS IN ENZYMOLOGY, VOL. 169

endothelium is to maintain the nonthrombogenic properties of the vessel wall, hence exposure of the subendothelium to the circulating blood following endothelial damage results in a hemostatic response.[2-4] Characteristic of this response is the adhesion of platelets to the subendothelium which is followed by morphological alterations, degranulation, thromboxane synthesis, and aggregation.[2-5] Thus, in contrast to the endothelium, the subendothelium provides a highly thrombogenic surface which plays an important role in the arrest of bleeding, and may be involved in pathological processes such as thrombosis and atherosclerosis.[6,7] The interaction of platelets with the subendothelial basement membrane was previously studied both in animal models, in which blood vessels were deendothelialized with a balloon catheter, and with denuded vascular segments which were interacted with blood in a specially designed perfusion chamber.[2-5] The major limitations to the wide use of these techniques is that they require a constant supply of fresh segments of blood vessels as well as some surgical skills and that with the use of deendothelialized arteries or veins, it is difficult, if not impossible, to discern the subendothelial basement membrane from other constituents of the vessel wall.

Cultured vascular or corneal endothelial cells, which closely resemble their *in vivo* counterparts, produce an extracellular matrix which is deposited exclusively beneath the cell layer. This ECM resembles in its organization and macromolecular composition (fibronectin, laminin, collagen types III and IV, elastin, heparan sulfate, and dermatan sulfate proteoglycans) the vascular subendothelial basement membrane.[8-10] Denudation of the endothelial cell monolayer by solubilization with Triton X-100 leaves the underlying ECM intact and firmly attached to the entire surface of the culture dish, while retaining its characteristic morphological appearance and biological properties.[8-10] This ECM has been used as an *in vitro* model for studies on platelet interaction with the subendothelium. The ECM was

[1] Y. Jaffe, C. R. Minick, B. Adelman, C. G. Becker, and R. Nachman, *J. Exp. Med.* **144,** 209 (1976).
[2] H. R. Baumgartner, M. B. Stemerman, and T. H. Spaet, *Experientia* **27**, 283 (1971).
[3] T. H. Spaet and B. V. Stemerman, *Ann. N.Y. Acad. Sci.* **201,** 13 (1972).
[4] H. R. Baumgartner and C. Haudenschild, *Ann. N.Y. Acad. Sci.* **201,** 22, (1972).
[5] H. R. Baumgartner and R. Muggli, *in* "Platelets in Biology and Pathology" (J. L. Gordon, ed.), p. 3. Elsevier/North-Holland, Amsterdam, The Netherlands, 1976.
[6] R. Ross and J. A. Glomset, *N. Engl. J. Med.* **295,** 369 (1976).
[7] M. B. Stemerman, in "Progress in Hemostasis and Thrombosis" (T. Spaet, ed.), p. 1. Grune & Stratton, New York, 1975.
[8] D. Gospodarowicz, I. Vlodavsky, and N. Savion, *J. Supramol. Struct.* **13,** 339 (1980).
[9] D. Gospodarowicz, I. Vlodavsky, and N. Savion, *Endocr. Rev.* **1,** 201 (1980).
[10] I. Vlodavsky, G. M. Lui, and D. Gospodarowicz, *Cell* **19,** 607 (1980).

shown to initiate the morphological and biochemical alterations occurring during platelet activation at sites of endothelial injury. The results obtained were virtually similar to those obtained using mechanically deendothelialized vascular segments.[11-13] It was demonstrated that upon incubation with ECM, platelets undergo shape changes (Fig. 1A and B) and aggregation (Fig. 1C and D) associated with release of vasoactive substances and thromboxane A_2 production (Table I).[11,12] Platelet interaction with the subendothelial ECM is dependent upon the presence of certain plasma proteins, especially fibrinogen and von Willebrand factor.[12,13] In the absence of fibrinogen, platelets adhere to the ECM, but do not form aggregates,[13] whereas von Willebrand factor is necessary for platelet adhesion under high shear forces.[14] Thrombasthenic platelets, which lack the fibrinogen receptor, were found to interact with the ECM in a manner similar to that shown with denuded vascular segments.[13,15] These platelets showed tight adhesion to the ECM and extension of typical filopodia, but failed to aggregate.[13]

The ECM can also provide a model to test the interaction of megakaryocytes with vessel wall components, and for the *in vitro* activation of megakaryocytes.[16,17] During the process of platelet liberation, the megakaryocytes penetrate the sinusoid walls. This process may require close association and interaction of the megakaryocytes with subendothelial matrix components on the abluminal surface of the sinusoid in order to attach to and penetrate the vessel wall. Unlike platelets, isolated megakaryocytes do not adhere to glass or collagen-coated surfaces.[18] In contrast, shortly after exposure to ECM, the megakaryocytes show a nonreversible adhesion, and extensive formation of filopodia with a tendency toward flattening.[16,17] The interaction of megakaryocytes with the ECM is associated with production of significant amounts of thromboxane A_2 (TXA_2).[16,17]

[11] I. Vlodavsky, A. Eldor, E. HyAm, R. Atzmon, and Z. Fuks, *Thromb. Res.* **28,** 179 (1982).
[12] K. S. Sakariassen, P.A.M.M. Aarts, P. G. de Groot, W. P. M. Houdijk, and J. J. Sixma, *J. Lab. Clin. Med.* **102,** 522 (1983).
[13] A. Eldor, I. Vlodavsky, U. Martinowicz, Z. Fuks, and B. S. Coller, *Blood* **65,** 1477 (1985).
[14] F. M. Booyse, S. Feder, and A. Quarfoot, *Thromb. Res.* **28,** 299 (1982).
[15] T. B. Tschopp, H. J. Weiss, and H. R. Baumgartner, *Experientia* **31,** 113 (1975).
[16] R. F. Levine, A. Eldor, H. Gamliel, Z. Fuks, and I. Vlodavsky, *Blood* **66,** 570 (1985).
[17] Y. G. Caine, I. Vlodavsky, M. Hersh, A. Polliack, D. Gurfel, R. Or, R. F. Levine, and A. Eldor, *Scanning Electron Microsc.* **3,** 1087 (1986).
[18] V. T. Nachmias, *Semin. Hematol.* **20,** 261 (1983).

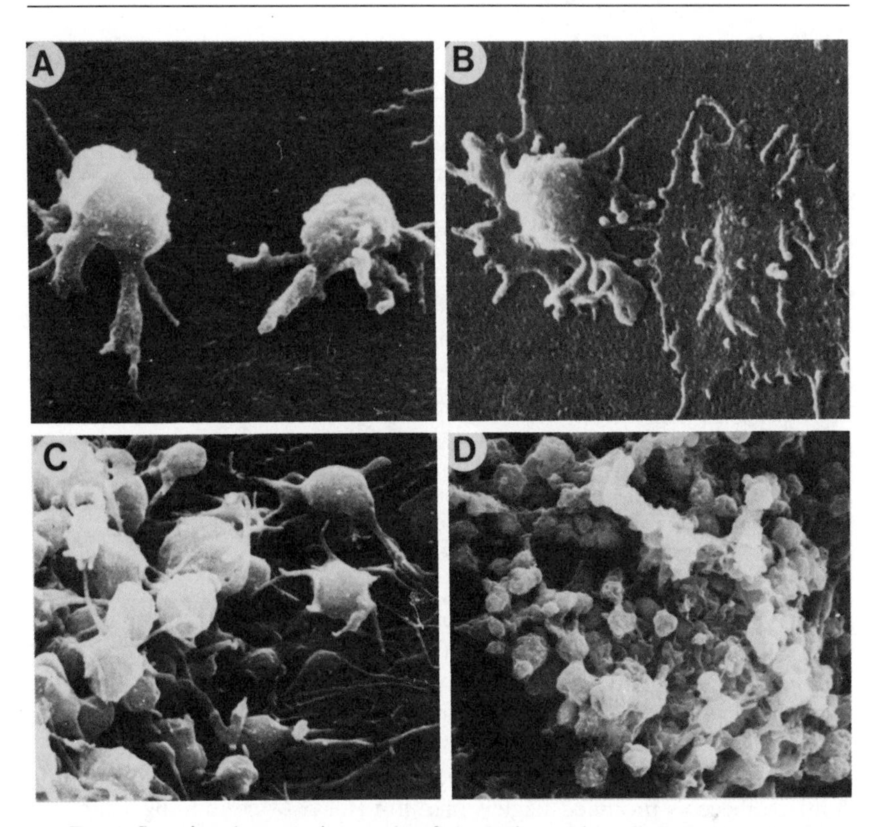

FIG. 1. Scanning electron micrographs of platelet interaction with subendothelial ECM. One milliliter of PRP was incubated (30 min, 37°, gentle shaking) on top of the subendothelial ECM. Unbound platelets were removed and the remaining firmly attached platelets fixed and processed for SEM. Note the extension of pseudopodia in panel A (×16,000), flattening and firm attachment of single platelets in panel B (×11,300), and the extensive aggregation of platelets on top of the ECM in panels C (×9000) and D (×1600).

Materials and Methods

Preparation of ECM-Coated Surfaces

Bovine corneal endothelial cells rather than vascular endothelial cells, are used routinely in our laboratory for the preparation of ECM-coated surfaces (tissue culture dishes, coverslips, or microcarrier beads), since these cells are easy to maintain in culture and produce a thick matrix similar in its organization, macromolecular constituents, and platelet reactivity to ECM produced by cultured vascular endothelial cells.[8-10]

TABLE I
INTERACTION OF PLATELETS WITH SUBENDOTHELIAL ECM[a]

Parameter	Empty plastic dish	Matrix-coated dish	Aortic endothelial cells on matrix
Aggregation			
Light microscopy	—	++++	—
Scanning electron microscopy	—	++++	—
Adherent platelets (% of total)			
^{111}In-labeled platelets	3.3	36.3	0
[^{14}C]Serotonin-labeled platelets	1.4	29.9	1.2
Thromboxane B_2 formation (ng/ml)	5.6 ± 1.9	32.4 ± 2.5	1.83 ± 0.1
[^{14}C]Serotonin release (%)	0	14.5	0

[a] Extracellular matrix produced by the bovine corneal endothelial cells was used. Human platelet-rich plasma obtained from different donors was used in five separate experiments in three to six culture dishes (35 mm) each incubated with 1 ml PRP for 30 min at 37°. Results are presented as mean ± SD.

Fresh cow eyes are obtained from the slaughterhouse and are either used immediately or can be stored for 12 hr at 4°. The cornea is washed with 95% ethanol and dissected out with fine scissors. The inner side of the cornea, which is covered with endothelial cells, is washed with phosphate-buffered saline (PBS) and is delicately scraped with a groove director. The groove director is then dipped into a 6-cm tissue culture dish containing 5 ml Dulbecco's modified Eagle's medium (DMEM containing 1 mg glucose/ml) supplemented with 10% fetal calf serum, 5% calf serum, and 50 μg/ml gentamicin. This procedure is repeated several times, and the plates which contain tissue fragments are then incubated at 37° in a CO_2 humidified incubator and left untouched for 5 days. During this period the endothelial cells will migrate from the explants, proliferate, and start to cover the culture dish surface. The proliferation of the endothelial cells is stimulated by the addition (every other day) of partially purified fibroblast growth factor (FGF; 100 ng/ml) prepared from bovine brain as described.[19,20] The primary cultures are treated with STV solution (0.9% NaCl, 0.01 *M* sodium phosphate, pH 7.4, 0.05% trypsin, and 0.02% EDTA) at 37° for 5 min and the detached cells are resuspended in DMEM supplemented with serum and antibiotics and passaged into 35-mm tissue culture dishes (Falcon Labware Division, Becton Dickinson Co., Oxnard, CA) at a split ratio of 1:10 (5×10^4–10^5 cells/35-mm dish). FGF (100

[19] D. Gospodarowicz, J. Moran, D. Braun, and C. Birdwell, *Proc. Natl. Acad. Sci. U.S.A.* **73,** 4120 (1976).

[20] D. Gospodarowicz, A. R. Mescher, and C. Birdwell, *Exp. Eye Res.* **25,** 75 (1977).

ng/ml) is added every other day during the phase of active cell growth, until the cultures are confluent. In order to obtain a better production of ECM, 4% dextran T-40 (Pharmacia, Fine Chemicals, Sweden) is included in the growth medium. Six to 8 days after the cells reached confluency, the subendothelial ECM is exposed by dissolving (3 min, 22°) the cell layer with a solution containing 0.5% Triton X-100 and 20 mM NH_4OH in phosphate-buffered saline (PBS), followed by four washes in PBS.[10] The ECM remains intact, firmly attached to the entire area of the tissue culture dish and free of nuclear and cellular debris. ECM-coated dishes containing PBS and antibiotics are stored at 4°C for periods up to 6 months without losing their platelet reactivity. Dishes can also be stored for short periods at 4° in the absence of any fluid.

For preparation of ECM-coated cover slips, each plastic cover slip (Thermanox, Lux Scientific Corporation, Newburg Park, CA) is placed in a 35-mm dish or into each well of a 4 well tissue culture plate. Corneal endothelial cells are then seeded and the cultures treated as detailed above.

For preparation of ECM-coated microcarrier beads (Cytodex 2 microcarriers, Pharmacia Fine Chemicals, Uppsala, Sweden) $0.5-1 \times 10^5$ beads are placed into each of 35-mm plastic Petri dishes; 1×10^5 corneal endothelial cells are then added in 2 ml of complete growth medium and the suspension incubated under constant stirring. Five to 7 days later the beads become covered with a monolayer of cells. Ten to 14 days after seeding the cell-coated beads are harvested by centrifugation at 600 rpm for 2 min and the beads are suspended in PBS containing 0.5% Triton X-100 and 0.025 N NH_4OH. After a 15-min incubation at room temperature with gentle shaking the cells are dissolved and the remaining ECM-coated beads are washed four times with PBS by sedimentation under gravity.[21] ECM-coated beads induce platelet aggregation when tested in a Born-type aggregometer. This aggregation has a 2-min lag period and is similar in shape to aggregation induced by collagen. It should be noted that under similar conditions, no platelet aggregation occurs with uncoated beads, or when ECM which is scraped from ECM-coated dishes is added to the aggregation cuvette.

Platelet Reactivity with ECM

The reactivity of platelets with ECM-coated surfaces can be assayed using ECM-coated culture dishes or in a perfusion system in which ECM-coated coverslips are introduced into a perfusion chamber. In both systems it is possible to use anticoagulated blood, platelet-rich plasma, or a suspension of washed platelets.[11-14]

[21] S. Mai and A. E. Chung, *Exp. Cell Res.* **152,** 500 (1984).

Platelet Interaction with ECM-Coated Dishes. One-milliliter aliquots of anticoagulated blood, citrated platelet-rich plasma, or washed platelet suspension is introduced into 35-mm ECM-coated dishes and incubated at 37° for various time periods. If the dishes are not agitated, platelets will adhere tightly to the ECM and will undergo shape change with the formation of long filopodia, lamellipodia, and subsequent flattening (Fig. 1A and B). This process is associated with TXA_2 synthesis and serotonin release. The number of adherent platelets can be evaluated by prelabeling the platelets with ^{111}In-oxine complex solution or ^{51}Cr.[22,23] Aliquots of the supernatant can be frozen and used to determine the amounts of TXA_2 (by radioimmunoassay) and other substances released from the platelets, such as [^{14}C]serotonin.[11-13] The morphological changes that occur in the adherent platelets can be visualized by scanning electron microscopy (SEM) following fixation (1 hr, 37°) with 2.5% glutaraldehyde in PBS, incubation with 2% guanidine hydrochloride and 2% tannic acid, postfixation, dehydration, and air drying.[24]

When ECM-coated dishes containing platelet-rich plasma are subjected to a gentle agitation (on an electric rocker platform), large platelet aggregates are formed on top of the ECM within a few minutes (Fig. 1C and D).[11,13] These aggregates can be seen with the naked eye, or by an inverted phase microscope and are composed of several layers of platelets forming a thrombus-like structure. Platelet aggregation is associated with TXA_2 synthesis and serotonin release in a manner similar to that observed *in vivo* or with a mechanically deendothelialized rabbit aorta (Table I). In this system no aggregation occurs with thrombasthenic platelets. These platelets which do not adhere to glass surfaces, adhere tightly to the ECM, undergo shape change as normal platelets, release serotonin, and produce TXB_2 in quantities similar or somewhat lower than those observed with normal platelets.[13] Aggregation on the ECM does not occur when washed platelet suspensions which are devoid of fibrinogen are interacted with this surface.[13] Platelet aggregation on the ECM is inhibited by pretreatment of the platelets with platelet inhibiting drugs such as aspirin, prostacyclin (PGI_2), or stable analogs of PGI_2, and by pretreatment of the platelets with a monoclonal antibody (10E5) which blocks the fibrinogen receptor (glycoproteins IIb and IIIa) on the platelet membrane, and produces a thrombasthenic-like defect.[13]

The potential of the ECM system and its relevance to the situation in the vascular intima can also be demonstrated by the protective, nonthrombogenic role of vascular endothelial cells plated on top of the ECM. As

[22] I. Vlodavsky, Z. Fuks, M. Ber Ner, Y. Ariav, and V. Schirrmacher, *Cancer Res.* **43,** 2704 (1983).

[23] N. Savion, I. Vlodavsky, and Z. Fuks, *J. Cell. Physiol.* **118,** 169 (1984).

[24] T. Murakami, K. Yamamoko, and T. Itoshima, *Arch. Histol. Jpn.* **40,** 35 (1972).

demonstrated in Fig. 2, incubation of platelets (under gentle agitation) with a confluent endothelial cell monolayer that was mechanically wounded so as to expose the subendothelial ECM resulted in platelet adhesion and activation, while no platelet aggregation occurred on exposed ECM next to vascular endothelial cells. Pretreatment of similar endothelial cell cultures with aspirin (100 μM) (Fig. 2B) resulted in the appearance of large platelet aggregates in the wound area, demonstrating that the inhibition of platelet aggregation by the untreated endothelial cell cultures was mediated by production of prostacyclin (PGI_2).[25] By measuring the quantities of PGI_2 and TXA_2 which were synthesized by the vascular cells and the platelets when interacted together, the known endoperoxide transfer between platelets and endothelial cells could be demonstrated. These results emphasize the validity of the *in vitro* endothelial ECM model for studies on the interaction of platelets with both intact and injured vessel walls.

Platelet Interaction with ECM in a Perfusion System

The role of plasma von Willebrand factor in the interaction of platelets with the vascular subendothelium has been elucidated in the investigations which used denuded vascular segments placed in annular perfusion chambers developed by Baumgartner.[5] Sakariassen and colleagues have developed a perfusion chamber which contained glass coverslips precoated with ECM produced by human umbilical vein endothelial cells.[12] Perfusions were carried out at 37° with constant flow obtained by gravity. Blood which contained ^{111}In-labeled platelets was perfused for 5 min through the chamber. After perfusion, the ECM-coated coverslips were removed from the chamber and counted for their radioactivity or fixed for morphological studies. It was demonstrated that adherence of platelets to the ECM had the same time course and resulted in deposition of the same number of platelets as found using subendothelium of human renal and umbilical arteries.[12] It was also demonstrated that platelet interaction with a coverslip coated with fibrillar equine collagen had different characteristics than the interaction with the ECM or the vascular subendothelium.[12] We have also used a perfusion system to investigate the interaction of megakaryocytes with the ECM under defined flow conditions. This system was developed by R. R. Stromberg, A. R. Koslow, and L. I. Friedman from the American Red Cross, Blood Services Research Laboratories, Bethesda, Maryland. The chamber, shown in Fig. 3, consists of a pair of parallel plates with a divergent entrance and a convergent exit channel. Two recessed wells in the base plate accommodate 22-mm-diameter coverslips precoated with ECM, so that the surfaces of the coverslips are flush with the base plate.

[25] A. Eldor, I. Vlodavsky, Z. Fuks, T. H. Muller, and W. G. Eisert, *Thromb. Haemostasis* **56,** 333 (1986).

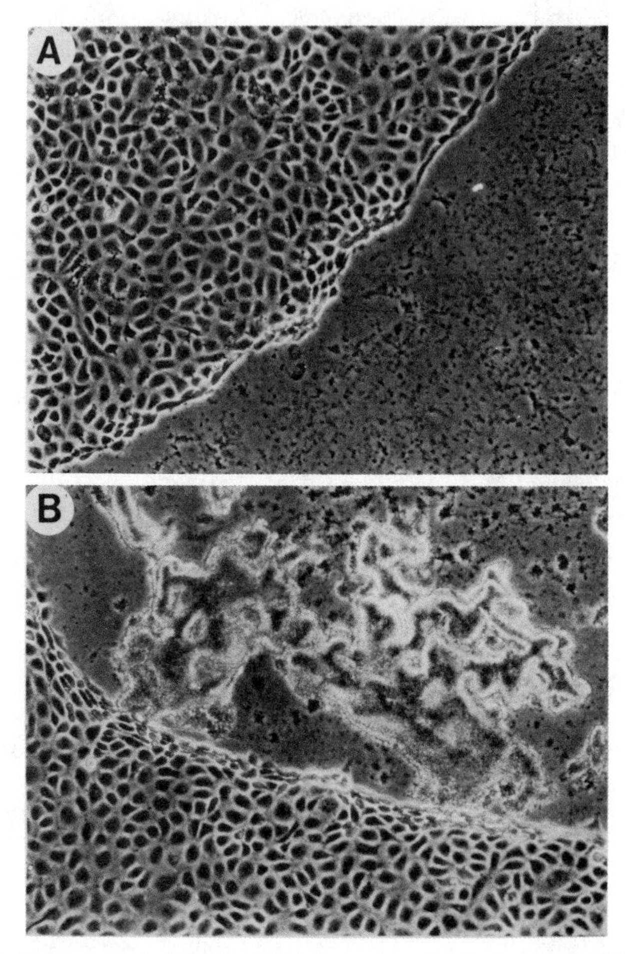

FIG. 2. Platelet-endothelial cells–ECM interaction. Confluent endothelial cell monolayers covering the entire subendothelial ECM were wounded with a rubber policeman to locally expose the underlying ECM. One milliliter PRP was then incubated (30 min, 37°) with (A) control cultures and (B) cultures which were pretreated with aspirin (100 μM, 60 min, 37°). Unbound platelets were removed and the dishes visualized by phase microscopy (×70). Note thrombus formation in (B).

The gap between the parallel plates is controlled by 0.6-mm shims. The base plate is type 316 stainless steel, the top plate is glass, and the gasket is a medical-grade reinforced silicone elastomer (Silastic, Dow Corning, MI). A 1-cm-diameter glass window in the base plate under the proximal coverslip permits continuous monitoring of the ECM surface during the entire experiment using video microscopy. The theoretical wall shear rates in the vicinity of the coverslips were experimentally verified using laser Doppler

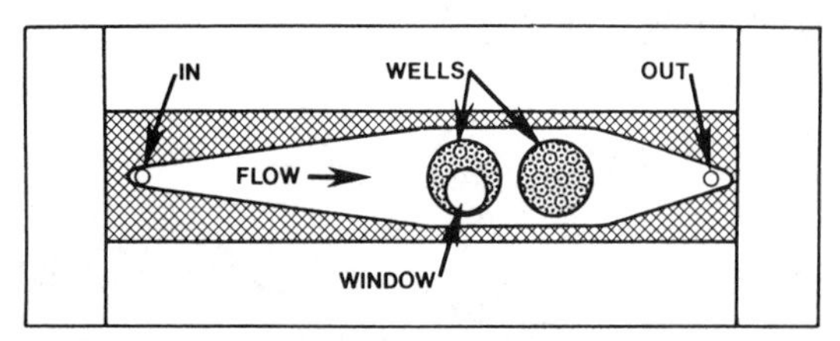

FIG. 3. Schematic representation of a perfusion chamber (top view). ECM-coated coverslips are introduced into the recessed wells and are flush with the surface of the base plate. A "window" permits microscopic observation of cellular interactions with the ECM. Laminar flow is developed in this chamber between the inlet and the coverslips.

anemometry and by an electrochemical technique. Streakline studies using dye demonstrated that laminar flow was present.

The chamber is incorporated into a continuous circuit, shown in Fig. 4, which allows steady, nonpulsatile flow at a rate dependent on the height difference between two reservoirs. Connecting tubing is silicone rubber (Sil-Med., Taunton, MA) and the reservoirs are 50-ml polypropylene centrifuge tubes (Corning Glass, Corning, NY). The chamber is precoated with polyethylene oxide. A roller pump (model SK10, Sarns, Ann Arbor, MI) maintains the fluid levels in the reservoirs, thus providing a constant head of pressure and constant flow through the perfusion chamber. The entire flow circuit is maintained at 37°. Desired flow rates are attained by adjusting the relative heights between the reservoirs. Microscopic observations of events occurring on the ECM can be recorded using an inverted microscope and video camera connected to a recorder. In the same manner, one can replace the microscope with a gamma irradiation counter, and use platelets or other cells prelabeled with a radioactive tracer. The system can be modified as desired, and the effects of various flow rates on the interaction of platelets (in plasma, whole blood, or buffer) with the ECM can be conveniently monitored. The system can be used to study the effects of various pharmacological agents, and can be used with ECM covered with confluent or subconfluent vascular endothelial cells.

Degradation of ECM Components by Human Platelets

We have recently reported that heparan sulfate in the subendothelial ECM is susceptible to cleavage by the platelet heparitinase.[26] Thus, incubation of a metabolically $^{35}SO_4^{2-}$-labeled ECM with platelet-rich plasma or washed platelets, but not with platelet-poor plasma, resulted in degradation of its heparan sulfate-containing proteoglycans into labeled fragments four to five times smaller than intact glycosaminoglycan side chains. These fragments were sensitive to deamination with nitrous acid and were not produced in the presence of heparin,[26] indicating their specificity as hepar-

[26] J. Yahalom, A. Eldor, Z. Fuks, and I. Vlodavsky, *J. Clin. Invest.* **74,** 1842 (1984).

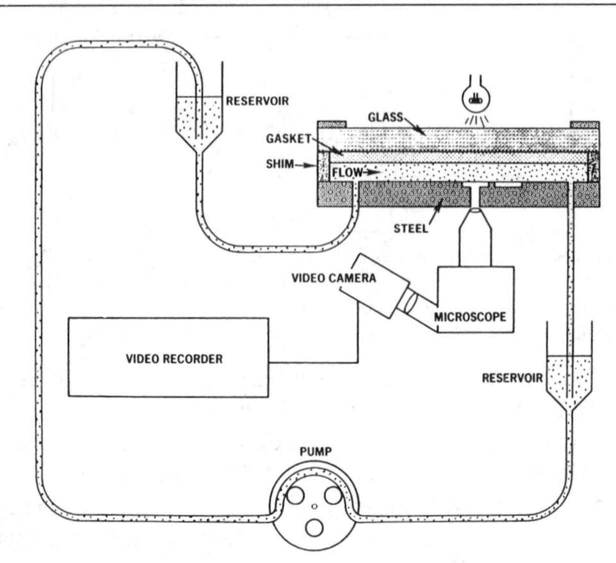

FIG. 4. Schematic representation of a flow circuit. The ECM-coated coverslips are exposed to steady nonpulsatile flow at a rate determined by the height difference between the two reservoirs. The perfusate fluid containing the cells is returned to the upper reservoir by the roller pump. The shear rate is determined by the gap (shim thickness) between the upper and lower plates and by the flow rate. A window in the bottom plate of the chamber permits continuous observation using time-lapse video microscopy.

itinase degradation products.[22,23,26-30] The production of a metabolically labeled ECM offers experimental approaches for studies that are difficult to perform with the currently available subendothelial tissue in isolated blood vessels.

Preparation of Sulfate-Labeled ECM

Subconfluent cultures of bovine corneal endothelial cells are metabolically labeled with $Na_2{}^{35}SO_4$ (540–590 mCi/mmol) added to the growth medium on days 4 and 8 after seeding (40 μCi/ml each time). Cultures are continuously incubated with radiolabeled sulfate without medium change. Eight days after confluence is reached the cell layer is solubilized with 0.5% Triton X-100 and 0.025 *N* NH_4OH, as described before, to prepare dished coated with ${}^{35}SO_4^{2-}$-labeled ECM. Previous studies have shown that 70–80% of the ${}^{35}SO_4^{2-}$ radioactivity in the labeled ECM is incorporated into the glycosaminoglycan moieties of heparan sulfate.[22,27] The release of heparan

[27] R. H. Kramer, K. G. Vogel, and G. L. Nicolson, *J. Biol. Chem.* **257,** 2678 (1982).

[28] Y. Naperstek, I. R. Cohen, Z. Fuks, and I. Vlodavsky, *Nature (London)* **310,** 241 (1984).

[29] M. Bar-Ner, A. Eldor, L. Wasserman, Y. Matzner, Z. Fuks, and I. Vlodavsky, *Blood* **70,** 551 (1987).

[30] A. Eldor, M. Bar-Ner, J. Yahalom, Z. Fuks, and I. Vlodavsky. *Semin. Thromb. Haemostasis* **13,** 475 (1987).

sulfate degradation products is analyzed following incubation (3–24 hr, 37°) of the labeled ECM with 1 ml platelet suspension at pH 6.5 (optimal pH for heparitinase activity). The incubation medium is then collected, centrifuged (10,000 *g*, 5 min), and applied for gel filtration on Sepharose 6B columns (0.7 × 35 cm). Samples are eluted with PBS or under dissociating conditions (4 *M* guanidine–HCl in 0.1 *M* sodium acetate, pH 5.5) and 0.2-ml fractions are collected at a flow rate of 5 ml/hr and counted in Biofluor scintillation fluid. The excluded volume V_0 is marked by blue dextran and the total included volume V_t by phenol red, which was shown to comigrate with free $^{35}SO_4^{2-}$.[22] Depending on the incubation conditions, heparan sulfate degradation fragments are eluted with a K_{av} value of 0.63 to 0.73 corresponding to an M_r of $6-10 \times 10^3$ (Fig. 5). This value is four to seven times smaller than intact heparan sulfate side chains released from the ECM by either papain digestion or cleavage with alkaline borohydride.[22] In contrast, incubation of the ECM with platelet-poor plasma results in a release of only high M_r degradation products as demonstrated by >90% of the released radioactivity eluted with and next to V_o (M_r 10^6) (Fig. 5).[26] Figure 5 also demonstrates an experiment in which the metabolically labeled [^{35}S]ECM was incubated with the supernatant of platelets treated with thrombin. Degradation of the heparan sulfate proteoglycans by heparitinase released from the thrombin-treated platelets is evident by the appearance of typical degradation products eluted with a K_{av} of 0.63 to 0.73. No heparitinase activity is detected in the supernatant obtained from platelet-poor plasma treated with thrombin. We have also shown that the release of heparitinase from thrombin-treated platelets can be inhibited by prostacyclin.[26] Heparitinase-mediated degradation of heparan sulfate occurred also during incubation of washed platelets or aspirin-treated platelets with ECM, indicating that plasma components or platelet cyclooxygenase are not necessary for expression of platelet heparitiniase.[26]

Studies on degradation of other macromolecular constituents of the ECM can be performed in a manner similar to that described above using labeled precursors (proline, glucosamine) suitable for the metabolic labeling of ECM. Alternatively, the ECM can be radioiodinated and the labeled degradation products analyzed by polyacrylamide gel electrophoresis and autoradiography.

Megakaryocyte Interaction with ECM

It has become apparent that megakaryocytes synthesize and assemble platelet components and organelles, are capable of shape change and degranulation, and respond with a prompt but reversible flattening reaction upon stimulation with known platelet agonists (ADP, thrombin, and arachidonic acid).[18] However, unlike platelets megakaryocytes do not adhere to or undergo shape change when exposed to glass or collagen-coated

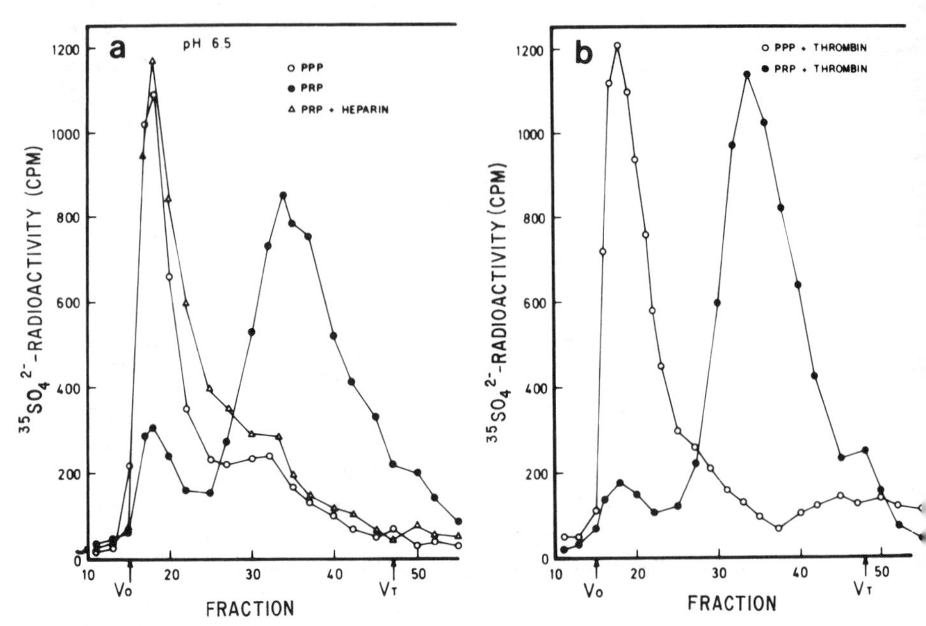

FIG. 5. Degradation of sulfate-labeled ECM by human platelet heparitinase. (a) Platelet-rich plasma vs platelet-poor plasma. One milliliter platelet-rich plasma (PRP, ●) or platelet-poor plasma (PPP, ○) was adjusted to pH 6.5 and incubated (24 hr, 37°) on top of sulfate-labeled ECM. The ECM was also incubated with PRP in the presence of 10 μg/ml heparin. (b) PRP (●) and PPP (○) were treated (30 min, 37°) with thrombin (1 unit/ml). One milliliter of the serum obtained after centrifugation was incubated (24 hr, 37°, pH 6.5) with the labeled ECM. Degradation products released from the ECM were analyzed by gel filtration on Sepharose 6B.

surfaces.[18] Thromboxane A_2 formation has also been reported in megakaryocytes in response to added arachidonic acid.[31] We have recently shown that shortly after exposure to ECM, isolated guinea pig, human, or rat megakaryocytes attach tightly and irreversibly to this surface and concomitantly produce TXA_2.[16,17] Adhesion and secretion occur within the first hour and in the next few hours a spreading typical of platelets occurs associated with the formation of long filopodia extending like the rays of the sun from a veil of cytoplasm in the periphery of the cell body (Fig. 6). These filopodia are generally linear with attenuated tips and are larger than but resembling the filopodia of rat or guinea pig platelets exposed to ECM. Ninety percent of the cells became adherent to the ECM at 30 min of incubation, and typical platelet-like flattening and shape change is observed in more than 50% of the cells.

For these experiments megakaryocytes are harvested from marrow

[31] R. E. Worthington and A. Nakeff, *Blood* **58,** 175 (1981).

tissue obtained from the femora and humeri of rats and guinea pigs and separated from other marrow cells by centrifugal elutriation and velocity sedimentation.[16,17] Resulting suspensions of megakaryocytes in DMEM (H-16) containing 2.3% bovine serum albumin, penicillin (50 units/ml), and streptomycin (50 μg/ml) exhibit 40–75% purity by number, and contain 3×10^5 to 1.2×10^6 megakaryocytes. The experiments are performed by prewarming the ECM-coated dishes in a 37° incubator, and placing the suspension of megakaryocytes into the dishes without moving them out of the incubator. We have found that the typical flattening of cells does not occur if the dishes are disturbed during the first 5 min of incubation with the megakaryocytes. As with platelets, aliquots of medium withdrawn from the dish can be used to monitor the formation of TXA_2 and release of other cellular constituents.

Using the perfusion system described above, we were able to show that megakaryocytes, circulated at shear rates of 10–200 sec^{-1}, adhered to the ECM surface, and that at low shear rates (10 sec^{-1}) megakaryocytes in the perfusate attached to adherent cells, forming megakaryocyte aggregates. This interaction was specific, since no adhesion of megakaryocytes to glutaraldehyde-fixed ECM, glass, or plastic coverslips occurred. Furthermore, many megakaryocytes, which attached to the ECM and were subjected to shear forces, subsequently underwent elongation and exhibited formation of pseudopods, flagelliform processes, and fragmentation of the cytoplasm.

The major advantage of the ECM model is that ECM-coated dishes can be easily prepared in large quantities and stored at 4° for more than 6 months without losing their platelet-activating properties. Studies on platelet interaction with this thrombogenic surface are reproducible, easy to perform, and do not require any surgical skills. Since the ECM, unlike the vascular intima, is a transparent structure, the events occurring on its surface (platelet adhesion, shape changes, etc.) can be monitored by direct visualization with a phase microscope. The ECM is also suitable for flow studies, and it has been demonstrated that platelets interacted with ECM-coated coverslips placed in a perfusion chamber in the same manner as with denuded vascular segments.[12] The ECM-coated surfaces can be prepared from cultured endothelial cells obtained from different species (human, bovine, porcine) and the reactivity of the different ECMs with human platelets was found to be similar.[11–14]

The ECM produced by vascular endothelial and smooth muscle cells has also been used to study the binding of proteins involved in coagulation or fibrinolysis. Plasminogen binds specifically and saturably to the ECM.[32]

[32] B. S. Knudsen, R. C. Silverstein, L. L. K. Leung, P. C. Harpel, and R. L. Nachman, *J. Biol. Chem.* **261,** 10765 (1986).

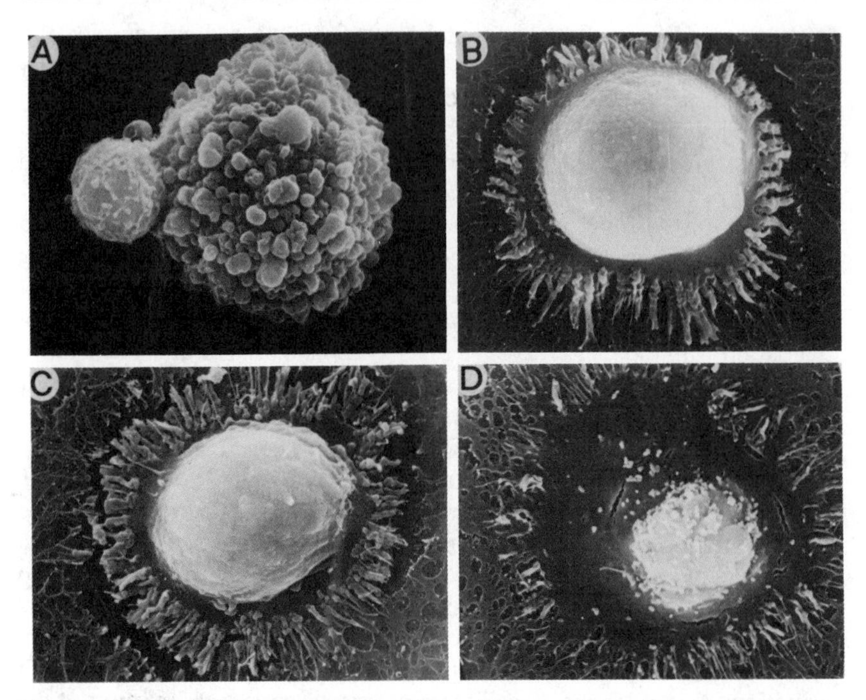

FIG. 6. Scanning electron micrographs of megakaryocyte interaction with ECM. One milliliter of megakaryocyte suspension (1.2×10^6 cells, 90% purity, 90% viability) harvested from guinea pig marrow was incubated (1 hr., 37°) on top of an empty plastic dish dish (A), or subendothelial ECM (B–D). Megakaryocytes do not adhere to the plastic dish and retain their characteristic "bumpy" appearances (A: ×3000, a megakaryocyte and adjacent lymphocyte). Megakaryocytes adhere tightly to the ECM, and undergo shape change with formation of long filopodia and concentric flattening (B: ×3000; C: ×2600; D: ×2100).

The plasmin generated on the ECM, in marked contrast to the fluid phase, was protected from α_2-plasmin inhibitor. Likewise, we have found that human α-thrombin binds to ECM through its "loop B" mitogenic domain in a specific, saturable, and irreversible manner.[33,34] The ability of ECM-bound thrombin to form a complex with its inhibitor anti-thrombin III was markedly decreased compared to complex formation in fluid phase.[33]

Another unique potential of the ECM system, as compared to deendothelialized vascular segments, is the possibility of producing a metabolically radiolabeled ECM. Thus, for example, by growing the ECM-producing endothelial cells in the presence of $Na_2{}^{35}SO_4$, it is possible to label sulfated glycosaminoglycans in the ECM.[22,23,26–30] This ECM has been used to characterize the degradation of ECM heparan sulfate by the platelet heparitinase—a lysosomal endoglycosidase which hydrolyzes glucuronyl-

[33] R. Bar-Shavit, A. Eldor, and I. Vlodavsky, *Blood* **70,** Suppl 1 398a (1987).

[34] R. Bar-Shavit, A. J. Kahn, K. G. Mann, and G. D. Wilner, *Proc. Natl. Acad. Sci.* U.S.A. **83,** 976 (1986).

glucosamine linkages.[26] Sulfate-labeled ECM was also used to investigate the requirements for heparan sulfate degradation by various normal and malignant blood-borne cells and its involvement in cell invasion and extravasation.[22,23,26–30]

Recently, we have found that endothelial cell growth factors (i.e., basic FGF) are bound to heparan sulfate in ECM.[35] These factors are released in an active form when the ECM heparan sulfate is degraded by the platelet heparitinase. Interaction of platelets with the subendothelial ECM is also associated with release of platelet-derived growth factor (PDGF) and heparan sulfate degradation products, which may stimulate or inhibit[36] smooth muscle cell proliferation, respectively.

A number of connective tissue constituents, in particular fibrillar collagen, have been used as an *in vitro* model for the initiation of platelet adherence, aggregation, and the release reaction.[37–39] The ECM, as a natural structure closely resembling the subendothelium *in vivo,* provides a convenient, *in vitro* model for studies of the interaction of platelets, megakaryocytes, and vascular cells with the subendothelium.

[35] I. Vlodavsky, J. Folkman, R. Sullivan, R. Fridman, R. I. Michaeli, J. Sasse, and M. Klagsbrun, *Proc. Natl. Acad. Sci.* U.S.A. **84,** 2292 (1987).

[36] J. J. Castellot, L. V. Favreau, M. J. Karnovsky, and R. D. Rosenberg, *J. Biol. Chem.* **257,** 11256.

[37] G. D. Milner, H. C. Nossel, and E. C. LeRoy, *J. Clin. Invest.* **47,** 2616 (1968).

[38] V. L. Leytin, E. V. Ljubimova, D. D. Sviridov, O. S. Zakharova, V. S. Repin, and V. N. Smirnov, *Thromb. Res.* **20,** 509 (1980).

[39] M. J. Barnes, A. J. Bailey, J. C. Gordon, and D. E. McIntyre, *Thromb. Res.* **18,** 375 (1980).

[8] *Ex Vivo* Measurement of Platelet Adhesion to Polymeric Surfaces

By KINAM PARK, DEANE F. MOSHER, and STUART L. COOPER

Introduction

Although significant progress has been made in the application of artificial materials as substitutes for natural organs, the long-term success of blood-contacting prosthetic devices is still very poor. Better understanding of the material behavior at the blood–biomaterial interface and subsequent precise control of material design are necessary. It is the platelet adhesion and subsequent thrombus formation which results in the potential danger of using artificial materials *in vivo.* Such defined thrombogenicity of biomaterials is profoundly affected by the initial adsorption of plasma proteins. Protein adsorption precedes other events when an artificial material first contacts blood.[1] The rate and the amount of protein adsorption and the type of the protein molecule adsorbed are thought to depend on the characteristics of the material.[2]

Adherence of individual platelets to a specific surface depends on receptor-mediated recognition of specific sites or chemical moieties on adsorbed proteins.[3] Some proteins, such as fibronectin,[4] von Willebrand factor (vWF),[5] fibrinogen,[6] and γ-globulin[7] are thrombogenic, i.e., they enhance platelet adhesion when preadsorbed on surfaces. Albumin, on the other hand, minimizes platelet adhesion and thrombogenesis.[8,9] Of the thrombogenic proteins, fibrinogen has been studied most widely, probably due to its abundance in the blood and its ability to clot. It has been suggested that the rate of platelet adherence onto a polymer surface is directly related to the amount of fibrinogen adsorbed onto the surface.[2,10] Some *in vitro* and *in vivo* experiments indicated that platelet adhesion and aggregate formation was dependent on the amount of fibrinogen on the surface.[2,11] Based on those observations, attempts have been made to correlate the amount of thrombogenic protein, especially fibrinogen, adsorbed onto a surface, but this approach has met with little success.

Rationale for the ex Vivo Test

Since *in vitro* measurements of thrombogenic proteins on surfaces to determine surface thrombogenicity have been inconclusive, platelet adhesion to surfaces has been examined as an alternative. Clotting times upon

[1] R. E. Baier, G. I. Loeb, and G. T. Wallace, *Fed. Proc., Fed. Am. Soc. Exp. Biol.* **30,** 1523 (1971).

[2] H. V. Roohk, J. Pick, R. Hill, E. Hung, and R. H. Bartlett, *Trans. Am. Soc. Artif. Intern. Organs* **22,** 1 (1976).

[3] R. H. Bartlett and J. C. Anderson, *in* "Biologic and Synthetic Vascular Prostheses" (J. C. Stanley, ed.), p. 63. Grune & Stratton, New York, 1982.

[4] T. A. Barber, L. K. Lambrecht, D. F. Mosher, and S. L. Cooper, *in* "Scanning Electron Microscopy III" (R. P. Becker and O. Johari, eds.), p. 881. Scanning Electron Microscopy, Inc., O'Hare, IL, 1979.

[5] P. A. Bolhuis, K. S. Sakariassen, H. J. Sander, B. N. Bouma, and J. J. Sixma, *J. Lab. Clin. Med.* **97,** 568 (1981).

[6] H. V. Roohk, M. Nakamura, R. L. Hill, E. K. Hung, and R. H. Bartlett, *Trans. Am. Soc. Artif. Intern. Organs* **23,** 152 (1977).

[7] M. A. Packham, G. Evans, M. F. Glynn, and J. F. Mustard, *J. Lab. Clin. Med.* **73,** 686 (1969).

[8] D. F. Mosher, *in* "Interaction of the Blood with Natural and Artificial Surfaces" (E. W. Salzman, ed.), p. 85. Dekker, New York, 1981.

[9] E. W. Salzman, *Blood* **38,** 509 (1971).

[10] D. J. Lyman, L. C. Metcalf, D. Albo, Jr., K. F. Richards, and J. Lamb, *Trans. Am. Soc. Artif. Intern. Organs* **20,** 474 (1974).

[11] H. Lagergren, P. Olsson, and J. Swedenborg, *Surgery* **75,** 643 (1974).

exposure of blood to test materials has been used as a complementary measure of platelet adhesion. However, it is necessary to distinguish platelet adhesion from blood clotting. Blood will clot with equal facility whether or not platelets are present. The data of Rodman and Mason[12] suggested that primary platelet adhesion and aggregation might be a response to factors other than activation of the clotting mechanism.

Although the measurement of platelet adhesion is a direct test of surface thrombogenicity, the number of platelets adhered on surfaces varies widely depending on experimental conditions such as the species of platelets used, the type of anticoagulant employed, the presence of plasma proteins, and whether *in vitro* or *in vivo* assays are used. Such variations make it difficult to compare results obtained in different laboratories. The variation in the number of adhered platelets sometimes differs by an order of magnitude on the same material, especially in comparison between *in vitro* and *in vivo* or *ex vivo* experiments. Transformation of adherent platelets to the spread form and subsequent mural thrombus formation invariably occurs *in vivo* or *ex vivo* upon exposure of artificial materials to flowing blood.[13] In contrast, only a partial surface coverage with single round platelets is observed in *in vitro* studies and the study of Turitto *et al.*[14] suggests that it is due to the inhibition of platelet spreading by chelators. This raises questions whether surface thrombogenicity can be measured by *in vitro* studies which use platelets in chelated media. Environments which allow platelet activation and thrombi formation under *in vivo* conditions are necessary to accurately evaluate the interaction of blood with biomaterials. *Ex vivo* animal tests satisfy such conditions.

Materials and Methods

The following method to study the thrombogenicity of various surfaces has been developed by our group over the last decade.

Preparation of Plasma Proteins

Canine fibrinogen is prepared from fresh citrated plasma by β-alanine precipitation.[15] Fibronectin is removed on a gelatin–Sephadex column (Bio-Rad). The clottability of the purified fibrinogen is at least 97%. Fibrinogen is desialylated using insoluble neuraminidase (*Clostridium per-*

[12] N. F. Rodman and R. G. Mason, *Thromb. Diath. Haemorrh.* **42,** 61 (1972).

[13] E. W. Salzman, J. Lindon, D. Brier, and E. W. Merrill, *Ann. N.Y. Acad. Sci.* **283,** 114 (1977).

[14] V. T. Turitto, R. Muggli, and H. R. Baumgartner, *ASAIO J.* **2,** 28 (1979).

[15] E. Jakobsen and P. Kierulf, *Thromb. Res.* **3,** 145 (1973).

fringens attached to agarose beads, Sigma). Canine albumin (Nutritional Biochemicals or Sigma) is used after further purification by column chromatography (Cellex D, Bio-Rad), or as received. Canine α_2-macroglobulin is purified according to Kurecki *et al.*[16] Canine and human von Willebrand factor are purified by the method of Sodetz *et al.*[17]

Human fibrinogen and human and canine fibronectin are prepared using methods described by Mosher and Johnson[18] from fresh frozen human plasma. The purified fibronectin at its plasma concentration contains less than 1% of the plasma concentration of von Willebrand factor and less than 0.05% of the plasma concentration of fibrinogen. Crude preparations of human and canine fibrinogen are also used and at their physiological concentration of 2 mg/ml, they contained approximately 50% of the plasma concentration of von Willebrand factor and fibronectin (impure fibrinogen). Human thrombospondin obtained from platelet releasate (see below) by a modification of the procedure of Margossian *et al.*[19] is further purified on Affi-Gel–heparin (Bio-Rad).

Platelet releasate is obtained using the following procedure.[20] Platelets are washed and reacted with thrombin as described by Margossian *et al.*[19] Thrombin is inhibited with hirudin after 2 min of reaction. After centrifugation, the supernatant is frozen in a dry ice–ethanol bath and then thawed in a 37° water bath. The small fibrin clot formed in the thawing solution is removed with a stirring rod. The remaining solution is referred to as the platelet releasate.

Radiolabeling of Platelets and Proteins

Proteins are labeled with ^{125}I (New England Nuclear) using either Enzymobeads (Bio-Rad) or Iodo-Beads (Pierce Chemical). Free iodide is separated from labeled protein on a gel column (BioGel P-30, 0.7 × 30 cm Econo Column, Bio-Rad) equilibrated with phosphate-buffered saline. Autologous platelet are labeled with ^{51}Cr (New England Nuclear) by the method of Abrahamsen.[21] ^{51}Cr-Labeled platelets and ^{125}I-labeled protein are injected into the animal 15 and 1 hr prior to surgery, respectively. Alternatively, canine platelets can be labeled with ^{111}In.[22]

[16] T. Kurecki, L. F. Kress, and M. Laskowski, Sr., *Anal. Biochem.* **99,** 415 (1979).

[17] J. M. Sodetz, S. V. Pizzo, and P. A. McKee, *J. Biol. Chem.* **252,** 5538 (1977).

[18] D. F. Mosher and R. B. Johnson, *Ann. N.Y. Acad. Sci.* **408,** 583 (1983).

[19] S. S. Margossian, J. W. Lawler, and H. S. Slayter, *J. Biol. Chem.* **256,** 7495 (1981).

[20] B. R. Young, M. J. Doyle, W. E. Collins, L. K. Lambrecht, C. A. Jordan, R. M. Albrecht, D. F. Mosher, and S. L. Cooper, *Trans. Am. Soc. Artif. Intern. Organs* **28,** 498 (1982).

[21] A. F. Abrahamsen, *Scan. J. Haematol.* **5,** 53 (1968).

[22] W. A. Heaton, A. P. Heyns, and J. H. Joist, this volume [14].

Preparation of Polymer Shunts and Protein Preadsorption

The polymers evaluated include plasticized poly(vinyl chloride) (Tygon R3603, Norton Plastics, 0.125-in. i.d.), segmented polyether-urethane-urea (Biomer, extruded grade, Ethicon, Inc., 0.10-in. i.d.), polyethylene (Intramedic, Clay-Adams, 0.125-in. i.d.), silicone rubber (Silastic, Dow Corning, 0.078-in. i.d.), polyacrylamide (Polysciences), and poly(hydroxyethyl methacrylate) (Hydron Laboratories) in the form of grafts polymerized to silicone rubber using ^{60}Co-initiated gamma radiation.

Polymer tubings are cut into 70-in. lengths for the single shunt experiment (much smaller sizes are used for the series shunt method; see below) and washed with running distilled deionized water for 2 hr. Poly(vinyl chloride) tubing is washed at room temperature with 200 ml of 0.1% Ivory detergent and rinsed like the other tubings. The tubings are filled with Tyrode's solution (pH 7.35) and left at 4° overnight. After equilibration overnight with the buffer solution, the tubing is implanted in the dog as described below.

For protein preadsorption, Tyrode's solution in the tubing is replaced with protein solution prior to the surgery. Bulk protein concentrations used for preadsorption are 0.12 mg/ml for von Willebrand factor; 0.3 mg/ml for purified fibrinogen, albumin, γ-globulin, α_2-macroglobulin, and fibronectin, and 0.5 mg/ml for crude fibrinogen. Protein is preadsorbed for 2 hr at room temperature. Shortly before implanting the tubing into the animal, the protein solution is removed by flushing with 50 ml of Tyrode's solution. When necessary, the protein concentration on the tubing is measured by using ^{125}I-labeled protein.

Animal Selection

Adult male mongrel dogs weighing 18–30 kg are selected following screening for normal blood counts and platelet aggregation in response to ADP and epinephrine. Each dog used for surgery has a platelet count within the range of 150,000–400,000/μl, a fibrinogen level of 1–3 mg/ml, and a hematocrit of 35–50%. Prior to surgery, the dog is fasted for 15 hr. Adult rhesus monkeys *(Macaca mulatta)* are also used after the same screening procedure.

Animal Surgery

Figure 1A schematically shows the experimental arrangement used for the single shunt *ex vivo* experiment. The animal is anesthetized with intravenous sodium thiamylal (20 mg/kg initial dose) and put on a respirator. The femoral artery and vein in one leg are exposed and ligated (at

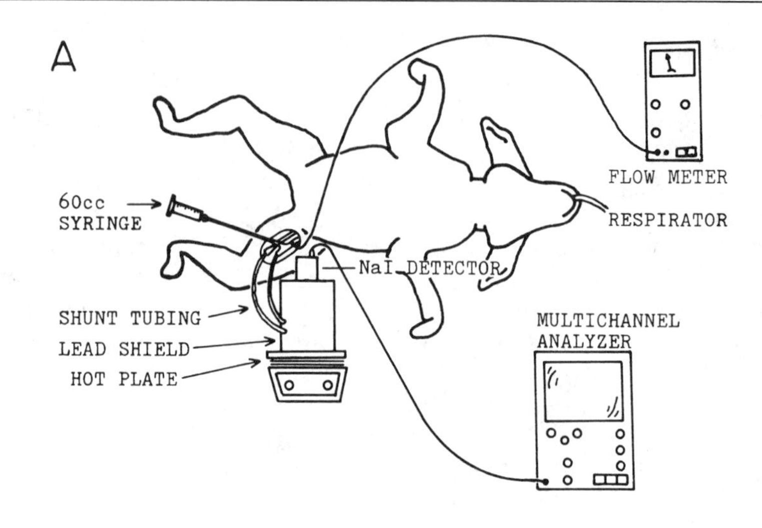

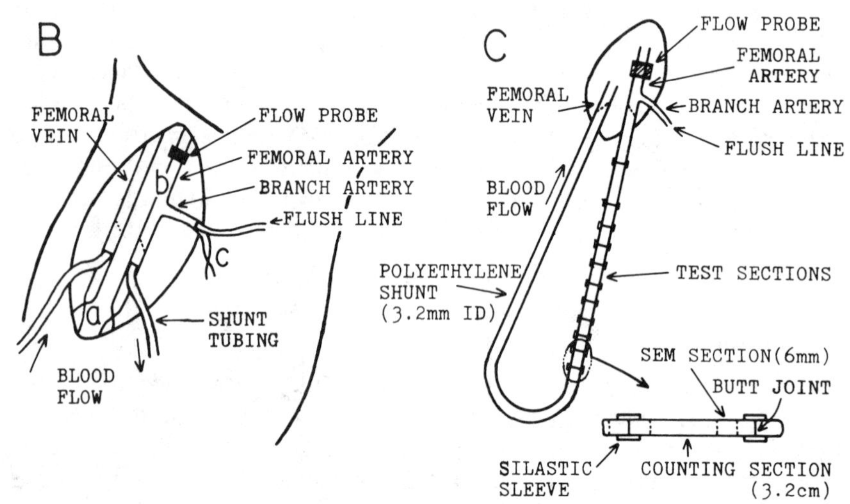

FIG. 1. Schematic presentation of the canine *ex vivo* shunt model. (A) Experimental arrangement for the single-shunt test. (B) Detail of the shunt cannulation site. (C) Detail of the series shunt.

point "a" in Fig. 1B). During cannulation of the shunt, arterial blood flow is interrupted by placing an atraumatic vascular clamp proximal to the cannulated site (point "b" in Fig. 1B). The artery and vein are cannulated with the test polymer shunt. In the case of the series shunt, the artery and vein are connected to the polyethylene in a series of test polymer segments connected to the polyethylene shunt (Fig. 1C). The shunt is initially filled with Tyrode's solution to prevent blood–air contact. Blood is allowed to flow through the shunt by removing the clamp. A small branch artery, proximal to the shunt cannulation site, is also cannulated (point "c" in Fig. 1B) with a polyethylene 16-gauge catheter and connected to a syringe containing Tyrode's solution. The branch line is used to inject buffer solution to flush the shunt. The flushing procedure is necessary to detect the small number of platelets deposited on the shunt compared to circulating platelets.

Blood flow is continuously monitored using a electromagnetic flow transducer (Gould SP2202) connected to a Physiograph Mark III recorder (Narco Bio-system). The flow rates range from 150 to 300 ml/min. At regular intervals when platelet deposition is measured, a small volume of blood is removed via an indwelling jugular catheter to determine the whole blood radioactivity, platelet count, fibrinogen level, hematocrit, thrombin time, and euglobulin lysis time. Platelet counting is done either manually by phase microscopy using a Unopette chamber (Becton Dickinson and Co.) or by using a Coulter counter.[23] Fibrinogen concentration is calculated by comparing thrombin clotting time of diluted plasma with those of standard fibrinogen solutions. Platelet aggregability to ADP, collagen, and epinephrine is monitored at the start and at the end of each experiment. Experimental animals are sacrificed at the end of surgery by a lethal dose of sodium thiamylal.

Single Shunt. A 24-in. segment of the 70-in. shunt is wrapped around a solid crystal detector which is placed inside a lead shield and the radioactivities of ^{51}Cr and ^{125}I are counted directly using a multichannel analyzer (Tektronix, Ino-Tech). At selected time points, the blood flow is stopped and the blood in the shunt is immediately flushed with 60 ml of Tyrode's solution into the venous circulation. The buffer solution is introduced at a rate of about 1 ml/sec. The number of platelets and the amount of protein deposited on the shunt surface are measured by determining the level of ^{51}Cr and ^{125}I radioactivity on the tubing. A small piece of the shunt is removed and fixed with 2% glutaraldehyde in 0.1 *M* phosphate buffer for scanning electron microscopy (SEM). Blood flow is then resumed through the same shunt until the next counting time. The blood exposure time of the shunt is measured cummulatively. The shunt is sampled for platelet deposition after 2, 5, 10, 15, 60, 90, and 120 min of blood–polymer contact time. A merit of the single shunt is the large surface area available

[23] B. S. Bull, M. A. Schneiderman, and G. Brecher, *Am. J. Clin. Pathol.* **44,** 678 (1965).

for the measurement of thrombus deposition. Since the initial development of a thrombus does not occur uniformly over a surface,[24] a large surface area produces more accurate average values.

Series Shunt. The radioactivities on series shunt materials are counted after removing the shunt from the animal. Thus, a solid crystal detector and a multichannel analyzer used for a single shunt experiment (Fig. 1A) are not necessary. At selected time points, the blood flow is stopped, and the tubing is flushed with 60 ml of Tyrode's solution. Immediately following the flushing, the joined test section is removed and fixed with 2% glutaraldehyde solution without surface–air contact. A 3.2-cm segment of each test surface is counted in an automated dual channel gamma counter (Beckman 5500). A 0.6-cm segment is examined by SEM. A new set of identical test surfaces joined as a shunt section is inserted for each time period and the blood contact and flushing procedures repeated. Blood exposure time for a series shunt is not cummulative. The advantage of a series shunt is that a large number of materials, up to 10, can be tested simultaneously in the same animal. It has been shown that, at least at the high flow rates and relatively short times used in the series shunt experiments, the deposition of thrombus material on proximal segments does not affect the response on other sections of the shunt.[25]

Merits of the *ex Vivo* Shunt Method

Homeostatic Compensatory Mechanism

The most distinct feature of the *ex vivo* method is that the homeostatic compensatory mechanism is maintained. Blood circulation through the test polymer shunt used in the *ex vivo* shunt method can also be achieved in *in vitro* systems,[26–29] but the homeostatic mechanism of a live animal is not maintained. Thrombosis and embolization on artificial surfaces are the results of a balance between thrombogenic and fibrinolytic events. The delicate balance between fibrinolytic and thrombogenic mechanisms is not expected to be maintained in *in vitro* studies. The development of such

[24] E. F. Leonard, Y. Butrille, S. Puszkin, and S. Kochwa, *Ann. N.Y. Acad. Sci.* **283,** 256 (1977).

[25] M. D. Lelah, L. K. Lambrecht, and S. L. Cooper, *J. Biomed. Mater. Res.* **18,** 475 (1985).

[26] J. N. Lindon, R. Rodvien, D. Brier, R. Greenberg, E. Merrill, and E. W. Salzman, *J. Lab. Clin. Med.* **92,** 904 (1978).

[27] I. A. Feuerstein, J. M. Brophy, and J. L. Brash, *Trans. Am. Soc. Artif. Intern. Organs* **21,** 427 (1975).

[28] V. T. Turitto and H. J. Weiss, *Science* **207,** 541 (1980).

[29] T. Karino and H. L. Goldsmith, *Microvasc. Res.* **17,** 217 (1979).

experiments as the *ex vivo* shunt is important for the evaluation of new biomaterials for clinical applications and the understanding of mechanisms involved in artificial surface-induced thrombosis.

Thromboembolization in the Presence of Shear Flow

In the *ex vivo* shunt, platelet adhesion to the test surface is measured under flow conditions. The flow field not only provides a continuous supply of platelets to form thrombi, but it also removes thrombi from the surface by the shear force generated at the blood–polymer interface. Thus, the growth rate of thrombi and the size of emboli are expected to be influenced by hemodynamic factors.[30] If the interaction force between the thrombus and the surface is low, the flowing blood will remove the thrombi from the prosthetic surface more readily, resulting in the formation of microemboli. Large emboli can cause transient ischemic attacks or strokes while no clinically detectable effect is produced by small emboli.[31] Since the size of the emboli is important, thromboembolization measured in the presence of shear flow gives more information on surface biocompatibility than *in vitro* static platelet adhesion tests.

Transient Platelet Deposition

Since thrombus formation is followed by embolization, the time of observation is important in determining the thrombogenicity for a surface. In this sense, the measurement of transient and dynamic thrombus formation in an *ex vivo* system has advantages over the *in vitro* systems and systems using *in vivo* end-point determinations.[32] It was shown by Ihlenfeld *et al.*[33] that transient platelet deposition is not due to the experimental procedure of flushing the blood from the polymer shunt with buffer solution. In addition, the results of a single shunt are the same as those for series shunts which do not require flushing.

Comments on the Limitations of the Model

There is always the question of the validity of an animal model to interpolate the results to human clinical situations. The dog is known to

[30] H. R. Baumgartner, *Microvasc. Res.* **5,** 167 (1973).

[31] R. G. Mason, F. Mohammad, and J. U. Balis, *in* "Perspectives in Hemostasis" (J. Fareed, H. L. Messmore, J. W. Fenton II, and K. M. Brinkhous, eds.), p. 71. Pergamon, New York, 1981.

[32] J. S. Schultz, A. Ciarkowski, J. D. Goddard, S. N. Lindenauer, and P. A. Penner, *Trans. Am. Soc. Artif. Intern. Organs* **22,** 269 (1976).

[33] J. V. Ihlenfeld, T. R. Mathis, L. M. Riddle, and S. L. Cooper. *Thromb. Res.* **14,** 953 (1979).

have a low susceptibility to cardiovascular disease and pulmonary embolism, very active hemostatic and fibrinolytic mechanisms, and hyperplastic vascular responses to incitants resulting in rapid occlusion of vessels.[34-36] Deposition of thrombi on various polymer shunts in the canine model is an order of magnitude greater than those in the monkey model (Table I).[37] Despite these problems, the canine model has been used rather widely to evaluate biomaterials.[34] Successful implantation in humans is expected if vascular prostheses exhibit minimal thrombogenic potential in dogs.[38] In addition, it was observed that the relative order of human platelet and fibrinogen deposition on five materials was similar to that observed in the canine system.[39]

Relationship to Long-Term Patency

The ultimate test of blood compatibility is the use of biomaterials for long-term implantation in the body. Thus, *in vivo* tests for long-term patency are necessary, but it is difficult to monitor blood interactions with the test materials using an end-point determination. In addition, *in vivo* tests are expensive and time consuming. The question of relationship between acute tests and long-term results is not limited to the *ex vivo* shunt method. Rather, it is a general problem for all acute experimentation. If endothelial cell growth on prosthetic vascular grafts is necessary for long-term patency and if it is facilitated by trapping endothelial cells on or in a fibrin matrix as suggested by Graham *et al.*,[40] maximal fibrinogen adsorption would be beneficial. Although valid predictive information regarding the long-term patency is not readily available from the acute tests, an understanding of the initial events occuring at the blood–polymer interface is essential to the design of truly biocompatible artificial materials.

[34] "Guidelines for Blood–Material Interactions," p. 11. U.S. Department of Health and Human Services, Bethesda, MD, 1980.

[35] E. F. Grabowski, K. K. Herhter, and P. Didisheim, *J. Lab. Clin. Med.* **88,** 368 (1976).

[36] W. J. Dodds, *in* "Proceedings on Platelets: A Multidisciplinary Approach," p. 45. Raven, New York, 1978.

[37] L. K. Lambrecht, M. D. Lelah, C. A. Jordan, M. E. Pariso, R. M. Albrecht, and S. L. Cooper, *Trans. Am. Soc. Artif. Intern. Organs* **29,** 194 (1983).

[38] F. A. Pitlick, *in* "Biologic and Synthetic Vascular Prostheses" (J. C. Stanley, ed.), p. 3. Grune & Stratton, New York, 1982.

[39] P. Didisheim, J. Q. Stropp, M. K. Dewanjee, and H. W. Wahner, *Thromb. Haemostasis.* **50,** 60 (1983).

[40] L. M. Graham, D. W. Vinter, J. W. Ford, R. H. Kahn, W. E. Burkel, and J. C. Stanley, *Arch. Surg.* **115,** 929 (1980).

Animal-to-Animal Variation

Tests of biomaterials are based on the assumption that the conditions remain exactly the same except for the text material. However, animal-to-animal and day-to-day variations in the same animal are the characteristic of all such studies. Although dogs are selected for the *ex vivo* experiment after examining parameters such as hematocrit, platelet count, platelet aggregability, fibrinogen level, and fibrinolytic activity (see Material and Methods), these parameters are not controllable within a predefined allowable range. Variations within the allowable ranges might be the reason for the relatively large deviations in the platelet deposition profiles shown in Fig. 2. In this regard, the excellent reproducibility of the *in vitro* column technique of Lindon *et al.*[26] still showed a large donor-to-donor variation, and the same blood sample has to be used for evaluation and comparison of the blood compatibility of various polymers. In the canine *ex vivo* model, small differences in thrombogenicity of various materials may not be observed due to relatively large deviations resulting from the use of different animal subjects, but the apparent difference in platelet deposition between nonthrombogenic and thrombogenic surfaces can still be determined.

Use of Radiolabeled Platelets and Anesthetics

The magnitude of platelet adhesion, thrombus formation, and embolization is measured by ^{51}Cr-labeled platelets and ^{125}I-labeled fibrinogen. It is still not clear if radiolabeling alters the behavior of platelets or proteins *in vivo* or to what extent. Limited evaluations suggest that the effect of radiolabeling is minimal.[41] The use of anesthetics during an *ex vivo* experiment may influence the results. It has been shown that on glass coverslips sodium thiamylal blocks platelet transition from the spreading form to the fully spread form and inhibits ADP-induced aggregation in a dose-dependent manner.[42,43] Although the actual concentration of barbiturates to induce anesthesia in experimental animals is lower than that used in the *in vitro* studies, the effects of anesthetics should be taken into account when the results are compared with those of studies which do not use anesthetics.

[41] B. R. Young, Ph.D. thesis. University of Wisconsin, Madison, Wisconsin, 1984.

[42] S. T. O'Rourke, J. D. Folts, and R. M. Albrecht, *Scanning Electron Microsc.* **3,** 1337 (1984).

[43] F. N. McKenzie, E. Svensjo, and K. E. Arjors, *Microvasc. Res.* **4,** 42 (1972).

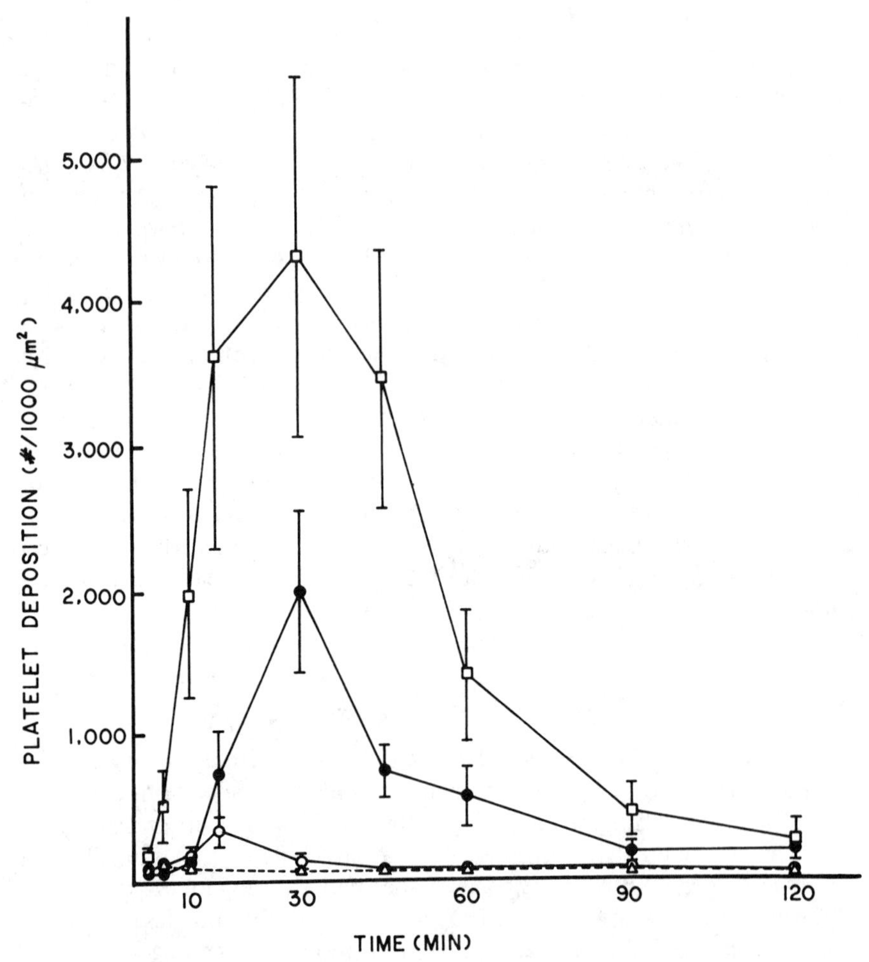

FIG. 2. Transient platelet deposition (mean ± SEM) on poly(vinyl chloride) (○), and poly(vinyl chloride) preadsorbed with vWF (□), fibrinogen (●), or albumin (△).

Limited Number of Experiments

As mentioned in the section on experimental methods, animals are used only after *in vitro* screening tests are completed, and the number of materials that can be tested in one surgery is limited. The *ex vivo* shunt experiment is therefore relatively costly and time consuming to carry out in large numbers.

TABLE I
THROMBUS DEPOSITION ON CANINE *ex Vivo* SHUNT SURFACES

Polymer surface and Precoated proteins	Source	Maximum platelet deposition: Time (min)	Maximum platelet deposition: Density (number/1000 μm^2)	Ref.
Poly(vinyl chloride)				
Control		15	320 ± 170 (20 ± 10)[a]	44, 45
Albumin	Canine	15	50 ± 10	20, 45
γ-Globulin	Canine	30	1500 ± 200	45
Crude fibrinogen	Canine	30	3600 ± 1800	41
Fibronectin	Canine	15	1750 ± 360	41
Fibrinogen	Canine	30	2000 ± 550	
Desialylated fibrinogen	Canine	30	2400 ± 500	
von Willebrand factor	Canine	30	4300 ± 1200	41
Crude fibrinogen	Human	30	2070 ± 1100	45
Fibronectin	Human	30	1600 ± 440	20, 45
Fibrinogen	Human	15	1780 ± 1000	45
von Willebrand factor	Human	30	2100 ± 570	20, 45
Platelet releasate	Human	60	600 ± 150	20
Thrombospondin	Human	15	3400 ± 300	20
α_2-Macroglobulin	Human	30	1160 ± 200	45
Silicone rubber				
Control		15	90 ± 40 (40 ± 10)[a]	4, 45
von Willebrand factor	Canine	15	120 ± 20	4
γ-Globulin	Canine	120	600 ± 300	45
Fibrinogen	Canine	45	600	
Fibronectin–antifibronectin	Human	30	900 ± 70	4
Antifibronectin	Human	15	1700 ± 170	4
Fibronectin	Human	30	1800 ± 900	4, 45
α_2-Macroglobulin	Human	5	70 ± 30	45
Acrylamide grafted		15	490 ± 190	44
HEMA[b] grafted		15	650 ± 50	44
Polyethylene				
Control		10	760 ± 210 (60 ± 10)[a]	41, 46
Chitosan		30	70 ± 20	46
Chitosan–heparin		5	70 ± 10	46
Fibrinogen	Canine	120	2000	
Chromic acid-oxidized polyethylene		30	3100 ± 300 (80 ± 20)[a]	41
Polyether-urethane-urea				
Control		5	50 ± 10	33

[a] Platelet deposition values in parentheses are for the monkey model.[37]

[b] HEMA, Hydroxyethylmethacrylate.

Data Obtained with the *ex Vivo* Shunt Model

Figure 2 shows transient platelet deposition profiles on poly(vinyl chloride) precoated with various proteins. The summary of the peak platelet deposition times and the platelet density on various polymer surfaces tested at the University of Wisconsin is presented in Table I.[44-46] Platelet deposition profiles, thrombus size, fibirn deposition, and leukocyte adhesion depend on the nature of the polymer substrate and the precoated protein layer. The influence of precoated proteins largely depends on the nature of the polymer substrates. von Willebrand factor showed the largest platelet deposition when precoated on poly(vinyl chloride), but such an effect was not observed when it was precoated on silicone rubber (Table I). The effect of vWF was negligible on silicone rubber and only fibronectin resulted in platelet deposition which is comparable to that on poly(vinyl chloride). On albumin- or serum-coated polymer surfaces, the number of platelets was consistently less than 100/1000 μm^2 and no platelet aggregates were observed in SEM micrographs. On all surfaces tested the radiolabeled fibrinogen deposition – release profiles closely paralleled the platelet deposition curves, suggesting fibrin stabilization of the platelets in thrombi.

[44] T. A. Barber, T. Mathis, J. V. Ihlenfeld, S. L. Cooper, and D. F. Mosher, *in* "Scanning Electron Microscopy II" (R. P. Becker and O. Johari, eds.), p. 431. Scanning Electron Microscopy, Inc., O'Hare, IL, 1978.

[45] B. R. Young, L. K. Lambrecht, S. L. Cooper, and D. F. Mosher, *in* "Biomaterials: Interfacial Phenomena and Applications" (S. L. Cooper and N. A. Peppas, eds.), p. 371. American Chemical Society, Washington, D.C., 1982.

[46] L. K. Lambrecht, B. R. Young, D. F. Mosher, A. P. Hart, W. J. Hammar, and S. L. Cooper, *Trans. Am. Soc. Artif. Intern. Organs* **27**, 380 (1981).

[9] Platelet Interaction with Artificial Surfaces: *in Vitro* Evaluation

By JACK N. LINDON, LESLIE KUSHNER, and EDWIN W. SALZMAN

Introduction

Recent advances in the development of artificial organs have demonstrated the need for blood-compatible surfaces and for test systems to evaluate and guide their development. Although the ultimate test of any artificial material will be its performance *in vivo, in vivo* testing is expensive, subject to numerous variables that are difficult or impossible to control, and in most cases does not allow direct comparison of different materials. Also, *in vivo* tests must almost always be conducted in species other than humans, and species differences in behavior and composition of the blood may cloud the results. *In vitro* tests, on the other hand, suffer from the fact that significant alterations to the blood occur during bloodletting, anticoagulation, and other processing steps prior to actual contact with the test material. Nevertheless, *in vitro* methods are generally the first level of testing employed since human blood can be used, direct comparisons of different materials can be made, and a large number of variables can be examined simultaneously and throughout the course of the tests. In addition, *in vitro* testing more readily permits study of the basic mechanisms of interaction of the blood with artificial surfaces.

Thrombosis induced *in vivo* when blood comes in contact with a prosthetic device is usually due to the adhesion and activation of platelets by the surface, since fibrin formation can generally be prevented by proper fluid dynamic design that minimizes zones of stasis or recirculating eddies, or through the use of anticoagulant drugs. For this reason, most *in vitro* testing concentrates on platelet adhesion and activation induced by blood/surface contact. It is important, however, not to rely on a single parameter when evaluating surfaces. Tests of coagulation and complement activation and fibrinolysis are also important in determining the hemocompatibility of a surface.

When an artificial surface is exposed to blood, platelets presumably do not interact directly with the surface, but rather with a plasma-derived, adsorbed coating of protein and perhaps lipid, the composition and molecular conformation of which are determined by the chemical topography of the surface.[1-3] Adhesion of platelets to a surface that has developed such a coat (formation of which takes a matter of seconds) may lead to profound structural and functional changes in the platelet.[4,5] Activated platelets change from their native discoid shape to a more spherical configuration and develop numerous pseudopods. Their surfaces become "sticky" to

[1] R. E. Baier, *Ann. N.Y. Acad. Sci.* **283,** 17 (1977).

[2] D. E. Scarborough, R. G. Mason, P. G. Dalldorf, and K. M. Brinkhous, *Lab. Invest.* **20,** 164 (1969).

[3] J. L. Brash, *in* "Interaction of the Blood with Natural and Artificial Surfaces" (E. W. Salzman, ed.), p. 37. Dekker, New York, 1981.

[4] P. Clagett, *in* "Hemostasis and Thrombosis" (R. W. Colman, J. Hirsh, V. J. Marder, and E. W. Salzman, eds.), 2nd Ed. Lippincott, Philadelphia, Pennsylvania, 1987.

[5] H. R. Baumgartner, R. Muggli, T. B. Tschopp, and V. T. Turitto, *Thromb. Haemostasis.* **35,** 124 (1976).

other platelets, aggregates of platelets form, and the contents of platelet storage granules are secreted (the "release reaction") into the surrounding media. In a prosthetic device, formation of such platelet thrombi usually leads to adverse effects, including impairment of blood flow, compromised function of the device (e.g., incomplete closure of a heart valve), detachment of thrombi and subsequent embolization, thrombocytopenia, or alteration in platelet function.[4]

Many of the techniques developed to investigate platelet–surface interactions are listed in Table I and have recently been reviewed.[6-8] The earliest tests by Wright using a rotating glass flask[9] and by Moolten and Vroman using glass wool filters[10] were reported to quantitate platelet "adhesiveness," but such interactions have since been shown to reflect the entire process of platelet activation, including aggregation.[11] The glass bead column test, introduced by Hellem,[12] was modified by Salzman,[13] who showed that platelet aggregation occurred on the glass beads, so that retention of platelets in a bead column was a product of both adhesion (platelet–surface interaction) and aggregation (platelet–platelet interaction). Other tests have been developed to measure platelet adhesion alone. Some involve the use of chambers which are filled with blood or a platelet suspension and are allowed to stand until platelets have settled onto the base of the chamber[11,14,15] or are centrifuged[16-19] and the adherent platelets are counted by microscopy. Other tests involve surfaces which are dipped

[6] J. M. Anderson and K. Kottke-Marchant, *CRC Crit. Rev. Biocomp.* **1,** 111 (1985).

[7] "Guidelines for Blood–Material Interactions," NIH Publ. No. 80-2185, Chap. 9. Devices and Technology Branch, Division of Heart and Vascular Diseases, National Heart, Lung and Blood Institute, 1980.

[8] G. P. McManama, P. C. Johnson, and E. W. Salzman, *in* "Platelet Function and Metabolism" (H. Holmsen, ed.). In press.

[9] H. P. Wright, *J. Pathol. Bacteriol.* **53,** 255 (1941).

[10] S. E. Moolten and L. Vroman, *Am. J. Clin. Pathol.* **19,** 701 (1949).

[11] R. L. Page and J. R. A. Mitchell, *Thromb. Haemostasis* **42,** 705 (1979).

[12] A. J. Hellem, *Scand. J. Clin. Lab. Invest. Suppl.* **51,** 1 (1960).

[13] E. W. Salzman, *J. Lab. Clin. Med.* **62,** 724 (1963).

[14] K. Breddin and H. Langbein, *Thromb. Diath. Haemorrh.* **10,** 29 (1963).

[15] E. Ruckenstein, A. Marmur, and S. R. Rakower, *Thromb. Haemostasis* **36,** 334 (1976).

[16] J. N. George, *Blood* **40,** 862 (1972).

[17] S. F. Mohammad, M. D. Hardison, C. H. Glenn, B. D. Morton, J. C. Bolan, and R. Mason, *Haemostasis* **3,** 257 (1974).

[18] S. F. Mohammad, M. D. Hardison, H. Y. K. Chuang, and R. G. Mason, *Haemostasis* **5,** 96 (1976).

[19] D. L. Coleman, D. E. Gregonis, and J. D. Andrade, *J. Biomed. Mater. Res.* **16,** 381 (1982).

TABLE I

In Vitro TESTS OF PLATELET–SURFACE INTERACTION[a]

Test system	Variable examined	Ref.
1. Adhesion induced by sedimentation onto a test surface	Platelet adhesion by light microscopy	11, 14, 15, 41
2. Immersion of test material into whole blood or PRP	^{111}In-labeled platelet deposition	42
	Platelet adhesion by light microscopy or SEM	20, 21, 43–45
	ADP-induced platelet aggregation	46
3. Adhesion induced by centrifugation onto a test surface	Platelet adhesion by light microscopy or SEM	16–19
4. Blood or PRP passed through columns of glass beads, polymer beads, or polymer-coated beads	Platelet retention calculated from change in platelet count across column. Platelet adhesion and aggregation by SEM	32, 37, 47–54
	[^{14}C]5HT release	32
5. Blood or PRP passed through parallel-plate chamber	Platelet adhesion by light microscopy or SEM	55–58
	Platelet adhesion by videodensitometry	59–62
	^{111}In-labeled platelet deposition	58
	ADP-induced platelet aggregation	46
6. Blood or platelet suspension passed through glass, polymer, or polymer-coated tubing	^{111}In-labeled deposition and [^{14}C]5HT release	63–65
	Platelet deposition by fluorescent videomicroscopy	22
7. Perfusion circuit	^{51}Cr-labeled platelet deposition	66–68
	Platelet adhesion by light microscopy	57
	Platelet adhesion and aggregation by TEM cross-section or SEM	58, 69–71
	[^{14}C]5HT release	68
	PF4, βTG, and 5HT release	72
	Collagen-induced platelet aggregation	73
	Platelet retention by glass bead column (after exposure to perfusion circuit)	72, 73
8. Cylindrical device rotated in platelet suspension	^{51}Cr-labeled platelet deposition and [^{14}C]5HT release	74–76
9. Polymer microspheres added to PRP	Platelet aggregation	77

[a] TEM, Transmission electron microscopy; SEM, scanning electron microscopy; PF4, platelet factor 4; βTG, β-thromboglobulin; 5HT, serotonin; ADP, adenosine diphosphate. Abridged from Anderson and Kottke-Merchant[6] (Table 5).

into blood or PRP, washed, and examined with a microscope.[20,21] In addition, tests have been devised to measure platelet adhesion in flowing systems which allow the effects of shear stress to be examined.[22-24]

The test described here, derived from the Hellem glass bead column method, allows surface-induced platelet adhesion, aggregation, and secretion to be examined for any material which can be prepared in bead form or coated onto glass beads by deposition from solution. Bead columns are exposed to flowing whole blood, anticoagulated with sodium citrate, and maintained at 37°. Platelet retention is determined by comparing the number of platelets entering and leaving the columns; platelet secretion is assayed by measuring extracellular serotonin[25] and β-thromboglobulin[26] levels in the effluent blood; platelet lysis is monitored by measuring extracellular lactate dehydrogenase levels,[27] and the relative effects of adhesion and aggregation are assessed by scanning electron microscopy.

Passage of blood through columns packed with small polymer-coated glass beads is a severe test of platelet–surface compatibility because of the large surface area to blood volume ratio. A 2-ml column packed with 0.3-mm-diameter beads provides $\sim 300\ cm^2$ of surface area which, at a point in time, is exposed to ~ 0.5 ml of blood. This is equivalent to spreading a film of blood 17 μm thick on a flat surface and represents approximately one-tenth the surface-to-volume ratio of a capillary bed (assuming a mean capillary diameter of 8 μm).[28]

Column Preparation

Bead columns are prepared by filling heat-shrinkable polyethylene columns with acid-washed glass beads, shrinking the columns, and then coating the beads with polymer by deposition from solution.

[20] S. W. Kim, S. Wisniewski, E. S. Lee, and M. L. Winn, *J. Biomed. Mater. Res. Symp.* **8,** 23 (1977).

[21] A. W. Neumann, O. S. Hum, D. W. Francis, W. Zingg, and C. J. van Oss, *J. Biomed. Mater. Res.* **14,** 499 (1980).

[22] I. A. Feurstein and J. Kush, *Trans. Am. Soc. Artif. Intern. Organs* **29,** 430 (1983).

[23] Y. A. Butruille, E. F. Leonard, and R. S. Litwak, *Trans. Am. Soc. Artif. Intern. Organs* **21,** 609 (1975).

[24] V. T. Turitto, H. J. Weiss, and H. R. Baumgartner, *J. Rheol.* **23,** 735 (1979).

[25] T. H. Spaet and M. B. Zucker, *Am. J. Physiol.* **206,** 1267 (1964).

[26] C. A. Ludlam, S. Moore, A. E. Bolton, D. S. Pepper, and J. D. Cash, *Thromb. Res.* **6,** 543 (1975).

[27] E. Amator, L. E. Dorfman, and W. E. C. Wacker, *Clin. Chem.* **9,** 391 (1963).

[28] C. Busch, P. A. Cancilla, L. E. DeBault, J. C. Goldsmith, and W. G. Owen, *Lab. Invest.* **47,** 498 (1982).

Short lengths of heat-shrinkable polyethylene tubing (10.4 mm i.d. × 6 cm long; High Voltage Engineering, Inc., Burlington, MA; part #ECC P/N 2400410CRZ) are rinsed with ethanol and preshrunk on a Teflon mandrel (Fig. 1) using a high-temperature heat gun (Milwaukee Machine Co., model #500X). The resultant flared ends facilitate the mounting of the column tips which are pushed into the preshrunk tubes as shown in Fig. 2. Polyethylene column tips (Kontes, Inc., Vineland, NJ, #K-420160-9006) are rinsed with ethanol and fitted with 15-gauge stainless steel tubing and polypropylene mesh disks (Small Parts, Inc., Miami, FL, part #CMP-125) as shown in the detail of Fig. 2. Short lengths of stainless steel tubing are cut by scoring the perimeter of the tubing using a thin emery grinding disk (Dremel part #409) mounted on a Dremel tool (Emerson Electric Co., Racine, WI; model #380-5) and rotated at slow speed. If the scoring is done uniformly and to the proper depth, the lengths of tubing can be snapped apart easily. Care should be taken to prevent the grinding wheel from cutting through the wall of the tubing as this will cause the lumen to become contaminated with fine corundum particles. Disks of polypropylene mesh ($\frac{1}{4}$-in. diameter) are soaked overnight in chloroform and placed by friction fit into the base of the column tips (Fig. 2).

Glass beads (0.3-mm diameter, Ferro Corp., Cataphote Div., Jackson, MS, class 5A microspheres, $-45 + 50$ mesh) are acid washed to remove adventitious material prior to loading into the columns. Acid washing also improves adhesion of the polymer coatings to the beads. Beads are heated in concentrated nitric acid to 80° and allowed to cool overnight. The beads are then rinsed five times with distilled, deionized water, placed in a 1 *M* NaCl solution for 10 min, and rinsed five more times with distilled, deionized water. Following rinsing, the water is aspirated thoroughly and the beads are dried at 80°. Properly washed and dried beads are free of visible clusters.

The acid-washed glass beads are poured, dry, into preshrunk column bodies with a column tip assembly mounted on the bottom. Packing of the beads is facilitated by gently tapping the side of the column body. Filled columns are placed in a rack overnight to allow the beads to settle. Beads are then added to the top of the column and the top column tip assembly is pressed down into place with care to avoid squeezing the column body and disturbing the bead packing.

After loading, the columns are heat shrunk again to stabilize the bead packing and to minimize the space between the beads and the inside wall of the column body which, in turn, minimizes the chance that blood flow will channel along the wall during testing. A column is held vertically and rotated slowly while air from the heat gun is trained on the bottom half of the column. This is a delicate procedure—the column must be heated

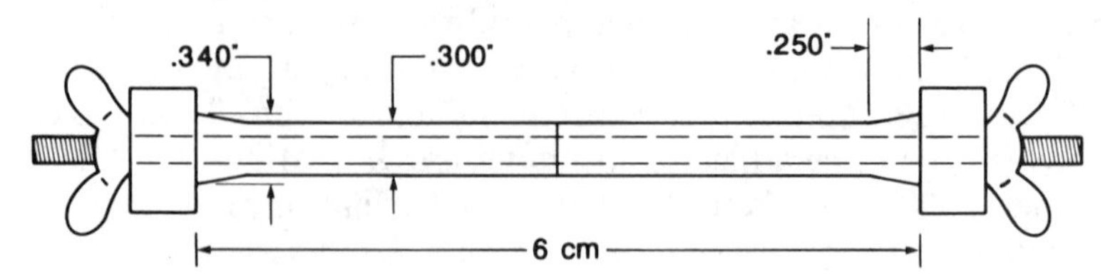

FIG. 1. Teflon mandrel used to preshrink heat-shrinkable polyethylene column bodies (see text for detailed description).

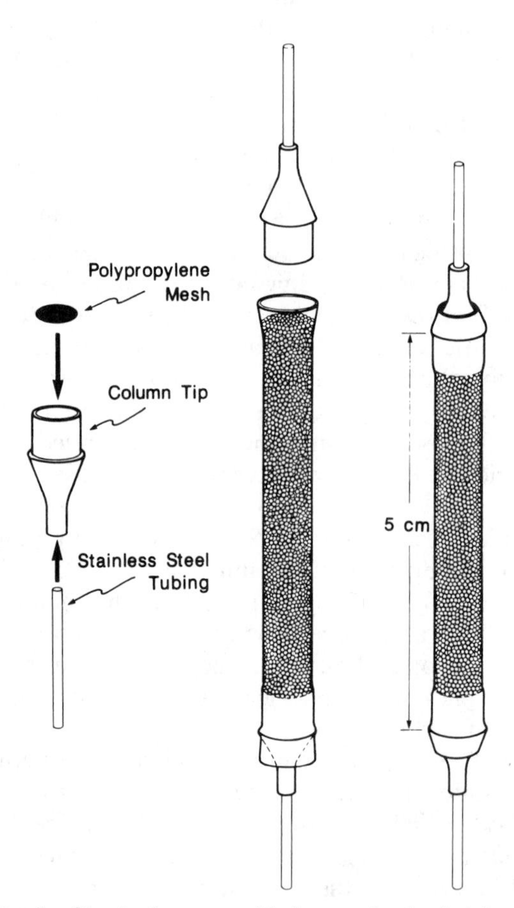

FIG. 2. Details of bead column assembly (see text for detailed description).

sufficiently to shrink the column body without melting the polyethylene tip assembly. The column is then inverted and the top of the column (now on the bottom) is shrunk in the same manner. When cut open and emptied of glass beads, properly shrunk columns will exhibit a well-defined, uniform dimple pattern on the inner surface of the heat-shrinkable tube.

Assembled glass bead columns are coated with polymer by filling from below with a polymer solution, draining the columns, and then purging with nitrogen until the solvent has been removed. Polymers must be dissolved in an organic solvent which will not affect the polyethylene or polypropylene column parts. Chloroform is most often employed, but a number of other solvents can be used, including dichloroethane, dimethylformamide, and dimethylacetamide. Polymer solutions are usually prepared at a concentration of 2 g/100 ml. The coating process begins by attaching the top of the column to a syringe and drawing the polymer solution up into the column until the liquid–air meniscus reaches the top of the column. This must be done slowly to avoid trapped air spaces between the beads, which, when they occur, are easily visible. After filling, the column is allowed to drain from the bottom by gravity (the drained polymer solution is discarded) and the top of the column is attached to a manifold which supplies dry, filtered nitrogen. Nitrogen is allowed to flow through the columns at a rate of ~5 cm^3/min until all of the solvent has evaporated (usually overnight). Polymer coatings are examined for thickness and uniformity by scanning electron microscopy.

Test Protocol

Prior to blood exposure, columns are placed in a 37° constant-temperature bath and washed with degassed, pyrogen-free saline (0.15 *M* NaCl) at a flow rate of 1 ml/min. This washing procedure prevents formation and trapping of microbubbles on the polymer surfaces which may alter platelet reactivity.[29,30] During prewashing, pressure measurements are taken at the column inlets to check for patency: for a flow of 1 ml/min, pressure readings should not exceed 10 mmHg.

Blood is drawn by venipuncture into 1/10 vol of 3.8% (w/v) sodium citrate from donors who have taken no medication for at least 10 days. Immediately after the blood is drawn, the tube is capped, mixed gently, and placed in a 37° constant-temperature bath for 30 min. If serotonin release is to be assessed, [^{14}C]serotonin (Amersham Corp., Arlington

[29] C. A. Ward, B. Ruegsegger, D. Stanga, and W. Zingg, *Trans. Am. Soc. Artif. Intern. Organs* **20,** 77 (1974).

[30] H. L. Goldsmith, P. L. Rifkin, and M. B. Zucker, *Microvasc. Res.* **9,** 304 (1975).

Heights, IL; final concentration 20 μM) is added prior to incubation at 37° to allow nearly complete uptake of the [^{14}C]serotonin by the platelets. Following serotonin uptake, imipramine is added (final concentration, 2.5 μM) to reduce potential reuptake of serotonin released by platelets retained in the columns. It is important that the blood be incubated at 37° for 30 min prior to testing: blood incubated for less than 20 min after phlebotomy yields irreproducible results; blood incubated for more than 60 min, like blood at less than 37°, loses much of its reactivity.

After incubation, blood is placed in 12-cm^3 polypropylene syringes, each containing a Teflon-coated, stainless steel ball for mixing, and transferred to a prewarmed (37°) heating block mounted on a modified multiple syringe pump (Harvard Apparatus Co., Millis, MA; model #935) which rests upon a rocking platform (10 oscillations/min). The syringes are connected to the column inlets via saline-filled polyethylene tubing. Blood is pumped through the columns, from below, at a flow rate of 0.25 ml/min. Five aliquots are collected from each column, beginning at the first appearance of red cells from the efferent tubing. Aliquot size depends on the column size: each aliquot can conveniently be taken equal to two column void volumes. A column void volume is about one-fourth of the total column volume. For columns described above (10.4-mm i.d. × 6 cm long, filled with coated glass beads), two void volumes is approximately 1 ml. Collection vials contain 12 μM indomethacin and 1.6 mM EDTA (final concentrations) to minimize further activation (release or aggregation) of the effluent blood by manipulations during platelet counting and other studies. Aliquot 1, containing blood diluted with saline, is discarded and aliquots 2 through 5 are retained for platelet counting and release studies.

Platelets are counted by standard techniques[31] immediately after passage through the columns. Three aliquots of blood, not exposed to the bead columns, are also counted as controls. Ratios of platelet counts in the effluent blood samples to the control counts in whole blood are obtained for aliquots 2 through 5 for each column. Reduction in this ratio reflects reactions induced by passage of blood through the column, which result in either platelet adhesion to the beads or the formation of platelet aggregates. Measurement of the products of platelet release, [^{14}C]serotonin,[25] β-thromboglobulin (BTG),[26] and platelet factor 4,[31] and the assay of lactate dehydrogenase (LDH),[27] a marker of platelet lysis, are performed by standard techniques. BTG and LDH are assayed using commercially available kits (BTG: Amersham Corp, Arlington Heights, IL; product code IM.88; LDH: Sigma Chem. Co., St. Louis, MO; kit #288-UV).

[31] K. L. Kaplan and J. Owen, this series, Vol. 84, p. 83.

Scanning Electron Microscopy (SEM)

Columns exposed to whole blood are fixed with glutaraldehyde immediately upon cessation of blood flow. First, a 0.1% solution of glutaraldehyde in 0.1 *M* sodium cacodylate buffer (pH 7.4), maintained at 37°, is allowed to flow by gravity through the columns, entering the columns at the top. When the blood has been displaced by glutaraldehyde, the tubing is clamped and the columns are allowed to incubate at 37° for 1 to 2 hr. The 0.1% glutaraldehyde is then displaced with 2% glutaraldehyde (in 0.1 *M* sodium cacodylate; pH 7.4, 37°) and the columns are incubated overnight. After fixing, the columns are purged of fixative by pumping 50 ml of distilled, deionized water through each column at the rate of 0.5 ml/min. Fixed columns are dried either by passage of dry nitrogen or by critical point drying. Fixed, dried beads are prepared for mounting by binding the beads into small clusters using epoxy. Depending on the portion of the column that is of interest, the epoxy is injected either directly into the top or bottom of the column or into the side of the column (using a small-gauge hypodermic needle). After the epoxy has hardened, the bead clusters are carefully removed and mounted on SEM pedestals with the orientations of the clusters noted. This technique allows examination of the "tops," "bottoms," and "sides" of the beads and allows assessment of the effects of flow characteristics (e.g., moderate shear on the upstream side and low shear on the downstream side of the beads) on platelet deposition. Examination of the beads by SEM before and after exposure to blood provides information regarding the topography and possible imperfections of the polymer surface, the relative contributions of platelet adhesion and aggregation to overall platelet retention, and the extent to which the spreading of adherent platelets is induced.

Statistical Analysis

The induction of platelet retention and release by polymer-coated beads in this system has proved to be highly reproducible when blood from a single donor is employed.[32] The day-to-day and column-to-column coefficient of variation with blood from one donor is usually less than 10%. Aliquot-to-aliquot variation is also small for most materials, implying that the processes by which platelets are retained in the columns tend to remove a constant fraction of the platelets presented to the beads. When platelet retention and/or release vary significantly from aliquot to aliquot, comparisons of materials should be made using equivalent aliquots. Donor-to-

[32] J. N. Lindon, R. Rodvien, D. Brier, R. Greenberg, E. Merrill, and E. W. Salzman, *J. Lab. Clin. Med.* **92,** 904 (1978).

donor variation for both platelet retention and release, however, is usually much larger. For this reason, when materials are being compared, each experiment involves the simultaneous exposure of blood from a single donor to each of the materials and experiments are repeated using at least six donors. This allows the use of a "blocked" statistical analysis as described below.

Multiple comparison tests[33] are employed to detect significant differences among the surfaces tested. Multiple factor analysis of variance must first be performed to ensure that variations among surfaces are significantly greater than the donor-to-donor, aliquot-to-aliquot, or replicate variation. Once such a test demonstrates that surfaces are the greatest source of variation, the appropriate multivariant test can be used to compare the platelet reactivities of the materials. Selection of the appropriate multivariant tests[33] depends on whether the data are blocked (an extension of the notion of paired data to more than two treatments), and whether the underlying population is normally distributed. Selection from among the available tests of the normality of sample distribution (e.g., the D'Agostino or Wilk–Shapiro test) depends on sample size.[33] If the data are normally distributed, blocked or unblocked Newman–Keuls tests can be performed. If the data are nonparametric (not normally distributed), the Friedman test is used for blocked data and the Kruskal–Wallis test is used for unblocked data.

Critique of the Method

As with all *in vitro* tests, this method is subject to the limitation that its relevance to the behavior of specific artificial materials *in vivo* must be established, not assumed. Materials currently thought to be most nearly nonthrombogenic, polyurethane and silicone rubber,[34] showed minimal induction of platelet retention or release with this test.[35,36] Two known thrombogenic materials, polystyrene and glass, show significant platelet

[33] J. H. Zar, "Biostatistical Analysis." Prentice-Hall, Englewood Cliffs, New Jersey, 1974.

[34] E. W. Salzman and E. W. Merrill, *in* "Hemostasis and Thrombosis" (R. W. Colman, J. Hirsh, V. J. Marder, and E. W. Salzman, eds.), 2nd Ed. Lippincott, Philadelphia, Pennsylvania, 1987.

[35] E. W. Merrill, E. W. Salzman, V. Sa da Costa, D. Brier-Russell, A. Dincer, P. Pape, and J. N. Lindon, *in* "Biomaterials: Interfacial Phenomena and Applications" (S. L. Cooper and N. A. Peppas, eds.), Adv. Chem. Ser. 199. American Chemical Society, Washington, D.C., 1982.

[36] E. W. Merrill, V. Sa da Costa, E. W. Salzman, D. Brier-Russell, L. Kushner, D. F. Waugh, G. Trudel III, S. Stopper, and V. Vitale, *in* "Biomaterials: Interfacial Phenomena and Applications" (S. L. Cooper and N. A. Peppas, eds.), Adv. Chem. Ser. 199. American Chemical Society, Washington, D.C., 1982.

activation *in vitro*.[35,37] However, a more extensive direct comparison of this *in vitro* test with *in vivo* tests remains to be performed.

Two characteristics of this *in vitro* test system are of concern in this regard: the use of citrated blood, and the measurement of surface-induced activation over a short time period. Surfaces that rapidly and efficiently activate platelets in an *in vitro* system are, intuitively, unlikely to perform well *in vivo*. However, surfaces that activate few platelets during a 10-min *in vitro* exposure will not necessarily remain benign during a long-term *in vivo* application.

The effects of citrate in this system are unclear. In most *in vitro* tests of platelet activation, anticoagulation with citrate produces a response intermediate to that seen with heparin, which produces a greater response, or EDTA, which eliminates the response.[8] Divalent cations are apparently essential for platelet activation by surfaces, but whether this is due to direct involvement of calcium and/or magnesium in the adhesion and aggregation processes or to a requirement for thrombin generation is not known. In this regard, replacement of citrate as an anticoagulant with hirudin or D-phenylalanyl-L-propyl-L-arginine chloromethyl ketone (PPACK) will facilitate delineation of the role of divalent cations.

Despite these apparent limitations, the bead column retention test has proved a valuable tool for evaluating the short-term thromboresistance of numerous polymers (e.g., polyurethanes,[36] heparin-coated surfaces,[38] polyacrylates, and polymethacrylates)[39] and for experiments of more fundamental aspects of blood/surface interactions (e.g., the role of adsorbed fibrinogen[40] and the role of internal redistribution of platelet Ca^{2+}).[8] The test is rapid, convenient, and relatively inexpensive. It is also highly versatile, being applicable to any polymer which can be prepared in the form of beads or can be coated onto glass beads by deposition from solution. It is useful, in conjunction with tests for surface-induced activation of the clotting cascade, complement activation, and fibrinolysis, to predict mate-

[37] C. R. Robertson and H. N. Chang, *Ann. Biomed. Eng.* **2,** 361 (1974).

[38] J. N. Lindon, E. W. Salzman, E. W. Merrill, A. K. Dincer, D. Labarre, K. A. Bauer, and R. R. Rosenberg, *J. Lab. Clin. Med.* **105,** 219 (1985).

[39] D. Brier-Russell, E. W. Salzman, J. N. Lindon, R. Handin, E. W. Merrill, A. K. Dincer, and J. S. Wu, *J. Colloid Interface Sci.* **81,** 311 (1981).

[40] J. N. Lindon, G. McManama, L. Kushner, E. W. Merrill, and E. W. Salzman, *Blood* **68,** 355 (1986).

rials appropriate for more demanding *in vivo* testing. The reader is once again referred to Table I[41-77] for an overview of the test systems mentioned here.

Acknowledgment

This work is supported by Grant NIH HL20079.

[41] A. W. Neumann, M. A. Moscarello, and W. Zing, *J. Biomed. Mater. Res.* **14,** 499 (1980).

[42] P. Didisheim, J. Q. Stropp, M. K. Dewanjee, and H. W. Wahner, *Thromb. Haemostasis* **50,** 60 (1983).

[43] Y. Mori, S. Nagaoka, H. Takiuchi, T. Kikuchi, N. Noguchi, H. Tanzawa, and Y. Noishiki, *Trans. Am. Soc. Artif. Organs* **28,** 459 (1982).

[44] S. W. Kim and E. S. Lee, *J. Polym. Sci.: Polym. Symp.* **66,** 429 (1979).

[45] C. D. Ebert, E. S. Lee, and S. W. Kim, *J. Biomed. Mater. Res.* **16,** 629 (1982).

[46] J. McRea and S. W. Kim, *Trans. Am. Soc. Artif. Intern. Organs* **24,** 746 (1978).

[47] J. Aznar, P. Villa, and E. Rueda, *Haemostasis* **8,** 38 (1979).

[48] T. Okano, S. Nishiyama, I. Shinohara, T. Akaike, Y. Sakurai, K. Kataoka, and T. Tsuruta, *J. Biomed. Mater. Res.* **15,** 393 (1981).

[49] E. W. Merrill, E. W. Salzman, S. Wan, N. Mahmud, L. Kushner, J. N. Lindon, and J. Curme, *Trans. Am. Soc. Artif. Intern. Organs* **28,** 482 (1982).

[50] E. S. Lee and S. W. Kim, *ASAIO J.* **8,** 11 (1979).

[51] K. Kataoka, T. Okano, Y. Sakurai, T. Nishimura, M. Maeda, S. Inoue, and T. Tsuruta, *Biomaterials* **3,** 237 (1982).

[52] K. Kataoka, T. Okano, T. Akaike, Y. Sakurai, M. Maeda, T. Nishimura, Y. Nitadori, T. Tsuruta, A. Shimada, and I. Shinohara, *Biomaterials* **3,** 493 (1982).

[53] E. W. Merrill, E. W. Salzman, V. Sa da Costa, D. Brier-Russell, A. Dincer, P. Pape, and J. N. Lindon, *in* "Biomaterials: Interfacial Phenomena and Applications" (S. L. Cooper and N. A. Peppas, eds.), Adv. Chem. Ser. 199. American Chemical Society, Washington, D.C., 1982.

[54] K. Kataoka, T. Tusurata, T. Akaike, and Y. Sakurai, *Makromol. Chem.* **181,** 1363 (1980).

[55] J. McRea and S. W. Kim, *Trans. Am. Soc. Artif. Intern. Organs* **24,** 746 (1978).

[56] J. Olijslager, *in* "The Development of Test Devices for the Study of Blood Material Interactions" (J. Olijslager, ed.), Chap. 3, p. 43. Delft University Press, Delft, Holland, 1982.

[57] E. F. Grabowski, P. Didisheim, J. C. Lewis, J. T. Franta, and J. Q. Stropp, *Trans. Am. Soc. Artif. Intern. Organs* **23,** 141 (1977).

[58] P. Didisheim, M. K. Dewanjee, C. S. Frisk, M. P. Kaye, and D. N. Fass, *in* "Contemporary Biomaterials: Materials and Host Response: Clinical Applications, New Technology and Legal Aspects" (J. W. Boretos and M. Eden, eds.), Chap. 10. Noyes Publications, Park Ridge, New Jersey, in press, 1984.

[59] K. Bodziak and P. D. Richardson, *Trans. Am. Soc. Artif. Intern. Organs* **28,** 426 (1982).

[60] P. D. Richardson, R. L. Kane, and K. Agarwal, *Trans. Am. Soc. Artif. Intern. Organs* **27,** 203 (1981).

[61] P. D. Richardson, S. F. Mohammad, and R. G. Mason, *Artif. Organs* **1,** 128 (1977).

[62] P. D. Richardson, S. F. Mohammad, R. G.Mason, M. Steiner, and R. Kane, *Trans. Am. Soc. Artif. Intern. Organs* **25,** 147 (1979).

[63] G. A. Adams and I. A. Feuerstein, *Trans. Am. Soc. Artif. Intern. Organs* **27**, 219 (1981).
[64] G. A. Adams and I. A. Feuerstein, *Am. J. Physiol.* **240,** H99 (1981).
[65] G. A. Adams and I. A. Feuerstein, *ASAIO J.* **4,** 90 (1981).
[66] H. V. Roohk, J. Pick, R. Hill, E. Hung, and R. H. Bartlett, *Trans. Am. Soc. Artif. Intern. Organs* **22,** 1 (1976).
[67] H. V. Roohk, M. Nakamura, R. L. Hill, E. K. Hung, and R. H. Bartlett, *Trans. Am. Soc. Artif. Intern. Organs* **23,** 152 (1977).
[68] R. Malmgren, R. Larsson, P. Olsson, and K. Radegran, *Haemostasis* **8,** 400 (1979).
[69] H. R. Baumgartner, R. Muggli, T. B. Tschopp, and V. T. Turitto, *Thromb. Haemostasis* **35,** 124 (1976).
[70] V. T. Turitto, R. Muggli, and H. R. Baumgartner, *Trans. Am. Soc. Artif. Intern. Organs* **24,** 568 (1976).
[71] R. G. Mason, W. H. Zucker, B. A. Shinoda, and H. Y. Chaung, *Lab. Invest.* **31,** 143 (1974).
[72] K. Kottke-Marchant, J. M. Anderson, A. Rabinovitch, and R. Herzig, *Proc. World Congr. Biomater., 2nd* p. 90 (1984).
[73] C. N. McCollum, M. J. Crow, S. M. Rajah, and R. C. Kester, *Surgery* **87,** 668 (1980).
[74] J. L. Brash and S. J. Whicher, *in* "Artificial Organs" (R. M. Kenedi, J. M. Courtney, J. D. S. Gaylor, and T. Gilchrist, eds.), p. 263. University Park Press, Baltimore, Maryland, 1977.
[75] S. J. Whicher, S. Uniyal, and J. L. Brash, Trans. Am. Soc. Artif. Organs **26,** 268 (1980).
[76] I. A. Feuerstein, J. M. Brophy, and J. L. Brash, *Trans Am. Soc. Artif. Inter. Organs* **21,** 427 (1975).
[77] G. A. Herzlinger, D. H. Bing, R. Stein, and R. D. Cummings, *Blood* **57,** 764 (1981).

[10] Platelet Aggregation Measured by the Photometric Method

By MARJORIE B. ZUCKER

Aggregation of platelets into a mass is responsible for their main physiological function, that is, the formation of a hemostatic plug sealing off the break in an injured blood vessel. Whereas formation of platelet aggregates (thrombi) *in vivo* is difficult to measure, the process of platelet aggregation *in vitro* has been quantitatively analyzed and employed in many studies of platelet function. The value of platelets in modern studies in cell biology and the subject of platelet aggregation have been reviewed elsewhere.[1-3]

[1] B. S. Coller, *CRC Handb. Ser. Clin. Lab. Sci.* p. 381 (1979).
[2] R. L. Kinlough-Rathbone, M. A. Packham, and J. F. Mustard, *in* "Methods in Hematology: Measurements of Platelet Function" (L. A. Harker and T. S. Zimmerman, eds.), p. 64. Churchill-Livingstone, New York, 1983.
[3] M. B. Zucker and V. T. Nachmias, *Arteriosclerosis* **5,** 2 (1985).

General Characteristics of Aggregation

Platelets are not "sticky" until they are stimulated by an agonist. Agonists are very diverse, consisting of small-molecular-weight compounds, such as ADP, serotonin, and epinephrine; enzymes, such as thrombin and trypsin; particulate material, such as collagen and antigen–antibody complexes; lipids, such as platelet-activating factor (PAF-acether); and ionophores, such as A23187. These agonists induce changes in the membrane of platelets, evoking their aggregation under appropriate conditions.

For aggregation to occur, the platelets must come into contact with one another, which is usually achieved by stirring the suspension. Aggregation in a suspension of platelets is detected on a macroscopic level by the development of visible clumps and clearing of the suspension. More sensitive measurements can be made in a platelet aggregometer, which is simply a photometer that records the clearing of a stirred suspension of platelets.

Aggregation of platelets consitutes a multistep process which can be analyzed by recording the tracing of changes in light transmission. Following the addition of certain agonists there is a decrease in light transmission due to a change in the shape of platelets from discoid to spherical. This is followed by a gradual increase in light transmission as the platelets aggregate and clear the way for the light to be transmitted through the platelet suspension. The initial phase of platelet aggregation is reversible unless it is followed by secretion of "proaggregatory" factors from the dense granules (e.g., ADP and serotonin) and α-granules (adhesive proteins such as fibrinogen, von Willebrand factor, thrombospondin, and fibronectin). This phase of platelet aggregation is irreversible and reaches a plateau which reflects the maximal level of light transmission.

Requirement for Fibrinogen and Divalent Cations

Platelet "stickiness" develops when the platelet membrane acquires the ability to bind fibrinogen. This dimeric molecule probably acts as a molecular "glue," bridging the gap between platelets.[4] With stimuli that can induce secretion without prior aggregation, e.g., thrombin or collagen, the fibrinogen can be secreted from the platelet α-granules, whereas with stimuli that require aggregation before secretion occurs, e.g., ADP and epinephrine, fluid-phase fibrinogen must be present.[5] Fibrinogen binding to platelets, and hence aggregation, also requires divalent cations (see below). It is important to note that certain responses, e.g., secretion induced by ADP, and the association of glycoproteins IIb/IIIa with the

[4] E. I. B. Peerschke, *Semin. Hematol.* **22,** 241 (1985).

[5] I. F. Charo, R. D. Feinman, and T. C. Detwiler, *J. Clin. Invest.* **60,** 866 (1977).

cytoskeleton, only occur when platelets have actually aggregated, not when they have been simply stimulated or even when they have bound fibrinogen.

Preparation of Platelet Suspensions

Precautions for Donors

Severe lipemia decreases the apparent response to aggregating agents in platelet-rich plasma because the plasma will remain turbid even if all of the platelets have aggregated. Hence a fatty meal should not be ingested within 6–8 hr before blood donation.

Ideally, donors should not have taken any drugs for a week before blood donation. Nonsteroidal antiinflammatory drugs inhibit the enzyme cyclooxygenase and thus prevent secretion via the arachidonate pathway. The effect of most such drugs does not persist when the drug disappears, except for that of acetylsalicyclic acid (aspirin). This drug irreversibly blocks the enzyme and hence affects platelets as long as they survive (about 10 days.) Secretion appears to recover in 4 to 5 days, since there are enough young uninhibited platelets that can synthesize endoperoxides.[6,7] An enormous number of drugs contain aspirin, including Alka-Seltzer and Dristan.[8,9] Some antibiotics, tranquilizers, and tricyclic antidepressants inhibit platelet function.[6,7]

Anxiety is reported to cause abnormally large increases in the primary aggregation responses to ADP and norepinephrine during the hour after preparation of citrated platelet-rich plasma.[10] It also abolishes second-wave aggregation and secretion induced by epinephrine.[11] Inhaling smoke from a single cigarette increases the response to low concentrations of ADP 10 to 20 min later.[12]

Blood Drawing

Disposable plastic syringes and tubes are usually used with either a needle or butterfly; often the first few milliliters of blood are discarded and blood for study is taken with a clean syringe. It is important to execute a

[6] H. J. Weiss, *Prog. Hemostas. Thromb.* **1,** 199 (1972).
[7] M. A. Packham and J. F. Mustard, *Circulation* **62,** (Suppl. 5), V-26 (1980).
[8] E. R. Leist and J. G. Banwell, *N. Engl. J. Med.* **291,** 710 (1974).
[9] J. C. Selner, *N. Engl. J. Med.* **292,** 372 (1975).
[10] M. J. G. Harrison, P. R. Emmons, and J. R. A. Mitchell, *J. Atheroscler. Res.* **7,** 197 (1967).
[11] Y. S. Arkel, J. I. Haft, W. Kreutner, J. Sherwood, and R. Williams, *Thromb. Haemostasis* **38,** 552 (1977).
[12] P. H. Levine, *Circulation* **48,** 619 (1973).

clean venipuncture with no difficulty in drawing the blood. The primary response to ADP is the same whether a 14-gauge needle and free flow or a syringe and 19-gauge needle is used.[13]

Citrated Platelet-Rich Plasma: Centrifugation and Characteristics

Citrated platelet-rich plasma is widely used for studying platelet aggregation. It is usually prepared from blood added to 1/9 vol of 0.109 to 0.129 *M* trisodium citrate (3.2 to 3.8% of the dihydrate salt).

To prepare platelet-rich plasma, blood is usually centrifuged at 150 to 200 *g* for 10–15 min. The maximum *g* force in a centrifuge tube is measured at the bottom of the tube. However, a higher *g* force or a longer time will be necessary for preparing platelet-rich plasma in tubes that are completely filled with blood than in tubes that are only partly filled. Calculation of the average *g* force in the blood sample is most useful, and centrifugation can be carried out for longer or shorter periods, provided that the product of the average $g \times$ time is constant. The most satisfactory harvest of platelet-rich plasma is achieved at $g_{av} \times$ time in minutes = 1750–2400. The g_{av} is calculated as $1.118 \times 10^{-5} \times$ (revolutions per minute)$^2 \times$ radius in centimeters (i.e., the distance from the center of the rotor to the midpoint of the column of blood).[14]

The red cells and most of the leukocytes sediment without tight packing, and the supernatant, about one-third of the volume, consists of plasma containing platelets. The platelet-rich plasma should not be tinged with red; the presence of erythrocytes reduces the sensitivity with which aggregation can be recorded, since they remain suspended in the plasma after the platelets have aggregated. Platelet-poor plasma is conveniently prepared by centrifuging the residual blood at a higher *g* force.

Platelet-rich plasma kept at 37° shows a "swirl" when it is agitated. This is characteristic of the presence of asymmetric particles, in this case discoidal platelets. When agents such as ADP or chilling cause the platelets to change their shape and become symmetrical "spiny spheres," the swirl disappears.

Factors that Affect Aggregation of Platelets in Citrated Platelet-Rich Plasma

pH. Aggregation reaches an optimum at approximately pH 8.0.[15] Since the major buffer in plasma is HCO_3^-/H_2CO_3, plasma pH readily increases

[13] B. P. Woods, A. Dennehy, and N. Clarke, *Thromb. Haemostasis* **36,** 302 (1976).

[14] M. V. Vickers and S. G. Thompson, *Thromb. Haemostasis* **53,** 216 (1985).

[15] L. Skoza, M. B. Zucker, Z. Jerushalmy, and R. Grant, *Thromb. Diath. Haemorrh.* **18,** 713 (1967).

toward this value if its dissolved CO_2 is allowed to escape. The rise in pH is enhanced by letting blood or plasma flow down the side of a tube, by frequent agitation, or by the presence of a large air interface. An easy way to maintain the pH is to draw platelet-rich plasma into a syringe and expel it only as needed. It can also be kept under oil, in a full tube which is capped, or in a box through which 5% CO_2 is continually passed. In contrast to aggregation, agglutination by ristocetin plus von Willebrand factor decreases above pH 7.6.[16]

Divalent Cations. Platelet aggregation is prevented by extreme reduction of divalent cation concentration in the medium. Plasma contains approximately 2.5 m*M* calcium and 1 m*M* magnesium. About half of each is ionized and most of the remainder is complexed with plasma proteins. Citrated plasma contains enough ionized calcium (about 40 μM) to support aggregation, whereas plasma anticoagulated with 5 m*M* EDTA does not.[17] Magnesium can support platelet aggregation,[18] but a trace of calcium is also necessary; 7 n*M* has an optimal effect in the presence of 1 m*M* magnesium.[19]

The sensitivity and magnitude of primary ADP-induced aggregation are increased when the concentration of citrate is less than that usually employed.[20] The second phase of aggregation, which is associated with secretion activated by the arachidonate pathway, only occurs when the concentration of ionized calcium in the medium is low, as it is in citrated plasma.[2,17,21,22] Men require slightly higher concentrations of ADP and epinephrine to elicit second-phase aggregation than women because the higher hematocrits in men result in a higher plasma citrate concentration.[23] Therefore, the plasma concentration of citrate should be controlled in routine studies on patients with abnormal hematocrits, or in careful determinations of threshold responses in normal subjects. If 1 ml of 0.11 *M* sodium citrate is added to 9 ml of blood with a hematocrit of 45%, the plasma citrate concentration will be $1 \times 0.11/[(9 \times 0.55) + 1]$ or

[16] B. S. Coller, *Blood* **47,** 841 (1976).

[17] B. Lages and H. J. Weiss, *Thromb. Haemostasis* **45,** 173 (1981).

[18] M. B. Zucker and R. A. Grant, *Blood* **52,** 505 (1978).

[19] L. F. Brass and S. J. Shattil, *J. Clin. Invest.* **73,** 626 (1984).

[20] E. C. Rossi and G. Louis, *Thromb. Haemostasis* **37,** 283 (1977).

[21] J. F. Mustard, D. W. Perry, R. L. Kinlough-Rathbone, and M. A. Packam, *Am. J. Physiol.* **228,** 1957 (1975).

[22] D. E. Macfarlane, P. N. Walsh, D. C. B. Mills, H. Holmsen, and H. J. Day, *Br. J. Haematol.* **30,** 457 (1975).

[23] R. M. Hardisty, R. A. Hutton, D. Montgomery, S. Rickard, and H. Trebilcock, *Br. J. Haematol.* **19,** 397 (1970).

0.0185 *M*. To maintain this concentration with different hematocrits, the volume of 0.11 *M* citrate solution (X) added to 9 ml of blood can be varied according to the formula:

$$(X)(0.11) = (0.0185)[9(1\text{-hct}) + X]$$

$$X = 1.82(1\text{-hct})$$

Anticoagulants Other than Citrate

Heparinized plasma contains a more physiologic concentration of divalent cations than citrated plasma. However, heparinized platelet-rich plasma is not easy to prepare from heparinized blood (5 – 10 U/ml) because heparin can induce aggregation, and the platelet clumps sediment with the erythrocytes or adhere to the walls of the tube. Although it can be prepared from citrated platelet-rich plasma by adding heparin and 15 m*M* $CaCl_2$ (0.1 vol), it is of questionable value for studying physiologic functions because of the stimulating effect of heparin itself on platelets.[17,24] Hirudin, the anticoagulant from leech salivary glands, can be used at a concentration of 200 U/ml[17] or 5 – 10 U/ml[21] to prevent blood from clotting without altering the concentration of divalent cations or affecting platelets.

Platelet Count

Platelet-rich plasma usually contains 350,000 to 500,000 platelets/μl. In comparing the response in different subjects, many investigators adjust the platelet count to 250,000 or 300,000/μl by adding the patient's platelet-poor plasma. Others feel that even this simple adjustment introduces artifacts. It is possible to correct the measured response without resetting the aggregometer by reference to a previously prepared curve relating response and platelet count.[25] For effects of differences in platelet size, see the section on "Assessing Aggregation."

Study of platelet responses in thrombocytopenic patients poses a real problem because aggregation is impaired at a low platelet count, presumably because the likelihood of platelet collisions is decreased. Although it has been reported that responses are maintained with 50,000 platelets/μl with a sensitive aggregometer,[26] most investigators have difficulty when the count is below 100,000/μl. If the count is low, it should be possible to concentrate the platelets by lowering the pH of the platelet-rich plasma to 6.5 with citric acid or by adding 5 m*M* EDTA, centrifuging, resuspending

[24] M. B. Zucker, *Thromb. Diath. Haemorrh.* **33,** 63 (1975).
[25] J. R. O'Brien, J. B. Heywood, and J. A. Heady, *Thromb. Diath. Haemorrh.* **16,** 752 (1966).
[26] P. H. Levine, *Am. J. Clin. Pathol.* **65,** 79 (1976).

the platelets in a smaller volume of citrated platelet-poor plasma, and incubating the sample at 37° until the platelets regain their discoidal shape. A similar procedure should be carried out with a normal sample to ensure that abnormalities are not introduced. Again, however, this treatment may introduce changes in responsiveness.

Temperature

ADP and epinephrine induce secretion and its accompanying second phase of aggregation at 37° but not at room temperature.[27,28] In contrast, thrombin and collagen can cause secretion at room temperature.[28]

Storage

Cold temperature can stimulate platelets and, on the other hand, responsiveness is lost more rapidly at 37° than at lower temperatures. It is, therefore, advisable to keep platelet-rich plasma at room temperature. Even better, in the author's experience, is to keep it in a tube inserted into an insulated flask at 18–20° since modern laboratories are often considerably warmer than this. When citrated platelet-rich plasma is kept at room temperature, the rate of ADP-induced primary aggregation is reported both to decrease[29] and to increase[30] with time, even when a pH change is prevented.

Secretory responses to epinephrine are usually better in platelet-rich plasma that has been kept on the bench for 30–60 min than in freshly prepared samples.[31] This enchancement may occur because centrifugation releases ADP from the red cells or platelets,[32] so that the platelets are initially slightly refractory, or because of a progressive increase in the pH of the platelet-rich plasma. Secretory responses to ADP and epinephrine gradually deteriorate so that it is advisable to use platelet-rich plasma within 2 ½ to 3 hr.

Platelet Suspensions Free of Plasma

The presence of plasma proteins or other components may interfere with some studies on platelet aggregation, so that gel-filtered or washed

[27] D. C. B. Mills, I. A. Robb, and G. C. K. Roberts, *Am. J. Physiol.* **195,** 715 (1968).
[28] J. F. Valdorf-Hansen and M. B. Zucker, *Am. J. Physiol.* **220,** 105 (1971).
[29] C. Warlow, A. Corina, D. Ogston, and A. S. Douglas, *Thromb. Diath. Haemorrh.* **31,** 133 (1974).
[30] W. Terres, B. F. Becker, M. A. A. Kratzer, and E. Gerlach, *Thromb. Res.* **42,** 539 (1986).
[31] E. C. Rossi and G. Louis, *J. Lab. Clin. Med.* **85,** 300 (1975).
[32] I. Aursnes and V. Vikholm, *Thromb. Haemostasis* **51,** 54 (1984).

platelets must be used. The ability of platelets to aggregate with ADP, however, is readily lost when platelets are isolated, probably largely because ADP released during the isolation procedure induces refractoriness to this agonist. Responsiveness can be maintained by adding apyrase to the suspension medium (see below and Section I in this volume).

Preparation of washed platelets is detailed elsewhere (see chapters [1] and [2] in this volume). Important aspects are noted here. Platelets centrifuged out of citrated platelet-rich plasma are almost impossible to resuspend. To avoid this, 5 m*M* EDTA is added or the pH is reduced to 6.5 by addition of 0.1 *M* citric acid. Incubation of platelets with EDTA at 37° should be avoided as this can cause irreversible impairment of aggregation.[18]

In the method developed by Mustard *et al.*,[33] heparin is added in the first resuspension medium to prevent formation of thrombin. Since the heparin may cause the platelets to clump, for the first wash we use a buffer like that developed by Mills *et al.*[34] with the following composition: 140 m*M* NaCl, 2.5 m*M* KCl, 1 m*M* $MgCl_2$, 2 m*M* $CaCl_2$, 0.5 m*M* NaH_2PO_4, 10 m*M* $NaHCO_3$, 22 m*M* trisodium citrate, 1 mg/ml glucose, 3.5 mg/ml bovine serum albumin, 0.1 mg/ml apyrase (Ec 3.6.1.5) (grade VIII, Sigma Chemical Co., St. Louis, MO) brought to pH 6.5 with 0.11 *M* citric acid. We resuspend the platelets in a small volume of this buffer, then add these platelets to a large volume of buffer in a clean tube, incubate them for 10 to 15 min at 37°, centrifuge again at room temperature after restoring the pH to 6.5 if necessary, and resuspend the platelets in buffered Tyrode's solution. Because bicarbonate is an inadequate buffer in the absence of CO_2, 10 m*M* HEPES (*N*-2-hydroxyethylpiperazine-*N*′-2-ethanesulfonic acid) buffer is added to help maintain the pH at 7.4. TES [*N*-tris(hydroxymethyl)methyl-2-aminoethanesulfonic acid] is preferred if one wishes to measure platelet proteins by the Lowry method with which HEPES interferes, but Tris, a similar compound, inhibits some platelet responses.[35] Incidentally, 10^9 platelets contain about 2 mg protein.

Albumin is usually added to platelet suspension media at 0.1 to 0.35%, i.e., less than 1/10 of the plasma concentration. It helps to prevent interaction of the platelets with foreign surfaces, be they glass or silicone, and traps small amounts of arachidonic acid or its products that may be liberated from the platelets.

[33] J. F. Mustard, D. W. Perry, N. G. Ardlie, and M. A. Packham, *Br. J. Haematol.* **22,** 193 (1972).

[34] D. C. B. Mills, W. R. Figures, L. M. Scearce, G. J. Stewart, R. F. Colman, and R. W. Colman, *J. Biol. Chem.* **260,** 8078 (1985).

[35] M. A. Packham, M. A. Guccione, M. Nina, R. L. Kinlough-Rathbone, and J. F. Mustard, *Thromb. Haemostasis* **51,** 140 (1984).

For studying aggregation with ADP, a low concentration of potato apyrase should be added to hydrolyze ADP released from the platelets during storage of the suspension and thus prevent refractoriness.[33] Responsiveness to so-called strong stimuli[5] is more easily preserved, therefore apyrase is not needed. Fibrinogen must also be added if ADP or epinephrine is used as the stimulus. Kabi (grade L) fibrinogen is a satisfactory preparation that can be obtained from Helena Laboratories (see below), dialyzed against 140 m*M* NaCl and 1.7 m*M* sodium citrate and frozen in small aliquots, or purified further.[2] Concentrations of about 0.25 mg/ml are adequate.

Recording Aggregation

Initially, the increase in light transmission caused by platelet aggregation was measured by placing samples in a spectrophotometer after aggregation had occurred.[36] Then the widely used method was developed in which light transmission through platelet-rich plasma is continuously recorded.[37,38] The photometer is equipped with a stirring motor below the light path so that a stir bar can be placed in the bottom of the cuvette.

Sources of Aggregometers

Some machines are best for routine study of patients, and other more adaptable machines are better suited for research. Aggregometers are available in the United States from several companies. These are (in alphabetical order): Bio/Data (Hatboro, PA), Chrono-Log (Havertown, PA), Logos Scientific (Henderson, NV), Helena Laboratories (Beaumont, TX), Payton Scientific (Buffalo, NY), and Sienco (Morrison, CO). Chrono-Log and Payton have been established for many years and each provides a number of models. Since the characteristics of the machines change and new features (e.g., computer ability) are added, individual models will not be described. Instead, important characteristics of aggregometers will be discussed so that a buyer can choose the machine best adapted to his/her needs.

Characteristics of Aggregometers

Recording. All machines record on a moving sheet of paper the transmission of light through a stirred platelet suspension. In some models, the

[36] T. H. Spaet, J. Cintron, and M. Spivack, *Proc. Soc. Exp. Biol. Med.* **111,** 292 (1962).
[37] J. R. O'Brien, *J. Clin. Pathol.* **15,** 452 (1962).
[38] G. V. R. Born and M. J. Cross, *J. Physiol.* **168,** 178 (1963).

recorder is an integral part of the aggregometer. If it is not, recorders can be purchased from the aggregometer vendors or from other sources. Most recorders travel at a speed of 1 in./min, but for research purposes it is convenient to have a recorder that can reach higher speeds, for example for studying changes in platelet shape.

Number of Channels. Some machines provide only a single channel whereas others have two or four channels that can record simultaneously. There are many advantages to an aggregometer with multiple channels: one sample can be prepared with an inhibitor, another without; one can be used to measure shape change, the other aggregation; or the several channels can simply be used to speed up the study.

Adaptability. Chrono-Log and Payton have models that can be adapted to quantify ATP secretion by measuring light emission with luciferin/luciferase (Lumi-aggregometer). In fact, some machines can record light transmission in two channels and luminescence measurements in both channels as well. Payton has models in which the platelets can be stimulated by application of ultrasound or ultraviolet light. A number of companies now have machines with a computer and limited memory.

Sample Volume and Containers. In some machines, adapters and different sized cuvettes are available so that different volumes of platelet-rich plasma can be used. A large volume is desirable for measuring the presence of secreted products in the supernatant after transferring the sample to a test tube, then chilling and centrifuging it, whereas very small cuvettes are useful for studying platelet responses in pediatric patients. Some investigators siliconize the cuvettes used in studying platelet aggregation. Since platelets may adhere more readily to siliconized surfaces than to glass under certain conditions,[39] there is no rationale for this procedure unless contact activation of blood coagulation is to be avoided.

Setting. White light is often used in studying light transmission through platelet-rich plasma. However, differences are most pronounced with a red light that is used in some machines. The aggregometer is ordinarily adjusted so that platelet-rich plasma has 0 to 10% light transmission and platelet-poor plasma from the same subject has 90 to 100% transmission. In some machines, the gain (or amplitude of deflection for a given difference in light transmission) and the zero (or the place on the scale at which this difference records) are adjusted automatically by placing platelet-rich plasma in one well, platelet-poor plasma in another, and pressing a button. In other machines the setting is more laborious but also more flexible.

When selecting an aggregometer, it is important to determine whether the machine can be set when the platelet-poor plasma is not transparent,

[39] R. L. Page and J. R. A. Mitchell, *Thromb. Haemostasis* **42,** 705 (1979).

for example if it is lipemic. The ability to set the machine with turbid "platelet-poor plasma" is also important in testing samples of platelet-rich plasma that still contain erythrocytes; platelet aggregation can be much more accurately measured in these samples by using as the "platelet-poor plasma" a sample of erythrocyte- and platelet-rich plasma in which the platelets have been maximally aggregated. The agglutination of fixed platelets with von Willebrand factor and ristocetin is also much easier to measure (see details in chapter [12] in this volume) if a higher gain is used. This can be achieved by setting the 90–100% transmission with a 1:1 dilution of the platelet suspension. A similar adjustment would be useful for measuring shape change or weak aggregation. In summary, when selecting a machine the prospective purchaser should consider whether to sacrifice versatility for convenience.

Temperature. In some machines, the temperature of the wells is fixed at 37°, whereas in others it can be varied and displayed on a dial. With a new machine, it is advisable to check the temperature in the well. The time required for a sample stored at room temperature to reach 37° when placed in the well should also be determined.

Stirring. Some machines have a fixed stirring rate of 1200 rpm whereas others stir at a rate that can be varied from 300 to 1100 rpm. The pH varies little when samples are stirred at 1000 rpm for 5 min in 7×45 mm cuvettes. Some manufacturers provide reusable, Teflon-coated stir bars that must be retrieved after each test. Others provide thinner, disposable, nickel-coated wire stir bars. The size and shape of the stir bar affect aggregation responses; thicker stir bars are more effective. In general bars that stir more effectively give greater aggregation, and particularly lower the concentration of ADP and epinephrine required to induce the second phase of aggregation.[40] Rough ends of cut wire also reduce the necessary concentration of epinephrine.[40] Some machines have an observation port to view the contents of the cuvettes.

Analysis of Aggregation Responses

Typical Responses to ADP. Figure 1 shows typical tracings of ADP-induced aggregation in citrated platelet-rich plasma at 37°. (Unless otherwise noted, concentrations represent final values.) In most aggregometers, the tracing shows considerable oscillation before the aggregating agent is added. This is due to the passage of the discoidal platelets across the light path (the "swirl"). Addition of the reagent causes brief obstruction of the light when the pipette or Hamilton syringe used to add the reagent is placed

[40] B. S. Coller and H. R. Gralnick, *Thromb. Res.* **8,** 121 (1976).

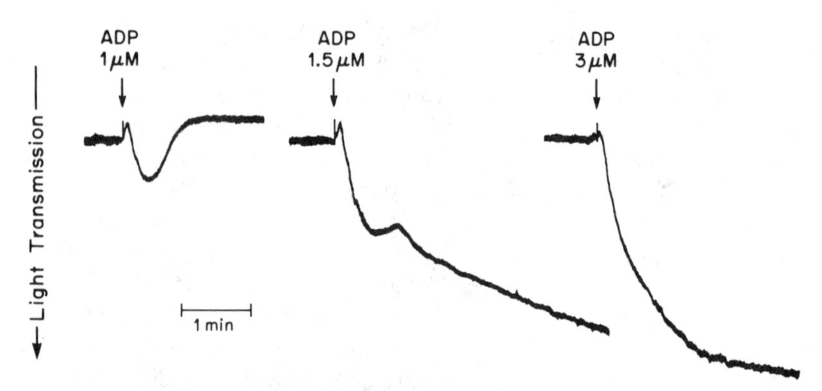

FIG. 1. Aggregometer recordings of platelet aggregation induced in citrated platelet-rich plasma by three concentrations of ADP. Note the prompt shape change, and the biphasic aggregation produced by a critical concentration of ADP.

in the suspension (as it should be when the volume of reagent is small). If a large volume of reagent is added, light transmission increases abruptly due to dilution. Next, light transmission decreases and its oscillation disappears due to the change in platelet shape from disks to spiny sheres. These changes may not be very noticeable because they are superseded by the increase in light transmission due to aggregation. If aggregation is prevented by adding EDTA before the ADP, the shape change is very clearly observed, especially if the gain on the aggregometer is increased. This is a recognized method of measuring platelet shape change.[41]

Light transmission increases progressively as aggregation begins, and reaches a plateau when aggregation is maximal. Even with maximal aggregation, light transmission never reaches the value recorded by platelet-poor plasma. When large aggregates pass the light beam, light transmission and hence the recording has large oscillations. If a reagent lyses the platelets, light transmission increases and the oscillations disappear. If secretion is not produced and the concentration of ADP is not too high, aggregation reverses because plasma contains enzymes that break down ADP. At a critical ADP concentration, usually 1 to 2 μM, two-phased aggregation is seen. The second phase is associated with secretion from the platelet dense granules. Both second phase aggregation and secretion depend upon the formation of endoperoxides and thromboxane A_2 from liberated arachidonic acid, and secretion of ADP from the dense granules. They are abolished by nonsteroidal antiinflammatory agents. With higher concentrations of ADP, the two phases of aggregation merge, and secretion must

[41] S. Holme and S. Murphy, *J. Lab. Clin. Med.* **96,** 480 (1980).

be detected by monitoring the appearance in the supernatant of nucleotides, Ca^{2+}, or radiolabeled serotonin from the dense granules, or of proteins from the α-granules (see Chapters [16] and [17] in this volume). ADP-induced secretion and hence two-phased aggregation fail to occur at room temperature[27,28] and require close platelet contact[5] as well as a low concentration of divalent cations.[2,17,21,22] Thus, if high concentrations of ADP are added to platelet-rich plasma without stirring, neither aggregation nor secretion takes place.[5]

Assessing Aggregation

The magnitude of aggregation as detected in an aggregometer can be assessed by measuring the maximum change in light transmission (T_{max}) or the maximum rate of aggregation (i.e., the tangent to the curve measured in millimeters or other units per unit time, termed also "slope value"). Under usual circumstances these two measurements correlate well.[15] Attempts have been made to interpret the curves according to mass action law.[15,20,42] However, it must be recognized that the aggregometer does not record the formation of small aggregates, which must be detected microscopically or by other means. Furthermore, consolidation of aggregates as well as their formation increases light transmission.[43]

Aggregation of platelet-rich plasma from patients is ordinarily compared to the mean values obtained from a number of samples of normal platelet-rich plasma with the same platelet count. However, platelet size as well as platelet count influences the response; samples of platelet-rich plasma with a high platelet count and small platelets have the same initial light transmission as samples with a low platelet count and large platelets. Furthermore, under the circumstances, the tracings after addition of an aggregating agent are similar and they would obviously differ if the samples were adjusted to the same platelet count.[44] This observation indicates that differing aggregation responses of two samples adjusted to the same platelet count do not indicate altered "stickiness" if the platelets in one sample are abnormally large (e.g., May–Hegglin anomaly, some types of thrombocytopenia) or small (some patients with reactive thrombocytosis).[44]

In clinical work, it does not seem necessary to test a normal sample every day. However, if no sample gives a normal response on a given day, the reagent(s) should be checked by testing normal platelet-rich plasma.

[42] S. Cronberg, *Coagulation* **3,** 139 (1970).
[43] G. V. R. Born and M. Hume, *Nature (London)* **215,** 1027 (1967).
[44] S. Holme and S. Murphy, *J. Lab. Clin. Med.* **97,** 623 (1981).

Whole-Blood Aggregometers

The whole-blood aggregometer was developed by the Wellcome Research Laboratories in England[45] and is manufactured by Chrono-Log under license from them. Use of whole blood eliminates preparation time before aggregation responses can be measured and permits use of smaller volumes of blood. Measurement depends on changes in electrical impedence as the platelets aggregate onto two electrodes immersed in the blood sample. When an electric current (15 kHz, 0.08 mW) is passed through the sample, the electrodes are immediately coated with a monolayer of platelets. If no aggregating agent is added, there is no further accumulation of platelets and conductance between the electrodes is constant. When an aggregating agent is added, however, platelets aggregate on the adherent platelets and impair conduction. This increase in impedence is displayed as a function of time on the recorder.

Several advantages of the whole-blood aggregometer are claimed.[45] Studies can be made in the presence of lipemia and obviously in the presence of red cells. However, the hematocrit strongly affects the results. Large platelets such as those present in the rare congenital disorder, Bernard–Soulier syndrome, are not lost from the sample as they are during preparation of platelet-rich plasma by centrifugation. Least valid is the argument that tests can be made before the decay of labile modulators of aggregation such as thromboxane A_2 and prostacyclin, because the concentration of these materials would be changing rapidly during the test.

When the impedance method is used on platelet-rich plasma, it is more sensitive than the optical method for detecting aggregation. Responses with collagen are somewhat different, possibly because the collagen particles adhere to the electrodes. A clear second phase of aggregation with ADP is not detected, but the response to arachidonic acid can be used to determine whether the patient has an "aspirin-like" defect.[46]

The Chrono-Log company manufactures machines that can measure impedance, light transmission, and luminescence output simultaneously on a single sample of blood or platelet-rich plasma. Measurement of ATP secretion and aggregation of a 5-ml sample of whole blood can provide rapid diagnosis of platelet disorders.[47]

[45] D. C. Cardinal and R. J. Flower, *J. Pharmacol. Methods* **3,** 135 (1980).
[46] C. Ingerman-Wojenski, J. B. Smith, and M. J. Silver, *J. Lab. Clin. Med.* **101,** 44 (1983).
[47] C. M. Ingerman-Wojenski and M. J. Silver, *Thromb. Haemostasis* **51,** 154 (1984).

Responses to Aggregating Agents

Volumes. Reagent volumes of 5–10 μl cause minimal changes in light transmission and are thus advantageous. They must be injected into the platelet-rich plasma, not applied to the wall of the cuvette.

Commonly Used Aggregating Agents: Sources and Effects on Citrated Platelet-Rich Plasma

Many platelet aggregation agents can be obtained through companies that supply aggregometers, but this is an expensive source for the reagents that can easily be prepared in the laboratory.

ADP. Effects of this reagent have been described. The secretion that occurs when the platelets aggregate at 37° requires the low concentration of Ca^{2+} in citrated plasma and does not occur with physiologic levels of ionized calcium.[17,21,22] It increases with increasing pH.[48] ADP can be made 10 m*M* in isotonic saline, stored in 0.1-ml aliquots at −15 to −20°, and diluted daily as desired.

Epinephrine. Epinephrine does not cause platelet shape change or marked primary aggregation. Secondary aggregation due to secretion, however, can be very marked, is accompanied by shape change, and is prevented by nonsteroidal antiinflammatory agents. Epinephrine is maximally effective in concentrations between 0.5 and 7.5 μ*M* in different subjects; increasing the concentration does not increase the magnitude of the primary or secondary response.[31] The secondary response increases with increasing pH[31,48] and with time up to about 45 min.[31] Injectable epinephrine (e.g., Parke-Davis 1 : 1000 or 5.5 m*M*) is a valuable source for this reagent because it contains a preservative that prevents rapid oxidation.

Thrombin. Thrombin is a potent aggregating agent that acts in several ways—first, by a primary effect, second (to a minor extent), by secretion through the cyclooxygenase pathway, and third, by another mechanism, possibly activation of protein kinase C by diacylglycerol formed from polyphosphoinositides. Concentrations below about 0.15 U/ml are ineffective because plasma contains thrombin inhibitors. At concentrations above about 0.25 U/ml, aggregation is followed by clotting; the latter is signalled by wild oscillations of the recording. Commercially available bovine thrombin preparations (Parke-Davis; Sigma), though impure and rich in calcium, affect human platelets essentially like purified human thrombin.

Collagen. Collagen does not induce primary aggregation, but platelet adhesion to collagen particles induces secretion. Most of the aggregation

[48] A. B. Rogers, *Proc. Soc. Exp. Biol. Med.* **139,** 1100 (1972).

response depends upon the products of the cyclooxygenase pathway and is therefore inhibited by nonsteroidal antiinflammatory agents. Aggregation occurs after a lag period during which stimulation and secretion are taking place.

Collagen is difficult to quantitate since only its native insoluble form is active, and the size of the particles varies. Concentrated, largely soluble preparations are available at low pH (Hormon Chemie, Munich, West Germany or Helena Laboratories) or can be prepared.[2] The collagen polymerizes into fibrils in the platelet suspension medium and the volume of acid added is so small that it does not affect the pH. The time required for polymerization contributes to the lag period before aggregation. An inexpensive stable product for routine use can be prepared from human subcutaneous tissue, obtained at surgery, stored frozen in small aliquots, and diluted many-fold for studying platelet aggregation.[49]

Arachidonic Acid. Arachidonic acid causes aggregation which depends entirely on its transformation by cyclooxygenase to endoperoxides and thromboxane A_2. It is, therefore, completely abolished by nonsteroidal antiinflammatory agents which inhibit cyclooxygenase. It is available from Nuchek (Elysian, MN) and from Sigma Chemical Company.

Potentiation. Combinations of aggregating agents may be much more effective than either one alone.[2] This potentiation is particularly noted with epinephrine. The mechanisms by which this occurs are not understood.

Reversibility of Aggregation

It is difficult to reverse aggregation of platelets that have undergone secretion. Apparently, several mechanisms are responsible for sustaining the aggregation; one is a lectin-like association between platelet-bound secreted thrombospondin and platelet-bound fibrinogen.[50]

Differences between Responses in Plasma and Buffer. The effect of many aggregating agents is different in washed platelets than it is in platelet-rich plasma. The response to ADP is usually minimal in washed platelets unless fibrinogen as well as apyrase have been added. Apyrase destroys the ADP that is continuously lost from the platelets, leading to a refractory state. The added apyrase causes ADP-induced aggregation to reverse rapidly unless secretion has occurred.

Washed platelets aggregate with lower concentrations of thrombin than platelets in plasma because thrombin inhibitors are absent. Washed platelets also require much lower concentrations of arachidonic acid than plate-

[49] M. B. Zucker and J. Borrelli, *Proc. Soc. Exp. Biol. Med.* **109,** 779 (1962).

[50] R. L. Kinlough-Rathbone, J. F. Mustard, D. W. Perry, E. Dejana, J. P. Cazenave, and E. J. Harfenist, *Thromb. Haemostasis* **49,** 162 (1983).

lets in plasma because some of the arachidonate binds to plasma albumin and cannot be converted to endoperoxides and thromboxane A_2 by the platelets. PAF-acether and a number of inhibitors of aggregation also bind to albumin.

Response in Species Other than Man

Platelets of most species of vertebrates do not aggregate with epinephrine although epinephrine can potentiate responses to other aggregating agents. Responses vary in different dogs; some do not react with thromboxane A_2. Guinea pig and cat platelets secrete in response to ADP-induced aggregation, whereas rabbit platelets do not.[2] Responses of rabbit and pig platelets to ADP and collagen have been compared.[51]

Congenital Abnormalities in Platelet Aggregation

Thrombasthenia. In this rare congenital disorder, platelets are unable to aggregate in response to any stimulus because they lack glycoproteins IIb and IIIa which form the fibrinogen receptor. They do, however, show the primary agglutination response with ristocetin in the presence of von Willebrand factor.

Secretion Defects. The most common cause of abnormal secretion is ingestion of aspirin. Patients with storage pool syndrome lack normal dense granules from which ADP can be secreted. Even rarer patients have a deficiency of the enzymes that convert arachidonic acid to thromboxane A_2 or other less clearly defined abnormalities. All of those exhibit defective aggregation with arachidonate, collagen, ADP, and epinephrine. Patients with the Grey Platelet Syndrome lack α-granules which contain the secretable proteins including fibrinogen, von Willebrand factor, and thrombospondin. This causes some abnormalities in aggregation.

Bernard–Soulier Giant Platelet Syndrome. These platelets aggregate normally except for a defect in the response to thrombin. They lack glycoprotein Ib, and hence do not bind or agglutinate with von Willebrand factor in the presence of ristocetin.

[51] A. Galvez, L. Badimon, J.-J. Badimon, and V. Fuster, *Thromb. Haemostasis* **56,** 128 (1986).

[11] Platelet Aggregation Measured *in Vitro* by Microscopic and Electronic Particle Counting

By MONY M. FROJMOVIC, JOHN G. MILTON, and ADRIAN L. GEAR

Definitions and Assay Principles

Here we describe the measurement of platelet aggregation by the use of optical[1,2] and electronic particle-counting methods.[3,4] These methods have proved particularly useful for measuring platelet aggregation in whole blood,[5] for investigations of the kinetics of early platelet aggregation,[1-3] and for quantitative–theoretical studies of platelet aggregation due to Brownian motion,[6] none of which can be directly and primarily measured by the more traditional light-scattering (photometric) methods used for assessing aggregation.[1,4] In the discussion that follows we have abbreviated platelet aggregation measured by particle counting methods as PA in order to distinguish it from aggregation measured turbidometrically ("photometrically") from the increase in light transmission through stirred platelet suspensions, TA (see chapter [10] in this volume).[7]

The methods described here measure the decrease in single platelet count or in platelet particle count as an indication of increasing platelet aggregation. No consideration is given to aggregate size. Following a lag time in the onset of aggregation of ~1 sec, PA is typically complete in 8–10 sec.[1-3] Both theoretical and experimental observations indicate that some biochemical and structural changes may occur in less than 1 sec following addition of activator.[2] Thus a major requirement of the procedure is that mixing times be short (less than 200 msec in present techniques) and that the method be able to cover the time scale from milliseconds to seconds. Two mixing procedures are described. In the first method platelet suspensions are mixed in an aggregometry cuvette using an aggre-

[1] M. M. Frojmovic, J. G. Milton, and A. Duchastel, *J. Lab. Clin. Med.* **101,** 964 (1983).
[2] J. G. Milton and M. M. Frojmovic, *J. Lab. Clin. Med.* **104,** 805 (1985).
[3] A. R. L. Gear, *J. Lab. Clin. Med.* **100,** 866 (1982).
[4] A. R. Nichols and H. B. Bosman, *Thromb. Haemostasis* **42,** 679 (1979).
[5] U. Siversten, *Thromb. Haemostasis* **36,** 277 (1976).
[6] H. N. Chang and C. R. Robertson, *Ann. Biomed. Eng.* **41,** 151 (1976).
[7] M. B. Zucker, this volume [10].

gometer as a stirrer, and aggregation stopped by addition of a glutaraldehyde solution. This technique allows an earliest measurement of ~500–750 msec.[1,2,8] Its main advantage is that the equipment required is generally available in most platelet laboratories. This facilitates comparison between PA and TA. Its main disadvantages are that it is not suitable for measurement of events occurring in less than 500 msec, that it cannot be scaled up for batch analysis due to the restraints imposed by mixing times, and that flow conditions, though turbulent, are not rheologically defined. The second method describes a rapid-mix, quenched-flow method, where shear forces (~30 dyn cm^{-2} ≡ shear rates of ~2000 sec^{-1}) in laminar flow of PRP or blood suffice to ensure efficient aggregation without causing shear-induced platelet activation or lysis.[3] This procedure allows for accurate measurement of platelet events at times as short as 40 msec, under flow conditions reasonably close to physiological and is readily scaled up for batch analysis.[3,8]

The second major requirement for an effective PA procedure is the ability to measure, at any moment in time, the number of single platelets or total platelet particles in the presence of ongoing aggregation. This is readily accomplished with optical methods using the hemacytometer since the observer can directly pick out both the single platelets and the aggregates which can be counted as individual particles. However, this procedure is tedious and hence not readily adapted to analysis of a large number of samples. The use of electronic particle counters offers speed of sample handling and in principle can be automated. However, with the use of this method care must be taken in the choice of size discriminator since it is possible that small aggregates containing two to four platelets may be misidentified as single platelets or large single platelets, counted as aggregates. These two methods are compared. Both methods give qualitatively similar results, though differences do exist, particularly in the first 1–3 sec when at least half or more of the aggregates formed are doublets and triplets. It is suggested that electronic particle counting methods are suitable for most purposes, and will generally reflect changes in platelet particle number with parameters of PA dynamics and sensitivity being very similar to those measured by total particle counts, rather than singlet platelet counts, using the hemacytometer. If quantitative data of an absolute rather than relative nature are required, then the hemacytometer method is to be preferred.

It is important to note that PA and TA do not measure the same aggregating process.[1] Properties which distinguish PA from TA include the following: (1) half-maximal extent of PA is attained at significantly lower

[8] A. R. L. Gear and D. Burke, *Blood* **60**, 1231 (1982).

activator concentration than TA for a wide range of activators; (2) PA is minimally refractory to ADP under conditions causing maximal refractoriness of TA; and (3) PA is less sensitive to inhibitors of platelet aggregation than TA. On the basis of observations of this type, it has been suggested that underlying platelet aggregation there are at least two distinct processes: (1) the recruitment of single platelets into small aggregates (measured as PA) and (2) the build-up of larger aggregates from smaller ones (measured as TA). Some recent examples supporting this concept include the demonstration that a monoclonal antibody directed toward the major α-granule protein of human platelets, i.e., thrombospondin, can inhibit TA by $\geq 80\%$ with considerably smaller effects on PA,[9] that PA remains normal in the absence of TA in thrombasthenia,[10] and that PA but not TA appeared distinctly different for male versus female donors.[11]

We may note that the more recent electrode method (whole-blood aggregometer) capable of measuring aggregation in whole blood appears to measure a process most similar to TA[12] with no relevance to PA measurements here under consideration.

In general, theoretical and experimental evaluations of aggregating colloidal particles including blood platelets have been focused on the earliest singlet recruitment into doublets, or more generally on the overall changes in particle number occurring with doublet to multiplet formation.[6] The methods for PA described below allow for both of these approaches.

Method I: Platelet Aggregation in Stirred Suspensions

Materials and Equipment

Stopwatch

Hamilton syringe: 10 μl in 0.2-μl divisions

Automatic pipet capable of quickly dispensing 0.4-ml solution

Stirring device: The geometry of the sample container can affect the frequency of cell collisions, shear rates, and mixing times. We use a Payton aggregometer as a stirrer. With this apparatus the mixing time for a 0.1-ml platelet suspension contained in a siliconized cuvette (6.9 × 45 mm) with stir bar (6 × 1 mm) spun at 1000 rpm, 37°, has been estimated to be less than 200 msec. Increasing the volume of the platelet suspension in the cuvette significantly

[9] L. L. K. Leung, *J. Clin. Invest.* **74,** 1764 (1984).

[10] S. Heptinstall, W. M. Burgess, S. R. Cockbill, S. C. Fox, and T. Sills, *Thromb. Haemostasis* **50,** 216 (1983).

[11] M. M. Frojmovic, J. G. Milton, and A. Duchastel, *Thromb. Haemostasis* **50,** 36 (1983).

[12] D. C. Cardinal and R. J. Flower, *J. Pharmacol. Methods* **3,** 135 (1980).

lengthens the mixing time (~0.4 sec for 0.4 ml; ~5 sec for 0.8 ml). If a different stirring device and cuvette is utilized it is necessary to measure the mixing times by either thermal quenching or use of a colored dye

Fixation solution: Four volumes of 0.8–1.0% (v/v) glutaraldehyde in Ca^{2+}- and Mg^{2+}-free Tyrode's solution, pH 7.4, is added to the reaction mixture at 37°. Lower concentrations (<0.01%) are to be avoided since these can directly induce significant aggregation.[13] This solution should be freshly made from the highest grade of glutaraldehyde available. Occasionally the use of less than 4 vol glutaraldehyde solution results in the platelet–glutaraldehyde mixture assuming a gellike consistency. The use of greater than 4 vol glutaraldehyde offers no advantages, though 5 vol may be convenient and acceptable

Solutions

Diluting solutions: Modified Tyrode's or 0.9% (w/v) NaCl or 0.154 *M* NaCl

Glutaraldehyde: E. M. grade, 8% aqueous stock in sealed ampoules (Polysciences, Inc., Wavington, PA), or Sigma grade II, or Kodak, 50% aqueous stock

Adenosine diphosphate: Sigma or Pharmacia P-L Biochemicals, disodium salt

(ADP): 50 m*M* stock, pH 7.4 (kept at −20 to −40°)

Epinephrine: Bitartrate salt; 10 m*M*, made fresh (Sigma)

Procedure (for Stirred Suspensions, Method I)

1. One-tenth milliliter of platelet suspension (platelet-rich plasma, washed platelets, whole blood) is added to the cuvette and stirred at 1000 rpm, 37°, in an aggregometer. A sample of the platelet suspension is fixed with the glutaraldehyde solution and is used to determine the initial platelet count and the platelet aggregation present prior to the addition of activator [%PA = 4 ± 2% (1 SD) in citrated PRP for 34 healthy donors].

2. At zero time the solution containing the activating agent (freshly diluted with Tyrode's solution; 1–10 μl) is rapidly added to the stirred platelet suspension using a Hamilton syringe. In the case of citrated whole blood, stirring must be initiated ≤10 sec before addition of activator to avoid time-dependent stirring-induced PA.

[13] J. A. Zeller, R. E. Dayhoff, K. Eurenius, W. Russel, and R. S. Ledley, *Clin. Lab. Haematol.* **6**, 145 (1984).

3. At time *t*, aggregation is stopped by quickly adding 0.4 ml of the glutaraldehyde solution; the stirring is simultaneously stopped by removing the cuvette from the aggregometer. Mixing is ensured by the addition of the relatively large volume of fixing solution to the small volume of the platelet suspension.

4. Platelet suspensions were prepared for counting within 2–3 hr of fixation. Otherwise, they were diluted 4- to 5-fold with modified Tyrode's solution (i.e., 16 to 20 cm^3 of Tyrode's) to prevent further aggregation, which tended to occur on allowing the undiluted samples to stand (e.g., ~10% further aggregation overnight). Handled in this way, samples could stand at least 24 hr without influencing the extent of platelet aggregation. NaCl (0.05 *M*) may replace modified Tyrode's solution as the diluting solution.

5. Single platelet counts are determined by either the use of the hemacytometer or electronic particle counter and the data analyzed as described in the section below, Platelet Counting.

Method II: Platelet Aggregation in Laminar Flow (Quenched Flow)

Principle

This approach is an adaptation[3] of stopped flow or, more specifically, quenched-flow techniques previously developed to measure rapid enzymatic reactions.[14,15] The basic idea is shown in Fig. 1. Separate solutions containing platelets and activating agents are rapidly mixed and then pumped down a narrow-diameter tube of precise length. Addition of fixation solution stops the reaction. The reaction time is determined from the relation

$$\text{Reaction time} = \text{tube length/flow rate} \qquad (1)$$

Materials and Equipment

Variable-speed syringe pump, e.g. Harvard No. 944 or 935 (Harvard Apparatus Co., South Natick, MA)

Syringes: Commercial, disposable plastic syringes are appropriate; e.g., 1- to 5-ml capacity for PRP, inducer and aldehyde quenching agent. In this configuration, only three syringes of equal volume are needed. Alternatively, two 60-ml syringes in parallel can be set up on the pump. This provides massive dilution quenching for on-line

[14] Q. H. Gibson, this series, Vol. 16, p. 187; see also H. Gutfreund, this series, Vol. 16, p. 229.

[15] A. Fersht, *in* "Enzyme Structure and Mechanism," p. 103. Frieman, San Francisco, California, 1977.

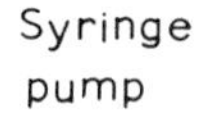

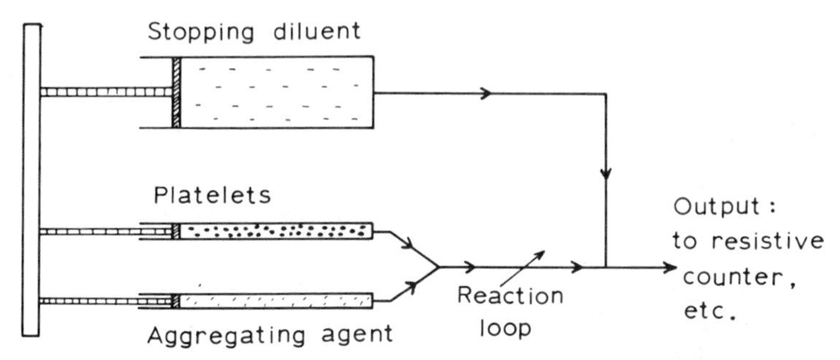

FIG. 1. Basic principle of quenched-flow aggregometry as described in method II. A platelet suspension is placed in one small plastic syringe, and an aggregating agent in another syringe. A dual-drive syringe pump forces these solutions through 0.8-mm i.d. Teflon tubing before entering a common reaction tube of 0.2- to 0.4-mm i.d. The extent of reaction is influenced by both the tubing length and pumping rate. The reaction may be stopped, for example, by massive dilution with 0.9% NaCl, if the diluted reaction mixture is to be continuously analyzed for platelet numbers and volume by a resistive-particle counter. Otherwise, glutaraldehyde is used to stop the reaction for batch-type counting (see method II).[3]

analysis of aggregation to the resistive-particle counter.[15] Employing 1-ml "milk pipets" (Falcon Plastics, Los Angeles, CA) as a syringe barrel with custom, stainless steel plungers and double O rings, and two 60-ml diluting syringes, the final dilution of PRP to the resistive counter becomes 190-fold.[3] The syringes are attached to the tubing (Fig. 1) via standard Luer and screw couplings (e.g., Altex Sci. Co., Berkely, CA; Durrum, Palo Alto, CA)

Tubing: Teflon tubing is used (same suppliers as couplings and T junctions) with 0.8-mm i.d. serving for inlet and outlets from the reaction loop which is assembled from 0.3-mm i.d. commercial microbore Teflon tubing. The reaction loops can be constructed either in a "resting" or "stretched" mode (0.30- and 0.25-mm i.d., respectively) according to the suppliers' instructions. The exact internal diameter is determined by calibration with a microliter syringe.

T junctions: Commercial ones may be used, but it is recommended to construct ones with minimal dead volumes out of nylon or appropriate plastic. The T going to the reaction loop should be drilled to 0.3-mm i.d., the other inlets are 0.8-mm i.d.

Water bath: The whole syringe pump and tubing assembly can be placed in a heated, hot-air box; or, alternatively the reaction loop and at least 10 cm of the inlet tubes to the first T junction are submerged in a small beaker kept at 37° by a circulating water bath. Reaction loop temperature is controlled to within 0.2°

Procedure

1. Different reaction times are obtained by varying the tube length and/or pump speed [see Eq. (1)]. The reaction time which works best for ADP as the activating agent ranges from 1.5 to 8 sec. For epinephrine, reaction times up to 10 sec are required.

2. The concentration of activating agent is twice that desired in the reaction loop since there is a 1:1 dilution of the platelet suspension on entering the loop.

3. The stopping diluent is either 2% glutaraldehyde (made up in 0.15 *M* NaCl) or 0.15 *M* NaCl. Two 60-ml syringes in parallel provide a 190-fold dilution (as described above). The aldehyde is employed for batch-type counting and the 0.15 *M* NaCl in the large syringes (which may contain aldehyde) for continuous-flow counting.[16]

4. For batch analyses, using glutaraldehyde as quencher, normally about 100 μl is collected from the reaction loop at each reaction time. Aliquots of 60 μl are then diluted with 10 cm^3 0.15 *M* NaCl to prevent plasma protein polymerization which may otherwise occur and result in electrical interference for electronic counting.

5. When on-line, continuous-flow analyses[3] are to be carried out, a control run is carried out first in the absence of activating agent to check for background noise. This may come from the diluting solutions. Commercial "saline for irrigation" is usually sufficiently free from particles; laboratory-made saline may need to be filtered to lower background counts. In the case of on-line analysis there may be electrical interference associated with the tubing connected to the electronic particle counter. Simple electrical grounds or use of the C-option (Particle Data) may be necessary.

6. The basic procedure is to slow the syringe pump stepwise, obtaining at each speed two or three steady 10-sec counts (continuous flow) or sufficient collected effluent (about 100 μl) in the batch procedure. A complete reaction profile should not take more than 5 min.

Quenching Problems

As previously pointed out,[3] dilution with saline or buffer rather than glutaraldehyde fixation is appropriate only if samples are observed directly

[16] A. R. L. Gear, *Anal. Biochem.* **72,** 322 (1976).

on-line, i.e., within 10 sec from aggregation onset to observation. Indeed, in method I (stirred suspension studies) it was found that PA at early times (1–5 sec) and moderate ADP concentration ($<2\ \mu M$) was reduced by 1.5- to 3-fold if samples were diluted with saline/buffer rather than with glutaraldehyde, even though observations with the resistive counter were made within a minute of activator addition. Considerably less platelet disaggregation was observed with longer reaction times and/or higher activator concentrations (conditions yielding $\geq 80\%$ PA).

Platelet counting

Hemacytometer Counting: Singlet Platelets (H_s) and Platelet Particles (H_p)

Materials and Equipment

Hemacytometer: Preferably one designed for use in conjunction with phase contrast microscopy, e.g., Neubauer improved hemacytometer (Fein-Optick, Blankenburg, GDR)

Humid chamber

Microscope equipped for phase contrast and/or differential interference contrast (DIC) microscopy. We have found that DIC is the most convenient method; however, with the use of proper lighting even bright-field microscopy can be used, although this is less preferable

Procedure

1. Samples for counting are applied to the hemacytometer in the usual manner and allowed to stand for 20 min to ensure platelet settling. To prevent evaporation from the samples over this time the hemacytometer is placed in a humid chamber.

2. Typically the platelet suspensions have undergone a 20- to 25-fold dilution prior to counting (see, for example, method I) and give 300 to 500 particles in 5/25 squares. Standard error in mean values for a given sample was $<7.4\%$, whereas variations between two different fixed samples from the same platelet suspension was typically $<3\%$ from the mean.

3. The number of singlet platelets is normally counted when estimating the properties for recruitment of singlets into aggregates (H_s method).[1] However, for primary studies of changes in total platelet particle number, doublets to multiplets are counted as single particles in addition to singlet platelets (H_p method—see below). The percentage of platelets in PRP present as aggregates prior to activator addition is $14 \pm 6\%$ ($n = 7$ donors).[1]

Electronic Particle Counting

Materials and Equipment

Electronic (resistive) particle counter (Particle Data, Inc., Coulter Electronics, Clay-Adams Co., etc.), possessing lower and upper discriminators
Sensing apertures: 30–70 μm with 50 μm being optimal
Counting vials: 20 cm^3, polystyrene, disposable
Diluent: 0.15 *M* NaCl or Hematall (Fisher Scientific; azide-free isotonic diluent: 0.15 *M* NaCl; 0.3 m*M* KCl; 15 m*M* PO_4, pH 7.5)
Repeater pipet (manual or automated) for 10.00-ml delivery

Procedure. The electronic (resistive) particle counter detects single platelets and particles of low-order aggregates by selection between lower and upper particle size discriminators. Depending upon gain and current settings, the upper discriminator is set such that greater than 50% of tetramers will be excluded from being sensed. The lower size discriminator is set to exclude particles smaller than about 1.8 fl. Consequently platelets counted by this method include the low-order aggregates: doublets, triplets, and some quadruplets.

Using a 48-μm orifice and 0.15 *M* NaCl or Hematall as blank, background counts should be less than 200/10 sec of analysis time (~50- to 100-μl samples). Counts with platelet suspensions (PRP or washed platelets) should be $\leq$7000/100 μl counted in 13 sec for stirred suspensions studied (see method I) or about 20,000 to 60,000/10 sec for quenched-flow method (see method II). If whole-blood platelet aggregation is being studied, then most of the erythrocytes must be removed by brief centrifugation,[3] or a sheath-flow counter such as the Ultra-Flo 100 (Clay-Adams Co.) must be employed.

Once the counts have been carried out, they are corrected for coincidence for particle count $\geq$7000/~10 sec. Dilutions in the 100- to 500-fold range normally provide enough counts without excessive coincidence corrections (the corrected count should not normally exceed twice the observed particle count). In the case in which platelet aggregation is being monitored in whole blood, accurate coincidence corrections for erythrocytes is essential.[3] Use of the total particle count, including erythrocytes, will provide the necessary correction for platelets being sensed in the platelet "window." Smaller diameter sensing orifices suffer less from the need for major coincidence corrections.

Data Analysis

Five parameters are typically used to characterize PA: %PA, onset time, maximal rate, maximal extent, and concentration of activator required to induce one-half maximal rate and/or extent, effective concentration (EC_{50}) for a given activator, e.g., ADP. Such concentration is also expressed as $[Act]_{1/2}$ in published reports from our laboratory.

The percentage of platelets aggregated at time t, %PA, is equal to

$$\text{Percentage PA} = [1\text{-}(N_t/N_0)](100) \tag{2}$$

where N_t, N_0 are, respectively, the singlet (H_s) or total particle (H_p) platelet count at time t and at time zero following addition of activators for hemacytometer counting: and, respectively, particle counts of "singlets" (including any doublets, triplets, and some quadruplets which are present) at times t and zero for the resistive particle counting technique (method II).

The onset time (t_0), initial rate (V), and maximal extent of PA (PA_{max}) are readily obtained from a plot of %PA (or N_t) versus time as shown in Fig. 2. The onset time (t_0) can be estimated from inspection of Fig. 2, or alternately in the case when the rate of aggregation is sufficiently high, as the point where the tangent to the maximum rate of aggregation, projected backward, intersects the 0% axis. The onset time is thought to reflect the occurrence of a Ca^{2+}-independent platelet activation step.[2] When the electronic counter is utilized to estimate the single platelet count, the onset time also includes the time required to form aggregates large enough (e.g., tetramers) not to be counted as singlets. Experience gained from the kinetics of consecutive chemical reactions emphasizes that the absolute value for the duration of the onset time has no quantitative significance, but rather is an indication of method's sensitivity.[17] However, the actual existence of an onset time is significant.

The maximum extent of PA is not 100%. When ADP is the activating agent ~10–20% of the platelets never aggregate and with epinephrine in the absence of released ADP, ≥40–50% of the platelets do not aggregate.[1-3]

In general, PA versus time is linear from ~1 to 3 sec (Fig. 2) and hence the initial rate V is readily determined.[1,3] For method I studies with ADP as activator, V is routinely measured from PA at 3 sec (PA_3) and maximal extent of aggregation from PA at 10 sec (PA_{10}). Maximal values for PA_3 and PA_{10} were routinely observed for ≥10 μM ADP. For method II, V is actually determined from the initial linear slope (linear in first 2–3 sec). V is expressed as the percentage platelet singlets aggregating per second relative to the control value. For method II, V can be normalized to a platelet count of 200,000 μl^{-1}. However, it has been observed in method I studies that PA_3 is not a simple linear function of initial platelet counts, so it is best to establish the V_{max} concentration dependence for distinct platelet suspensions and mixing conditions.

[17] T. M. Lowry and W. T. John, *Chem. Soc. Trans.* **97,** 2634 (1910).

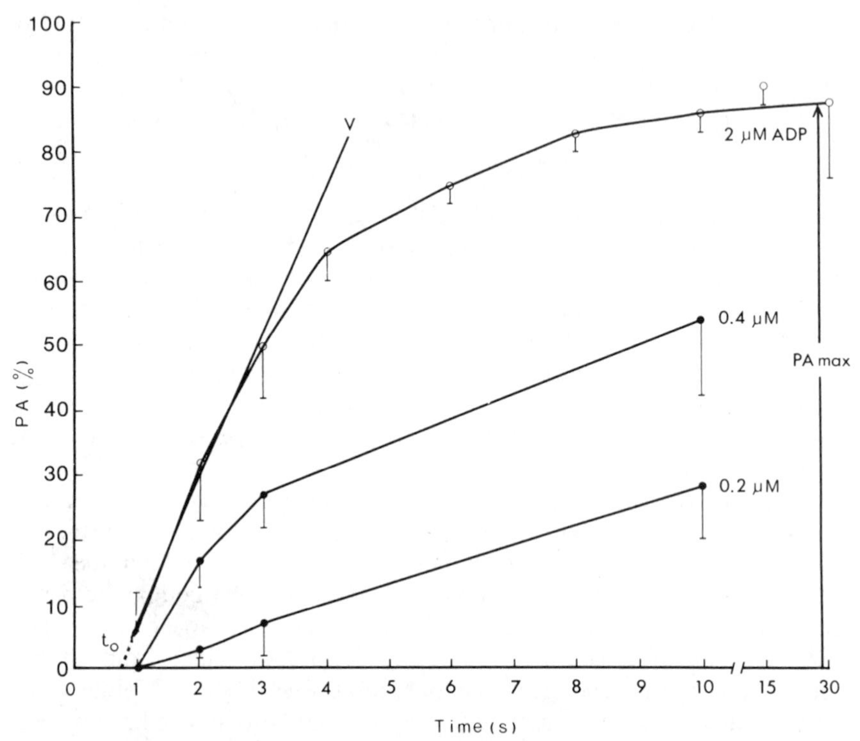

FIG. 2. Kinetics of microscopically measured ADP-induced platelet aggregation (PA). PA_t was determined as a function of time following ADP addition to stirred citrated PRP using singlet platelet counting with the hemacytometer (H_s) and calculated from Eq. (2). Values are shown for 12 donors (6 males and 6 females) determined at 2 μM ADP (○) and for 4 of the 12 donors (●) for 0.4 and 0.2 μM ADP. One-half of standard deviation bars are shown in all cases. t_0, V, PA_{max} are onset time, initial rate and maximal extent for PA, respectively. $N_0 = 339-687 \times 10^3\ \mu l^{-1}$; initial hematocrits = 42–48%.

The sensitivity of platelets (or platelet reactivity) in response to an agonist leading to early platelet recruitment (PA) can be determined from log dose–response curves as shown in Fig. 3. The law of mass action, as developed for drug–receptor interaction, is valid when applied to evaluations of platelet aggregation.[1] Thus, aggregation runs are conducted at a number of inducer concentrations covering the range of V and PA, generating a family of curves as shown for selected ADP concentrations in Fig. 2. The normalization of V relative to V_{max} (or PA relative to PA_{max}), and plotting against log ADP allows a direct read-out of EC_{50} (Fig. 3). In method II, analyses have been made from double-reciprocal plots of velocity of PA(V) against activator concentration, yielding apparent $K_x \equiv EC_{50}$

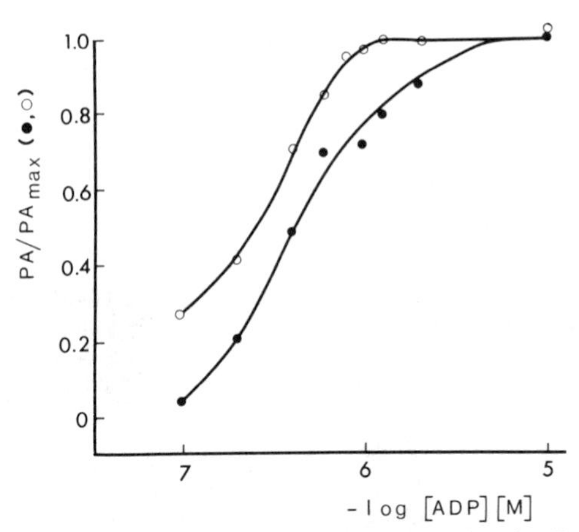

FIG. 3. Dose–response curves for ADP-induced aggregation for TA and PA. Typical curves are shown for relative changes in PA_3 (●) and PA_{10} (○) for citrated PRP, determined by singlet platelet counting with the hemacytometer (H_s).[1]

and V_{max} values.[3,18] In method I it has been found that the simplest analysis shown in Fig. 3, using PA_3 to reflect V, gives essentially identical results as for more sophisticated analyses using the double-reciprocal plot to determine maximal initial velocities and a Hill-type plot to determine EC_{50} ($\equiv K_x$). In addition, studies using PA_2 (aggregation at 2 sec) also yielded identical EC_{50} (K_x values). It should also be noted that EC_{50} values are independent of N_0 (initial platelet counts) over the normal range of N_0 studied for platelet suspensions ($\sim 200-600 \times 10^3\ \mu l^{-1}$), precluding any normalization requirements.

Comparison of Studies of Aggregation in Stirred Suspensions for Counting with Hemacytometer vs Electronic (Resistive) Particle Counter

It is seen from the results in Table I that EC_{50} values for both ADP-induced PA_3 (rate) and PA_{10} (extent) for early platelet recruitment are quantitatively distinct for the two methods, with significantly higher values observed with electronic counter measurements than for singlet platelet counting with the hemacytometer. This apparently arises because doublets, triplets, and many quadruplets are being counted as "single particles" by the electronic counting method. In this regard, it is instructive to directly compare the kinetics and ADP sensitivities for PA measured by electronic particle counting (E) and by singlet platelet counting with the hemacyto-

[18] V. M. Haver and A. R. L. Gear, *Thromb. Haemost.* **48,** 211 (1982).

TABLE I
EC_{50} FOR ADP USING HEMACYTOMETER (H_s)[a] VS ELECTRONIC (RESISTIVE) COUNTER (E)

Apparatus	EC_{50} for PA_3	EC_{50} for PA_{10}
Hemacytometer	0.68 ± 0.05[b]	0.51 ± 0.04 (7, 7)[c]
Resistive counter	1.00 ± 0.10	0.61 ± 0.06 (8, 11)

[a] H_s, Singlet platelet determinations (see Platelet Counting).

[b] Mean ± standard error in mean.

[c] Numbers in parentheses indicate the number of donors (male only) and the number of observations. The same donors were observed for PA_3 and PA_{10} for a given method but not necessarily for the different methods.

meter (H_s), with PA measured by total platelet particle counting with the hemacytometer (H_p), i.e., where multiplets are counted as single particles. Such a comparison, made in parallel for each of the three counting methods for any given donor, is summarized in Table II. Estimates of EC_{50} values for both ADP-induced PA_3 and PA_{10} obtained from measurements of particle number per unit volume by electronic (E) or by hemacytometer (H_p) counting yield on average essentially identical results, though some scatter is observed for PA_3, also shown in Fig. 4 for linear regression analyses of all the data points. This approximate identity for E and H_p also holds for PA_{max} values (see Table II). Comparisons of ADP-induced changes in E and H_s (singlet platelets measured with the hemacytometer), show ~20 and ~10% higher EC_{50} values but comparably lower PA_{max} values for E in measurement of PA_3 and PA_{10}, respectively (Table II). Therefore, it can be expected that electronic counting (E) will closely reflect changes in platelet particle number, distinct from changes in singlet platelet counts (H_s) measured directly with the hemacytometer. These differences (E vs H_s) may vary especially where early shifts in small multiplet distributions are occurring. Such varying quantitative differences between EC_{50} values derived from E vs H_s have been found when comparing PRP samples from eight male vs five female donors, with the latter appearing most sensitive to ADP by method E (1.0 ± 0.1 μM vs 0.59 ± 0.08 μM, respectively, for male vs female donors), in contrast to method H_s (0.68 ± 0.05 μM vs 0.54 ± 0.04 μM).

TABLE II
EC_{50} AND PA_{max} VALUES FOR ADP-TREATED SAMPLES IN ELECTRONIC (E) VS HEMACYTOMETER SINGLET (H_s) AND TOTAL PARTICLE (H_p) COUNTS

Measurements for	Mean values[a] from electronic counts (E)	Relative values[b] for	
		E/H_p	E/H_s
PA_3 ($n = 7$)[c]			
EC_{50} (μM)	0.7 ± 0.2 (0.5–0.9)	1.0 ± 0.1++ (0.8–1.1)	1.2 ± 0.3* (0.8–1.7)
PA_3 (maximum)[d](%)	49 ± 4 (45–54)	1.00 ± 0.04++ (0.86–1.04)	0.75 ± 0.07** (0.66–0.82)
PA_{10} ($n = 5$)			
EC_{50} (μM)	0.4 ± 0.2 (0.2–0.7)	0.94 ± 0.06* (0.75–1.00)	1.1 ± 0.1* (1.0–1.2)
PA_{10} (maximum)[d](%)	77 ± 5 (72–85)	0.97 ± 0.05+ (0.92–1.03)	0.89 ± 0.03** (0.85–0.92)

[a] Mean ± SE for pooled donors; range of values shown in parentheses.
[b] These were calculated for each donor and then pooled. Significance for differences between pairs of values for each donor were calculated using a Student's t test, shown as **, $p < 0.01$; *, $p < 0.1$; +, $p < 0.2$; ++, $p < 0.4$, respectively.
[c] Number (n) of different donors evaluated in parallel for all three parameters (~1 : 1 male and female).
[d] PA (maximum) values obtained at 10 μM ADP.

Blood and Platelet Preparations: Effects of Anticoagulants

Human blood collected by standard phlebotomy[1,3,18]
Trisodium citrate: 3.8 g/100 ml H_2O
Acid–citrate–dextrose (ACD): 12 g glucose, 25 g trisodium citrate, and 8.8 g citric acid per liter of solution
Heparin: 1000 U/ml from porcine intestine

Human venous blood is collected into polypropylene tubes normally containing either sodium citrate (1 vol 3.8% citrate plus 9 vol blood: final citrate ≈ 13 mM, with adjustments possible for hematocrit variations) or ACD (1 : 10/11). Heparin can also be used (1.5 U ml^{-1} final), but it is not recommended because it may induce significant aggregation depending on source and batch. The citrated whole blood may be studied directly in methods I and II, with studies best made within 10–30 min of blood collection. Between 30 min and about 3 hr, aggregation rates may be apparently abnormal, while between 3 and 5 hr they seem stable with less loss in reactivity.[3] Platelet-rich plasma (PRP) is isolated after low-speed

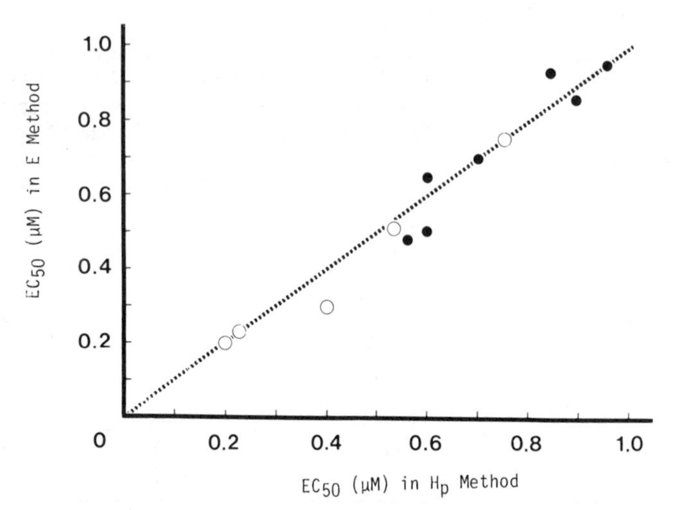

FIG. 4. A comparison of EC_{50} values determined for PA_3 and PA_{10} for electronic counting (E) vs hemacytometer total particle counting (H_p). Donors and averaged data are summarized in Table II. Data is shown for PA_3 (●) and PA_{10} (○). The theoretical dotted line is for a slope = 1.0.

centrifugations by standard procedures,[1,3,19] gassed with 5% CO_2–95% air to preserve pH and function, stored at room temperature or 37° (if platelet morphology is to be preserved),[20] and normally used within 4 hr without major loss in aggregability. Indeed, major loss in aggregability may occur as measured turbidimetrically[1] (see also this volume [10])[7] with little change in kinetics or sensitivity for PA.[1]

It is important to note that different anticoagulants can yield different kinetics and platelet sensitivities to aggregating agents such as ADP. Thus, EC_{50} ($\equiv K_x$) is ~10–30× and 2–3× smaller, i.e., more sensitive, respectively, for heparinized PRP and citrated PRP, than for ACD-PRP (see Table III).

The calcium concentrations in the platelet suspensions can be decreased with EDTA (added preferably a few seconds[21] before activator addition) or citrate, while $CaCl_2$ can be added just prior to studies[1-3] without concern for clotting complications in the presence of fibrinogen in cases where early kinetics ($\leq$10 sec) are being studied.

[19] M. M. Frojmovic, *Prog. Hemost. Thromb.* **4,** 279 (1978).

[20] M. M. Frojmovic and J. G. Milton, *Physiol. Rev.* **62,** 185 (1984).

[21] L. Skoza, M. B. Zucker, Z. Jerushalmy, and R. Grant, *Thromb. Diath. Haemorrh.* **18,** 713 (1967).

TABLE III
TYPICAL ADP SENSITIVITIES FOR PRP FOR DIFFERENT ANTICOAGULANTS

Anticoagulant	EC_{50} for	
	Stir (method I)[a]	Quenched-flow (method II)[b]
ACD	2.5 ± 0.5 (2)[c]	1.8 ± 1.0 (5)
Citrate	0.6 ± 0.0 (2)	0.9 ± 0.4 (5)
Heparin	0.12 ± 0.07 (2)	0.32 ± 0.16 (5)

[a] Determined for PA_3 for the same two donors for all three anticoagulants, using a hemacytometer and singlet counts (H_s).
[b] Determined from V (see Fig. 2) using the resistive counter.
[c] Numbers in parenthesis refer to number of donors studied.

Acknowledgments

We are grateful to the Medical Research Council of Canada (Grant 248-59 for M.M.F.) the Canadian Heart Foundation (M.M.F and J.G.M), and the National Institute of Health (U.S.) (Grant HL27014 for ALR).

[12] Fixation Platelets and Platelet Agglutination/Aggregation Tests

By K. M. BRINKHOUS and MARJORIE S. READ

For quantitative assays of the rate of platelet aggregation and the study of factors that determine this cell-to-cell interaction, standard platelet preparations are needed. For many platelet aggregation tests, isolated washed platelets are prepared daily by differential centrifugation to separate them from plasma and other blood cells. Variation in the reactivity of such preparations may be considerable. In a search for a stable platelet preparation that would avoid many of the problems inherent in the use of fresh platelets, Allain *et al.*[1] found that appropriate fixation of platelets with a common histopathologic fixative, the polymer of formaldehyde, paraformaldehyde, yielded preparations highly suitable for use in platelet aggregation tests dependent on plasma von Willebrand factor (vWF). In

[1] J. P. Allain, H. A. Cooper, R. H. Wagner, and K. M. Brinkhous, *J. Lab. Clin. Med.* **85,** 318 (1975).

these initial studies, fixed human platelets were used in two different types of vWF tests. One was an assay for vWF in human plasma using ristocetin, a vWF-dependent platelet-aggregating reagent. The other was an assay for vWF in bovine plasma, in which the animal plasma directly aggregates fixed human platelets without ristocetin. This property of vWF of certain animal plasmas is also known as the platelet-aggregating factor (PAF). The utility of fixed human platelets as a reagent for vWF testing was quickly confirmed.[2,3] In our studies, paraformaldehyde as a fixative appeared to give more reactive fixed platelets than those fixed with either unpolymerized formaldehyde or glutaraldehyde. Fixed human platelets were found to be likewise effective for vWF testing using a snake venom activator, venom coagglutinin (VCA),[4] also known as botrocetin. Unlike ristocetin or PAF testing for which human platelets are needed, botrocetin testing can be done with either human or animal platelets and with plasmas of many animals.[5] Fixed platelets may be frozen or lyophilized for long-term storage.[6]

Fixed platelets have been used in other types of test procedures besides vWF assays. Antibody inhibitors of vWF-dependent platelet aggregation have been recognized in two clinical situations: (1) in some multitransfused patients with homozygous von Willebrand disease and (2) in the autoimmune disorder, acquired von Willebrand syndrome. Tests for screening for these inhibitors and for quantitation of antibody titers using fixed platelets have been reported.[7,8] Fixed platelets are also used for detection of platelet agglutinins in autoimmune thrombocytopenic purpura.[9] Another use of fixed platelets is in mixtures with fresh platelets to study mechanisms of platelet aggregation.[10,11] In studies with a mixture of fresh and fixed platelets, enhanced platelet aggregation has been observed, in which the fixed platelet contribution may be dependent on the presence of fresh platelets. Fixed platelets are not suitable for aggregation reactions

[2] E. P. Kirby and D. C. B. Mills, *J. Clin. Invest.* **56,** 491 (1975).

[3] D. E. Macfarlane, J. Stibbe, E. P. Kirby, M. B. Zucker, R. A. Grant, and J. McPherson, *Thromb. Diath. Haemorrh.* **34,** 306 (1975).

[4] M. S. Read, R. W. Shermer, and K. M. Brinkhous, *Proc. Natl. Acad. Sci. U.S.A.* **75,** 4514 (1978).

[5] M. S. Read, J. Y. Potter, and K. M. Brinkhous, *J. Lab. Clin. Med.* **101,** 74 (1983).

[6] K. M. Brinkhous and M. S. Read, *Thromb. Res.* **13,** 591 (1978).

[7] K. M. Brinkhous and H. R. Culp, *Thromb. Res.* **8** (Suppl. 2), 125 (1976).

[8] W. A. Fricke, K. M. Brinkhous, J. B. Garris, and H. R. Roberts, *Blood* **66,** 562 (1985).

[9] A. E. G. Kr. von dem Borne, F. M. Helmerhorst, E. F. van Leeuwen, H. G. Pegels, E. von Riesz, and C. P. Engelfriet, *Br. J. Haematol.* **45,** 319 (1980).

[10] T. K. Gartner, J. M. Gerrard, J. G. White, and D. C. Williams, *Nature (London)* **289,** 688 (1981).

[11] G. Agam and A. Livne, *Thromb. Haemostasis* **51,** 145 (1984).

dependent on energy metabolism or the release reaction of the platelet. Fixed platelets are not aggregated by ADP, epinephrine, collagen, or thrombin, but are aggregated by poly(L-lysine).[1]

In this chapter we describe our methods for preparation of fixed platelets and for the preparation of botrocetin, as well as our various tests in which they are used for assay of vWF and for detection and titrating of vWF inhibitors. We use the general term platelet aggregation for these tests. The term platelet agglutination is often used for those aggregation reactions of fresh platelets that also occur with fixed platelets and do not require any metabolic capacity on the part of the platelets.[12]

Preparative Procedures

1. *Fixed Platelets*

Materials

50-ml round-bottom centrifuge tubes
Disposable pipets
Centrifuge
Phase microscope
Hemacytometer

Tubes and pipets should be made of plasticware such as polycarbonate, polypropylene, or other types to which platelets do not adhere.

Reagents

Acid–citrate–dextrose (ACD): 0.084 *M* sodium citrate, 0.064 *M* citric acid, and 0.11 *M* dextrose, pH 4.5
Citrate–saline: 0.0054 *M* sodium citrate, 0.146 *M* NaCl, pH 6.8, adjusted with 1.0 *N* HCl
Paraformaldehyde, 1.8%: 9 parts of stock solution of 4% paraformaldehyde in 0.135 *M* NaH_2PO_4, pH 6.8, 1 part ACD, and 10 parts 0.135 *M* NaH_2PO_4
Imidazole-buffered saline: 0.084 *M* imidazole, 0.154 *M* NaCl, pH adjusted to 6.8 with 1.0 *N* HCl
Paraformaldehyde, 0.8%: Prepared by mixing 9.0 ml of 1.8% paraformaldehyde, 1.0 ml ACD, and 10.0 ml 0.135 *M* NaH_2PO_4
Imidazole buffer: 0.084 *M* imidazole dissolved in distilled water and the pH adjusted to 7.35 with 1.0 *N* HCl

[12] E. P. Kirby, *Thromb. Haemostasis* **38,** 1054 (1977).

Fixed human platelets[1] are prepared from blood drawn directly into anticoagulant (7.5 parts ACD to 42.5 parts blood). The blood is divided into 25-ml aliquots in 50-ml round-bottom tubes and centrifuged for 8 min at 200 *g*. The platelet-rich plasma (PRP) is removed for processing. All procedures are carried out at 25°. The PRP is centrifuged for 3 min at 200 *g* to remove the remaining red blood cells. The supernatant is decanted and the platelets pelleted by centrifugation for 8 min at 800 *g*. The supernatant is discarded and the platelet pellet is resuspended in 5–10 ml of citrate–saline. This wash procedure is repeated twice. After the third wash, the platelet pellets are resuspended in 5.0 ml 1.8% paraformaldehyde and allowed to stand for 2 hr. After fixation, 5.0 ml imidazole-buffered saline is added and the mixture is centrifuged for 8 min at 800 *g*. The supernatant is discarded and the platelets are suspended in 5–10 ml imidazole-buffered saline. This wash is repeated three times. After the final wash, the platelets are resuspended in citrate–saline containing 5% bovine serum albumin (BSA), and are counted in a hemacytometer. The platelet concentration is adjusted to 800,000/mm^3 and 1.0-ml aliquots are placed in 20-ml vials. The platelets may be stored at 4° or frozen at −20°.

Human platelets may also be prepared from 72-hr-old platelet concentrates obtained from a blood bank. Four- and 5-day-old stored platelets are less reactive, but may be used. The procedure described above is used for processing.

Fixed animal platelets are prepared from freshly drawn blood with ACD as anticoagulant. The procedure is as described for human platelets, except for the fixation process. The platelets are resuspended in 5.0 ml 0.8% paraformaldehyde and placed at 4° for 24 hr. Higher concentrations of fixative with animal platelets often cause spontaneous platelet aggregation. An equal volume (5.0 ml) of imidazole-buffered saline is added to each tube and the platelets pelleted by centrifugation for 8 min at 460 *g*. The platelets are then washed and further processed as described for the human platelet preparation.

Fixed platelet preparations may also be lyophilized.[6] Lyophilized platelets are stable for months or even years.

2. *Botrocetin*

Materials

Glass chromatography columns, 1.6 × 20 cm
DEAE-cellulose (DE-52)

Reagents

Dried *Bothrops* venom. We use *Bothrops jararaca* venom. Several other species are equally useful

Starting chromatography buffer contains 0.015 *M* imidazole, 0.154 *M* NaCl dissolved in distilled water. The pH is adjusted to 7.35 with 1.0 *N* HCl

Elution buffers are (a) 0.015 *M* imidazole, 0.175 *M* NaCl and (b) 0.015 *M* imidazole, 0.5 *M* NaCl, dissolved in distilled water and the pH adjusted to 7.35 with 1.0 *N* HCl

Botrocetin is prepared from the crude venom obtained from any of several different suppliers. Venom (20 mg) is dissolved in 5.0 ml starting chromatography buffer. The venom is chromatographed on DEAE-cellulose with chromatography buffers with increasing salt concentration for separation of botrocetin. Briefly, chromatography columns are packed with DEAE-cellulose, washed according to manufacturer's instructions, and equilibrated with the starting chromatography buffer. The venom solution is pumped onto the column, followed by 50 ml of elution buffer (a), until the optical density returns to baseline. Elution of botrocetin is with elution buffer (b), containing 0.5 *M* NaCl. Fractions are tested for the presence of botrocetin-aggregating activity (see below). The aggregating fractions are combined and dialyzed at 4° for 4 hr against starting buffer. Bovine serum albumin (BSA) is added to reach final concentration of 0.1%. The botrocetin is assayed as described below.

Bioassay of Botrocetin Activity. The bioassay of botrocetin is based on the observation that the rate of platelet aggregation under standard conditions with a constant amount of plasma vWF is dependent on the concentration of botrocetin (Fig. 1).[13] On the arithmetic plot, the segment of the curve between 12 and 16 sec approaches linearity and was selected for measuring relative potency of botrocetin solutions. One unit of botrocetin is defined as that amount in 1 ml test mixture which causes a platelet aggregation time of 14 sec in a four-part test system, one part reference human PPP, one part buffered saline, pH 7.35, one part fixed platelets, and one part botrocetin. From the data of Fig. 1 (inset) Table I was developed, which indicates the aggregation times resulting with varying botrocetin concentrations in the range 0.75–1.35 units/ml.

[13] K. M. Brinkhous, D. S. Barnes, J. Y. Potter, and M. S. Read, *Proc. Natl. Acad. Sci. U.S.A.* **78,** 3230 (1981).

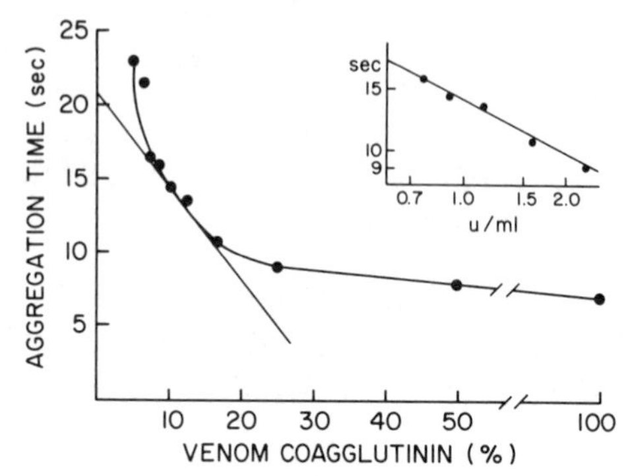

FIG. 1. Derivation of a botrocetin unit: Relationship of botrocetin to platelet aggregation time. Test system consisted of equal parts of (1) normal human plasma, reference pool stored at $-70°$, designated as 1 unit of vWF/ml; (2) buffer; (3) fixed human platelets, 800,000/mm^3; and (4) botrocetin solution. Inset shows logarithmic replot of data in the range 9–16 sec, with conversion of botrocetin concentration to units/ml: 1 unit of botrocetin/ml is equivalent to aggregation time of 14 sec in a 1.0-ml standard test mixture.[13]

An illustrative example of bioassay of an unknown preparation for botrocetin potency follows. The botrocetin solution was diluted such that at least two of the diluted samples caused aggregation in the range 12–16 sec (Table I). The samples diluted 1 : 128 and 1 : 175 met this criterion with aggregation times of 13.0 and 15.5 sec, respectively (each an average of three replicate determinations). These time values are equivalent to botrocetin concentrations of 1.16 and 0.78 units/ml in the final test mixture. The botrocetin value based on the 1 : 128 sample would be 128 (reciprocal of the initial dilution) × 4 (reciprocal of 1 : 4 dilution in the test mixture) × 1.16 (botrocetin in final test mixture from Table II) = 594 units/ml. For the 1 : 175 sample, the value would be 546 units/ml for a mean of 570 units/ml.

After bioassay 100 units of botrocetin is aliquoted into 2.0-ml wide-mouth vials, frozen at $-70°$, and lyophilized for 8 hr, or to dryness. The lyophilized botrocetin is stored at $-20°$ until used. Botrocetin containing 0.1% BSA may also be stored frozen, but we have found that after several months, frozen products tend to lose activity. It is more convenient to store this botrocetin reagent as a lyophilized powder. Reconstitution is with 1.0 ml distilled water. It is recommended that the venom be assayed after lyophilization, since there may be some loss of activity during drying. Further dilution of the botrocetin is with imidazole-buffered saline, pH 7.35, containing 0.1% BSA, to provide solutions of desired strength.

TABLE I
DERIVED CONVERSION FACTORS FOR BIOASSAY OF BOTROCETIN

Platelet aggregation time (sec)	Botrocetin (units/ml)
12.0	1.35
12.5	1.25
13.0	1.16
13.5	1.08
14.0	1.00
14.5	0.92
15.0	0.86
15.5	0.78
16.0	0.75

Bioassay Procedures for von Willebrand Factor

Macroscopic Tests

1. Flocculation Titer Plate Test

Materials

2 × 4 in. clear glass titer plates containing three rows of four wells each. The wells are 1 cm in diameter
Stop watches
Microliter pipets and tips

Reagents

Ristocetin, 2.0–2.4 mg/ml, dissolved in saline
Botrocetin
Fixed platelets: Fixed frozen or fixed lyophilized platelets may be used
Normal human reference plasma (control plasma)
Citrated platelet-poor plasma (PPP) (test plasma): PPP is prepared by centrifuging whole blood (eight parts blood, one part 3.2% sodium citrate anticoagulant) at 4° for 20 min at 2000 *g*
Imidazole buffer, pH 7.35 (0.084 *M* imidazole, pH adjusted with 1.0 *N* HCl)
Imidazole-buffered saline, pH 7.35 (0.084 *M* imidazole, 0.154 *M* NaCl, pH adjusted with 1.0 *N* HCl)
0.1% bovine serum albumin (BSA) in imidazole-buffered saline
Acid–citrate–dextrose (ACD) (0.084 *M* sodium citrate, 0.064 *M* citric acid, and 0.11 *M* dextrose, pH 4.5)

Citrate–saline, pH 6.8 (0.0054 *M* sodium citrate, 0.154 *M* NaCl, pH adjusted with 0.1 *N* HCl)

Protocol

1. Both the test plasma and the control plasma are serially diluted with buffered saline to a maximum dilution of 1:32.

2. If lyophilized platelets are used, they are reconstituted in 1.0 ml imidazole buffer and allowed to stand undisturbed for 5 min.

3. The lyophilized botrocetin is reconstituted in 1.0 ml water and diluted to 50 botrocetin units/ml with buffered 0.1% BSA for assay of vWF of human plasma with human platelets. For assay of canine plasma vWF with either human or animal platelets, the botrocetin is diluted to 5.0 units/ml, and for porcine vWF assays, the botrocetin is diluted to 8 units/ml. For other animal plasmas a preliminary test should be run to determine optimal botrocetin concentration. The reconstituted botrocetin is allowed to stand 10–15 min before testing begins.

4. Ristocetin concentration of 2.0–2.4 mg/ml provides a final concentration of 0.5–0.6 mg/ml in the four-part test mixture. Higher concentrations may precipitate plasma proteins. (Ristocetin is used only for assay of human plasma vWF with human platelets.)

5. To each of the four wells in row one is added 0.025 ml buffer and 0.025 ml platelet suspension. Then undiluted test plasma (0.025 ml) is added to well one; 0.025 ml 1:2 diluted test plasma to well two; 0.025 ml 1:4 diluted test plasma to well three; 0.025 ml 1:8 diluted test plasma to well four. Finally, 0.025 ml botrocetin is added to each well, beginning at well four and proceeding to well one. A watch is started for each well upon addition of botrocetin to the well. The titer plate is gently swirled under a light and at the first appearance of platelet clumping in each of the four wells, the appropriate watch is stopped and the time of onset of aggregation is recorded (Fig. 2).

2. Tap-Tube Test

Materials

10 × 75 mm glass tubes, clean and free of lint and scratches
Stop watches
Microliter pipets and tips
12 × 75 mm plastic test tubes

Reagents. The reagents for this test are the same as for the titer plate test (see above).

Protocol

1. Steps 1–4 of the titer plate test are followed.

2. Platelet aggregation is determined as follows: The reagents are mixed

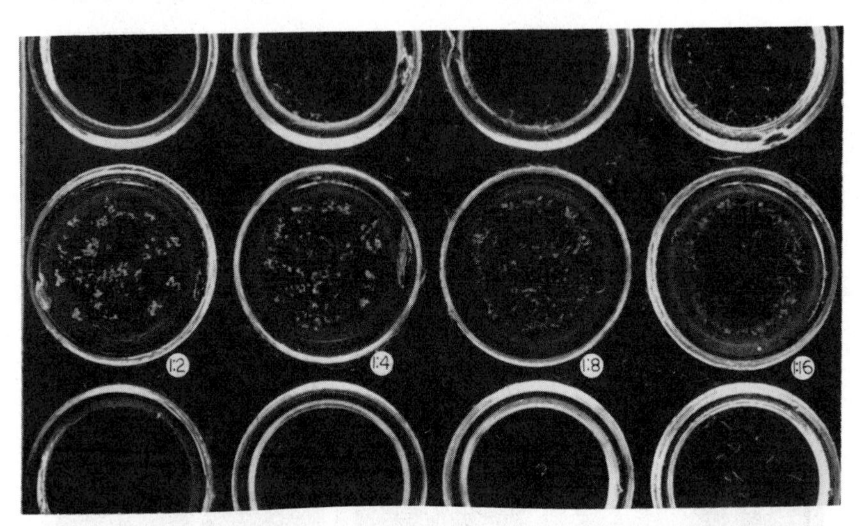

FIG. 2. Flocculation titer plate test for vWF, illustrating macroscopic platelet aggregation with human reference plasma. Test system: 0.025 ml imidazole buffer, 0.025 fixed human platelet suspension (800,000/cmm), 0.025 ml serially diluted reference plasma (1 : 2 to 1 : 16), 0.025 ml botrocetin solution (50 units/ml) in each well. Macroscopic platelet aggregation times determined. Note size of aggregates is a function of plasma vWF concentration.

in a 10 × 75 mm tube (0.025 ml buffer, 0.025 ml plasma, 0.025 ml platelets). Mix gently, then add 0.025 ml botrocetin or ristocetin.

3. Upon the addition of botrocetin or ristocetin, a watch is started and the tube is tapped gently. The film of fluid flowing down the wall of the tube is observed. The time of the first appearance of macroscopic platelet aggregates in the film is the end point. The watch is stopped at the first sign of aggregation and the time recorded.

4. Gentle tapping of the tube is continued for 60 sec, at which time the size of platelet aggregates is judged (Fig. 3). The degree of aggregation is expressed on a scale of 1+ to 4+, with 1+ being the smallest detectable aggregate with 3–5 platelets per aggregate; 2+, with 5–20 platelets per aggregate; 3+ with 20–50 platelets per aggregate; and 4+ with 50 or more platelets per aggregate. Macroscopically, 1+ aggregation is difficult to judge reliably, but 2+ to 4+ aggregates are easily seen in both the tap-tube and titer plate methods, and with experience microscopic examination is not needed.

Calculations. Platelet aggregation times for each dilution of normal reference plasma and of test plasma are plotted on semilog graph paper as illustrated in Fig. 4. In this example, using botrocetin as the aggregating

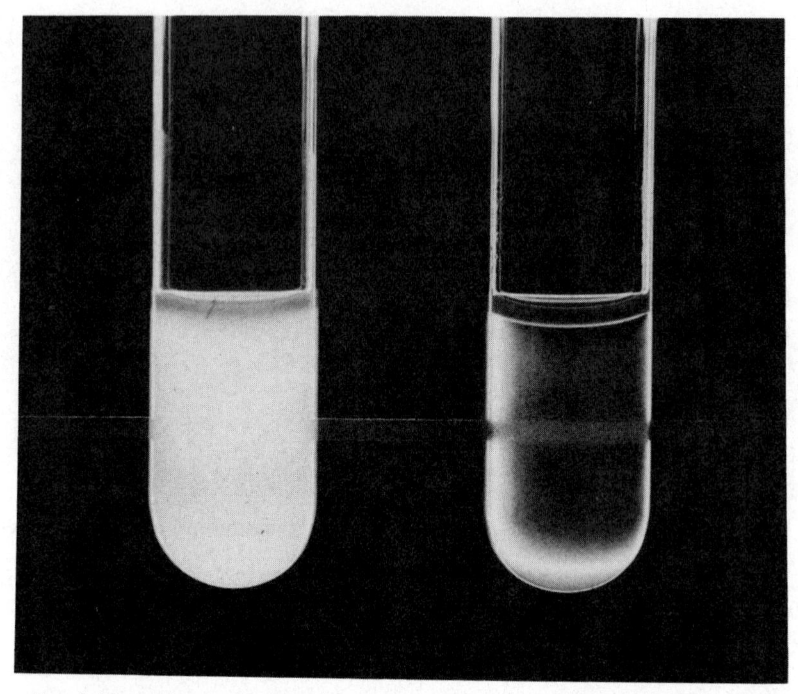

FIG. 3. Tape-tube test for vWF, illustrating macroscopic platelet aggregation with undiluted human reference plasma. Test system as in Fig. 2. Tube on left illustrates opacity of mixture with nonaggregated platelets, prior to addition of botrocetin. Platelets in test mixtures aggregated with botrocetin are shown in tube on right. Photo taken several minutes after platelet aggregation had occurred, with aggregates settled on bottom of tube.

agent, the test plasma was that of a patient with mild von Willebrand disease. The testing was done with both fixed human platelets and fixed dog platelets. The values for the test plasmas are interpolated from the graph of aggregation times of reference plasma arbitrarily assigned a value of 100%, or 1 unit vWF/ml. A horizontal line is drawn from observed values along the patient line to determine interpolated points of intercept on the reference plasma line. The plasma concentration (on abscissa) is ascertained for each interpolated point on the normal plasma line. This normal plasma concentration is divided by the corresponding test plasma concentration similarly obtained, multiplied by 100, to give the relative percentage vWF concentration in test plasma. The average of the individual vWF values is obtained.

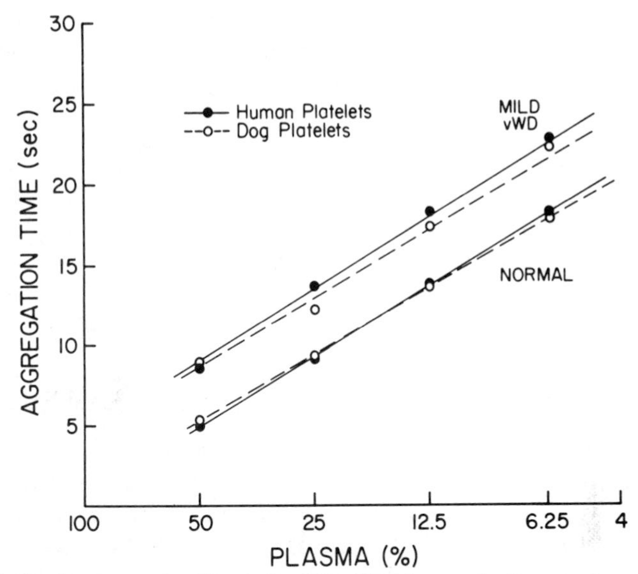

FIG. 4. Platelet aggregation times of serially diluted normal reference plasma and mild von Willebrand disease plasma, using either human or dog fixed platelets, on a semilog plot. See text for method of calculating relative vWF concentration in test plasma.

Aggregometry Test Materials

Aggregometer: We use either Chronolog or Payton instruments
A good recorder
Microliter pipets and tips
37° water bath

Reagents. The reagents used for the aggregometer testing are the same as those given for the titer plate test.

Protocol

1. Plasma dilutions are prepared the same as for the titer plate test.
2. The botrocetin is reconstituted in a volume of distilled water sufficient to yield 625 botrocetin units/ml.
3. The ristocetin concentration is 30.6 mg/ml in buffer for a final concentration in the test mixture of 0.6 mg/ml.
4. The mixture consists of 0.3 ml imidazole buffer, 0.1 ml fixed platelets, 0.1 ml plasma sample, 0.01 ml botrocetin or ristocetin.
5. The testing is performed at 37° and the stir rate is 1000 rpm. The recorder is set at 0.1 V. The baseline setting for the 0% light transmission is established with a suspension of platelets in the test mixture above, with

buffer in place of botrocetin or ristocetin. The maximal light transmission (100%) is set using a maximally aggregated suspension of platelets in undiluted normal reference plasma containing either botrocetin or ristocetin, depending upon which aggregating agent is being used in the assay.

Calculation of vWF values. We use the slope of a line for calculating plasma vWF. Briefly, one determines the steepest part of the curve, constructs a tangent to that part of the curve, and determines the slope of the tangent line. The slope indicates rate of aggregation.

The percentage vWF in the test plasma is determined from the ratio of the slopes of the test plasma dilutions to the slopes of the respective dilutions of the reference plasma, as follows:

[slope (test plasma)/slope (reference plasma)] × 100 = % vWF test plasma

The aggregometric tracing is also observed to determine the lag phase and the maximal deflection. The lag phase is the time from addition of aggregating agent to the beginning deflection in the tracing due to platelet aggregation. The lag phase is roughly proportional to the vWF concentration. The magnitude of each deflection observed at maximal platelet aggregation on the aggregometry tracing is an index of the size of the platelet aggregate—the larger the aggregate size, the greater the deflection.

Special Applications of vWF Testing with Fixed Platelets

1. *Screening Test for vWF Multimer Size by Combined Use of Ristocetin and Botrocetin Tests*

By employing either the macroscopic flocculation macrotiter plate test or the tap-tube test and determining simultaneously the platelet aggregation times on test material and reference plasma samples, a qualitative index of vWF multimer size range may be obtained provided that the vWF antigen level of the reference plasma is adjusted to be approximately that of the test samples. The basis of this test is that ristocetin is reactive with only the larger vWF multimers whereas botrocetin is reactive with a wide range of sizes of vWF multimers (all except for the small sizes).[14] An example of this type of testing is illustrated in Table II. It will be noted that plasmas of patients with variant type IIA von Willebrand disease, commercial high-potency factor VIII concentrates, and cryosupernatant, all lacking high-molecular-weight forms of vWF, were unreactive in the ristocetin test, whereas platelet aggregation was prompt with botrocetin.

[14] K. M. Brinkhous, M. S. Read, W. A. Fricke, and R. H. Wagner, *Proc. Natl. Acad. Sci. U.S.A.* **80**, 1463 (1983).

TABLE II
COMPARISON OF REACTIVITY OF LOWER MOLCULAR WEIGHT MULTIMERS OF vWF ANTIGEN (vWF:Ag) IN THE RISTOCETIN AND THE BOTROCETIN TESTS FOR PLASMA vWF

Plasma or plasma fractions	vWF:Ag (%)	Macroscopic vWF test: Ristocetin (sec)	Macroscopic vWF test: Botrocetin (sec)	Multimeric distribution of vWF:Ag: Higher M_r	Multimeric distribution of vWF:Ag: Lower M_r
vWD plasmas, type IIA					
Subject W.R.	43	Negative	14	—	+
Subject V.N.	78	Negative	14	—	+
Plasma fractions					
Factor VIII concentrate	100	Negative	18	Trace	+
Cryosupernatant	25	Negative	30	—	+
Human reference plasma					
Undiluted	100	8	8	+	+
Diluted 1:4	25	14	14	+	+

2. *Measurement of Anti-vWF Antibody (Ab) Titers*

These measurements are performed in multitransfused von Willebrand disease subjects with vWF inhibitor and in acquired von Willebrand syndrome. Several test procedures are used to measure anti-vWF titers. The anti-vWF Abs may be identified by their inhibitory activity on platelet aggregation as demonstrated with any of the several vWF tests,[7,15] or by their ability to bind vWF in the absence of inhibitory activity.[8]

Anti-vWF Inhibitor Tests. The inhibitor tests may be performed with human fixed platelets and human plasma in both the ristocetin and botrocetin platelet aggregating tests, or with bovine or porcine plasma (PAF inhibitor test).

vWF inhibitor test with ristocetin and/or botrocetin. Either of the macroscopic test procedures, the titer plate or the tap-tube test, may be used. The vWF inhibitor test consists of mixing equal parts (0.025 ml) of serially diluted test plasma (the plasma being assayed for antibody titer), normal reference plasma diluted 1:4, fixed platelet suspension, and aggregating agent, either ristocetin or botrocetin. The time between addition of aggregating agent and onset of aggregation is recorded as the macroscopic aggregation time. The highest dilution of inhibitor plasma which causes an

[15] R. D. Stratton, R. H. Wagner, W. P. Webster, and K. M. Brinkhous, *Proc. Natl. Acad. Sci. U.S.A.* **72**, 4167 (1975).

aggregation time at least 3 sec longer than the aggregation time of the 1:4 dilution of reference plasma is given as the antibody titer of the inhibitor plasma.

The bovine or porcine PAF inhibitor test. This test consists of addition of equal parts (0.025 ml) of imidazole buffer, human inhibitor test plasma or dilution thereof, platelet suspension, and bovine plasma diluted 1:32 or porcine plasma diluted 1:8 with buffered saline. The aggregation time is the time between the addition of animal plasma and the appearance of macroscopic aggregates of platelets. The antibody is given as the dilution of inhibitor plasma which causes an aggregation time at least 3 sec longer than that obtained when buffer is added in place of inhibitor test plasma.

vWF-Binding Antibody Test. Some subjects with autoimmune or acquired von Willebrand disease have a circulating Ab which depletes the patient's plasma of its factor VIII macromolecular complex, with absence of factor VIIIR:Ag, vWF activity as determined with either ristocetin or botrocetin, and antihemophilic factor or F.VIII:C. *In vitro* a seeming paradox is observed. The patient's plasma exhibits little or no anti-vWF inhibiting activity. Pathophysiologically, the anti-vWF Ab is believed to bind to a vWF epitope unrelated to the functional domain(s) responsible for vWF-dependent platelet aggregation as observed in vWF bioassays. The Ag–Ab complex, on removal from the circulation, removed the factor VIII complex with its two components, the vWF and the F.VIII:C molecules. A test has recently been devised for measuring the titer of the vWF-binding capacity of anti-vWF Abs.[8] This test employs a suspension of staphylococci with protein A on the cell surface (SpA) (Pansorbin, Calbiochem-Boehring, La Jolla, CA) to isolate the anti-vWF binding antibody from the patient's plasma. The Ab of IgG class 1, 2, or 4 is attached to SpA (SpA–Ab). The principle of the test is that the isolated Ab (SpA–Ab) added to normal plasma binds its factor VIII–vWF complex which is then removed by centrifugation. Little or no factor VIII–vWF complex activities remain in the plasma. By serially diluting test plasma before addition of SpA–Ab, the titer of the binding Ab can be determined. The amount of vWF, vWF:Ag, or F.VIII:C removed from the normal plasma is a function of antibody concentration. The test is performed as follows:

The patient plasma is serially diluted with buffered saline 1:4, 1:8, 1:16, 1:32, 1:64, 1:128, 1:256, 1:512. Eight aliquots of the SpA suspension (0.2-ml aliquots) diluted with 0.5 ml of buffered saline is sedimented by centrifugation (1620 *g* for 8 min). The sedimented SpA aliquots are then resuspended in 0.1 ml of diluted patient plasma being tested and incubated at 23° for 20 min. The suspension is diluted with 0.5 ml of the above buffer and the SpA sedimented as above. The pellet is resuspended in 0.2 ml normal reference plasma and is incubated for an additional 20 min at 23°. The Ag–Ab complex attached to SpA is removed by centrifugation as above and the supernatant plasma is assayed for residual FVIII–vWF complex activities. Controls are run in parallel by substituting normal reference plasma incubated with SpA for patient plasma. The titer of antibody is chosen to be the greatest dilution of test plasma showing removal of at least 25% of a given F.VIII complex activity from the normal plasma in comparison to a control of reference plasma incubated with SpA suspension. As a rule we use only the ristocetin and the botrocetin macroscopic test procedures for vWF in this test.

Accuracy and Reproducibility

The vWF values obtained with macroscopic tests are reproducible to approximately ± 10% in the hands of experienced operators. For the beginning operator, the end point in the microtiter plate test appears to be more easily observed and replicated than that in the tap-tube test. The latter test when mastered allows rapid and accurate testing. Small differences in vWF concentration detectable with the macroscopic tests have not been detectable with the aggregometer. The time of beginning aggregation of replicate samples in the macroscopic tests may vary with operators. Regardless of this, between operators the vWF values for test samples in comparison to reference plasma are within approximately ± 10%.

Precautions

The main precaution to be taken in the preparation of fixed platelets is to start with carefully collected blood. Prompt mixing with the anticoagulant is essential. If blood is collected by syringe, air bubbles from leaks at the needle junction should be avoided. "Spontaneous aggregation" of platelets during processing often occurs with poorly collected blood samples.

The macroscopic platelet aggregation tests require practice to obtain replicate values for the platelet aggregation time. A good light and a dark background to view the aggregates is desirable. With aggregometric testing with fixed platelets there is usually an initial period of decreased light transmission, especially with ristocetin.[1] The lag phase then is the time from the addition of the last reagent to the beginning of increased light transmission.

[13] Determination of Platelet Volume and Number

By DAVID BESSMAN

The platelet count in circulating blood is an important parameter that reflects the balance between their production by bone marrow megakaryocytes, on the one hand, and their utilization, destruction, or loss, on the other hand. The normal number of platelets in human blood varies from 150,000 to 450,000/μl of blood depending in part on the method used in their counting. A decrease in the number of circulating platelets indicates a relative or absolute quantitative deficiency of platelets (thrombocytopenia). An increase in the number of circulating platelets above 500,000/μl indicates a quantitative excess of platelets (thrombocytosis or thrombocythemia).

The platelet is a colorless, discoid body with a diameter of 1 – 5 μm. Especially after *in vitro* stimulation, it may easily change its shape to a spiny sphere extending filamentous pseudopods, often longer than the body of the platelet. Such a change is accompanied by an increase in the surface area of the platelet. The particle's volume may not actually change, for its mass does not, but its *apparent* volume, as measured currently, does change (see below).[1]

Most of the studies of platelet functional attributes presented in this volume require determination of platelet count for comparative controls, to monitor the yield of platelets in separation procedures, and for utilization of platelets in biological experiments. Therefore, platelet count may be performed in samples of whole blood, platelet-rich plasma (PRP), or platelets separated from plasma proteins.

The most suitable methods of platelet counting for the purpose of experimental work are direct methods with electronic particle counters. Phase microscopy is preferable only when electronic counting is impractical. These methods will be presented below. Questions of the relation of platelet size vs platelet density are reviewed elsewhere.[2,3]

[1] D. D. Mundschenk, D. R. Connelly, J. G. White, and R. D. Brunning, *J. Lab. Clin. Med.* **88,** 301 (1976).

[2] S. Karpatkin, *Blood* **51,** 307 (1978).

[3] C. B. Thompson, J. A. Jakubowski, P. A. Quinn, D. Deykin, and C. R. Valeri, *J. Lab. Clin. Med.* **101,** 205 (1983).

Platelet Counting with Phase Microscope

Whole blood is diluted with ammonium oxalate solution (10 g/liter) in a standard red cell-counting pipet and agitated for 5–10 min.[4] Half of the mixture is introduced into each side of a hemacytometer chamber and all 25 squares are counted in duplicate under phase microscopy (×45). The principle is that platelets are the only blood cells remaining intact to be counted. The results are expressed as the average of the two scores $\times 10^3/\mu l$. Usually a few hundred platelets at most are counted. Therefore, the duplicate error just from counting is substantial, from 11 to 23%.[5,6] This method is recommended only in two circumstances: when a particle counter is not available, and when data from the counter are suspect, as when the particle size distribution is abnormal.

Platelet Counting with Electronic Particle Analyzers

Several commercially available instruments permit platelet measurements as part of an automated blood count differential. These instruments distinguish particles on the basis of impedance or optical density. They allow quantitation of several cell types, including platelets, from a small blood sample and yet offer the statistical soundness of a 10,000-cell count. This is the method of choice when such an instrument is available and the platelets are from species with suitable platelet size. They include platelet count, mean platelet volume (MPV), and a platelet volume histogram. Some instruments also offer an assay of the heterogeneity of platelet size (PDW).

Platelet Count. In any automated particle counter the particles detected within a defined size range are presumed to be platelets and are so counted, although they are not necessarily platelets. For human subjects, the range that is used in most automated instruments is 2–20 fl. From the raw count of these particles the data are extrapolated to make a smooth curve and it is from the extrapolated curve that both platelet count and MPV are calculated. Note that sometimes the extrapolation posits "platelets" <2 or >20 fl (Fig. 1).[7] This range has been chosen to optimize the correlation between the number of particles of a given size (automated technology)

[4] G. Brecher and E. P. Cronkite, *J. Appl. Physiol.* **3,** 365 (1958).

[5] G. Brecher, M. Schneiderman, and E. P. Cronkite, *Am. J. Clin. Pathol.* **23,** 15 (1953).

[6] B. S. Bull, M. Schneiderman, and G. Brecher, *Am. J. Clin. Pathol.* **44,** 678 (1965).

[7] J. D. Bessman, "Automated Blood Counts and Differentials." Johns Hopkins University Press, Baltimore, Maryland, 1986.

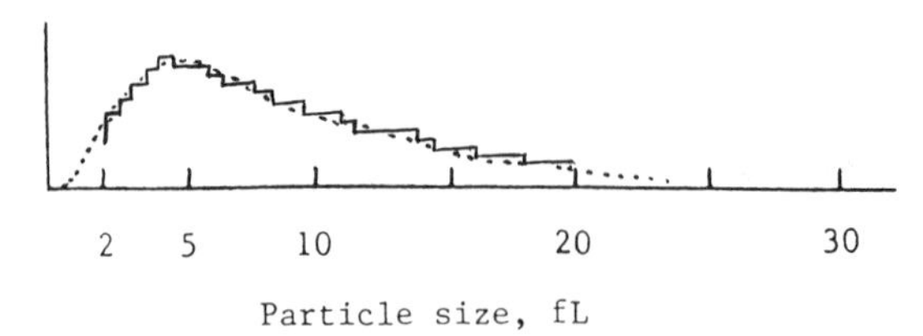

FIG. 1. Platelet size distribution curve generated by a Coulter counter model S-Plus IV. Solid line, raw data; dotted line, extrapolation to a smoothed curve from which the platelet count and size are calculated. From Bessman.[7]

and the number of morphologically recognizable platelets (manual technology). The platelet count is usually expressed as a number $\times 10^3/\mu l$ (or $\times 10^9$/liter in the SI terminology).

Mean Platelet Volume (MPV). Within the range designated for platelet size, the extrapolated histogram of platelet volume distribution is then integrated for mean platelet volume. Again, this is calculated from a curve that has been smoothed. The scale for mean platelet volume depends on the calibrating standard. Each manufacturer has such a calibrating standard, but they are not equivalent. That is, calibrators for two manufacturers will not yield similar calibrations in a given instrument, nor will one calibrator give the same result in two different instruments.[8] The user should determine the calibrated values for the instrument being used with the calibrator provided by the manufacturer. Nonplatelet materials, such as plastic beads, unfortunately do not resemble platelets in either shape or deformability. Therefore, using such beads as calibrators of a size scale is advised only if the size scale desired is nominal and relative rather than absolute.

There has been criticism of the use of mean platelet volume as a value not independent of platelet volume distribution and, therefore, not appropriate for serious consideration.[9] The relative mathematical merits of mean, median, or mode can be debated. However, in our experience, the distribution of the platelet volumes in 98% of specimens with platelets $>250{,}000/\mu l$ is right skewed in an approximately log-normal pattern.[10-12] The mean value, therefore, is statistically valid. The modal value, being a single point, would be more subject to vagaries of duplicate error and curve

[8] V. Lippi and P. Cappellitti, *Am. J. Clin. Pathol.* **79,** 648 (1983).

[9] J. M. England, *in* "Laboratory Hematology" (J. C. Koepke, ed.), p. 927. Churchill Livingstone, New York, 1984.

[10] S. K. Garg, E. L. Amorosi, and S. Karpatkin, *Ann. Intern. Med.* **77,** 361 (1972).

[11] Z. Zeigler, S. Murphy, and F. H. Gardner, *Blood* **51,** 479 (1978).

[12] J. Levin and J. D. Bessman, *J. Lab. Clin. Med.* **101,** 295 (1983).

shape. For the 2% of specimens with an abnormal shape and distribution of platelet volume (see Pitfalls), the mean, median, and mode are equally unreliable.

Platelet Distribution Width (PDW). This is an index of platelet volume heterogeneity, similar in concept to the red cell distribution width (RDW). This too is calculated from the extrapolated platelet volume distribution (Fig. 1). Since all platelet volume distributions are right skewed, the heterogeneity is calculated as the geometric standard deviation.[9,13]

Platelet Volume Distribution Histogram. This is displayed and/or printed by many automated or semiautomated counters.[7] A standard method of analysis to describe normal or abnormal shape has not yet been agreed upon. However, review of such a histogram is valuable to check against the artifacts described below.

Instruments. As is discussed above, the calibration and, therefore, the value reported differs from instrument to instrument. Also, several instruments measure particle size differently. Coulter-type instruments measure impedance of particles crossing an electronic current. Examples are all Coulter instruments (Coulter Electronics, Hialeah, FL) and Sysmex (Toa Medical, Del Amo, CA), Cellect (Instrumentation Laboratories, Cambridge, MA), and Sequoia-Turner (Sunnyvale, CA) instruments. The H-6000/H*1 (Technicon, Tarrytown, NY) instruments measure particles by light scatter. Both impedance and light scatter are influenced by cytoplasmic constituents as well as by cell size and shape. However, in neither technology have specific disorders of platelet cytoplasm been shown to cause artifacts of platelet size on that basis.

In all instruments, the blood sample is diluted in an isotonic solution to produce a cell concentration of about $5-10 \times 10^3/\mu l$. The dilute cell suspension then is focused into a thin stream in which particles presumably are single. The stream crosses an interrogatory probe beam, which measures light scatter or impedance as noted above, records the amount of the characteristic in each particle measured, and accumulates individual particle data for a histogram and analysis.

For the above reasons, the mean platelet volume should be viewed as a relative, rather than an absolute, value. The scale will vary depending on calibration and instrument. However, as long as the same technique is used, normal values can be established and tested specimens compared to them.

Accuracy of Variables. Several studies have shown linear and essentially unitary correlation between manual and automated platelet counts. Accu-

[13] G. A. Threatte, C. Adrados, S. Ebbe, and G. Brecher, *Am. J. Clin. Pathol.* **81**, 769 (1984).

racy of mean platelet volume is discussed elsewhere.[7,12] There are no reference methods for platelet distribution width or the platelet volume distribution histogram.

Precision is the degree of unavoidable technical error. For automated platelet counting, the duplicate error is approximately 10,000/μl.[6,9,10] This means that platelet counts < 10,000/μl are statistically not distinguishable, and platelet counts of 11,000–20,000/μl, barely so. The absolute duplicate error remains similar at higher platelet counts. However, the higher the platelet count, the lower the proportion of duplicate error to reported value. As noted above, this is greater precision than is available with manual counting. The duplicate error for automated determination of MPV is 0.5 fl, and 0.8% for PDW. The duplicate error for the platelet volume distribution histogram has not been well established.

Sample Handling. These instruments require only that the sample be properly anticoagulated. Most now routinely use whole blood, which is automatically diluted. However, platelet-rich plasma or washed platelets also may be used. Care must be taken for the following points. First, the specimen must be fully agitated to produce a homogeneous cellular suspension. This is usually complete with five mechanical or manual inversions. Second, if whole blood is used, the specimen must be effectively anticoagulated. Run a thin wooden stick around the inside perimeter of the tube, withdraw it, and examine for small clots. This will detect incomplete anticoagulation better than visual inspection of the whole tube. Incomplete anticoagulation will yield false—usually low—counts. Third, the specimen age is important. If less than 1 hr has elapsed since venipuncture, the platelet count will be accurate but MPV will vary. EDTA anticoagulation causes a change in platelet shape and therefore in perceived platelet size, amounting to an increase of 20% over the first hour.[12–14]

Anticoagulant. If only platelet count is to be determined, either a citrate (e.g., acid–citrate–dextrose, ACD) or EDTA anticoagulation (liquid or powder) can be used. If ACD is used, its volume is significant in reference to the specimen, hence the count must be adjusted by the manually measured ratio of specimen/specimen plus anticoagulant. If EDTA is used, its volume is negligible to materially affect the result.

Animal Studies. Automated instruments usually are set to measure human platelets. Animal platelets are 20–80% the size of human platelets. Rat and dog platelets generally fall between the automated thresholds. In contrast, mouse and guinea pig platelets are too small for the thresholds. Many animal platelets will be <2 fl and extrapolation of the 2- to 20-fl portion of the distribution will be unreliable. A semiautomated technique

[14] B. S. Bull and M. V. Zucker, *Proc. Soc. Exp. Biol. Med.* **120,** 269 (1965).

is more useful in such animals. Results should be interpreted by comparison to an established normal range. For humans, the normal values for platelet count and MPV are best expressed as a nomogram (see below, Normal Values). For animals, a similar nomogram has yet to be established.

Semiautomated Platelet Counter (Coulter Counter Type)

Semiautomated counters provide easy adjustment of amplification and scale to allow sizing and counting of particles outside the preset thresholds on automated instruments (e.g., mouse platelets). The sample requires more preparation than does the sample for automated evaluation and may require manual calculation of the platelet count and MPV. Some instruments will develop a particle-volume size distribution histogram and the user must integrate under it; others will integrate automatically. The lower thresholds level be set above particle and instrument noise, and the upper threshold, must be set to exclude red cells. These thresholds can be determined initially by measurement of whole blood and adjustment of thresholds from the whole-blood histograms.

Platelet-Rich Plasma. Semiautomated counters do not eliminate red cells from platelets by "smoothing" a platelet-volume histogram. Since red cells are 10- to 20-fold more numerous than platelets, a cleaner platelet volume distribution is obtained by eliminating at least most red cells. The cost of this is the loss of a few platelets in the discarded red cells. Whether these platelets differ in some variables (perhaps those of interest to the particular investigator) is not well established. The investigator must consider on an individual basis whether the benefit of less red cell artifact outweighs the risk of selective platelet loss. Over 95% of platelets can be isolated as platelet-rich plasma by spinning at 400 *g* for 4 min, and retaining the supernatant. Spinning too hard will redistribute some or all platelets into the buffy coat at the plasma-red cell interface. Should this happen, resuspension of the cells and centrifugation at a lower speed can provide platelet-rich plasma (PRP). For small volumes (e.g., a capillary tube) gravity alone usually will suffice—as it will for regular tubes as well. Allow the tube or capillary to stand on end for 60-100 min. For capillary tubes, break off the tube at the buffy coat and aspirate the sample from the platelet-rich plasma. Occasionally, subjects with abnormal red cell sedimentation rates will not have satisfactory PRP for sedimentation. In such cases centrifugation is required.

Washed Platelets. The platelet distribution should be freer of nonplatelet particles than is platelet-rich plasma. Assuming that the initial dilution is known, handling is similar to platelet-rich plasma.

Dilution and Counting. The dilution must be precise to assure accurate platelet count. One part of plasma to 1000–5000 parts of diluent is a satisfactory range. The diluent for Coulter instruments is usually commercial Isoton, Isoton-Plus, or Isoton-III (Coulter Electronics, Hialeah, FL). Other instruments may have other recommended diluents. Reagents may be less than stringently controlled for the presence of small particles. Each lot should have a control histogram run. Further, the effect of reagents' tonicity on platelet size is not well characterized except for the Isoton diluents. Platelet-rich plasma is added by micropipet into a cuvette already containing 10 ml diluent, and mixed by inversion. The instrument reading should be corrected for coincident cell passage from the tables provided by the manufacturer. Platelet count in platelet-rich plasma should be adjusted to the volume of whole blood by multiplying the corrected platelet count in PRP by the hematocrit factor derived from the chart provided by the manufacturer.

Calibration. In the past, latex particles of known volume were used to calibrate the semiautomated counter. Although it may seem straightforward, the latex particle has shape and deformation properties different from blood cells, hence providing quite artificial values. Calibration with latex particles, typically 2.02 fl in volume (Sigma), fixes only one or a few points on a scale that rarely is strictly linear. Platelets have a wide range of size, 10-fold in a normal sample. An alternative method is calibration by human platelets. If the platelet volume histogram is available from an automated instrument, it can be compared with the histogram of volume on the semiautomated counter. Since normal values for healthy human subjects have been published,[7] the modal platelet size of 10 normal subjects can be expected to be approximately 7 fl on a Coulter instrument. If one is using an instrument that does not calculate mean platelet volume, the model channel for 10 normal subjects thus can be used as a practical calibrator.

Note that an automated instrument's scale for the histogram for platelet volume often does not correspond to the reported MPV, but itself is scaled to latex particles. Even so, use of a histogram of human platelet volume derived from normal subjects as a reference scale of platelet size in semiautomated instruments, as described above, is preferable to the use of latex particles. Accuracy, as described above, is relative rather than absolute in any system.

Generation of the Histogram. The cuvette containing the diluted PRP is raised so that the aspirating tip of the instrument is fully into the fluid. If the tip is fully plugged, no data will be recorded. However, a partial plug will obviously or subtly increase the perceived volume of a given particle. Subtle changes are especially troublesome. Therefore, the aperture should be brushed between each specimen and rinsed thoroughly. Ten thousand particles in the resultant histogram are sufficient for a smooth curve that can be well analyzed. The instrument's amplification controls should be adjusted to yield a volume–distribution histogram that is contained in the center of the volume scale.

Data Evaluation. Inspect the histogram to be certain the artifacts described below are not present. If they are, data interpretation will be compromised. Integration under the curve allows mean platelet volume. Modal platelet volume can be used, but is more subject to statistical variation. The scale of values will differ for the mode vs mean platelet volume; in a right-skewed curve, almost universally seen, the mean will exceed the mode value.

Normal Values

Mean platelet volume is related inversely but nonlinearly to platelet.[9,11,12] Therefore, it is not advantageous in presenting the data to give simply a normal range for each. Instead, the nonlinear relation of platelet size and count should be shown as a nomogram (e.g., Ref. 7). This relation has been shown (for human platelets only) by several independent groups and has the same characteristics regardless of the technology used for measuring MPV. The values for individual subjects should be compared to such a nomogram. As is the case with MPV, there is no single range of normal for PDW. Instead, it is directly and nonlinearly related to MPV. Thus, the higher the MPV value, the higher the PDW that can be expected.

The platelet volume histogram is ordinarily right skewed, though not necessarily log-normal. Since there is a variable range of "normal" for both MPV and PDW, many normal histograms are possible. However, the right-skewed shape remains, even when the platelet count falls below 20,000/μl (Fig. 1). What is distinctly abnormal is a deviation from the right-skewed shape. Artifacts are the principal cause for nonplatelet particles that are the size of platelets counted as platelets and included in the histogram. The more numerous such particles are relative to platelets, the more they will distort the platelet volume histogram. Electronic noise appears in the left (smaller size) of the histogram, while red cell fragments appear primarily on the right (largest size). These will be discussed more extensively below.

Pitfalls

Red Cell Fragments. These generally are larger than platelets and so appear on the right hand of the histogram, creating a plateau. If the plateau is high enough (in Coulter instruments, $>5\%$ of the modal value) the histogram may not fit the instrument's criteria for calculation, and neither MPV nor platelet count may be obtained.[9,15] The higher the platelet count, the less influence such red cell fragments have on the histogram. Less often, similar effects are caused by fragments of leukemic blasts.

Extreme Thrombocytopenia. When the platelet count is $<10{,}000/\mu l$, there may be insufficient platelets to allow a statistically valid histogram that is distinguishable from background noise. The histogram will not appear right skewed and unimodal, but rather will have a left margin peak or will be indefinitely irregular. Such results suggest that the actual platelet number is lower as compared to a low automated platelet count, accompanied by a right-skewed, unimodal platelet volume histogram. In the latter cases, platelet signals exceed noise sufficiently to allow the platelet volume histogram.

Platelet Clumping. EDTA occasionally induces *in vitro* platelet clumping. The reported platelet count falls, and the MPV rises.[16] The increase in MPV is variable, and does not reflect the MPV–platelet count relationship seen in normals. In such cases an ACD-anticoagulant specimen should be used for a platelet count. The MPV will be about 20% less than the sample containing EDTA as an anticoagulant. This artifact can be detected by examination of the peripheral blood smear made from EDTA-anticoagulated blood.

[15] P. J. Cornbleet and S. Kessinger, *Am. J. Clin. Pathol.* **83,** 78 (1985).
[16] D. P. Shreiner and W. R. Bell, *Blood* **42,** 541 (1973).

[14] Measurement of *in Vivo* Platelet Turnover and Organ Distribution Using ^{111}In-Labeled Platelets

By W. ANDREW HEATON, ANTHON DU P. HEYNS, and J. HEINRICH JOIST

Introduction

Until recently, chromium −51 (^{51}Cr) as labeled sodium chromate was considered the most suitable isotope for studies of platelet survival and organ distribution.[1] However, sodium [^{51}Cr] chromate has several major shortcomings as a platelet-labeling compound, i.e., low labeling efficiency (6–15%), a relatively long half-life (27 days), and γ-photon emission

[1] International Committee for Standardization in Hematology: Panel on Diagnostic Applications of Radioisotopes in Hematology, *Blood* **50,** 1137 (1977).

(9% at 320 keV) too low for external detection.[2,3] With the introduction by Thakur *et al.*[4] in 1976 of Indium-111 (^{111}In) oxine (or 8-hydroxyquinoline) complex for labeling of platelets, many of the shortcomings of ^{51}Cr as a platelet label were overcome. ^{111}In is a cyclotron-produced radionuclide, available essentially carrier free, with a physical half-life of 2.8 days. It decays by electron capture and thus emits no high-energy particulate radiation that would increase radiation exposure of organ tissues. Furthermore, ^{111}In emits two γ photons (172 and 247 keV) in high abundance (90 and 94%, respectively) yielding a high flux of radiations that are readily collimated and efficiently detected by current scintillation cameras.

As reviewed in detail elsewhere,[3] ^{111}In is taken up by platelets rapidly with high labeling efficiency and there is little elution of the label from platelets upon prolonged *in vitro* storage. ^{111}In is not lost from human platelets upon stimulation with strong release-inducing agents. ^{111}In uptake, adequate to allow both survival and organ-distribution studies, can be achieved without the induction of measurable changes of *in vitro* function or *in vivo* survival of the labeled platelets.

The method of labeling of platelets with ^{111}In-oxine still has several drawbacks. First, ^{111}In-oxine is a lipid-soluble complex that is stable above pH 4.0 and appears to be rapidly and nonspecifically taken up by all blood cells by a mechanism that at least with human platelets is temperature dependent but not dependent on metabolic energy.[3] Thus, for specific platelet labeling with ^{111}In-oxine, platelets must be isolated from other blood cells and any such isolation procedure may be associated with platelet damage. Second, ^{111}In-oxine has a high affinity for a variety of plasma proteins, in particular, transferrin. Thus, to achieve high labeling efficiency it is necessary to isolate platelets from plasma and label them in a nonplasma medium (balanced electrolyte solution) and this procedure also may be associated with substantial platelet damage.[3] Third, both oxine and ethanol (used as solvent for oxine) are potentially (above certain concentrations) toxic to platelets.[3] Fourth, although human platelets do not seem to lose appreciable amounts of ^{111}In radioactivity when stimulated *in vitro* with strong release-inducing stimuli such as collagen or thrombin,[5] substantial loss of ^{111}In radioactivity may occur from rabbit platelets,[6] and perhaps other species,[3] upon *in vitro* exposure to such agents. Such a loss of ^{111}In radioactivity may complicate the interpretation of platelet survival

[2] M. L. Thakur, *Semin. Thromb. Hemostasis.* **9,** 79 (1983).

[3] J. H. Joist, R. K. Baker, and M. J. Welch, *Semin. Thromb. Hemostasis.* **9,** 86 (1983).

[4] M. L. Thakur, M. J. Welch, J. H. Joist, and R. E. Coleman, *Thromb. Res.* **9,** 345 (1976).

[5] J. H. Joist, R. K. Baker, M. L. Thakur, and M. J. Welch, *J. Lab. Clin. Med.* **92,** 829 (1978).

[6] J. R. K. Baker, K. D. Butler, M. N. Eakins, G. F. Pay, and M. A. White, *Blood* **59,** 351 (1982).

data. Subsequently, a method of labeling of platelets with ^{111}In-(tropolone) in plasma was developed, which results in labeling efficiency comparable to that achieved in nonplasma medium, and possibly improved preservation of platelet function.[7]

Methods

Preparation of ^{111}In-Oxine

Appropriate amounts (0.2–1.5 ml) of ^{111}In-chloride (Mediphysics, Inc., Emeryville, CA) in 0.02 *N* HCl are added to 50 μl of oxine in 95% ethanol (1 mg/ml) in a sterile conical glass tube. To achieve optimal platelet labeling, the concentration of oxine in the labeling solution should be 6.5 μg/ml or greater but should not exceed 50 μg/ml to avoid platelet damage[3] and minimize ^{111}In elution (due to oxine carryover) in the preparation of labeled platelets.[3] Since preformed ^{111}In-oxine is commercially available (Amersham, Arlington Heights, IL; Diagnostic Isotopes, Bloomfield, NJ) there seems to be little if any advantage in preparing the complex in one's own laboratory. In some early studies,[4,5] ^{111}In-oxine was extracted using 2 vol of chloroform or dichloromethane to increase purity of the preparation. However, this procedure was subsequently omitted when it was found to be not only unnecessary but also associated with variable loss of ^{111}In due to adsorption of the complex to glass surfaces.[3] The ^{111}In-oxine is added to 4 ml of a mixture of one part of acid–citrate–dextrose (ACD) solution (citric acid, 8.0 g/liter; sodium citrate, 22.4 g/liter; anhydrous dextrose, 2.0 g/liter) and six parts of isotonic saline, and the pH adjusted to 6.5 with 0.1 *N* NaOH.

Preparation of ^{111}In-Labeled Platelets

Variable amounts of whole blood [43 ml in subjects with platelet counts > 100,000/μl, 86 ml with platelet counts > 50,000 < 100,000/μl, and 129 ml with severe thrombocytopenia (< 20,000/μl)] are collected by clean venipuncture (using a 19-gauge needle and a two-syringe technique) into a sterile, disposable 50- or 60-ml syringe containing 7 ml of ACD solution, and carefully mixed. An additional 18 ml of blood is collected into a 20-ml disposable plastic syringe containing 2 ml of 3.8% sodium citrate solution, to be used for preparation of citrated platelet-poor plasma (C–PPP) for resuspension of ^{111}In-labeled platelets and measurement of *in vitro* function of labeled platelets. All subsequent procedures are carried

[7] M. K. Dewanjee, S. A. Rao, and P. Didisheim, *J. Nucl. Med.* **22**, 981 (1981).

out at room temperature under aseptic conditions in a laminar flow hood. Starting solutions are sterilized by heat or Millipore filtration. Using sterile techniques, the ACD–blood is transferred (using sterile plastic pipets) to a sterile plastic 50-ml screw-cap, conical, plastic centrifuge tube and centrifuged in a swing-out bucket centrifuge at 180 *g* for 15 min to yield ACD–platelet-rich plasma (ACD–PRP) with minimum loss of platelets and minimum red blood cell (RBC) contamination. The upper two-thirds of the supernatant ACD–PRP are transferred into a 15-ml sterile, disposable, conical, plastic tube and centrifuged at 2000 *g* for 10 min. The supernatant ACD–platelet-poor plasma (ACD–PPP) (to be used for washing of ^{111}In-labeled platelets) is carefully transferred to a sterile, disposable, plastic tube and the tube is capped. The platelet sediment is immediately aspirated, avoiding gross contamination with red blood cell sediment, and resuspended in 4 ml of prewarmed (37°), sterile ACD–saline solution, pH 6.5, containing the desired amount of ^{111}In-oxine, and the mixture is incubated for 3–5 min with intermittent agitation. An aliquot (0.5 ml) is removed and centrifuged in a microcentrifuge at 12,000 *g* for 2 min. The supernatant PFP is removed from the platelet sediment carefully and as completely as possible and both are counted in a gamma counter for determination of labeling efficiency according to the formula:

$$\text{Labeling efficiency} = [P/(P + M)]\,(100)$$

P is the platelet-associated radioactivity and *M* is the radioactivity in the suspending medium. The remainder of the labeled platelet suspension is centrifuged at 2000 *g* for 10 min at room temperature, the supernatant solution is discarded, and the platelet sediment is immediately resuspended in 4 ml of ACD–PPP and incubated for 3 min at room temperature to remove non-platelet-bound ^{111}In. The mixture is again centrifuged at 2000 *g* for 10 min, the supernatant PPP is removed, and the ^{111}In-labeled platelet sediment is resuspended in 5 ml of C–PPP. One milliliter of the ^{111}In-labeled PRP is removed for platelet counting and determination of the extent of contamination by other blood cells (phase-contrast microscopy), and examination of *in vitro* platelet function (aggregation in response to ADP and collagen using a turbidimetric method, aggregometry).[8] With this method, concentrations of ADP and collagen should be used that cause slightly less than maximal changes of light transmission in the aggregometer in intact C–PRP (not subjected to the labeling procedure). The changes in light transmission observed at 5 min after addition of the stimulus using the labeled C–PRP (diluted with C–PFP to give a platelet concentration similar to that in the control C–PRP) should be at least 50%

[8] G. V. R. Born and M. J. Cross, *J. Physiol. (London)* **168,** 178 (1963).

of those found with the control C–PRP. This turbidimetric method can be used reliably only with a platelet concentration in PRP > 70,000/μl. A simpler method for monitoring the functional integrity of the labeled platelets (which can be used with PRP preparations having a platelet concentration of < 70,000/μl), based on examination of platelet clumping after mixing of a drop of labeled C–PRP with 10 μl of an ADP solution on a glass slide, has also been suggested.[9] The labeled platelets are used if aggregates are observed on visual inspection or—if this is not the case—if on subsequent examination by phase-contrast microscopy, aggregates consisting of 10 or more platelets are seen.[9] The criteria for adequate preservation of platelet function with this method and with aggregometry are somewhat arbitrary and perhaps overly conservative since platelets appear to have the capacity to recover from certain types of physical and chemical injury with time both *in vitro* and *in vivo.* However, if a preparation of labeled C–PRP does not meet these quality control standards, more extensive "collection injury" (platelet damage during preparation) should be suspected. In this case, step-by-step reevaluation of the entire labeling procedure, including checks on preparation of solutions and reagents, should be undertaken. Major platelet damage during the labeling procedure may lead to rapid trapping of the reinjected platelets in the lungs and spleen with poor platelet recovery and shortened survival. Occasionally, both with normal volunteers and patients, markedly impaired platelet aggregation may be observed in both the labeled and control C-PRP. This can usually be attributed to inadvertent ingestion or administration of drugs (aspirin, penicillin-like antibiotics) or a clinical condition known to be associated with platelet dysfunction. If the labeled C–PRP is found to contain more than one red blood cell (RBC) or white blood cell (WBC) per 100 platelets, the C–PRP should be centrifuged at 200 *g* for 5 min to remove the contaminating blood cells.

Injection of Labeled Platelets

The labeled C–PRP is aspirated into a 5-ml plastic, sterile disposable syringe, the syringe is weighted (to 0.01 g), and the amount of radioactivity in the syringe is determined using a dose calibrator. For studies of platelet survival and organ distribution 30–50 μCi of injected radioactivity is adequate to yield radioactivity in the circulating blood sufficient for accurate counting. Standards are prepared to simulate dilution of the injected labeled platelet preparation by adding 200 μl of labeled C–PRP to 100 ml of distilled or deionized water.

[9] A. du P. Heyns, P. N. Badenhorst, H. Pieters, M. G. Lötter, P. C. Minnaar, L. J. Duyvene de Wit, O. R. van Reenen, and F. P. Retief, *Thromb. Haemostasis.* **42,** 1473 (1980).

A clean venipuncture is performed (using a 19-gauge butterfly infusion set) and 4.5 ml of blood is removed and added to an EDTA tube for determination of background radioactivity. The syringe containing the labeled C–PRP is then attached to the infusion set, the labeled C–PRP is injected, the empty syringe is removed, 5 ml of sterile saline is injected through the infusion set to ensure infusion of all of the labeled C–PRP in the set, and the infusion set is removed. The syringe used for injection of the labeled C–PRP is weighted again to determine the weight of the injected volume of C–PRP, which is divided by a correction factor of 1.027 and used for determination of the exact volume of the injectate.

Blood Sampling and Simple Analysis

Following injection of the labeled C–PRP, 4-ml blood samples (to be added to EDTA-tubes) are obtained by clean venipuncture at various time intervals for a total of at least 10 samples or until the residual blood radioactivity is less than 20% of the starting radioactivity. The time intervals are chosen according to the expected degree of shortening of platelet survival as indicated by the extent of thrombocytopenia and clinical or experimental conditions and factors. Thus, if more severe shortening of platelet survival is expected, blood samples are collected at 10, 15, 30, 45, 60, 90, 180, 300, 600, and 1200 min after injection of labeled C–PRP, whereas time intervals of 2, 3, 4, 24, 48, 72, 96, 120, 144, and 170 hr may be appropriate if minor to moderate shortening of platelet survival is anticipated. Blood sampling times should always be included in reports of platelet survival studies. If blood sampling at shorter intervals is required, samples may be collected through an indwelling intravenous catheter which may be kept open with sterile saline containing 5 units(u)/ml of heparin. Under such circumstances, a sufficient volume of blood (3–5 ml) must be withdrawn and discarded before each blood sampling to prevent dilution of the blood sample due to admixture of heparinized saline.

Within 4 hr after blood sampling, the samples are gently but thoroughly mixed and duplicate 1-ml aliquots are removed from each sample and added directly to separate counting vials. A third aliquot is centrifuged at 2000 *g* for 20 min, the supernatant PPP is decanted as completely as possible into a counting tube, and the sediment itself is resuspended in 0.5 ml of saline and added to another counting tube for estimation of platelet-associated radioactivity according to the formula:

$$\text{Platelet-associated radioactivity} = [\text{cpm platelets}/(\text{cpm platelets} + \text{cpm PPP})](100)$$

All blood samples, together with the appropriate standards, are counted on day 7, or with more rapid decrease of blood radioactivity on the last day of sample collection, in a NaI gamma well counter set to include both photopeaks of ^{111}In (130–270 keV). The minimum cpm per sample should be such that the counting error is less than 2% (generally greater than 5000 cpm). If the total time required for counting of a given set of samples exceeds 3 hr (approximately 5% of $t_{1/2}$ of ^{111}In), it is necessary to correct each sample count for decay in radioactivity according to the formula:

$$N_0 = N_T e^{kt}$$

N_0 is the number of cpm corrected to zero time; N_T is the actual cpm at time t; k is the ln 2/2.8 ($t_{1/2}$ of ^{111}In); and t is the elapsed time from counting first sample (in days). After estimation of the blood volume of the subject under study (usually by a simple method based on height and weight),[10] recovery of radioactivity for a given sample is calculated using the formula:

$$\text{Recovery (\%)} = \frac{\text{cpm/ml blood sample} \times \text{estimated blood volume (ml)}}{\text{cpm/ml standard} \times 500 \times \text{volume of injectate (ml)}}$$

Note that the cpm/ml blood sample must be corrected to yield platelet-associated radioactivity before value is entered into formula. Percentage recovery of injected platelet-associated radioactivity and survival of labeled platelets may be determined by a variety of methods,[1] including linear regression analysis, exponential mode, weighted mean of linear and semilogarithmic least-squares estimate, and gamma function analysis, using appropriate computer programs. Each of these methods may be particularly appropriate for a certain set of data. Thus, after plotting the percentage recovery values on the y axis versus time after injection of labeled platelets on the x axis using both an arithmetic scale and a semilogarithmic scale [log(y) vs x), it will be apparent whether the data fit more closely a linear or exponential model. Outliers and data errors may also be detected. The estimate of the mean survival time based on the exponential model is strongly affected by systematically large or small values of y for times near 0. This is important to note since very early data points (< 20 min) may be quite variable because of variable initial sequestration in the reticuloendothelial system of labeled platelets that were damaged during the labeling procedure. After the labeled platelets reach equilibrium with blood (20–30 min), variability of recovery values should decrease. Since platelet survival

[10] International Committee for Standardization in Hematology Report on Standard Techniques for the Measurement of Red Cell and Plasma Volume, *Br. J. Haematol.* **25,** 801 (1973).

curves generally are neither strictly linear (consistent with age-related destruction) nor strictly exponential (consistent with random destruction), the multiple hit model has been proposed.[11] This model includes the strictly linear and strictly exponential model as extremes and fits the data points to curves in between depending on the number of hits which may be estimated.[12] The topic of analysis of platelet survival data is discussed more extensively elsewhere.[11-14]

There appears at present to be no general agreement on a uniformly optimal method of analysis of platelet survival data. Nevertheless, to facilitate comparison of study groups[12] and interlaboratory comparison of platelet survival data, it seems useful as has been recommended[13,15] that the gamma function method be regularly employed. Data obtained with the use of this method—which is based on the "multiple hit" concept[11]—is reported together with those obtained with the use of other methods deemed more suitable in a particular clinical or experimental situation.

A comparison of data on ^{111}In-labeled platelet survival in normal subjects published by different groups of investigators is shown in Table I. Recovery ranged from 58 to 76% and survival varied from 162 to 216 hr depending on the method used for data analysis. Correction for changes in plasma radioactivity did not appreciably affect platelet survival estimates in normal subjects since radioactivity in plasma increased only slightly (6.6–9.6%) during 7 days of study. However, in cases where platelet survival is short, plasma radioactivity may rise more rapidly and to a greater extent than in normal subjects.[16] This, together with a more rapid decline in platelet-associated radioactivity, may lead to appreciable errors in the estimation of platelet survival unless appropriate corrections are made.

Data Interpretation and Clinical Applications

The validity of platelet survival estimation using ^{111}In-labeled platelets was first established by comparative studies using ^{111}In- and ^{51}Cr-labeled platelets in the same normal subjects.[17,18] Whereas survival and decay

[11] E. A. Murphy and M. E. Francis, *Thromb. Diath. Haemorrh.* **25,** 53 (1971).

[12] V. Fuster, J. H. Chesebro, L. Badimon, K. P. Offord, and H. W. Wahner, *Methods Hematol.* **8,** 189 (1983).

[13] A. M. Peters and J. P. Lavender, *Semin. Thromb. Hemostasis.* **9,** 100 (1983).

[14] E. A. Murphy, *Transfusion* **26,** 22 (1986).

[15] E. L. Snyder, G. Moroff, T. Simon, A. Heaton, and Members of the Ad Hoc Platelet Radiolabeling Study Group, *Transfusion* **26,** 37 (1986).

[16] H. H. Davis, A. Varki, W. A. Heaton, and B. A. Siegel, *Am. J. Hematol.* **8,** 81 (1980).

[17] W. A. Heaton, H. H. Davis, M. J. Welch, C. J. Mathias, J. H. Joist, L. A. Sherman, and B. A. Siegel, *Br. J. Haematol.* **42,** 613 (1979).

[18] K. G. Schmidt, J. W. Rasmussen, A. D. Rasmussen, and H. Arendrup, *Scand. J. Haematol.* **30,** 465 (1983).

TABLE I
RECOVERY AND SURVIVAL OF AUTOLOGOUS ^{111}INDIUM LABELED PLATELETS IN NORMAL SUBJECTS

Number of subjects (n)	Method of analysis: Linear		Weighted mean		Gamma function		
	Recovery (%)	Survival (hr)	Recovery (%)	Survival (hr)	Recovery (%)	Survival (hr)	Re
10	71 ± 4	214 ± 5	76 ± 4	172 ± 11	74 ± 4	180 ± 12	12
	67 ± 4[a]	212 ± 5[a]	72 ± 5[a]	162 ± 9[a]	71 ± 5[a]	176 ± 11[a]	
7	72 ± 6	216 ± 6					18
5	74 ± 10	200 ± 4			74 ± 10	196 ± 4	19
27	58 ±3 [a]	209 ± 3[a]	60 ± 3[a]	190 ± 4[a]	60 ± 3[a]	187 ± 5[a]	14

[a] Data corrected for plasma radioactivity. Figures represent means ± SEM.

patterns were similar, mean recovery of ^{111}In-labeled platelets was higher in one study.[17] In the other study,[18] a small increase in circulating radioactivity was noted in the first few hours following injection of labeled platelets. This may have reflected reentry into the circulation of a small fraction of platelets which were reversibly damaged during the labeling process. More recently, results of similar comparative studies in patients with shortened platelet survival, immune thrombocytopenic purpura (ITP), have also been reported.[19] Survival of homologous ^{51}Cr-labeled platelets was consistently shorter than that of autologous ^{111}In-labeled platelets, particularly in patients with predominant splenic sequestration, which could be attributed at least in part to increased elution of ^{51}Cr from platelets, consistent with previously reported *in vitro* findings.[20] The method described here has been successfully employed for estimation of autologous platelet survival in patients with severe thrombocytopenia.[16,19,21] In the latter study[21] several modifications (isolation of platelets from 170 ml of blood; repeated washing of red blood cells and buffy coat with balanced salt solution) were introduced to increase the total number and recovery of platelets available for labeling. Although preliminary data seem to indicate that ^{111}In-labeled platelet survival and organ distribution studies may be useful in predicting the severity of the platelet destruction process and the response to splenectomy, this remains to be definitively established before such studies can be recommended for routine clinical practice.

[19] A. du P. Heyns, H. Pieters, P. Wessels, P. N. Badenhorst, H. F. Kotze, and M. G. Lötter, *Thromb. Haemostasis.* **54,** 127 (Abstr.) (1985).

[20] J. H. Joist and R. K. Baker, *Blood* **58,** 350 (1981).

[21] A. du P. Heyns, P. N. Badenhorst, P. Wessels, H. Pieters, H. F. Kotze, and M. G. Lötter, *Thromb. Haemostasis.* **52,** 226 (1984).

Platelet survival using ^{111}In-labeled platelets has also been studied in patients with atherosclerosis and venous thromboembolic disease.[22,23] If the method described here is used for experimental studies in animals, data interpretation may be complicated by the fact that platelets from certain animal species may lose substantial amounts of ^{111}In upon induction of the release reaction.[3]

In Vivo Distribution of ^{111}In-Labeled Platelets

The distribution of ^{111}In-labeled platelets following intravenous injection into the donor is detected by an imaging device with a recording system. The scintillation camera is an excellent imaging device and scintigraphy of the whole body, an organ or organs, or region(s) of interest (ROI) is possible. Images of the distribution of radioactivity may be permanently recorded in black and white or color. It is useful to store the acquired images with the computer. The data may subsequently be analyzed and manipulated to obtain the maximum information from an "ideal" image. It should be recognized that tissue is not transparent to gamma rays and quantification of organ ^{111}In-radioactivity must be corrected for the influence of attenuation and geometry.[24] For instance, if only anterior images are acquired, the radioactivity in the spleen, which is posteriorly situated, is underestimated.[24,25]

Instrumentation

Typically, the components of the system are the scintillation camera, a collimator, and a computer. A scintillation camera consists of a large sodium iodide crystal viewed by an array of photomultiplier tubes. A γ photon interacts with the crystal, light is emitted, and this energy is converted into an electronic signal by the phototubes. The signals are summed and analyzed by a pulse-height spectrometer. A computerized circuit converts the output so that the signals received by the individual phototubes are spatially arranged to give an image of the region viewed by the camera.

[22] H. H. Davis, W. A. Heaton, B. A. Siegel, C. J. Mathias, J. H. Joist, L. A. Sherman, and M. J. Welch, *Lancet* **1,** 1185 (1978).

[23] H. H. Davis, B. A. Siegel, L. A. Sherman, W. A. Heaton, T. P. Naidich, J. H. Joist, and M. J. Welch, *Circulation* **61,** 982 (1980).

[24] O. van Reenen, M. G. Lötter, A. du P. Heyns, F. de Kock, C. Herbst, H. Kotze, H. Pieters, P. C. Minnaar, and P. N. Badenhorst, *Eur. J. Nucl. Med.* **7,** 80 (1982).

[25] A. du P. Heyns, M. G. Lötter, P. N. Badenhorst, O. R. van Reenen, H. Pieters, P. C. Minnaar, and F. P. Retief, *Br. J. Haematol.* **44,** 269 (1980).

A large-field-of-view camera is preferred. Some mobile scintillation cameras, designed for imaging of the 140-keV photons of ^{99m}Tc, may be adapted by modifying the collimator and are useful for studies on patients at the bedside. A camera with digital image processing capabilities is necessary for quantification of *in vivo* distribution of ^{111}In-labeled platelets. The crystal of the camera should be 12.5 mm thick; thinner crystals are less sensitive for the 247-keV energy photons of ^{111}In and the image quality suffers. A dual spectrometer is advantageous. Each of the two windows is set at 20% to accept either the 173- or the 247-keV photon peak. If only one window is available, the 247-keV peak should be used.

The collimator directs the photons emitted from the object to the crystal. Collimators generally consist of multiple parallel channels separated by lead septa. Therefore, only those photons directed in parallel with the channels reach the crystal. A medium-energy, high-sensitivity, parallel hole collimator is best. Suitable collimators are not available for mobile scintillation cameras. This may be overcome by superimposing two collimators, a high-resolution type and a high-sensitivity type, and shielding their rims with a 5-mm-thick lead strip to prevent lateral leakage of radiation.

Quantitative information of the analog image obtained with the scintillation camera may be derived with a computer interfaced to the camera. The signals are digitized and stored in the memory of the computer either as a sequential list of events, or as an image on a matrix. The data are displayed as an image in shades of gray, or in color. The standard utility programs of the computer software can handle general data acquisition, display of images, ROI selection, subtraction of background activity, smoothing of the image, and contrast enhancement. A group of images may be generated, replayed at different gray levels or in color, and the optimum method of display selected. The persistence oscilloscope of the camera is used for interactive-13-quality control and to position the patient correctly. A permanent record of the image is obtained on Polaroid film or, preferably, on X-ray film.

Image Acquisition

An image may encompass a whole body scan illustrating the *in vivo* distribution of all the ^{111}In-labeled platelets if scanning facilities are available. The radioactivity counts acquired should not exceed the acquisition rate capabilities of the computer. This may arise if the scanning rate over the spleen is too slow. Imaging of a ROI is usually done in the static mode. The collimator is placed as close as possible to the organ or region. Counts are accumulated over the area; 500,000 counts are necessary for an excellent image, but this is usually not practical with ^{111}In-labeled platelets. Counts are therefore usually acquired for a fixed time, often about 15 min, depending on the radioactivity in the ROI relative to that of the background. Thus the relative merits of practical considerations and the excellence of the image required have to be weighed.

Quantification of whole body and ROI radioactivity is much more time consuming. The process is conveniently divided into two phases: an equilibrium phase and a platelet-survival phase analysis. The equilibrium phase analysis involves the dynamic measurement of radioactivity accumulation within a selected organ or region during the first 30 to 90 min after reinjection of labeled platelets. In the platelet-survival phase, platelet redistribution in various organs or deposition in a ROI is serially measured, sometimes until the end of platelet life span.

Images of the whole body and ROI are acquired in the anterior and the posterior mode. Correction for attenuation and geometry is made by the geometric mean method.[24] The whole-body scan image is displayed in TV video format and the ROI is selected with the computer. ROI's are selected with the standard computer software, but the images obtained may be heavily influenced by observer bias. This may be minimized by having all selections made by a single observer, or averaging multiple measurements. Background counts are subtracted and counts are corrected for isotope decay.

The geometric mean counts of the whole body as well as the ROI are calculated by the formula:

$$G = (AP)^{1/2}$$

where G is the geometric mean counts and A and P are anterior and posterior counts.

In the equilibrium phase, images can be acquired only in a single projection, usually the anterior view. The geometric mean counts of a ROI can therefore not be calculated directly. The data may later be normalized by using the geometric mean estimate of ^{111}In-labeled platelet radioactivity in the ROI at equilibrium.

In the equilibrium phase, data are presented as time–radioactivity curves; this represents changes in counts relative to time. Serial images may be selected and recorded for documentation of events. In the platelet survival phase, whole-body and ROI images are acquired daily or every other day. Organ or ROI radioactivity is expressed as a percentage of whole-body radioactivity according to the formula:

$$\text{Organ } (i) \text{ radioactivity } (\%) = (A_i P_i / W)(100)$$

A_i and P_i are anterior and posterior counts for organ or region of interest and W is the geometric mean of whole-body counts. The time-radioactivity curves for the whole body and the ROI may be plotted with a linear regression technique by the method of least squares. The method has been validated by *in vitro* correlation.[26] The results reported in normal subjects are summarized in Table II.

Alternative methods for measuring ^{111}In-radioactivity in a single organ have been described.[27,28] Radioactivity is sometimes expressed relative to the administered dose of ^{111}In. The radioactivity of an ^{111}In standard is determined in an "organ" within a phantom constructed to correspond to the patient's body dimensions. Alternatively, radioactivity counts of an organ, measured in the patient, are corrected for attenuation with the geometric mean method and the results expressed as a percentage of administered radioactivity. The latter is mathematically corrected for attenuation, assuming that attenuation is monoexponential, or measured with a transmission flood source.

The scintillation camera does not provide a perfect image of the object being examined because it compresses a three-dimensional image into two planes. Resolution suffers and is primarily determined by the size of the ROI or "lesion" (e.g., thrombus) and also by the radioactivity in the lesion or ROI relative to that in the background and the surrounding tissues. This limitation may be minimized either by dual-isotope scintigraphy or emission-computed tomography. In the latter, sectional images, similar to those obtained with X-ray transmission-computed tomography, are obtained. Single-photon-emitting radionuclides or positron-emitting isotopes may be used as cell labels.[29] Although this is theoretically ideal, the major limitation is that these short-lived isotopes may not be appropriate to demonstrate the accumulation of sufficient numbers of platelets in the lesion, which often takes 24 to 48 hr. In dual-isotope scintigraphy, images of a second radiotracer distributed in the blood pool, usually ^{99m}Tc-labeled red cells, are acquired at the same time as those with ^{111}In-labeled platelets. The physical characteristics of ^{99m}Tc and ^{111}In permit easy separation of the two isotopes with a scintillation camera fitted with a medium energy collimator. The windows are adjusted with a 10% window for ^{99m}Tc (140 keV) and 20% window for ^{111}In (247 keV). It is possible to correct for the

[26] A. du P. Heyns, M. G. Lötter, H. F. Kotze, H. Pieters, and P. Wessels, *J. Nucl. Med.* **23,** 943 (1982).

[27] E. D. Williams, M. V. Merrick, and J. P. Lavender, *Br. J. Radiol.* **48,** 275 (1975).

[28] I. Klonizakis, A. M. Peters, M. L. Fitzpatrick, M. J. Kensett, S. M. Lewis, and J. P. Lavender, *Br. J. Haematol.* **46,** 595 (1980).

[29] M. E. Phelps, *Semin. Nucl. Med.* **7,** 337 (1977).

TABLE II
EARLY AND LATE ORGAN DISTRIBUTION OF AUTOLOGOUS ^{111}INDIUM-LABELED PLATELETS IN NORMAL SUBJECTS[a]

	Equilibrium phase			End-of-survival phase		
Distribution	Heart	Liver	Spleen	Heart	Liver	Spleen
Mean	7.3	9.6	31.1	1.9	28.7	35.6
± SD	1.4	1.2	6.1	1.2	8.3	9.7

[a] Data represent ^{111}In-related organ radioactivity expressed as a percentage of whole-body radioactivity.

blood content of the ROI, and the net deposition of platelets in a lesion may be determined accurately by subtracting the contribution due to ^{111}In-labeled platelets in the circulating blood. This method is especially appropriate for the study of small lesions in areas with high circulating blood activity such as heart, lungs, or large vessels. It has been used successfully for the demonstration of atherosclerotic lesions and small arterial thrombi.[30]

Interpretation of Images

The interpretation of gamma camera images is similar to that of radiographs. The general quality of the image should be assessed and aberrations noted. Relevant clinical, radiographic, and angiographic information enhances the clinical value of the interpretation of the image and makes it easier to recognize anatomical variations. An image may simply be interpreted by visual inspection. A positive scintigram is a discrete area of increased radioactivity, clearly greater than that of the background blood pool, and increasing with time if compared to the radioactivity in the blood as assessed in an organ such as the heart or large vessels. Images displayed in black and white on Polaroid film generally allow lesions with a 20% increase in radioactivity to be easily seen.

The image may be semiquantitatively analyzed by determining the ratio between radioactivity in a ROI and the background. The areas of the regions should be similar. The total counts per area may be used, or the number of counts per pixel per unit time may be calculated. A change in radioactivity in an organ may be expressed as a time–activity curve, or as the ratio between initial and final radioactivity.

[30] W. J. Powers and B. A. Siegel, *Semin. Thromb. Hemostasis.* **9,** 115 (1983).

ROI radioactivity may also be quantitated and expressed relative to that of a standard, or of ^{111}In radioactivity in the whole body. The ROI radioactivity may then be compared to that of a reference population, or dynamic changes in ^{111}In radioactivity in a ROI may be studied by generating and comparing time–radioactivity curves.

Applications

The techniques described above have been applied for the study of normal platelet kinetics and quantifications of their sites of sequestration[26,27,31] (Table II). The dynamics of the splenic platelet pool, splenic blood flow, and the transit time of platelets in the spleen have been determined.[26,27,32] Platelet kinetics and patterns of destruction (sequestration) of labeled platelets have been quantitatively assessed with autologous platelets in patients with ITP, even in the presence of severe thrombocytopenia.[16,19,33] However, the clinical usefulness of studies of ^{111}In-labeled platelet survival and organ distribution remains currently uncertain. The presence of accessory spleens may be demonstrated with the use of ^{111}In-labeled platelets[16] but it is not clear whether this method is more sensitive than that using heat-damaged ^{99m}Tc-RBC.

The radiation absorbed by various organs has also been assessed.[34] Normalized radiation absorbed dose estimates expressed in rem/mCi in normal subjects were for spleen 29.33, liver 2.48, testes 0.19, ovaries 0.33, red bone marrow 0.48, and total body 0.54, given a total body effective dose equivalent of 2.32 rem/mCi.

With the use of ^{111}In-labeled platelet imaging, venous thrombi, particularly those high in the ileofemoral venous system, and carotid artery atherosclerosis and arterial thrombosis may be visualized, the latter two lesions especially well with the dual-isotope technique.[30] Platelet deposition in aortic aneurysms and prosthetic arterial grafts may be dynamically and quantitatively studied and the effect of antiplatelet drugs on platelet deposition tested.[35,36] Thrombi in aneurysms of the left ventricle may be de-

[31] U. Scheffel, M.-F. Tsan, T. G. Mitchell, E. E. Camargo, H. Braine, M. D. Ezekowitz, E. L. Nickoloff, R. Hill-Zobel, E. Murphy, and P. A. McIntyre, *J. Nucl. Med.* **23,** 149 (1982).

[32] A. M. Peters, I. Klonizakis, J. P. Lavender, and S. M. Lewis, *Br. J. Haematol.* **46,** 587 (1980).

[33] A. du P. Heyns, M. G. Lötter, P. N. Badenhorst, F. de Kock, H. Pieters, C. Herbst, O. R. van Reenen, H. Kotze, and P. C. Minnaar, *Am. J. Hematol.* **12,** 167 (1982).

[34] A. du P. Heyns, M. G. Lötter, and P. N. Badenhorst, *Methods Hematol.* **9,** 216 (1983).

[35] J. L. Ritchie, J. R. Stratton, B. Thiele, G. W. Hamilton, L. N. Warrick, T. W. Huang, and L. A. Harker, *Am. J. Cardiol.* **47,** 882 (1981).

[36] A. du P. Heyns, M. G. Lötter, P. N. Badenhorst, F. de Kock, H. Pieters, C. Herbst, C. J. C. Nel, and P. C. Minnaar, *Arch. Surg.* **177,** 1170 (1982).

tected.[37] ^{111}In-labeled platelet imaging may also be of value in the diagnosis of early, acute renal transplant rejection and monitoring of the response to therapy.[38] Other applications have been the study of the redistribution of platelets after injection of protamine sulfate[39] and quantification of platelet consumption during cardiopulmonary bypass surgery.[40] The method of labeling platelets with ^{111}In-oxine clearly offers major advantages over that using sodium [^{51}Cr] chromate and has become an important tool for basic and clinical investigation in hemostasis and thrombosis and related areas. However, the role of and indication, for the use of ^{111}In-labeled platelets in clinical practice have yet to be determined.

[37] M. D. Ezekowitz, J. C. Leonard, E. O. Smith, E. W. Allen, and F. B. Taylor, *Circulation* **63,** 803 (1981).

[38] A. Fenech, A. Nicholls, and F. W. Smith, *Br. J. Radiol.* **54,** 325 (1981).

[39] A. du P. Heyns, M. G. Lötter, P. N. Badenhorst, H. Kotze, F. C. Killian, C. Herbst, O. R. van Reenen, and P. C. Minnaar, *Thromb. Haemostas.* **44,** 65 (1980).

[40] A. F. Hope, A. du P. Heyns, M. G. Lötter, O. R. van Reenen, F. de Kock, P. N. Badenhorst, H. Pieters, H. Kotze, J. M. Meyer, and P. C. Minnaar, *J. Thorac. Cardiovasc. Surg.* **81,** 880 (1981).

Section III

Platelet Secretion

Article 15

A. Dense Granule Constituents
Articles 16 and 17

B. α-Granule Constituents
Articles 18 through 28

C. Lysosomal Enzymes
Articles 29 and 30

[15] Platelet Secretory Pathways: An Overview

By JACEK HAWIGER

Platelets are carriers for a rich assortment of biologically active molecules packed into storage granules. Their content is secreted after membrane stimulation with different secretagogues. Secreted molecules interact with other platelets, with plasma proteins, and with cells and the extracellular matrix components in the vessel wall. The secretory response from platelets constitutes an effector mechanism for the formation of platelet thrombi, the regulation of fibrinolysis, the promotion of wound healing, and the proliferation of atherosclerotic lesions.

Four types of storage organelles are discerned in platelets (Fig. 1).[1] Dense bodies, also called δ granules, contain a nonmetabolic pool of adenine nucleotides (ATP and ADP) and guanine nucleotides (GTP and GDP), serotonin, Ca^{2+} and Mg^{2+}, and inorganic phosphate. The neurotransmitter, serotonin, is actively taken up against a concentration gradient, a process utilized to load dense bodies with a ^{14}C-labeled amine that along with adenine nucleotides serves as a marker for dense granule secretion. A second type of storage organelle, α granule, contains a wide array of small and large proteins: heparin-neutralizing platelet factor 4 and β-thromboglobulin, utilized as markers of α-granule proteins; adhesive molecules, fibrinogen, von Willebrand factor (vWF), fibronectin, and thrombospondin; coagulation factor V; and cellular mitogens, such as platelet-derived growth factor (PDGF) and transforming growth factor beta. They also contain protease inhibitors, such as plasminogen activator inhibitor (PAI) and α_2-antiplasmin, that regulate fibrinolysis. The third type of storage organelle, lysosome, is in the form of small vesicles distinct from α granules, and contains the hydrolytic enzymes, acid phosphatase (EC 3.1.3.2), β-glucuronidase, *N*-acetylglucosaminidase, arylsulfatase (EC 3.1.6.1), cathepsin, and heparitinase, among others. Lysosomes are primary rather than secondary since little endocytosis can be demonstrated in platelets. Therefore, lysosomal enzymes seem to function in the extraplatelet setting, their substrates are either in plasma or in the subendothelial extracellular matrix (e.g., heparan sulfate). The fourth type of storage

[1] M. E. Bentfeld-Barker and D. F. Bainton, *Blood* **59,** 472 (1982).

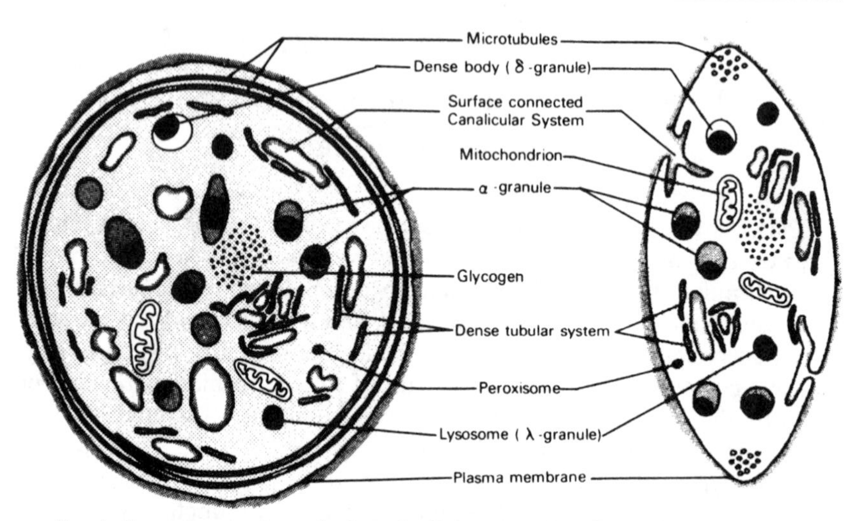

FIG. 1. Secretory granules and other subcellular organelles of the human platelet. In this diagram, based on electron microscopic and cytochemical visualization, the secretory granules are depicted in two planes. Among then, α granules predominate, interspersed with dense bodies (δ granules), lysosomes (λ granules), and peroxisomes. Note that the surface-connected canalicular system meanders around the granules which discharge their content therein, to be exported to the extraplatelet space. Reproduced with permission from Bentfeld-Barker and Bainton.[1]

organelle is the peroxisome, characterized by the presence of catalase.[2] The four types of storage organelles can be identified and studied in platelets and their progenitors, megakaryocytes, using ultrastructural, cytochemical, and immunocytochemical techniques, subcellular fractionation, and secretion kinetics of marker constituents. Selective or combined deficiencies of dense bodies and α granules ("storage pool deficiency") provide an opportunity to assess the contribution of granule content to platelet function *in vitro* and *in vivo*.[3,4] Measurement of secretion from platelet storage organelles constitutes the main thrust of this section. Methods describing the measurement of β-thromboglobulin and platelet factor 4 were published in Vol. 84 of this series.[5,6]

Secretion is initiated in response to agonists, such as thrombin, collagen, epinephrine, and ADP, acting on specific membrane receptors. At least three stimulatory pathways in human platelets that can be initiated all

[2] J. Breton-Gorius and J. Guichard, *J. Microsc. Biol. Cell* **23,** 197 (1975).

[3] H. J. Weiss, L. D. Witte, K. L. Kaplan, B. A. Lages, A. Chernoff, H. L. Nossel, D. S. Goodman, and H. R. Baumgartner, *Blood* **54,** 1296 (1979).

[4] J. G. White, *Hum. Pathol.* **18,** 123 (1987).

[5] K. Kaplan and J. Owen, this series, Vol. 84, p. 83.

[6] K. Kaplan and J. Owen, this series, Vol. 84, p. 93.

at once, or individually, depending on the nature and the potency of the agonist-evoked stimulus. The phospholipases A_2 and C, specific for phosphatidylcholine and phosphatidylinositol, play a key role in the generation of messenger molecules derived from phospholipids, archidonic acid, diacylglycerol, and inositol- 1, 4, 5-triphosphate. The latter mobilizes calcium from its membrane-bound pool. Another stimulatory molecule derived from phospholipids is platelet activating factor (PAF). GTP-binding proteins (G proteins) regulate phospholipase C activity. The Na^+/H^+ pump influences phospholipase A_2 activity. Among the known stimulatory pathways, shown schematically as solid lines in Fig. 2, the first pathway involves archidonic acid, which is transformed by an enzyme, cyclooxygenase, into biologically active endoperoxides (prostaglandin G_2 and H_2). Endoperoxides are converted enzymatically into thromboxane A_2 that triggers the secretory response through a poorly understood mechanism. The second stimulatory pathway is controlled by diacylglycerol, another phospholipid-derived messenger molecule. In the presence of Ca^{2+} and phospholipid, diacylglycerol activates protein kinase C, phosphorylating a number of intraplatelet proteins, most notably a protein of M_r 40,000 or 47,000. Whether protein kinase C-phosphorylated proteins mediate the fusion of membranes between storage organelles and the open canalicular system and whether the annexins (calpactin, lipocortin I, endonexin, endonexin II, and p70) are involved therein awaits examination. The third pathway involves the Ca^{2+}/calmodulin complex that activates myosin light chain kinase. The ensuing phosphorylation of myosin light chain is associated with early shape change of platelets and precedes their secretory response.

The nature of the secretory stimulus is relevant because certain agonists such as ADP require close contact between platelets before secretion can be initiated. Thrombin, on the other hand, is a very potent agonist, inducing secretion irrespective of cell–cell interaction. The major feature of stimulation is that the agonists (ADP, epinephrine, thrombin, collagen) interact with their own receptors evoking an intracellular signal. This signal is carried by messenger molecules to the granules, fusing them with the membrane of the surface-connected open canalicular system through which stored molecules are disgorged. Phospholipid-derived messenger molecules that mobilize Ca^{2+} or activate protein kinase C may work indirectly, or directly, as membrane fusogens promoting redistribution of the contents of storage organelles into a system of surface-connected open canaliculi. Thus, phospholipid-derived messengers exercise positive control of the secretory response.

On the other hand, inhibitory pathways exercise negative control imposed by the cyclic nucleotides, cyclic adenosine monophosphate (cAMP)

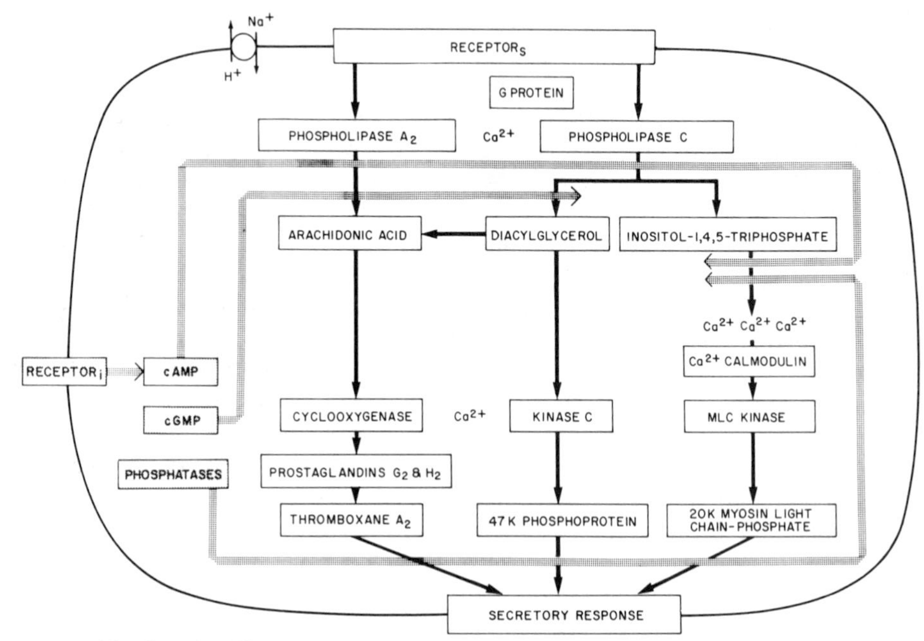

FIG. 2. A simplified scheme of the stimulatory and inhibitory pathways that modulate the secretory response in human platelets. Stimulation of platelets by agonists (ADP, epinephrine, collagen, thrombin), interacting with their specific receptors (receptor$_s$), generates a signal for activation of phospholipases A_2 and C. The former is influenced by the Na^+/H^+ pump, the latter by GTP-binding protein (G protein). Phospholipid-derived messenger molecules (arachidonic acid, diacylglycerol, and inositol-1,4,5-triphosphate) proceed along separate pathways represented by the solid lines to cyclooxygenase, protein kinase C, and myosin light chain kinase. Their enzymatic activity results in the formation of prostaglandin endoperoxides and thromboxane A_2 (cyclooxygenase pathway), or phosphorylated proteins (protein kinase C and myosin light chain kinase pathways). Ca^{2+} as an intracellular messenger is involved in several key steps in each stimulatory pathway. Individually, or collectively, they culminate in the secretory response. Inhibitory pathways involve receptor$_i$ for inhibitors (prostaglandin I_2, E_2, and adenosine) linked to adenylate cyclase. Soluble guanylate cyclase is independently activated by endothelial-derived relaxing factor (EDRF). Cyclic nucleotides (cyclic AMP and cyclic GMP) proceed along the stippled lines to interrupt the stimulatory pathways. Phosphatases exert negative control by dephosphorylating inositol 1,4,5-triphosphate and phosphoproteins involved in the stimulatory pathways.

and cylic guanosine monophosphate (cGMP). The former is elevated when inhibitors such as PGI_2 and adenosine interact with their receptors. Cyclic GMP is generated when endothelial-derived relaxing factor (EDRF), identified as nitric oxide, interacts with platelets. Phosphorylated molecules involved in intraplatelet stimulatory pathways are dephosphorylated by phosphatases, e.g., IP_3-specific 5′-monophosphoesterase. The methods concerning studies of intracellular and intercellular messenger molecules that exercise positive (stimulatory) and negative (inhibitory) control on the secretory response in platelets are presented in Section IV of this volume.

The secretory mechanisms in other blood cells such as neutrophils and macrophages have many similarities to platelets, but the latter have unusual features that presently appear to be unique in terms of the repertoire of molecules stored in dense bodies and α granules and the plurality of secretagogues. Much more information will be required before the mechanism of secretagogue-induced membrane fusion between platelet organelles and the surface connected open canalicular system is understood fully.

Detailed references are given in the chapters that follow. The analysis of platelet secretory pathways and their regulation can be found in a number of recent reviews.[7-12]

[7] J. G. White, *Blood Cells* **9,** 237 (1983).
[8] H. Holmsen and H. J. Weiss, *Annu. Rev. Med.* **30,** 119 (1979).
[9] P. W. Majerus, T. M. Connolly, H. Deckmyn, T. S. Ross, T. E. Bross, H. Ishii, V. S. Bansal, and D. B. Wilson, *Science* **234,** 1519 (1986).
[10] Y. Nishizuka, *Science* **233,** 305 (1986).
[11] J. Hawiger, M. L. Steer, and E. W. Salzman, *in* "Hemostasis and Thrombosis: Basic Principles and Clinical Practice" (R. W. Colman, J. Hirsh, V. J. Marder, and E. W. Salzman, eds.), p. 710. Lippincott, Philadelphia, Pennsylvania, 1987.
[12] S. Moncada, R. M. J. Palmer, and E. A. Higgs, *in* "Thrombosis and Haemostasis" (M. Verstrade, J. Vermylen, R. Lijnen, and J. Arnout, eds.), p. 587. Leuven University Press, Leuven, The Netherlands, 1987.

[16] Measurement of Secretion of Adenine Nucleotides

By HOLM HOLMSEN and CAROL A. DANGELMAIER

Secretion is an agonist-induced response of platelets. The secretable substances are stored in the dense granules,[1-3] the α granules, and the acid hydrolase-containing granules (lysosomes)[4] of the platelet. Upon stimulation the contents of these granules are rapidly emptied into the surrounding medium by exocytosis, and this secretory step can be monitored by

[1] Among them are ATP, ADP, serotonin, and divalent metal ions, e.g., Ca^{2+} (man) and Mg^{2+} (pig) (K. M. Meyers, H. Holmsen, and C. L. Seachord, (unpublished observations).
[2] K. M. Meyers, H. Holmsen, and C. L. Seachord, *Am. J. Physiol.* **243,** R451 (1982).
[3] H. Holmsen and H. J. Weiss, *Annu. Rev. Med.* **30,** 119 (1979).
[4] H. Holmsen and C. A. Dangelmaier, this volume [29].

measuring the secreted substances ATP or Ca^{2+} continuously[5,6] in the extracellular fluid of the platelet suspension, or discontinuously by stopping the secretion followed by separation of cells and medium and determining the secreted substances in the latter. Only the discontinuous methods will be described herein. Several steps in the procedure are common for secretion of adenine nucleotides, serotonin,[7] and acid hydrolases[4] and they will be described here.

The separation of platelets and medium has created problems because use of filters both activates and disrupts platelets and centrifugation of activated platelets causes further secretion.[8,9] The centrifugation-induced secretion can, however, be eliminated by fixing the platelets with formaldehyde prior to centrifugation. Plasma contains enzymes that rapidly hydrolyze ATP and ADP as they are secreted; plasma also contains large amounts of other major secreted constituents, calcium and acid hydrolases in particular. Therefore, removal of plasma eliminates the instability of secreted ATP and ADP as well as the high background levels of the other constituents. The methods described for secretion of adenine nucleotides herein and for the secretion of acid hydrolases[4] and serotonin[7] are based on the use of platelets in a plasma-free system. The gel filtration method is used to transfer the cells from platelet-rich plasma to a suitable buffer. However, the measurement of secretion can be performed with any suspension of washed platelets, and several procedures are easily adapted for a platelet-rich plasma system; brief recommendations for such adaptations are given.

Common Steps for Determining Secretion of Adenine Nucleotides, Serotonin, and Acid Hydrolases (Glycosidases)

Unless otherwise noted in the individual procedures, the general outline for collection of samples is as follows:

Blood is collected from human volunteers by venipuncture into one of the two anticoagulants described below. To avoid activation of the platelets during blood collection, it is advisable (1) to use briefly a gentle stasis on the arm to expose the veins, (2) to collect blood by *free flow,* i.e., without stasis and using a wide-bore (16-gauge) needle which leads the blood

[5] R. D. Feinman, J. Lubowsky, I. Charo, and M. P. Zabinski, *J. Lab. Clin. Med.* **90,** 125 (1977).

[6] J. W. N. Akkerman, H. Holmsen, and M. Loughnane, *Anal. Biochem.* **79,** 387 (1979).

[7] H. Holmsen and C. A. Dangelmaier, this volume [17].

[8] H. Holmsen, *in* "Mechanisms of Hemostasis and Thrombosis" (Mielke and Rodvien, eds.), p. 91. Symposia Specialists, Miami, Florida, 1978.

[9] H. Holmsen and C. A. Dangelmaier, *J. Biol. Chem.* **256,** 9393 (1981).

directly, or via a short plastic tubing, into an open collecting tube (absolutely avoid any vacuum device) and (3) to discard the first 5 ml of blood before collecting it in the tube(s) containing the anticoagulant. This procedure should be used for volunteers in the laboratory. However, similar principles may be used for acquisition of blood from blood bank donors. Thus, *after* the donor has given a unit under standard blood bank conditions and the tubing leading from the cannula to the blood bag is clamped, this tubing is cut between clamp and cannula and some 40 ml of blood is collected in a tube containing anticoagulant.

The blood may be collected in 1/10 vol of 0.11 *M* trisodium citrate. It is then spun at 180 *g* for 15 min at room temperature. The supernatant, platelet-rich plasma, is carefully collected using a plastic pipet. Alternatively, the blood is collected in $\frac{1}{6}$ vol ACD (2 g trisodium citrate, 1.5 g citric acid, 2.0 g glucose in 100 ml H_2O) and centrifuged at 700 *g* for 6 min at room temperature; the platelet-rich plasma is collected as above and further centrifuged for 10 min at 1500 *g* (room temperature) and resuspended in a one-third volume of the supernatant plasma (concentrated ACD platelet-rich plasma). This latter procedure enables appropriate adjustment of the platelet concentration with Tyrode's buffer after gel filtration (see below).

Both types of platelet-rich plasma are then gel-filtered through Sepharose 2B equilibrated with a Ca^{2+}-free Tyrode's solution containing 5 m*M* glucose and 0.2% albumin, yielding gel-filtered platelets (GFP). When prepared from concentrated ACD platelet-rich plasma the GFP contains normally around 10^9 cells/ml and should be diluted to $3-4 \times 10^8$ cells/ml with the Tyrode's solution before use. Details for the gel-filtration procedure are given elsewhere (see also Chapter [2] in this volume).[10] Although gel filtration is a reliable and reproducible method for transfer of platelets from plasma to suspension media, other methods may also be used for secretion studies.[11-13] However, methods employing apyrase (EC 3.6.1.5) in the suspending medium cannot be used for measurement of adenine nucleotide secretion.

Aliquots of GFP are equilibrated at 37°. One-tenth volume of agonist solution is added to test samples and solvent for the agonist is added to control sample. (If thrombin is used as agonist, the samples may be incu-

[10] B. Lages, M. C. Scrutton, and H. Holmsen, *J. Lab. Clin. Med.* **85,** 811 (1975).

[11] J. F. Mustard, D. W. Perry, N. G. Ardlie, and M. A. Packham, *Br. J. Haematol.* **22,** 193 (1972).

[12] H. J. Day, H. Holmsen, and M. B. Zucker, *Thromb. Haemostasis* **36,** 263 (1976).

[13] J. W. N. Akkerman, M. H. M. Doucet-de Bruine, G. Gorter, S. DeGraff, S. Holme, J. P. Lips, A. Numeuer, J. Over, A. E. Starkenburg, A. C. Trieschnigg, J. V. Veen, H. A. Vlooswijk, J. Wester, and J. J. Sixma, *Thromb. Haemostasis* **39,** 146 (1978).

bated without stirring; use of collagen, ADP, or epinephrine requires stirring, for example, in an aggregometer with a magnetic stirrer at 900 rpm.) The reaction is terminated after a given incubation time by mixing the sample with $\frac{1}{5}$ vol of ice-cold 0.633 *M* formaldehyde in 0.05 *M* EDTA.

These formaldehyde-fixed aliquots may be kept on ice for up to 60 min before further treatment. The aliquots are then centrifuged at 12,000 *g* for 2 min at room temperature in an Eppendorf centrifuge 3200 or a similar desk-top centrifuge. The supernatants obtained contain the substances secreted from the platelets (S_{test}) or the "background" levels of these substances in the medium of solvent-treated platelets ($S_{control}$). Additional aliquots for the determination of the total content ($T_{control}$) of secretable substances must also be prepared: For serotonin and acid hydrolases, an aliquot of control (i.e., 1 vol solvent + 10 vol GFP) GFP is mixed with $\frac{1}{5}$ vol of the formaldehyde–EDTA solution (above) on ice; for the adenine nucleotides aliquots are mixed with $\frac{1}{5}$ vol of 0.15 *M* NaCl on ice. These samples ($T_{control}$) are *not centrifuged.* The supernatants, as well as the totals (noncentrifuged samples) are then prepared for analysis of adenine nucleotides (below), serotonin,[7] or acid hydrolases.[4] Percentage secretion is calculated by the formula:

$$\text{Percentage secretion} = [(S_{test} - S_{control})/(T_{control} - S_{control})]\ (100)$$

where *S* designates the amount of substance or enzyme activity in the supernatants and *T* the total content or enzyme activity.

Maximal secretion is defined as the percentage secretion obtained with 5 NIH U/ml of thrombin for 5 min at 37°.

Detailed methods for the determination of rate and extent of secretion of the individual granule-located constituents are not given below. This is because the time course of secretion depends on the agonist used, its concentration, and the particular circumstances under which platelets and agonists are incubated (stirring or not, geometry of reaction vessel, type of platelet suspension species, etc.). Time courses should therefore be established empirically after the following guidelines:

Incubate 1–2 ml GFP with a given concentration of agonist under the conditions used and remove 200-μl samples into 50 μl of the formaldehyde–EDTA mixture on ice at noted time intervals. Analyze for amounts of constituent in supernatants of test and control samples, as well as in the noncentrifuged (total) sample and calculate percentage secretion. For thrombin, 0.2 U/ml or more, secretion is rapid and terminated within 20 sec (adenine nucleotides and serotonin) to 60 sec (acid hydrolases). For ADP, epinephrine, and collagen time periods of up to 3 min are required. Plot percentage secretion as a function of time with several agonist concentrations for GFP from normal donors who have not taken medications for at least 10 days prior to blood collection. Typical time courses for thrombin-induced secretion of adenine nucleotides and β-hexosaminidase are shown in Fig. 1. These time courses constitute the "normal" secretion patterns which should be compared to that of platelets from patients or normal platelets exposed to various substances (drugs). Normal platelets should always be processed simultaneously with patient platelets exposed to drugs in order to eliminate differences in platelet reactivity due to day-to-day variations. The time elapsed from blood collection to actual test of secretion should be kept constant.

Determination of Adenine Nucleotides

About two-thirds of the adenine nucleotide content of platelets of most species is stored in the dense granules and secreted, together with serotonin, when platelets are properly stimulated. The stored or secreted adenine nucleotides consist of ATP and ADP in an ATP/ADP ratio that varies considerably among species.[2,3] In platelets of most species, however, more ADP than ATP is secreted, and the method for ATP and ADP determination described below includes determination of both nucleotides, in contrast to the continuous assay[5,6] with the lumiaggregometer which only measures ATP. The secreted adenine nucleotides, as well as secreted GTP and GDP, may also be determined by HPLC, and several methods have been described.[14-18] They are, however, more time consuming than the firefly luminescence method and since EDTA can elute together with ADP and absorbs light at 255 nm[17] the EDTA–formaldehyde solution used for rapid termination of secretion and EDTA–ethanol extraction (see below) cannot be used for determination of all nucleotide secretion to the medium. However, only one secreted nucleotide needs to be measured in order to monitor dense granule secretion. Therefore, rapid HPLC procedures (isocratic elution, automatic sample injection) may well be developed for sole measurement of a given nucleotide (ATP, ADP, GTP, or GDP)[1-3] without interference from EDTA.

[14] G. H. R. Rao, J. G. White, A. A. Jachimowicz, and C. J. Witkop, *J. Lab. Clin. Med.* **84,** 839 (1974).

[15] R. E. Parks, G. W. Crabtree, C. M. Kong, R. P. Agarwal, K. C. Agarwal, and E. M. Scholar, *Ann. N.Y. Acad. Sci.* **255,** 412 (1975).

[16] L. D. D'Souza and H. I. Glueck, *Thromb. Haemostasis* **38,** 990 (1977).

[17] J. L. Daniel, I. R. Molish, and H. Holmsen, *Biochim. Biophys. Acta* **632,** 444 (1980).

[18] G. H. R. Rao, J. D. Peller, and J. G. White, *J. Chromatogr.* **266,** 466 (1981).

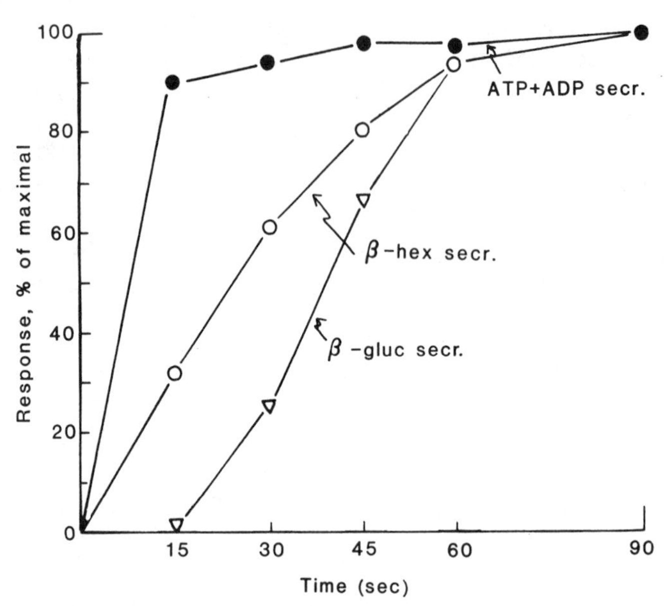

FIG. 1. Time course of secretion of platelet constituents. GFP was incubated with 0.2 U/ml of thrombin at 37°. Samples were taken out at the times indicated and mixed with formaldehyde–EDTA (see text). The samples were centrifuged and analyzed for secreted ATP + ADP (ATP + ADP secr.), β-hexosaminidase (β-hex secr.), and β-glucuronidase (β-gluc secr.). The secretions are given in percentages of the amounts of each constituent secreted at 90 sec.

Determination of ADP and ATP by Firefly Luminescence

Principle. The firefly luciferin–luciferase reaction emits light according to the equations:

$$\text{E (luciferase)} + \text{LH}_2 \text{ (luciferin)} + \text{ATP} \xrightarrow{\text{Mg}^{2+}} \text{LH}_2 \cdot \text{AMP} + \text{PP}_i \quad (1)$$

$$\text{E} \cdot \text{LH}_2 \cdot \text{AMP} + \text{O}_2 \rightarrow \text{E} + \text{product} + \text{CO}_2 + \text{AMP} + \text{light} \quad (2)$$

When a sample containing ATP is mixed with a suitably buffered reaction mixture of the firefly reagents, the intensity of the resulting light flash is directly proportional to the concentration of ATP. (Depending on the type and quality of the commercial firefly luciferin–luciferase preparation, the light intensity may be constant for several minutes or appear as an initial peak value from which it decreases with time. The characteristics of the light intensity–time relationship are usually given by the manufacturers.)

The ADP is converted to ATP by the pyruvate kinase system,

$$\text{ADP} + \text{phosphoenolpyruvate} \xrightarrow[\text{Mg}^{2+},\ \text{K}^+]{\text{pyruvate kinase}} \text{ATP} + \text{pyruvate}$$

which works optimally at pH 7.4 and requires the presence of Mg^{2+} and K^+.

Equipment

A Du Pont 760 luminescence biometer, a LKB 1250 Lumimeter, or a similarly sensitive, commercial, or self-constructed[19] instrument for luminescence measurement with dark injection.

Reagents

Ethanol–EDTA solution (EE)
0.2 *M* Tris–maleate buffer, pH 7.4
10 m*M* phosphoenolpyruvate (PEP solution) containing 0.4 *M* $MgSO_4$ and 1.3 *M* KCl
Pyruvate kinase (PK)—Boehringer Mannheim, 10 mg/ml
0.01 *M* morpholinopropanesulfonic acid (MOPS), pH 7.4, containing 0.01 *M* $MgSO_4$
Luciferin–luciferase powder (firefly lantern extract; FLE)
ADP stock standard solution (1 m*M*) in 0.15 *M* NaCl, aliquoted and kept frozen prior to use

Preparation of Solutions

I. *Inactive buffer* (to measure ATP): Mix 1.0 ml Tris–maleate buffer and 100 μl PEP solution, and make up to 25 ml with distilled water. Add 10 μl PK. Place in a boiling water bath for 30 min. Cool to room temperature.

II. *Active buffer* (converts ADP to ATP in order to measure ATP + ADP): This is the same mixture as (I) without boiling.

III. *FLE reagent:* The contents of one vial of the Du Pont luciferin–luciferase powder are dissolved in 10 ml MOPS by adding the enzyme–substrate powder to the MOPS. The reagent from LKB ("ATP monitoring reagent") is dissolved in 5 or 10 ml H_2O (depending on the sensitivity desired, according to manufacturer's label). Allow the mixture to stand 10 to 20 min at room temperature to dissipate inherent light. Do not shake. Aliquot 100-μl portions into disposable glass cuvettes (Du Pont Luminescence Biometer) or keep the FLE solution cold in the reservoir in the LKB instrument.

IV. *ADP standards:* A set of standards is always run with the samples. ADP is preferred over ATP because it provides an internal control for the

[19] H. Holmsen, I. Holmsen, and A. Bernhardsen, *Anal. Biochem.* **17,** 456 (1966).

conversion of ADP to ATP by the active buffer. The standards are diluted with normal saline to concentrations dependent on the platelet count. For analysis of adenine nucleotide secreted from platelets in "normal" concentrations in GFP or platelet-rich plasma ($3-4 \times 10^8$ cells/ml) or normal platelet-rich plasma ($2-7 \times 10^8$ cells/ml) the standard is diluted to 10–50 μM.

V. *EE solution:* This is a freshly prepared mixture of 1 vol of 0.1 *M* EDTA, pH 7.4, and 9 vol of 95% ethanol on ice.

Stability of Solutions and Reagents. The inactive buffer (I) is stable for a week if refrigerated between use. However, it should be checked periodically with the luminescence assay to avoid ATP contamination from bacterial growth. The active buffer (II) must be prepared immediately before use as the dilute pyruvate kinase activity decays rapidly. The Du Pont FLE reagent in MOPS (III) should be used within 24 hr, while the LKB preparation can be kept frozen for up to 2 months. It is recommended to run standards with each set of samples as the FLEs slowly lose their activity over time. It is imperative to keep the dried enzyme–substrate powder frozen and desiccated to minimize loss of activity.

The ADP standard (IV) is aliquoted and can be stored at $-20°$ in well-sealed containers (ampoules) for up to a year without decay. A new aliquot is prepared for each set of samples. Dilutions of the stock for working standards are kept on ice prior to use in the procedure.

The EE solution (V) is prepared fresh immediately before use. EDTA is salted out in this mixture, and becomes finely dispersed upon mixing ethanol and the EDTA solution. However, large clumps form during storage for 10 min. It is therefore desirable to distribute the freshly prepared EE immediately in 100- to 200-μl portions (in centrifuge tubes) placed on ice. On mixing the EE in the tubes with sample (or standard) as described below, the EDTA particles dissolve instantaneously.

Procedure

Preparation and Stability of Sample. To measure total ATP and ADP content, 1 vol (100–200 μl) of supernatants, "total samples" (see above), or ADP standard, or blank (0.15 *M* NaCl) is added to 1 vol (100–200 μl) of prechilled EE and vortexed immediately. The turbid mixture obtained is kept in ice until centrifuged for 1 min at $G_{max} = 12{,}000\ g$ at room temperature. The supernatant EE extract is kept in ice for immediate determination of ATP and ADP or stored at $-20°$ until analysis. Samples are stable up to 2 months at $-20°$ (liquid state) and 6 months at $-60°$ (frozen state).

Assay System

Stage I: Conversion of ADP to ATP

1. For each run an active blank, inactive blank, and standards are prepared.

2. Each sample is added to a tube with active buffer (II) and to a tube with inactive buffer (I). No more than nine samples should be measured in one run.

3. Pipet 1.0 ml of inactive buffer (I) into appropriate reagent tubes ("inactive" tubes).

4. Prepare active buffer (II) and immediately pipet 1.0 ml into appropriate reagent tubes ("active" tubes):

a. Add 25 μl of sample to one of the "active" tubes and start a stopwatch.
b. Add 25 μl of the remaining samples and standard to respective "active" tubes with a constant time interval (15 sec) between additions.
c. Incubate each of these mixtures for exactly 10 min at room temperature and place samples in 80–100° water bath for exactly 6 min.
d. At the end of the 6-min incubation place tubes in ice.

5. During the 10-min incubation at room temperature of the "active" samples, quickly add 25 μl of each sample (not standards) to respective "inactive" tubes. The addition of these samples does not have to be timed accurately since no reaction is taking place. They are also placed in the 80–100° water bath for exactly 6 min and then placed in ice.

Stage II: Measurement of ATP. For the Du Pont Luminescence Biometer inject 10 μl of the ice-cold mixtures resulting from steps 4 and 5 above into a cuvette containing 100 μl FLE reagent (III). For the LKB Lumimeter add 400 μl of the mixture to the cuvette and inject 100 μl FLE reagent (III). Record the readout, and repeat with a new cuvette each time, until three consecutive readouts show a variation of no more than $\pm$5%. Calculate the mean of the three injections to obtain the mean light response (MLR). All blanks, standards, "active," and "inactive" samples are read in this manner.

Calculations. The blank values are subtracted from all readings ("corrected MLR reading"). Since the peak intensity of light is directly proportional to the concentration of ATP, the (unknown) concentration C in the sample (C_{sample}, in μM) is

$$C_{sample} = (\text{MLR sample/MLR standard})\,(C_{standard})$$

where $C_{standard}$ is the known concentration (in μM) of ADP in the ADP standard that was mixed with EE.

When the corrected MLR readings from the inactive samples are used in the formula, the ATP concentration in the samples is obtained. The concentration of ADP is calculated as follows:

$$\mu M \text{ ATP} + \text{ADP (MLR from active reading)} - \mu M \text{ ATP (MLR from inactive reading)} = \mu M \text{ ADP}$$

The ATP and ADP contents are often expressed per 10^{11} cells, which may be obtained as shown:

$$[\mu M \text{ ATP or ADP}](10^{11}/\text{platelet count in tested sample})(10^{-3}) = \mu\text{mol}/10^{11} \text{ platelets}$$

where "platelet count" refers to the number of platelets in 1 ml of tested sample.

Normal values: The total content of ATP and ADP is 5.4 ± 0.32 and 3.1 ± 0.38 μmol/10^{11} platelets, respectively. During maximal secretion 30% of ATP and 80% of ADP is secreted.

Percentage adenine nucleotide secretion: For most secretion experiments released ATP and ADP in the supernatants are used to monitor dense granule secretion, and only the values for ATP + ADP (in "active" samples) are of interest. While calculating the amounts of ATP + ADP in the formula for percentage secretion, the unit terms disappear and the formula is reduced to

$$\text{Percentage secretion} = (\text{MLR of } S_{test} - \text{MLR of } S_{control}) / (\text{MLR of } T_{control} - \text{MLR of } S_{control})(100)$$

where MLR is the mean light reading of "active" samples only, corrected for the blank. Maximal percentage of secretion of ADP + ATP in GFP is 55–60%.

The exact concentration of ADP in the standard used above should be measured by the coupled pyruvate kinase–lactate dehydrogenase method as described elsewhere.[20] Sources of errors in the above procedure are (1) too high EDTA concentration in the sample, which will inhibit the conversion of ADP to ATP in the active sample, and (2) rapid decay of the pyruvate kinase activity upon dilution when preparing the active buffer. Both errors may be checked by processing ATP of known concentration together with the ADP standards.

[20] C. A. Dangelmaier and H. Holmsen, *in* "Measurement of Platelet Function" (L. A. Harker and T. S. Zimmerman, eds.), p. 98. Churchill-Livingstone, New York, 1983.

Adaptation for PRP

Secretion of ATP and ADP can be measured in platelet-rich plasma and the procedure above can be followed without modifications. However, ATPases and ADPase are present in plasma and will degrade ATP (half-life at 37° = 2–3 min) and ADP (half-life at 37° = 7 min) when they are released from the platelet. This degradation is abolished by the EDTA–formaldehyde treatment and can be prevented by the presence of 5 mM EDTA in the platelet-rich plasma.

[17] Measurement of Secretion of Serotonin

By Holm Holmsen and Carol A. Dangelmaier

The substances found in the secretory granules of circulating platelets are already packaged in these granules in the megakaryocyte. With the exception of serotonin in the dense granules, all other granule-stored substances do not exchange with their cytoplasmic or extracellular counterparts. Serotonin, however, readily crosses the plasma membrane with its transport system and is subsequently taken up by the granules through another system.[1] This ability of platelets to take up and store serotonin provides us with a simple, reliable method for determination of serotonin secretion (see Radioactive Assay, below). However, this ability also creates difficulties for secretion experiments because the secreted serotonin will rapidly reenter the same platelets that secreted it. This reuptake is promptly stopped by fixation with cold formaldehyde.[2,3] However, reuptake also takes place while serotonin is secreted, which results in underestimation of the amounts secreted.[4] Uptake during secretion can, however, be effectively blocked by specific inhibitors of serotonin transport across the plasma membrane, such as the tricyclic antidepressants which have IC_{50} values in the nanomolar range.[4-7] In the procedures outlined below, 2 μM

[1] J. F. Stoltz, *in* "Serotonin and the Cardiovascular System" (P. M. Vanhoutte, ed.), p. 37. Raven, New York, 1985.
[2] J. L. Costa and D. L. Murphy, *Nature (London)* **255,** 407 (1975).
[3] H. Holmsen and C. A. Dangelmaier, this volume [16].
[4] P. N. Walsh and G. Gangatelli, *Blood* **44,** 157 (1974).
[5] E. F. Marshall, G. Stirling, A. C. Tait, and A. Todrick, *Br. J. Pharmacol.* **15,** 35 (1960).
[6] A. Carlsson, K. Fluxe, and U. Ungerstedt, *J. Pharm. Pharmacol.* **20,** 150 (1968).
[7] M. Rehavi, Y. Ittah, K. C. Rice, P. Skolnick, F. K. Goodwin, and S. M. Paul, *Biochem. Biophys. Res. Commun.* **99,** 954 (1981).

imipramine is added to the platelet suspension before secretion is induced. Another drawback in using serotonin to monitor dense granule secretion is that serotonin may be released from platelets by other mechanisms than secretory exocytosis. Thus, other aromatic amines (tyramine, tryptamine) rapidly displace endogenous serotonin. Several ionophores may act as aminophores, causing rapid serotonin release. Inhibition of the serotonin uptake systems (both in the plasma membrane and the dense granule membrane) causes a slow serotonin release. These mechanisms are discussed in more detail elsewhere.[8]

A method detailed below is based on the use of platelets loaded with radioactive serotonin. A chemical (fluorimetric) assay is also described in detail below. Measurement of released, endogenous serotonin can also be done with HPLC methods.[9-11]

Radioactive Assay

Principle

Radiolabeled serotonin that has been taken up equilibrates rapidly and totally with the endogenous serotonin, and therefore behaves functionally as endogenous serotonin.[12] This allows quantitation of serotonin by simple measurement of the total radioactivity.

Preparation of Reagents and Stability of Solutions

5-Hydroxy[side chain-2-^{14}C]tryptamine creatinine sulfate or binoxalate (50–90 mCi/mmol). [^{14}C]Serotonin is usually in an aqueous solution containing 2% ethanol and is used without further preparation. It slowly decomposes when exposed to light and should be stored wrapped in aluminum foil or other light-proof containers at −20°. (Serotonin labeled with ^{3}H has been found particularly unsuitable for this "radioactive assay," since frequent freezing and thawing of [^{3}H]serotonin preparations leads to massive transfer of ^{3}H to the solvent H_2O)

Imipramine (1 m*M* in distilled water, stored at −20°)

[8] H. Holmsen, *in* "Mechanisms of Hemostasis and Thrombosis" (H. Mielke and R. Rodvien, eds.), p. 73. Symposia Specialists, Miami, Florida, 1978.

[9] B. Petrucelli, G. Bakris, T. Miller, E. R. Korpi, and M. Linnoila, *Acta Pharmacol. Toxicol.* **51,** 421 (1982).

[10] T. Fujimori, Y. Yamianishi, K. Yamatsu, and T. Tajima, *J. Pharmacol. Methods* **7,** 105 (1982).

[11] O. C. Ingebretsen, A. M. Bakken, and M. Farstad, *Clin. Chem.* **31,** 695 (1985).

[12] H. Holmsen, A. C. Ostvold, and H. J. Day, *Biochem. Pharmacol.* **22,** 2599 (1973).

Toluene/Triton X-100 (2:1 by volume) contains 2 g/liter of 2,5-diphenyloxazole or a corresponding scintillation cocktail that accepts 1% water. These scintillation liquids are stable at room temperature

Procedure

The procedure described elsewhere[13] for the general steps in secretion experiments should be followed with the following exceptions: (1) Platelet-rich plasma is incubated for 10 min at 37° for 30 min at room temperature with 1 μM of [^{14}C]serotonin before gel filtration; (2) prior to the secretion studies (in GFP or platelet-rich plasma) imipramine is added in a final concentration of 2 μM to prevent reuptake of [^{14}C]serotonin while it is being secreted.

After secretion studies have been done, the samples (formaldehyde-fixed supernatants, "total" samples) may be stored at −20° for several weeks or counted immediately.

Assay. Sample (100 μl) is counted in 10 ml of scintillation cocktail in a liquid scintillation counter.

Calculations. Because there are no chemical amounts measured in this assay, secretion is calculated as a percentage of the total radioactivity, i.e.,

$$\text{Percentage secreted} = (\text{cpm in } S_{\text{test}} - \text{cpm in } S_{\text{control}}) / (\text{cpm in } T_{\text{control}} - \text{cpm in } S_{\text{control}})$$

Normal Values for Serotonin Secretion. Under conditions of maximal secretion with 5 NIH units of thrombin/ml for 5 min at 37°, about 80–95% of the total radioactive serotonin is released extracellularly. During secondary aggregation (ADP, epinephrine), however, only up to 40% serotonin is secreted.

Adaptation for Platelet-Rich Plasma. Practically all the radioactive, authentic serotonin is taken up by the platelet so that very little radioactivity should be left in the plasma, which will constitute the supernatant of the control sample (S_{control} in the formula above). When the batch of radioactive serotonin is fresh, there may be 1–5% of contaminating radioactivity that is not associated with serotonin and which is not absorbed by the platelets and thus remain in S_{control}. With such low values of "cpm in S_{control}" secretion studies are easily done in platelet-rich plasma. However, during storage the radioactive serotonin will slowly be converted to compounds that are not taken up by the platelets, and thus the "cpm in S_{control}" term will increase to values that make the determination of serotonin

[13] H. Holmsen and C. A. Dangelmaier, this volume [16].

secretion increasingly inaccurate. This background radioactivity is removed during gel filtration, so this problem does not exist for GFP.

Sources of Error. Major sources of error are the failure of adding imipramine prior to secretion and the failure of using formaldehyde/EDTA to stop secretion (and other ways of serotonin release). It should be emphasized that the blocking of the plasma membrane transport system with imipramine causes a slow release of serotonin from the platelets. Therefore, imipramine should be added immediately before secretion is induced. Secretion results obtained with this radioactive assay will also be erroneous when the radioactive serotonin preparation contains radioactive contaminants that are absorbed by the platelets but not secreted and when serotonin undergoes substantial oxidation (to 5-hydroxytryptophol and 5-hydroxyindole acetate) once it has been taken up by the platelets, such as in various degrees of storage pool deficiency.[14-15] In both cases the denominator in the formula above will be artificially large and the percentage secreted becomes falsely low.

Chemical Method (Fluorimetric)

Principle and Sensitivity

Serotonin forms a fluorophore with *o*-phthalaldehyde (OPT) which can be easily detected down to 22 pmol under optimal fluorimetric conditions.[16] The content of serotonin varies in platelets from different species, from 300 nmol/10^{11} human platelets to 6930 μmol/10^{11} rabbit platelets.[17] The OPT method described below is therefore suitable for human platelet suspensions of $1-3 \times 10^8$ cells/ml.

Reagents.

HCl (1 and 8 *M*)
6 *M* Trichloroacetic acid (TCA)
o-Phthalaldehyde (OPT)
Chloroform
Serotonin creatinine sulfate

[14] J. W. Akkerman, H. K. Nieuwenhuis, M. E. Mommersteg-Leantand, G. Gorter, and J. J. Sixma, *Br. J. Haematol.* **55**, 153 (1983).

[15] K. M. Meyers, C. L. Seachord, K. Benson, M. Fukami, and H. Holmsen, *Am. J. Physiol.* **245**, H150 (1983).

[16] A. H. Drummond and J. L. Gordon, *Thromb. Diath. Haemorrh.* **31**, 366 (1974).

[17] M. DaPrada, J. G. Richards, and R. Vettler, *in* "Platelets in Biology and Pathology" (J. L. Gordon, ed.), p. 107. Elsevier/North-Holland, New York, 1981.

Preparation of Reagent Solutions and Their Stability. o-Phthalaldehyde (OPT), 0.5% (w/v) in ethanol, is mixed with 10 vol of 8 *N* HCl; this solution is stable for at least 2 months when stored in an amber bottle at +4°.

Serotonin creatinine sulfate (for standard) is dissolved to 1 m*M* in 1 *N* HCl and kept frozen and protected from light in small portions at −20°; no change in concentration takes place for at least 1 month. Working solutions (from 0 to 1 μM) are made in H_2O just before measurement.

Procedure

Treatment and Stability of Samples. After secretion has been performed, samples are transferred to $\frac{1}{10}$ vol 50 m*M* EDTA, pH 7.4, on ice. Formaldehyde must be omitted as it interferes with the assay. Samples are then centrifuged (except for the "total" sample). If serotonin analysis is not to be done immediately, add $\frac{1}{10}$ vol of 1 *N* HCl and store at −20°. Samples are stable for at least a month under these conditions.

Assay

1. To 600 μl sample add 120 μl 6 *M* TCA.
2. Centrifuge for 2 min at room temperature and 12,000 *g* in an Eppendorf centrifuge 3200 or a similar desk-top high-speed centrifuge.
3. To 500 μl of TCA extract add 2.0 ml of OPT reagent.
4. Place in boiling water bath for 10 min. Cool in ice.
5. Wash sample 2× with chloroform.
6. Read on a fluorimeter: activation = 360 nm; excitation = 475 nm.

Standard and blanks are processed as the samples.

Calculations. The blank values are subtracted from all readings.

$$C_{\text{sample}} = (\text{fluorimeter reading of sample}/\text{fluorimeter reading of standard})(C_{\text{standard}})$$

where C is the serotonin (μM).

$$\mu\text{mol 5-HT}/10^{11}\text{ platelets} = [\mu M\text{ 5-HT in tested sample}](10^{11}/\text{platelet count in tested sample})(10^{-3})$$

where "platelet count" refers to the number of platelets in 1 ml of tested sample.

Sources of Error. The formation of the OPT fluorophore under the conditions described is specific for serotonin in human platelets.

Sample readings must be within the 0–1.0 μM standard curve. If not, samples should be diluted prior to TCA extraction.

Since formaldehyde cannot be used, further secretion induced during centrifugation may occur with some agonists.[18] This unwanted effect may be prevented by separating cells and medium by centrifugation through silicone oil.[19]

Normal Values. Under the conditions of maximal secretion, some 80–90% of the platelet serotonin is released.

Adaptation for Platelet-Rich Plasma. The OPT method can be used without modification for PRP.

[18] H. Holmsen and C. A. Dangelmaier, *J. Biol. Chem.* **256,** 10449 (1981).

[19] H. Holmsen, C. A. Dangelmaier, and S. Rongved, *Biochem. J.* **22,** 157 (1984).

[18] Platelet-Derived Growth Factor: Purification and Characterization

By HARRY N. ANTONIADES and PANAYOTIS PANTAZIS

Introduction

Platelet-derived growth factor (PDGF) is the major mitogen of clotted blood serum. Its structural and functional properties are summarized below, and have been reviewed elsewhere.[1-5] PDGF is indispensable for the growth of mesenchyme-derived cells in culture, such as diploid fibroblasts, arterial smooth muscle cells, and brain glial cells. It is a heat-stable, cationic (p*I* 9.8–10.2) polypeptide consisting of two homologous polypeptide chains (PDGF-1 and PDGF-2) linked together by disulfide bonds. It circulates in blood stored in the α granules of platelets and it is released

[1] R. Ross and A. Vogel, *Cell* **14,** 203 (1978).

[2] C. D. Scher, R. C. Shepard, H. N. Antoniades, and C. D. Stiles, *Biochim. Biophys. Acta* **560,** 217 (1979).

[3] B. Westermark, C. H. Heldin, B. Ek, A. Johnsson, K. Mellstrom, M. Nister, and A. Wasteson, *in* "Growth and Maturation Factors" (G. Guroff, ed.), p. 73. Wiley, New York, 1984.

[4] H. N. Antoniades and A. J. Owen, *Horm. Proteins Pept.* **12,** 231 (1984).

[5] H. N. Antoniades, *Biochem Pharmacol.* **33,** 2823 (1984).

from platelets into the serum during blood clotting. *In vivo,* PDGF is apparently synthesized by megakaryocytes and it is stored and transported by platelets. These cellular elements adhere to the sites of vascular injury, aggregate there, and release their contents. In this manner, the platelets selectively deliver PDGF to sites of injury where it stimulates wound repair in connective tissue.

The primary *in vivo* function of PDGF is to induce mitosis in quiescent cells such as diploid fibroblasts, arterial smooth muscle cells, and brain glial cells. This would corroborate the proposed *in vivo* role of PDGF as a wound healing factor. The mitogenic effect of PDGF in target cells in culture occur at low concentrations (0.1 nM), similar to those required for the action of other hormonal polypeptides. In addition to its mitogenic activity, PDGF has been shown to be a potent chemoattractant for cultured fibroblasts, smooth muscle cells, and for human neutrophils and monocytes (see Chapter [20] in this volume). Other functions of PDGF include its ability to stimulate synthesis of protein, phospholipid and cholesterol esters, and prostaglandins, and to modulate receptor binding of several biologically important components including low-density lipoproteins (LDL) and serotonin.

The effects described above relate to possible physiologic functions of PDGF. However, PDGF has also been linked with abnormal states such as atherosclerosis and malignant transformation. The ability of PDGF to stimulate the proliferation and migration of arterial smooth muscle cells, and to stimulate cholesterol ester synthesis, prompted a hypothesis on its role in the pathogenesis of atherosclerosis.[6,7] According to this hypothesis, platelets that aggregate at the site of vascular injury release PDGF which initiates events involving the migration and proliferation into the intima of arterial smooth muscle cells. This process, along with an increase in cholesterol ester synthesis and the formation of connective matrix, may eventually lead to formation of the fibrous plaque that characterizes atherosclerosis.

Recent investigations established a prominent role for PDGF as a link to malignant transformation. These studies have shown that the PDGF-2 polypeptide chain and the transforming protein of the simian sarcoma virus (SSV), an acute transforming retrovirus of primate origin, derive from the same cellular gene. PDGF-2 and the SSV-transforming protein share a near identical amino acid sequence homology, have common antigenic determinants and structural conformation, including a disulfide-linked dimeric structure, and exert identical biological functions. It ap-

[6] R. Ross and J. Glomset, *N. Engl. J. Med.* **295,** 369 (1976).

[7] R. Ross, *Atherosclerosis* **1,** 293 (1981).

pears from these studies that the ability of the simian sarcoma virus to induce transformation derives from the incorporation of the PDGF gene within the retroviral genome. The resulting transforming *onc* gene (v-*sis*) region codes for a PDGF-like mitogen and is capable of inducing transformation by the continuous production of this potent mitogen, causing sustained cell proliferation. These studies have been extended to include the role of c-*sis* activation and PDGF synthesis and secretion in human malignant cells of mesenchymal origin such as fibrosarcoma, glioblastoma, and osteosarcoma. Detection of v-*sis*-related messenger RNAs (c-*sis* mRNA) has been reported in these malignant cell lines[8-11] and the synthesis and secretion of PDGF-like mitogen have been established.[9-11] These findings suggest that activation of *sis* transcription, accompanied by the synthesis and secretion of PDGF-like mitogen, may be involved in the processes leading normal cells of mesenchymal origin toward malignancy.

Localization of PDGF in the α Granules of Platelets

Platelets contain a number of organelles that store polypeptides and low-molecular-weight components. The platelet organelles release their contents during the clotting process. The platelet α granules, which are formed in the megakaryocytes, contain PDGF[12-15] along with platelet factor 4,[13] β-thromboglobulin,[14] and platelet fibrinogen.[13] The presence of these proteins in the α granules has been established by subcellular fractionation of disrupted platelets on a sucrose gradient. Consistent with the findings described above is the observation of reduced levels of PDGF, platelet factor 4, and β-thromboglobulin in a human patient with significantly reduced number of platelet α granules.[15]

[8] A. Eva, K. C. Robbins, P. R. Andersen, A. Srinivasan, S. R. Tronick, E. P. Reddy, N. W. Ellmore, A. T. Gallen, J. A. Lautenberger, T. Papas, E. H. Westin, F. Wong-Staal, R. C. Gallo, and S. A. Aaronson, *Nature (London)* **295,** 116 (1982).

[9] D. T. Graves, A. J. Owen, R. K. Barth, P. Tempst, A. Winoto, L. Fors, L. E. Hood, and H. N. Antoniades, *Science* **226,** 972 (1984).

[10] P. Pantazis, P. G. Pelicci, R. Dalla-Favera, and H. N. Antoniades, *Proc. Natl. Acad. Sci. U.S.A.* **82,** 2404 (1985).

[11] C. Betsholtz, B. Westermark, B. Ek, and C.-H. Heldin, *Cell* **39,** 447 (1984).

[12] D. R. Kaplan, F. C. Chao, C. D. Stiles, H. N. Antoniades, and C. D. Scher, *Blood* **58,** 1043 (1979).

[13] K. L. Kaplan, M. J. Brockman, A. Chernoff, G. R. Lesznik, and M. Drillings, *Blood* **53,** 604 (1979).

[14] L. D. Witte, K. L. Kaplan, H. L. Nossel, B. A. Lages, H. J. Weiss, and D. S. Goodman, *Circ. Res.* **42,** 402 (1978).

[15] H. J. Weiss, *N. Engl. J. Med.* **293,** 531 (1975).

Sources of PDGF Purification and Characterization

PDGF has been purified from human clotted blood serum, human platelets, and human platelet-rich plasma. The isolation and characterization of PDGF from human serum preceded its purification from platelets. These early studies revealed that the serum growth factor is a strongly cationic (p*I* 9.7) polypeptide, stable to heating (100° for 20 min), and sensitive to reducing agents.[16,17] It was shown subsequently that the serum growth factor activity resides in platelets[18-21] and it was named platelet-derived growth factor (PDGF). This, in turn, allowed the use of platelets as the main source for the large-scale purification and characterization of PDGF.

The procedures for the purification of PDGF, described below, take advantage of the unusual properties of PDGF. For example, its strongly cationic nature enabled the application of cation-exchange chromatography for the isolation and separation of PDGF from the bulk of other inactive proteins present in serum, platelet lysates, or plasma. Its "hydrophobic" nature allowed further purification using Blue Sepharose or phenyl-sepharose chromatography. The heat stability of PDGF provided additional purification, by precipitating contaminating proteins during heating at 100° for 10–20 min, and at the same time it served for the inactivation of proteolytic enzymes and may have contributed for the inactivation of viruses, such as the hepatitis virus, contaminating the derivative, from large pools of human blood.

Complications during purification derived primarily from the cationic nature of PDGF and its low concentration in the source material. The cationic nature of PDGF became the source of significant losses during purification, resulting from nonspecific adsorption on surfaces. This necessitated the introduction of conditions minimizing nonspecific adsorption of PDGF. Whenever possible the procedures were designed to utilize either high salt concentration (0.3 to 1.0 *M* NaCl) or strongly acidic conditions, such as 1 *M* acetic acid. Even under these extreme conditions significant losses during fractionation persist, accounting for the small yields of pure PDGF. These difficulties are compounded by the fact that PDGF is

[16] C. D. Scher, D. Stathakos, and H. N. Antoniades, *Nature (London)* **247,** 279 (1974).

[17] H. N. Antoniades, D. Stathakos, and C. D. Scher, *Proc. Natl. Acad. Sci. U.S.A.* **72,** 2535 (1975).

[18] S. D. Balk, J. F. Whitfield, T. Youdale, and A. C. Braun, *Proc. Natl. Acad. Sci. U.S.A.* **70,** 675 (1973).

[19] R. Ross, J. Glomset, B. Kariya, and L. Harker, *Proc. Natl. Acad. Sci. U.S.A.* **71,** 1207 (1974).

[20] N. Kohler and A. Lipton, *Exp. Cell Res.* **87,** 297 (1974).

[21] H. N. Antoniades and C. D. Scher, *Proc. Natl. Acad. Sci. U.S.A.* **74,** 1973 (1977).

present only in trace amounts in the source material. Its concentration in human serum was estimated at about 50 ng/ml by radioimmunoassay,[21,22] and at about 15 to 25 ng/ml by radioreceptor assay.[23] This necessitated fractionation of thousands of human platelet units/year, representing several thousand liters of human blood, for the production of adequate amounts of PDGF for characterization and functional studies.

Following is a description of the methods and techniques applied for the purification of PDGF from human blood serum, platelets, and platelet-rich plasma. A brief description is included of the cell culture techniques used for the identification and estimation of PDGF during purification.

Cell Culture Assays for PDGF

During purification, PDGF is detected and assayed for its effects on DNA synthesis in cultures of confluent BALB/c 3T3 cells. Autoradiography, or uptake of [^{3}H]thymidine by the cultured cells, is used for the evaluation of DNA synthesis.

For autoradiography, BALB/c 3T3 cells (clone A-31) are plated in 0.3-cm^2 microtiter wells (Falcon) in Dulbecco–Vogt's modified Eagle's medium (DME) supplemented with 10% calf serum (Biofluids) and incubated at 37°. When confluent monolayers are formed, the spent medium is aspirated and replaced with 0.2 ml of fresh DME supplemented with 5% human platelet-poor plasma, [^{3}H]thymidine (5 μCi/ml; 6.7 Ci/m*M*, New England Nuclear), and 10 μl of sample to be tested for PDGF activity. Platelet-poor plasma does not contain PDGF but it is required for the optimal replicative response.[2] After the sample addition, the microtiter cell cultures are incubated at 37° for 24 hr and then fixed with methanol and processed for autoradiography. Kodak nuclear tract photographic emulsion NTB2 is added (about 100 μl) into each microtiter well and then removed immediately using a Titertek 12-channel pipettor (Flow Laboratories). The plates are then kept in the dark in the presence of calcium sulfate desiccant for 24 hr;. The plates are developed by pipetting with the Titertek pipettor about 50 μl/well of Kodak Microdol-X developer and fixing with Kodak Rapid Fixer (100μl). The cultures are then stained with Giemsa. Percentage labeled nuclei is determined by counting at least 200 cells with microscopic grid while scanning approximately 10,000 cells. On

[22] J. S. Huang, S. S. Huang, and T. F. Deuel, *J. Cell Biol.* **97**, 383 (1983).

[23] J. P. Singh, M. A. Chaikin, and C. D. Stiles, *J. Cell Biol.* **95**, 667 (1982).

the basis of this assay, a unit of PDGF activity has been defined as the amount required to induce 50% (approximately 10^4) of the cells to synthesize DNA.

For estimation of the uptake of [^{3}H]thymidine, the 3T3 cells are incubated for 18 hr, as described above, and then they are pulse labeled for 6 hr with [^{3}H]thymidine (5 μCi/ml). At the end of the 6-hr incubation the cells are rinsed with cold saline, fixed in 200 ul of 10% trichloroacetic acid, and then rinsed with distilled water. The cells are solubilized in 200 μl of 1% SDS and the radioactivity is measured in a liquid scintillation counter.

It is recommended that assay samples of PDGF be diluted using 1% human serum albumin (reworked Cohn fraction V), to avoid losses of PDGF from surface adsorption.

Purification and Characterization of PDGF from Human Platelets

The direct isolation and characterization of PDGF from human platelets has been reported by our group in Boston,[24-27] and by Heldin and associates in Uppsala.[28-30] Our original procedure has been modified and improved as described below. The method developed by Heldin's group in Uppsala is described elsewhere in this series.[31]

Procedures and Modifications Applied by Antoniades *et al.*[24-27]

Source Material

Clinically outdated human platelet units (3–5 days old) are donated by the American Red Cross Blood Services, Northeast region. Each platelet unit derives from about 500 ml of human blood. It is stored in individual plastic bags and the platelets are suspended in about 50 ml of the donor's plasma.

[24] H. N. Antoniades, *Science* **127,** 593 (1958).
[25] H. N. Antoniades, C. D. Scher, and C. D. Stiles, *Proc. Natl. Acad. Sci. U.S.A.* **76,** 1809 (1979).
[26] H. N. Antoniades, *Proc. Natl. Acad. Sci. U.S.A.* **78,** 7314 (1981).
[27] H. N. Antoniades and M. W. Hunkapiller, *Science* **220,** 963 (1983).
[28] C.-H. Heldin, B. Westermark, and A. Wasteson, *Proc. Natl. Acad. Sci. U.S.A.* **76,** 3722 (1979).
[29] C.-H. Heldin, B. Westermark, and A. Wasteson, *J. Biochem. (Tokyo)* **193,** 907 (1981).
[30] A. Johnson, C.-H. Heldin, B. Westermark, and A. Wasteson, *Biochem. Biophys. Res. Commun.* **104,** 66 (1982).
[31] C.-H. Heldin, A. Johnsson, B. Ek, S. Wennergren, L. Rönnstrand, A. Hammacher, B. Faulders, A. Wasteson, and B. Wastermark, this series, Vol. 147, p. 3.

Step 1: Separation of Platelets from Plasma. The platelet units are pooled, concentrated by centrifugation at 3200 *g* for 30 min, and washed twice in 9 vol of 17 m*M* Tris–HCl, pH 7.5–0.15 *M* NaCl–0.1% glucose, and 1 vol of acid–citrate–dextrose buffer (ACD). The ACD buffer is prepared according to National Institutes of Health formula A: 8 g of citric acid monohydrate–22 g of dextrose (anhydrous)–26 g of sodium citrate (dihydrate), made up to 1 liter with distilled water. The washed platelets are concentrated by centrifugation, transferred into plastic containers with a minimal volume of 1 *M* NaCl–0.01 *M* sodium phosphate at pH 7.4, and stored at $-20°$. All further manipulations with platelets are conducted at 4° in plastic containers unless otherwise stated.

Step 2: Heat Extraction of PDGF from Platelets. Five hundred to 1000 units of washed platelets are thawed, and sufficient 1 *M* NaCl–0.01 *M* sodium phosphate buffer, pH 7.4, is added to bring the final volume to 2 ml/original platelet unit. The platelet suspension is heated at 100° for 15 min, with continuous gentle stirring. The supernatant fluid is separated by centrifugation and the precipitate is extracted twice with the 1 *M* NaCl solution (2 ml/original platelet unit).

Step 3: Cation-Exchange Chromatography. The combined platelet extracts from step 2 are dialyzed against 0.08 *M* NaCl–0.01 *M* sodium phosphate buffer, pH 7.4, and mixed overnight with 0.2 vol of CM-Sephadex C-50 (Pharmacia) equilibrated with the buffer. The mixture is poured into columns (5 × 100 cm), washed extensively with 0.08 *M* NaCl–0.01 *M* sodium phosphate buffer, pH 7.4, and eluted with 1 *M* NaCl while 10-ml fractions are collected.

Step 4: Blue-Sepharose Chromatography. Active fractions from step 3, corresponding to 500 platelet units, are pooled and dialyzed against 0.3 *M* NaCl–0.01 *M* sodium phosphate buffer, pH 7.4, centrifuged, and passed through a column (2.5 × 25 cm) containing 50 ml wet Blue Sepharose (Pharmacia) equilibrated with 0.3 *M* NaCl–0.01 *M* sodium phosphate buffer, pH 7.4. The flow rate is 20 ml/hr. The column is washed with 10 vol buffer and PDGF is eluted with a solution of 1 *M* NaCl and ethylene glycol (1 : 1), collecting 5-ml fractions. The active fractions are pooled, diluted (1 : 1) with 1 *M* NaCl, and dialyzed against 1 *M* acetic acid and lyophilized.

Step 5: CM-Sephadex Rechromatography. The lyophilized preparation from step 4 is dissolved in 0.08 *M* NaCl–0.01 *M* sodium phosphate, pH 7.4, centrifuged, and passed through a column (1.2 × 40 cm) of CM-Sephadex C-50 equilibrated with the same buffer. PDGF is eluted with a NaCl gradient (0.08 to 1 *M*). Active fractions with high specific activity are combined, dialyzed against *M* acetic acid, lyophilized, and dissolved in a small volume (about 3 ml) of 1 *M* acetic acid.

Step 6: BioGel P-150 Filtration in 1 M Acetic Acid. Samples (0.5 ml) from step 5 are applied to a column (1.2 × 100 cm) of BioGel P-150 (100 to 200 mesh; Bio-Rad) equilibrated with 1 *M* acetic acid at room temperature. The PDGF is eluted with 1 *M* acetic acid while 2-ml fractions are collected. Active fractions with high specific activity are pooled and lyophilized.

Step 7: High-Pressure Liquid Chromatography. Lyophilized fractions from step 6 containing 100 to 200 μg of protein are dissolved in 100 μl of 0.4% trifluoroacetic acid, and subjected to reversed-phase high-pressure liquid chromatography (HPLC) on a phenyl Bondapak column (Waters Associates), at room temperature. The PDGF is eluted with a linear acetonitrile gradient (0 to 60%). Pure PDGF preparations obtained from this step consist of two major bands corresponding to protein bands of 35,000 (PDGF-I) and 32,000 (PDGF-II) Da (Fig. 1). Separation of PDGF-I and II can be achieved by preparative SDS–PAGE.

Step 8: Separation of PDGF-I and PDGF-II by Preparative SDS–PAGE (Nonreducing Conditions). Electrophoresis is carried out as described by Laemmli[32] on 16% polyacrylamide gels. For preparative runs the PDGF obtained from HPLC is dialyzed overnight at room temperature against 0.05% aqueous sodium dodecyl sulfate (SDS). The dialyzed PDGF preparation is lyophilized and reconstituted in 2% SDS–0.06 *M* Tris–HCl, pH 6.8–15% sucrose, heated at 100° for 2 min, and applied on 16% polyacrylamide gels along with synthetic molecular weight standards (14,200–72,000, British Drug House, Poole, England). After electrophoresis, the gels are stained for 3–4 hr in 10% acetic acid–10% methanol (v/v) containing 0.5% Coomassie Brilliant Blue R-250 (Sigma). For destaining, the gels are placed in aqueous 10% acetic acid–10% 2-propanol (v/v). After destaining, the gels are rinsed with distilled water, incubated for 10–20 min in 0.1 *M* ammonium bicarbonate solution in order to neutralize the acid from the staining process and rinsed with distilled water. The stained protein bands in the region of PDGF-I (35,000) and PDGF-II (32,000) (Fig. 1) are sliced and placed in borosilicate culture tubes (Fisher Scientific) containing 1 ml 0.1 *M* ammonium bicarbonate–0.04% SDS. The gels are minced and allowed to elute for 8–12 hr at room temperature. The eluate is removed from the minced gel slices and stored at −20°. Elution of the gels is repeated once under the same conditions. The eluates are combined and aliquots are diluted in 1% human serum albumin for assay by cell culture. Preparation of PDGF-I and PDGF-II obtained by this procedure exhibited similar specific activities of about 3000 units/μg protein. The

[32] U. K. Laemmli, *Nature (London)* **227,** 680 (1970).

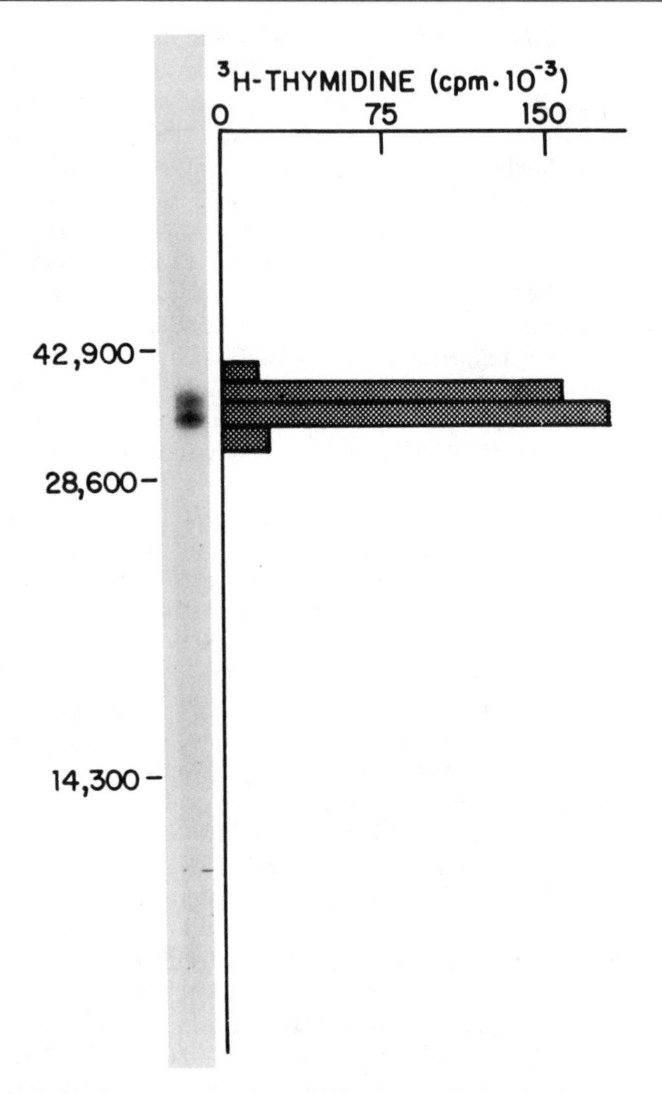

FIG. 1. Purified PDGF (unreduced) consists of two major polypeptides exhibiting molecular weights of about 35,000 (PDGF-I) and 32,000 (PDGF-II). PDGF activity was identified only with the two stained polypeptide bands corresponding to PDGF-I and PDGF-II.

specific activity of the original platelet lysate is about 0.03 unit/μg of protein.

The application of preparative SDS–PAGE for the separation of PDGF-I and II under nonreducing conditions, and of their reduced alkylated products (Fig. 1), proved effective in determining the amino terminal amino acid sequence of the PDGF homodimer.[27]

General Considerations

The procedures described above are applicable for the preparation of PDGF from relatively fresh platelet units. PDGF from platelet units stored for prolonged periods of time is present primarily in the plasma, due to progressive platelet degranulation during prolonged storage. For these preparations, we recommend the platelet-rich plasma as the source material of PDGF as described below.

Heat treatment of platelet lysates, described above, proved an important step, since this treatment largely destroyed the proteolytic enzymes released from platelets, stabilizing the PDGF preparations. In addition, heat treatment may contribute to inactivation of hepatitis virus and perhaps the Acquired Immunodeficiency Syndrome (AIDS) virus introduced to the large pools from potentially contaminated platelet units.

As mentioned previously, a serious problem during PDGF purification derives from the nonspecific adsorption of this cationic polypeptide by surfaces and contaminating proteins. For this reason, the techniques described above were designed to handle PDGF in high salt concentrations or in 1 *M* acetic acid. For assay for PDGF, where dilutions as great as 1 : 100,000 are required, we recommend that dilutions are made in 1% human, or bovine serum albumin–0.15 *M* NaCl solution.

Purification and Characterization of PDGF from Outdated Platelet-Rich Plasma

Purification of PDGF from platelet-rich plasma has been reported by Deuel *et al.*[33] and Raines and Ross.[34] Platelet-rich plasma derived from units of outdated platelets is a desired source for PDGF purification, since most of the PDGF activity is released from platelets into the plasma during prolonged storage. The procedures for PDGF isolation from plasma are largely similar to those applied for the purification of PDGF from platelets.

Procedure Reported by Deuel *et al.*[33]

Source Material

The source material consists of clinically outdated human platelets. The platelets from individual donors are stored in plastic bags and they are suspended at about 50 ml of the donor's plasma.

[33] T. F. Deuel, J. S. Huang, R. T. Proffitt, J. U. Baenzinger, D. Chang, and B. B. Kennedy, *J. Biol. Chem.* **256,** 8896 (1981).

[34] E. W. Raines and R. Ross, *J. Biol. Chem.* **257,** 5154 (1982).

Step 1: Preparation of Platelet-Rich Plasma. Sixteen liters of thawed human platelets with plasma derived from 400 pooled platelet units is stirred overnight at 4° in the presence of protease inhibitors at pH 6.1–6.5. The mixture is centrifuged at 13,000 *g* for 30 min and the clarified supernatant fluid is removed for further processing.

Step 2: Sulfadex G-50 Chromatography. The supernatant fluid derived from step 1 is mixed with 800 g (wet weight) of Sulfadex gel preswollen in 0.1 *M* NaCl–5 m*M* EDTA–0.01 sodium phosphate, pH 7.4. The mixture is allowed to stand overnight, the supernatant fluid is decanted, and the gel is packed into a column (5 × 90 cm). The column is washed with 0.5 *M* NaCl–5 m*M* EDTA–0.01 *M* sodium phosphate, pH 7.4, until the A_{280} is below 0.05. PDGF is eluted from the gel stepwise with 1.5 *M* NaCl–5 m*M* EDTA–0.01 *M* sodium phosphate, pH 7.4.

Step 3: CM-Sephadex Chromatography. The 1.5 *M* NaCl eluate of the Sulfadex gel is concentrated 10-fold by ultrafiltration with Amicon PM10 membranes, and heated in a bath of boiling water for about 10 min. The heated concentrate is centrifuged at 20,000 *g* to remove insoluble material and the supernatant fluid is dialyzed against 0.1 *M* NaCl–0.01 *M* phosphate buffer, pH 7.4. The dialyzed material is applied on a CM-Sephadex column (2 × 90 cm), equilibrated with the buffer described above. PDGF is eluted with 0.3 *M* NaCl.

Step 4: Blue-Sepharose Chromatography. PDGF preparation from step 3 is dialyzed against 0.3 *M* NaCl–0.01 *M* sodium phosphate buffer, pH 7.4, and is applied on a Blue-Sepharose column (2 × 25 cm), equilibrated with the same buffer. The column is washed extensively with the buffer, and PDGF is eluted with 50% (v/v) ethylene glycol in 1 *M* NaCl–0.01 *M* phosphate buffer, pH 7.4, dialyzed against 1 *M* acetic acid, and lyophilized.

Step 5: BioGel P-100 Chromatography in 1 M Acetic Acid. The lyophilized material from step 4 is dissolved in 0.5 ml of 1 *M* acetic acid and applied onto a BioGel P-100 column (1.5 × 90 cm), equilibrated with 1 *M* acetic acid. Active PDGF is localized in two overlapping protein peaks, corresponding to molecular weights of about 28,000 (PDGF-I) and 32,000 (PDGF-II).

Procedures Reported by Raines and Ross[34]

Source Material

Human platelet-rich plasma is derived from outdated platelet units (about 50 ml plasma/platelet unit).

Step 1: Preparation of Crude PDGF Extract. Outdated platelet-rich plasma from 80 units (about 4000 ml) is subjected to three cycles of freeze thawing, heat treated at 55–57° for 8 min to precipitate fibrinogen, transferred to large centrifuge tubes, and left at 4° overnight. The precipitated fibrinogen and platelet membranes are removed by centrifugation at 27,000 *g* for 30 min and the supernatant fluid containing the bulk of PDGF is collected and saved. The pellet, containing significant PDGF activity, is first washed with 200 ml of 0.39 *M* NaCl–0.01 *M* Tris, pH 7.4, followed by a second wash with 200 ml of 1.09 *M* NaCl–0.01 *M* Tris, pH 7.4. After each wash the pellet is removed by centrifugation at 27,000 *g* for 30 min. All supernatant fluids containing PDGF are combined and the pH is adjusted to 7.4 by the addition of 1.0 *M* Tris–base. The conductivity of the pool should not exceed 13.0 mmho, and should be adjusted with 0.01 *M* Tris, pH 7.4, if necessary, before further processing.

Step 2: Carboxymethyl-Sephadex Chromatography. The defibrinogenated platelet-rich plasma is stirred overnight at 4° with preswollen (1000 ml) CM-Sephadex C-50 equilibrated in 0.9 *M* NaCl–0.01 *M* Tris, pH 7.4. The slurry is layered on a column containing a 1.5-liter packed column bed of CM-Sephadex equilibrated in the same buffer and is loaded at a flow rate of about 150 ml/hr. The column is eluted by a step gradient which includes 0.09 *M* NaCl–0.01 *M* Tris, pH 7.4; 0.19 *M* NaCl–0.01 *M* Tris, pH 7.4; and 0.5 *M* NaCl–0.01 *M* Tris, pH 7.4. The last elution step removes the majority of the PDGF biological activity. The fraction eluted by 0.5 *M* NaCl is concentrated to 50 ml in an Amicon ultrafiltration unit using a PM10 filter. The concentrated material is then dialyzed against 1 *M* acetic acid using Spectrapor No. 3 dialysis tubing.

Step 3: Sephacryl S-200 Chromatography. The concentrated and dialyzed fraction from step 2 is centrifuged at 27,000 *g* for 30 min to remove particulate matter. After adjusting the pH to 3.5 with concentrated ammonium hydroxide, the fraction is applied at 4° on a Sephacryl S-200 column (5 × 92 cm, 1800 ml column volume) equilibrated in 1 *M* acetic acid with the pH adjusted to 3.5 with ammonium hydroxide. Maximum application volume is 60 ml containing a maximum of 150 mg protein, and the flow rate is 20 ml/hr. The column is eluted with 1 *M* acetic acid adjusted to pH 3.5, collecting 8-ml fractions.

Step 4: Heparin–Sepharose Chromatography. The PDGF fractions derived from step 3 are pooled, adjusted to pH 7 with concentrated ammonium hydroxide and to a conductivity of about 28 mmho (at 20°) by diluting with distilled water (at this step it is convenient to pool Sephacryl S-200 active fractions obtained from 160–240 units of platelet-rich plasma). The diluted Sephacryl fraction is applied on a column of heparin–Sepharose, previously equilibrated in 1 *M* acetic acid, and adjusted to pH 7 with ammonium hydroxide and a conductivity of about 28 mmho by the addition of distilled water. The flow rate of the column (1.5-cm diameter) is 20 ml/hr with a maximum load of 1 mg protein/ml of swollen gel. The column is eluted with a gradient of the same ammonium acetate buffer, with the conductivity increasing linearly from 28 to 80 mmho, at 20°. The majority of PDGF is eluted between conductivity of 39 and 53 mmho. The fractions containing PDGF are pooled for further purification.

Step 5: Phenyl-Sepharose Chromatography. The PDGF fraction derived from step 4, representing 160–240 units of platelet-rich plasma, is adjusted to a conductivity of 34 mmho (at 20°) with distilled water. The preparation is applied to a phenyl-Sepharose column at a flow rate of 15 ml/hr and with a maximum protein load of 1 mg/ml swollen gel. The column (0.9-cm diameter) has been previously equilibrated with 1 *M* acetic acid adjusted to pH 7.0 and conductivity of 24 mmho by the addition of distilled water (20°). The column is washed first with a buffer containing 10 ml 1 *M* acetic acid, adjusted to pH 7.0 with concentrated ammonium hydroxide, 60 ml distilled water, and 30 ml ethylene glycol. PDGF is eluted with a buffer containing 10 ml 1 *M* acetic acid, adjusted to pH 7.0 with concentrated ammonium hydroxide, 40 ml distilled water, and 50 ml ethylene glycol. Fractions eluted from the phenyl-sepharose column are tested for mitogenic activity after being dialyzed in the presence of 2 mg/ml bovine serum albumin.

Characterization Studies and Properties of PDGF

Purified PDGF obtained from platelets was shown to be a heat-stable, cationic (p*I* 9.8–10.2) polypeptide, sensitive to reducing agents.[24,27] These properties are identical to those described previously from the cationic polypeptide growth factor isolated from human serum.[16,17] Similar properties have been reported for PDGF derived from platelet-rich plasma[33,34] (Table I). The molecular weight of active, unreduced PDGF was estimated at 28,000–35,000. Upon reduction, the molecular weight appeared to be between 12,000 and 18,000, suggesting that biologically active, unreduced PDGF consists of two polypeptide chains. The amino-terminal amino acid sequence of human PDGF provided evidence that it consists of two homologous polypeptide chains linked together by disulfide bonds.[27] Recent studies have shown the presence of multiple molecular weight forms of active, unreduced PDGF obtained from platelets[26] or from platelet-rich plasma.[33,34] The two predominant forms obtained from platelets, named PDGF-I and PDGF-II, have molecular weights of about 35,000 and 32,000, respectively.[32] It appears that PDGF-II derives from PDGF-I by

TABLE I
PROPERTIES OF PDGF OBTAINED FROM HUMAN CLOTTED BLOOD SERUM, HUMAN PLATELETS, AND PLATELET-RICH PLASMA (PRP)

Properties	Serum[a]	Platelets[b]	PRP[c]
Heat (100° stability (10–20 min)	Stable	Stable	Stable
Isoelectric point	± 9.7	9.8–10.2	± 10.2
Molecular weight			
Unreduced	Not tested	28,000–35,000	28,000–32,000
Reduced	13,000	12,000–18,000	12,000–18,000
pH (2.0–9.6)	Stable	Stable	Stable
Reduction (2-mercaptoethanol)	Inactivated	Inactivated	Inactivated
Trypsin, chymotrypsin	Inactivated	Inactivated	Inactivated
Means of transport		Platelets	
Localized		α Granules	
Human blood serum concentration (ng/ml)			
Radioimmunoassay	50[d]		
Radioreceptor assay	15–25		

[a] From Refs. 16 and 17.
[b] From Refs. 21, 25, 27, and 29.
[c] From Refs. 33 and 34.
[d] From Refs. 21 and 22.

partial proteolysis, primarily at the carboxy-terminal region, which occurs during the storage of outdated platelets or during their handling and fractionation. Amino acid sequence data indicate that the amino-terminal regions of PDGF-I and -II are similar.[27]

General Comments and Precautions

Purification of PDGF during fractionation is assessed by monitoring its biologic activity as described above. It is important to relate the purification of PDGF to its activity at all stages of fractionation. Purified PDGF should be stored at −20°. In order to minimize nonspecific surface adsorption, it is advisable that pure PDGF solutions are made in 0.3 to 1.0 *M* NaCl or NH_4HCO_3, or in 1 *M* acetic acid, and stored at −20°.

Since PDGF is derived from human blood collected from thousands of donors, it is conceivable that this large pool contains units contaminated with hepatitis virus or Acquired Immune Deficiency (AIDS) virus. For this reason, it is extremely important that laboratories involved in PDGF purification establish all necessary precautions to avoid contamination. The handling and pooling of the bags containing the platelet units must take place in a segregated area. The personnel involved in the opening of the bags and pooling of the units must wear disposable gloves, gowns, and shoe covers. All disposable labware, including the empty plastic bags from the platelet units, should be collected, placed in autoclave containers, and autoclaved before disposal. Nondisposable labware should be autoclaved or soaked in 1 : 10 dilution of Clorox (5.25% sodium hypochlorite). The laboratory areas used for pooling of platelet units and subsequent fractionation must be washed on at least a daily basis with 1 : 10 dilution of Clorox.

The boiling step of PDGF purification, described above, may contribute to the inactivation of viruses, and its incorporation at some stage of the purification process may prove beneficial. Finally, careful records must be kept of the incoming platelet units, to allow the tracing of units possibly contaminated with hepatitis or AIDS virus.

Acknowledgments

We thank Erika Albert for preparing the manuscript. Work conducted in our laboratories was supported by National Institutes of Health Grants CA-30301, HL-29583 (to H.N.A.), CA-38784 (to P.P.), and grants from the American Cancer Society, and The Council for Tobacco Research, U.S.A. (to H.N.A.).

[19] Platelet Basic Protein, Low-Affinity Platelet Factor 4, and β-Thromboglobulin: Purification and Identification

By JOHN C. HOLT and STEFAN NIEWIAROWSKI

β-Thromboglobulin (βTG) is a platelet-specific secreted protein which was first described by Pepper and colleagues.[1] It binds to heparin with lower affinity than platelet factor 4, despite the high degree of amino acid sequence homology between the two proteins.[2] The plasma level of βTG is now widely measured by radioimmunoassay as an indicator of platelet secretion and hence activation. Continuing basic studies on this protein have shown, however, that the antigen detected may be in several different

[1] S. Moore, D. S. Pepper, and J. D. Cash, *Biochim. Biophys. Acta* **379,** 360 (1975).

[2] G. S. Begg, D. S. Pepper, C. N. Chesterman, and F. J. Morgan, *Biochemistry* **17,** 1739 (1978).

forms which are indistinguishable by current radioimmunoassays.[3,4] The different forms are related by proteolytic cleavage at the NH_2-terminal end of the molecule, as shown in Fig. 1. Normal platelets, obtained from freshly drawn blood under conditions which minimize proteolysis, contain predominantly low-affinity platelet factor 4 (LA-PF_4) together with small amounts of platelet basic protein (PBP) and, in our experience, even smaller amounts of βTG itself.[5] PBP appears to be the synthesized precursor, most of which undergoes proteolytic processing before platelets are released into the circulation.

Our purpose in this chapter is to describe the isolation of PBP and LA-PF_4 from platelets and a method for estimating the relative proportions of different βTG-antigens[6] in whole platelets.

Purification of PBP and LA-PF_4

The first step is to obtain a mixture of LA-PF_4 and PBP from platelets. Usually this has meant preparing a release supernatant from fresh, washed platelets as described in the next section. Recently we have developed a simple procedure for cell lysis with trichloroacetic acid (below). Not only is cleavage minimized, but also the ability to extract PBP and LA-PF_4 from the trichloroacetic acid pellet offers a significant purification step.

Preparation of Release Supernatant from Fresh, Washed Platelets

Washed platelets[7] are stirred with 1–10 units/ml α-thrombin (Dr. J. W. Fenton III, New York State Dept of Health, Albany, NY) for 2 min at 37°. Phenylmethylsulfonyl fluoride (3 m*M*) and soybean trypsin inhibitor (0.1 mg/ml) are added and aggregated platelets removed by centrifuga-

[3] B. Rucinski, S. Niewiarowski, P. James, D. A. Walz, and A. Z. Budzynski, *Blood* **53,** 47 (1979).

[4] S. Niewiarowski, D. A. Walz, P. James, B. Rucinski, and F. Kueppers, *Blood* **55,** 453 (1980).

[5] J. C. Holt, M. Harris, A. Holt, E. Lange, A. Henschen, and S. Niewiarowski, *Biochemistry* **25,** 1988 (1986).

[6] The Subcommittee on Platelets of the International Committee on Thrombosis and Hemostasis recently considered appropriate nomenclature for the proteins described here. While recognizing that βTG is a defined chemical species which is not generally present in platelets, members were reluctant to discard this established name. The consensus was that the term βTG-antigen or βTG-related antigen should be used when the protein detected by radioimmunoassay is not further identified. Investigators who identify the particular form of the antigen should take responsibility for defining the terms used in their reports.

[7] J. F. Mustard, D. W. Perry, N. G. Ardlie, and M. A. Packham, *Br. J. Haematol.* **22,** 193 (1972).

```
1               5                   10                  15
Ser-Ser-Thr-Lys-Gly-Gln-Thr-Lys-Arg-Asn-Leu-Ala-Lys-Gly-Lys-Glu-Glu-Ser-Leu-   PBP

                                    Asn-Leu-Ala-Lys-Gly-Lys-Glu-Glu-Ser-Leu-   LA-PF4

                                                Gly-Lys-Glu-Glu-Ser-Leu-   βTG
```

FIG. 1. NH_2-terminal amino acid sequence of PBP,[5] LA-PF_4[3] (D. A. Walz, personal communication), and βTG.[2]

tion at 10,000 g_{max} for 15 min. The supernatant is maintained at 100° for 5 min prior to storage at −18°. βTG-antigen constitutes about 10% of the protein in the release supernatant.

Preparation of Cell Lysate from Fresh Platelets

Platelets are pelletted from platelet-rich plasma (PRP) in acid citrate dextrose and 5 m*M* EDTA by centrifugation at 1000 g_{max} for 15 min. The supernatant plasma is discarded and the platelets resuspended in 0.15 *M* sodium chloride, 0.005 *M* EDTA (0.1 ml/ml of starting PRP). To the suspension at 0° is added $\frac{1}{5}$ vol of cold 60% (w/v) trichloroacetic acid. After mixing, the suspension is incubated 10 min at 0° and then the precipitate collected by centrifugation at 8000 g_{max} for 10 min. The supernatant is discarded and the precipitate resuspended in 0.1% trifluoroacetic acid (0.1 ml/ml PRP) containing 0.1% Triton X-100, first by repeated pipetting and then by sonication for up to 30 sec in an ice bath until a uniform suspension is obtained. This extract of the trichloroacetic acid pellet is clarified by centrifugation at 20,000 g_{max} for 15 min and stored at −18°.

Comments. This procedure yields a platelet extract containing LA-PF_4 and PBP with a minimum of proteolytic cleavage. Although neither protein is susceptible to thrombin,[5,8] some cleavage can occur either during the washing process (which may consume several hours at 20–37°), or during secretion, which may liberate proteolytic enzymes as well as the characterized secreted proteins. A second advantage is that samples as small as 100 μl PRP can be processed for analytical purposes, using an Eppendorf centrifuge operated for periods of 30 sec.

Separation of LA-PF_4 and PBP[5]

A column of CM-Sephadex C-50 (1.9 × 8 cm) is equilibrated in 0.08 *M* sodium chloride, 0.01 *M* sodium phosphate, pH 7.4. Samples prepared as described above are thawed, clarified, and adjusted in pH and ionic strength to match the column buffer. They are then applied to the column. Unbound protein is removed from the column by washing with 100 ml of starting buffer. A gradient is then applied, consisting of 100 ml each of starting buffer and 0.4 *M* sodium chloride, 0.01 *M* sodium phosphate, pH 7.4. Fractions of 3 ml are collected and analyzed for protein by measurement of absorbance at 210 nm. A typical elution profile is shown in Fig. 2. Peak I contains LA-PF_4 and peak II, PBP. The yield is 200–400 μg βTG-antigen from 100 ml PRP, whichever starting material is used. Since platelets contain about 20 μg βTG-antigen/10^9 cells, overall recovery is in the range of 30–60%.[3] Analysis by gel electrophoresis in sodium dodecyl sulfate shows both proteins are at least 85% pure. Further purification, including removal of platelet-derived growth factor from PBP, can be achieved by chromatography on heparin–agarose and Sephadex G-75.[5]

Preparation of βTG

βTG is prepared from LA-PF_4 by cleavage with trypsin.[8] Purified LA-PF_4 (0.1–1 mg/ml) in 0.15 *M* sodium chloride, 0.01 *M* Tris–chloride, pH 7.4, is incubated for 15 min at 37° with 1/100 weight of trypsin that has been treated with L-1-tosylamido-2-phenylethyl chloromethyl ketone (TPCK–trypsin, Cooper Biomedical, Cappel-Worthington Division). Trypsin is inactivated by addition of an experimentally determined amount of bovine lung trypsin inhibitor (aprotinin, Sigma Chemical Co.). The extent of cleavage is monitored using isoelectric focusing as described below, or disc gel electrophoresis.[8]

Identification of β-Thromboglobulin-Antigens

Since all forms of βTG-antigen cross-react immunologically, the essence of the method is to combine immunological specificity with a physical separation of the different forms. Equilibrium isoelectric focusing in a sucrose density gradient was first used for this purpose,[4] and was instrumental in revealing the existence of PBP and the distinctness of βTG and LA-PF_4 (Fig. 3). The procedure set out below is similar in principle to that originally described, but offers (1) increased sensitivity (10 ng versus 1μg), (2) increased resolution for basic proteins, and (3) the opportunity to analyze numerous single-donor platelet samples in a reasonable amount of time. These gains are achieved by using nonequilibrium isoelectric focusing in a polyacrylamide gel, with detection of antigen by incubating a nitrocellulose transfer of the gel first in anti-LA-PF_4 antibody and then in peroxidase-conjugated second antibody.

[8] J. C. Holt and S. Niewiarowski, *Biochim. Biophys. Acta* **632**, 284 (1980).

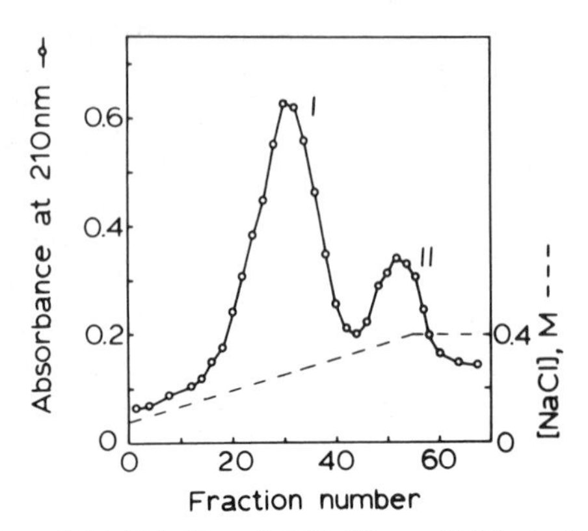

FIG. 2. Separation of LA-PF_4 (I) and PBP (II) on CM-Sephadex C-50. The sample, derived from 400 ml of blood, was prepared by lysis of platelets with trichloroacetic acid as described in the text.

Isoelectric Focusing

A vertical polyacrylamide slab gel (12 cm × 12 cm × 1.5 mm) is prepared, containing 6% polyacrylamide, 0.16% methylenebisacrylamide, 1.6% (w/v) ampholytes (Ampholines, pH 3.5–10, LKB Instruments), 0.6 mg/ml ammonium persulfate (freshly prepared) and 0.125% (v/v) N,N,N′,N′-tetramethylethylenediamine (TEMED). After the gel has polymerized at 20°, it is transferred to a cold room at 4°. Electrolytes are prepared as follows, and cooled to 4°: upper reservoir, anode connection, 9 m*M* phosphoric acid; lower reservoir, cathode connection, 20 m*M* sodium hydroxide. The volume of electrolyte required is determined only by the construction of the gel apparatus; we use 40 ml in the upper chamber and 100 ml in the lower chamber.

Samples are composed of 2–15 μl test material (which may contain up to 0.4 *M* sodium chloride or have an unbuffered pH as low as 3), 3 μl 25% aqueous glycerol, 5 μl 2 mg/ml cytochrome *c* (Sigma Chemical Co.), and water to 23 μl. Cytochrome *c* serves as a visible marker and as a carrier protein which is important when purified samples lacking endogenous carrier are analyzed. In our experience, 10 ng βTG-antigen is detectable, and 50 ng optimal, in a single band when antiserum that can be used at 1:10,000 dilution in radioimmunoassay is available.

When the gel and electrolytes are equilibrated at 4°, the apparatus is assembled and the sample wells, drained of any unpolymerized gel mix, are filled with a 1.6% (w/v) solution of ampholytes in water. Samples are layered in the individual wells under the ampholyte "cushion" with mini-

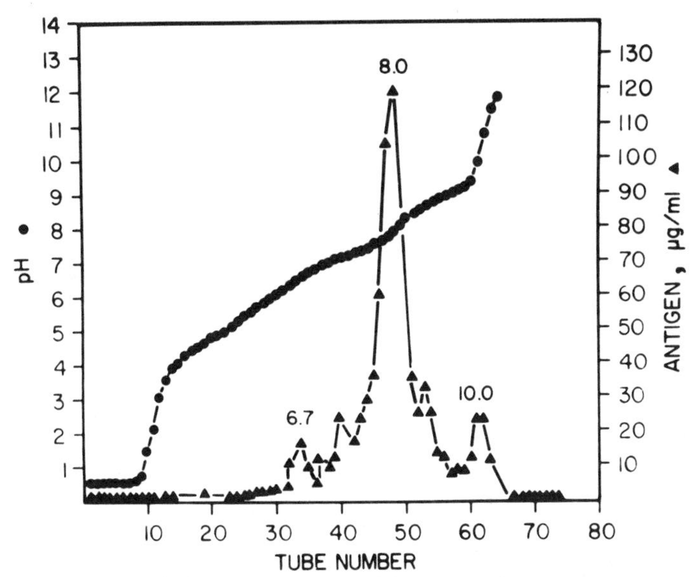

FIG. 3. Separation of βTG, LA-PF$_4$, and PBP by isoelectric focusing in a sucrose density gradient with detection by radioimmunoassay of LA-PF$_4$.[4] Washed platelets were stimulated with Ca^{2+}-ionophore A23187 and the supernatant, after removal of aggregated platelets, heated at 100°. The supernatant, dialyzed against 1% (w/v) glycine, was incorporated together with 1% (w/v) ampholytes (Ampholines, pH 3.5–10, LKB Instruments) in a sucrose density gradient formed in a glass focusing column (LKB Instruments). Focusing was carried out at 600 V for 24 hr. Fractions (1 ml) were displaced from the column by injection of a high-density sucrose solution. The pH and βTG-antigen level were determined in each fraction. Numbers beside the peaks in the βTG-antigen profile are isoelectric points deduced from the pH profile.

mum disturbance. Finally, the electrolyte chambers are filled and the electrodes connected, positive to the upper, acidic reservoir and negative to the lower, basic reservoir.

The process of focusing is carried out at 4° in a cold room. Provided current and voltage are carefully controlled, the circulating chilled air is sufficient to control the temperature within the gel and so prevent the convective disturbances to which focusing gels are particularly susceptible. The gel is initially run under constant current conditions at 2.5 mA. Every 30 min, the current is increased in increments of 0.5 mA to an upper limit of 5 mA. The voltage is allowed to rise to a limit of 450 V, which is maintained as the resistance of the gel continues to increase and the current falls. The run is ended when the visible cytochrome *c*, a highly basic protein with p*I* 10.5, is 1 cm from the gel bottom. At this point, 450 V produces a current of about 3 mA. A power supply designed for isoelectric focusing is helpful but not essential for this process. For a gel of the dimensions given, the elapsed time is about 6 hr. Gels as short as 3 cm can also be used ($2\frac{1}{2}$ hr to complete) with a slight loss of resolution for more basic components.

Comments. In liquid-phase focusing, proteins with p$I > 9$ are in or close to the junction between the pH gradient and the cathodic electrolyte, and may be poorly resolved from one another. In gel focusing, they will not be retained on the gel. Basic proteins applied at the anodic end of a focusing gel, however, migrate as a zone toward the same equilibrium position but can be retained on the gel by ending the run before equilibrium is reached.[9] Resolution of basic proteins is high, and is further enhanced when the run is ended at slightly (15–20%) shorter times.

Transfer to Nitrocellulose

The completed gel is placed in a shallow tray of 0.15 *M* sodium chloride, 0.01 *M* Tris–glycine, pH 8.2 (buffer A). When the gel is free floating, most of the buffer is removed by aspiration, leaving a moist gel of undistorted dimensions. A sheet of nitrocellulose paper (0.45-μm pore size, Schleicher and Schuell), wetted in buffer A, is placed on top of the gel with the careful exclusion of all air bubbles. Several layers of moistened filter paper are placed, one by one, on top of the nitrocellulose. The whole assembly, still on the tray, is enclosed without pressure in plastic wrap to prevent evaporation. Transfer of protein to paper is allowed to proceed by capillary action for 16 hr at 20–23° following the method of Reinhart and Malamud.[10]

Comments. Native LA-PF_4 binds poorly to nitrocellulose in a voltage gradient at pH values below 7, which is the necessary condition if βTG, LA-PF_4, and PBP are all to migrate in the same direction. Capillary transfer on the other hand does not involve forces that overcome the weak binding of LA-PF_4 to the paper. If desired, a second transfer may be prepared by placing a second nitrocellulose sheet beneath the gel.

Antigen Staining with Specific Antibody

The nitrocellulose transfer is incubated in the following solutions with constant mixing (e.g., Ames aliquot mixer, Miles Laboratories). At each stage, the volume is the minimum to ensure uniform wetting of the paper, i.e., 5–10 ml for a 12×12 cm sheet of paper in a box as nearly as possible the same size.

[9] P. L. Storring and R. J. Tiplady, *Biochem. J.* **171,** 79 (1978).
[10] M. P. Reinhart and D. Malamud, *Anal. Biochem.* **123,** 229 (1982).

5 min in buffer A

5 min in 0.3% gelatin, dissolved by boiling in buffer A for 20 min, and cooled to 20° before use. This step minimizes nonspecific binding to the paper

60 min in the same gelatin solution now containing 50–100 μl clarified rabbit anti-LA-PF_4 serum

25 min in five or more changes of buffer A to remove unbound IgG and serum proteins

60 min in 0.3% gelatin containing 40 μl peroxidase-conjugated goat anti-rabbit IgG (Cooper Biomedical, Cappel Worthington Division), reconstituted according to the supplier's directions

25 min in five or more changes of buffer A to remove unbound conjugate

2–10 min in a substrate solution composed of 10 mg 4-chloro-1-naphthol dissolved in 3 ml methanol and added to 47 ml buffer A containing 50 μl 30% hydrogen peroxide. Color development is stopped by washing with distilled water and air drying

Comments. Several preparations of gelatin should be screened to select one which gives minimum background staining. With βTG-antigen, Tween 20 cannot be used to prevent nonspecific binding to the paper because it promotes selective dissociation of LA-PF_4 from the paper. Ratios of LA-PF_4 to PBP are visibly altered when Tween 20 is used, and become inconsistent with ratios determined for the same samples by other methods. 4-Chloro-1-naphthol is a vastly superior substrate to *o*-dianisidine, and is not carcinogenic.

Applications of the Method to Estimate the Distribution of βTG-Antigens

The profile shown in Fig. 2 may be analyzed to confirm the separation of PBP and LA-PF_4 as well as to allow collection of the most homogeneous fractions for preparative purposes. Figure 4a shows a typical analysis.

Single-donor platelets are lysed with trichloroacetic acid (final concentration 10%, w/v) and the pellet extracted with 0.1% trifluoroacetic acid containing 0.1% Triton X-100. About 10^6 platelets are required for each gel sample. We found that, of 40 normal samples analyzed, 38 contained predominantly LA-PF_4 with PBP present as the minor component.[5] The remaining two samples were unusual in that they contained equal amounts of LA-PF_4 and PBP. Platelets from each of these two normal donors were obtained on two occasions, more than a year apart, and showed the same properties. The distribution of βTG-antigens in congenitally abnormal platelets was also studied. We examined two examples each of Bernard–Soulier syndrome and Glanzmann's thrombasthenia. All were normal in

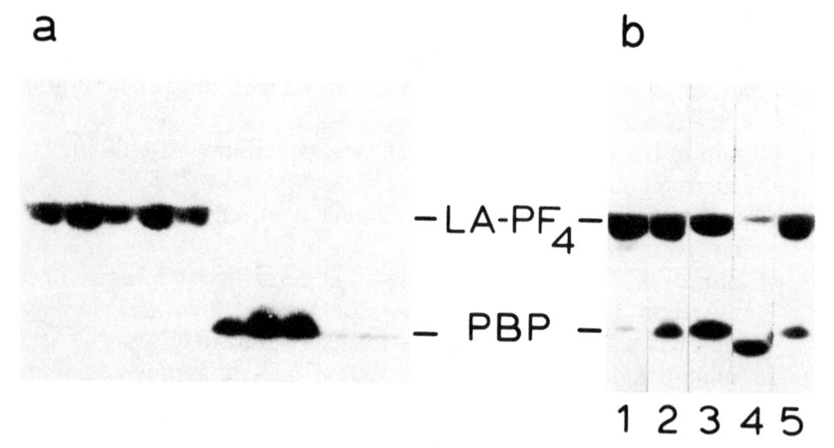

FIG. 4. Nonequilibrium focusing gel transfers stained with antibody to LA-PF_4. (a) Column fractions from the elution profile of Fig. 2 as follows: Starting from the left, 20, 25, 30, 35, 40, 48, 53, 57. Aliquots of 2–10 μl were analyzed, based on the concentration of protein implied by the absorbance at 210 nm. (b) Whole-platelet lysates showing low, average, and high ratios of PBP to LA-PF_4 (samples 1–3, respectively), the basic fragment seen in lysates obtained after sedimentation of red cells at 1 *g* (sample 4), and the same platelets as in sample 4, but after lysis with trichloroacetic acid and extraction with 0.1% trifluoroacetic acid (sample 5). A small amount of uncleaved LA-PF_4 is evident in this particular basic fragment sample. Each sample corresponds to about 3×10^6 platelets.

that LA-PF_4 was the major form of βTG-antigen detected. Figure 4b shows examples of platelets containing low, average, and high amounts of PBP.

Our initial analysis of βTG-antigen in platelets was qualitative in that we wished to determine whether platelets might in fact contain only precursor PBP, which was cleaved to LA-PF_4 during secretion or cell lysis. Since this was not the case,[5] we have recently begun quantitative estimation of the PBP:LA-PF_4 ratio. Antibody-stained transfers such as those shown in Fig. 4 are scanned with a reflectance densitometer. Peak areas are obtained over a 3-fold range of sample load where response is demonstrated to be linear. For five average normal samples (as sample 2 in Fig. 4), PBP was 28 ± 10% (SD) of the βTG-antigen. In the two unusual normal samples described above, the values were 47 and 48%. In bone marrow megakaryocytes, the average level of PBP was higher (49 ± 13% (SD), $n = 11$) than in platelets, supporting the status of PBP as the precursor form of βTG-antigen.[11]

PRP was occasionally obtained from anticoagulated blood by allowing red cells to sediment at 1 *g* for 0.5–2 hr. This was done to avoid potential loss of large platelets, e.g., in Bernard–Soulier syndrome, during the nor-

[11] J. C. Holt, E. M. Rabellino, A. M. Gewirtz, L. M. Gunkel, B. Rucinski, and S. Niewiarowski, *Exp. Hematol.* **16**, 302 (1988).

mal centrifugal separation of red cells. When platelets obtained at 1 *g* were pelletted and lysed in the absence of protease inhibitors, all βTG-antigen was detected in a novel, stable, highly basic form (Fig. 4b, sample 4). When the same platelets were lysed with trichloroacetic acid and the extract of the pellet analyzed, a normal distribution of LA-PF_4 and PBP was found (Fig. 4b, sample 5). Several lines of evidence, immunochemical and enzymatic, suggested that cathespin G from neutrophils was responsible for producing the highly basic species by cleaving 15 NH_2-terminal residues from LA-PF_4. Neutrophils are normally absent from PRP prepared by centrifugation, but could well be incompletely sedimented at 1 *g*.

Acknowledgment

This work was supported by grants from the NIH (HL14217, HL36579, and HL15226) and the W. W. Smith Charitable Trust, Ardmore, Pennsylvania.

[20] Platelet α-Granule Protein-Induced Chemotaxis of Inflammatory Cells and Fibroblasts

By ROBERT M. SENIOR, GAIL L. GRIFFIN, AKIRO KIMURA, JUNG SAN HUANG, and THOMAS F. DEUEL

Introduction

The migration of neutrophils and mononuclear phagocytes into tissues is a hallmark of inflammation. Under appropriate stimulation these cells function in the destruction of foreign debris and microorganisms, in the clearance of injured and autolysed endogenous materials, and in immune responses. Besides inflammatory cells, mesenchymal cells and endothelial cells also migrate to sites of inflammation and tissue injury, where they are important in the resolution of inflammation and subsequent tissue repair.

Both inflammation and tissue repair are complex responses that involve highly regulated and coordinated reactions and multiple extracellular and cellular components. In these responses molecules ordinarily confined intracellularly may gain access to the extracellular environment where they can be signals for cell migration. Chemotaxis refers to directed cell migration in response to a defined chemical stimulus. In recent years, chemotactic responses have been observed with many cell types and an ever-increasing number of chemotactic factors has been found.[1-8]

Inflammatory cells have been identified in close proximity to platelets in various pathological lesions and in experimental models of inflammation, immune complex disease, and atherosclerosis.[9-17] Stimulated neutrophils release platelet-activating factor[18] which, in turn, is capable of activating neutrophils themselves[19-21] and also of stimulating the release of secretory proteins from platelets.[22] Thus, a reciprocity of stimulatory functions between platelets and inflammatory cells may exist, and there may be important roles for platelets and proteins stored in platelet secretory granules in inflammation and tissue repair.[23-27]

Platelet factor 4 (PF_4) is a well-characterized secretory α-granule pro-

[1] P. C. Wilkinson, "Chemotaxis and Inflammation." Churchill-Livingstone, New York, 1982.

[2] H. U. Keller, P. C. Wilkinson, M. Abercrombie, E. L. Becker, J. G. Hirsch, M. E. Miller, W. S. Ramsey, and S. H. Zigmond, *Clin. Exp. Immunol.* **27,** 377 (1977).

[3] E. L. Becker and P. A. Ward, "Clinical Immunology" (C.W. Parker, ed.), p. 272. Saunders, Philadelphia, Pennsylvania, 1980.

[4] R. Snyderman and E. J. Goetzl, *Science* **213,** 830 (1981).

[5] E. Schiffman, *Annu. Rev. Physiol.* **44,** 553 (1982).

[6] S. H. Zigmond, "Cell Behavior." (R. Bellairs, A. Curtis and G. Dunn, eds.), p. 183. Cambridge University Press, New York, 1982.

[7] A. E. Postlethwaite, *Adv. Inflamm. Res.* **5,** 27 (1983).

[8] A. Albini, B. C. Adelmann-Grill, and P. K. Müller, *Collagen Relat. Res.* **5,** 283 (1985).

[9] A. A. Angrist and M. Oka, *J. Am. Med. Assoc.* **183,** 249 (1963).

[10] P. H. Levine, R. S. Weinger, J. Simon, K. L. Scoon, and N. I. Krinsky, *J. Clin. Invest.* **57,** 955 (1976).

[11] J. T. Prchal and J. Blakely, *N. Engl. J. Med.* **289,** 1146 (1973).

[12] M. Bachofen and E. R. Weibel, *Clin. Chest Med.* **3,** 35 (1982).

[13] M. Bachofen, E. R. Weibel, and B. Roos, *Am. Rev. Respir. Dis.* **111,** 247 (1975).

[14] P. M. Henson and C. G. Cochrane, *J. Exp. Med.* **133,** 554 (1971).

[15] H. Jellinek, *Adv. Exp. Med. Biol.* **82,** 324 (1977).

[16] L. Jørgensen, M. A. Packham, H. C. Rowsell, and J. F. Mustard, *Lab. Invest.* **27,** 341 (1972).

[17] I. Joris and G. Majno, *Adv. Inflamm. Res.* **1,** 71 (1979).

[18] P. M. Henson and R. N. Pinckard, *Monogr. Allergy* **12,** 13 (1977).

[19] E. J. Goetzl, C. K. Derian, A. I. Tauber, and F. H. Valone, *Biochem. Biophys. Res. Commun.* **94,** 881 (1980).

[20] J. T. O'Flaherty, R. L. Wykle, C. H. Miller, J. C. Lewis, M. Waite, D. A. Bass, C. E. McCall, and L. R. DeChatelet, *Am. J. Pathol.* **103,** 70 (1981).

[21] J. O. Shaw, R. N. Pinckard, K. S. Ferrigni, L. M. McManus, and D. J. Hanahan, *J. Immunol.* **127,** 1250 (1981).

[22] L. M. McManus, C. A. Morley, S. P. Levine, and R. N. Pinckard, *J. Immunol.* **123,** 2835 (1979).

[23] J. F. Mustard, H. Z. Movat, D. R. L. MacMorine, and A. Senyi, *Proc. Soc. Exp. Biol. Med.* **119,** 988 (1965).

[24] M. A. Packham, E. E. Nishizawa, and J. F. Mustard, *Biochem. Pharmacol. (Suppl.)* **17,** 171 (1968).

tein of platelets that has been shown to be a chemotactic factor for inflammatory cells. Highly purified PF_4 stimulated the migration of human polymorphonuclear leukocytes and monocytes at concentrations of PF_4 less than that in human serum. PF_4 was strongly chemotactic for fibroblasts also, but the maximum chemotaxis was observed at concentrations of PF_4 less than $\frac{1}{10}$ those required for maximum chemotaxis of inflammatory cells.[29] A second platelet secretory granule protein, the platelet-derived growth factor (PDGF), also was strongly chemotactic for human monocytes and neutrophils. Optimal concentrations of PDGF required for maximum activity were 20 and 1 ng/ml, respectively,[30] whereas PDGF concentrations in serum were ~ 50 ng/ml. Furthermore, PDGF was demonstrated to be a chemoattractant for fibroblasts and smooth muscle cells,[29,31,32] suggesting that PDGF also may be involved in wound healing through its influence in fibroblast movement and perhaps smooth muscle cells as well. These effects of platelet granule proteins on cell migration were extended by the observation that another platelet secretory α-granule protein, β-thromboglobulin (β-TG), was a highly active chemotactic factor for fibroblasts.[29] Other materials that have been found to exert fibroblast chemotactic activity include lymphokines, a peptide derived from the fifth component of complement, collagens and collagen-derived peptides, fibronectin, tropoelastin, and elastin-derived peptides.[33-37]

Besides stimulating directed migration other cellular effects have been observed with platelet α-granule secretory proteins. PDGF promoted acti-

[25] R. L. Nachman, B. Weksler, and B. Ferris, *J. Clin. Invest.* **49,** 274 (1970).

[26] R. L. Nachman, B. Weksler, and B. Ferris, *J. Clin. Invest.* **51,** 549 (1972).

[27] B. B. Weksler, *Clin. Lab. Med.* **3,** 667 (1983).

[28] E. A. Bauer, T. W. Cooper, J. S. Huang, J. Altman, and T. F. Deuel, *Proc. Natl. Acad. Sci. U.S.A.* **82,** 4132 (1985).

[29] R. M. Senior, G. L. Griffin, J. S. Huang, D. A. Walz, and T. F. Deuel, *J. Cell Biol.* **96,** 382 (1983).

[30] T. F. Deuel, R. M. Senior, J. S. Huang, and G. L. Griffin, *J. Clin. Invest.* **69,** 1046 (1982).

[31] G. R. Grotendorst, H. E. J. Seppä, H. K. Kleinman, and G. R. Martin, *Proc. Natl. Acad. Sci. U.S.A.* **78,** 3669 (1981).

[32] H. Seppä, G. Grotendorst, S. Seppä, E. Schiffmann, and G. R. Martin, *J. Cell Biol.* **92,** 584 (1982).

[33] R. M. Senior, G. L. Griffin, and R. P. Mecham, *J. Clin. Invest.* **70,** 614 (1982).

[34] A. E. Postlethwaite, R. Snyderman, and A. H. Kang, *J. Exp. Med.* **144,** 1188 (1976).

[35] A. E. Postlethwaite, R. Snyderman, and A. H. Kang, *J. Clin. Invest.* **64,** 1379 (1979).

[36] A. E. Postlethwaite, J. M. Seyer, and A. H. Kang, *Proc. Natl. Acad. Sci. U.S.A.* **75,** 871 (1978).

[37] H. E. J. Seppä, K. M. Yamada, S. T. Seppä, M. H. Silver, H. K. Kleinman, and E, Schiffman, *Cell Biol. Int. Rep.* **5,** 813 (1981).

vation of human polymorphonuclear leukocytes[38] and monocytes.[39] It also stimulated a dose-dependent, saturatable increase in the release of collagenase by fibroblasts in culture,[28] an effect that may be important in tissue repair.[40-44] These results may be interpreted to suggest that platelet secretory granule proteins, ordinarily sequestered within the cell but released to the extracellular compartment by platelet activation at sites of injury, may be recognized by phagocytic cells and by cells involved in wound healing, evoking chemotaxis and other responses. Recent studies with PDGF in *in vivo* models support these functional roles for platelet secretory proteins.[45]

This chapter describes methods for measuring chemotaxis of neutrophils, monocytes, and fibroblasts in response to purified PDGF, a potent chemotactic protein. The methods to be described have been used in several of the investigations cited above that demonstrated chemotactic responses of neutrophils, monocytes, and fibroblasts to PDGF. The methods are simple, yield highly reproducible data, and allow a high degree of sensitivity to small differences in concentrations of chemotactic factors.

Methods

The methods for cell collection and chemotaxis assay are similar for neutrophils and monocytes so these are described together. The procedures for fibroblast chemotaxis are described separately because there are a number of differences from the procedures for neutrophils and monocytes. All assays, however, are done with the same chemotaxis chamber.

Chemotaxis Chamber. The chemotaxis apparatus can be made, or alternatively, suitable chemotaxis microchambers may be purchased[46-48] through Neuro Probe (Cabin John, MD). In either case the chamber consists of two plastic plates with multiple aligned wells. If the apparatus is to be made the use follows specifications. The top plate is 114 mm long × 132 mm wide, 7 mm thick with 30 holes, each 6 mm in diameter and

[38] D. Y. Tzeng, T. F. Deuel, J. S. Huang, R. M. Senior, L. A. Boxer, and R. L. Baehner, *Blood* **64,** 1123 (1984).

[39] D. Y. Tzeng, T. F. Deuel, J. S. Huang, and R. L. Baehner, *Blood* **66,** 179 (1985).

[40] J. Gross, *in* "Biochemistry of Collagen" (G. N. Ramachandran and A. H. Reddi, eds.), p. 275. Plenum, New York, 1976.

[41] H. C. Grillo and J. Gross, *Dev. Biol.* **15,** 300 (1967).

[42] W. B. Riley Jr. and E. E. Peacock, Jr., *Proc. Exp. Biol. Med.* **124,** 207 (1967).

[43] A. Z. Eisen, *J. Invest Dermatol.* **52,** 442 (1969).

[44] R. B. Donoff, J. E. McLennan, and H. C. Grillo, *Biochim. Biophys. Acta* **227,** 639 (1971).

[45] G. R. Grotendorst, G. R. Martin, D. Pencev, J. Sodek, and A. K. Harvey, *J. Clin. Invest.* **76,** 2323 (1985).

[46] W. Falk, R. H. Goodwin, Jr., and E. J. Leonard, *J, Immunol. Methods* **33,** 239 (1980).

[47] L. Harvath, W. Falk, and E. J. Leonard, *J. Immunol. Methods* **37,** 39 (1980).

[48] K. L. Richards and J. McCullough, *Immunol. Commun.* **13,** 49 (1984).

7 mm deep (through the plate). The holes are surrounded by rubber gaskets to ensure a tight seal when the plates are fastened together. The bottom plate is 114 mm long × 132 mm wide, 9 mm thick. It has 30 wells, each 6 mm deep. The holes or wells on each plate are placed in five rows of six wells per row on 14-mm centers. Each well in the bottom plate holds 0.24 ml and each hole in the top plate (after the membranes are in place) holds 0.34 ml. There are four plastic nuts and four plastic screws, 25 mm long. Holes for the screws are near the corners of the plates. These screws hold the two plates together during the experiment. In a typical experiment, the sample (chemotactic factor or medium blank) is loaded in the bottom well, a 0.45-μm pore size, 13-mm diameter, Millipore filter is placed on top of the well followed by a Nucleopore membrane with the pore size determined by the experiment. Then, the top plate is placed on the bottom plate, the two plates are locked together with the screws, and the cell suspension is placed in the upper compartments.

Chemotaxis of Monocytes and Neutrophils. Neutrophils and monocytes are harvested from freshly collected venous blood. The cells are isolated using Ficoll-Hypaque gradients following procedures originally described by Böyum.[49] Routinely the procedure is done with 60 ml of blood. The quantities described below are for that amount of blood.

Ficoll-Hypaque Gradient. The gradient is prepared as follows: Mix 42 ml Hypaque (Sodium Hypaque 50%, Winthrop Labs, NY), 20 ml H_2O, 150 ml 9% (w/v) Ficoll (Sigma, type 400, Cat. F4375). As a convenient alternative one can use sterile, ready-made mixtures (Sigma, Histopaque #1077). Put 20-ml aliquots of the Ficoll-Hypaque mixture into 50-ml round-bottom plastic centrifuge tubes.

Blood Collection and Application to the Ficoll-Hypaque. Blood is drawn into a heparinized (1000 units/ml, 1–2 ml) plastic syringe using a small vein infusion set with a 19-gauge needle (McGraw Laboratories, Sabana Grande, PR, Cat. V1100). The blood is diluted 1 : 1 with 0.9% (w/v) saline and then the mixture is gently swirled. Twenty milliliters of the saline–blood mixture is carefully layered on top of the Ficoll-Hypaque. The interface between the Ficoll and the blood should be sharp. The tubes are then centrifuged at 2000 rpm, at room temperature, for 30 min.

Isolation of Mononuclear Cells. After centrifugation, the cloudy layer at the interface of the Ficoll-Hypaque and serum is carefully removed. This layer contains the monocytes and lymphocytes, which in a normal individual are in a ratio of about 1 : 4. As the cells are removed they are put into a 50-ml centrifuge tube. The cells are diluted with saline 1 : 1 and spun at 2000 rpm at room temperature for 10 min. The cells are then washed twice

[49] A. Böyum, *Scand. J. Clin. Lab. Invest.* **21** (Suppl. 97), 77 (1968).

by resuspending them in saline or medium, centrifuging at 1000 rpm for 10 min after each wash. Finally the cells are resuspended in medium containing 0.5 μg/ml human serum albumin and adjusted to a concentration of 2.5×10^6 cells/ml.

Isolation of Neutrophils. After removal of the mononuclear cells from the top layer of Ficoll-Hypaque, the Ficoll is aspirated down to the red cell pellet on the bottom of the centrifuge tube with care to leave 2–3 ml of Ficoll on top of the pellet. The tube is mixed to loosen the pellet and the mixture is poured into a plastic beaker. Then 8.0 ml of a 2.5% (w/v) dextran (Sigma, M_r 250,000) solution in 0.9% (w/v) saline is added to each tube. The tube is agitated and the contents transferred to the plastic beaker containing the cell pellet. This washing procedure is repeated using 4 ml of dextran solution. The plastic beaker is gently swirled so that the cells are suspended evenly in the dextran solution. The solution is divided in half and each half is put into a 50-ml disposable syringe. The syringes are then put in a 37° incubator for 15 min in an upright position resting on the plunger. The buffy coat, containing neutrophils, is collected by gently pushing the buffy coat out of the syringe through a plastic catheter of the vein infusion set (without the needle) into a 50-ml centrifuge tube. The syringe should be kept vertical at all times. An equal volume of saline is added to each tube to lower the dextran concentration. The tubes are then spun at 2000 rpm for 10 min at room temperature. The red cells remaining in the neutrophil pellet are lysed with a solution made up as follows: 8.3 g NH_4Cl, 1.0 g $KHCO_3$, 0.37 g NaEDTA, per liter, pH 7.2–7.3, kept at 4°. The cells are suspended in cold lysing solution and spun for 10 min at 1000 rpm. After this, the cells are washed three times with saline or medium and finally resuspended in DME medium containing 0.5 mg/ml human serum albumin, at a concentration of 1.5×10^6 cells/ml.

Monocyte and Neutrophil Chemotaxis Assays

Monocytes. The multiwell chemotaxis chamber is set up with double membranes. The bottom membrane is a 0.45-μm (Millipore, Bedford, MA) and the top membrane is a 5.0-μm type (Nucleopore, Pleasanton, CA). The cell concentration in the upper compartment is 2.5×10^6 cells/ml. After the upper and lower compartments are loaded, the plates are incubated for 120 min at 37° in 5% CO_2–95% air. The plates are taken apart, making sure not to separate the two filters, and the membrane pairs are stained with hematoxylin (Table I).

Neutrophils. The multiwell chemotaxis chamber is set up with double membranes. The bottom membrane is a 0.45-μm (Millipore) and the top membrane is a 2.0-μm type (Nucleopore). The cell concentration in the upper compartment is 1.5×10^6 cells/ml. After the upper and lower com-

TABLE I
PROCEDURE FOR STAINING MEMBRANES

Put staining rack containing membranes from chemotaxis assay in the following solutions[a]	
100% ethanol	15–30 sec
Filtered hematoxylin	6 min
Rinse in H_2O bath	15 sec
70% 2-propanol with 1–2 drops HCl	30 sec
Rinse in H_2O bath	15 sec
Bluing solution	3 min
10 g $MgSO_4$ and 1 g $NaHCO_3$ in 500 ml H_2O	
Rinse in H_2O bath	15 sec
Remove as much water as possible	
70% 2-propanol	5 min
95% 2-propanol	5 min
100% 2-propanol	5 min
100% 2-propanol	5 min
100% Xylene	5 min

[a] The above solutions can be used more than once (except H_2O baths). The hematoxylin must be filtered before each use.

partments are loaded, the chambers are incubated for 60 min at 37° in 5% CO_2-95% air. The plates are taken apart, making sure not to separate the two filters, and the membrane pairs are stained using the same procedure as that for monocytes.

Fibroblast Chemotaxis

Cells. Highly responsive fibroblasts are obtained from fetal calf neck ligament explants as described in detail in ref. 50. Human skin fibroblasts can also be used. The method for obtaining these cells is described below. In either case, the maintenance medium for the cells is DME with 10% fetal calf serum and the cells are grown in plastic tissue culture dishes.

Explant Culture of Human Skin Fibroblasts. Breast reduction tissue is used. Care is taken to keep the tissue sterile throughout the procedure. The tissue specimen is washed in medium containing penicillin/streptomycin (1×) and Fungizone, 1 μg/ml. After washing, the skin side is laid down on a

[50] R. P. Mecham, G. Lange, J. Madaras, and B. Starcher, *J. Cell Biol.* **90,** 332 (1981).

sterile board and the fat layer is shaved off with a scalpel. The specimen is then placed in L-15 medium 0.25% (w/v) trypsin overnight at 37°. After the overnight incubation, the epidermis is readily stripped off and the remaining dermis is then chopped with a scalpel into small explants, less than 0.5 mm. The tissue suspension is then diluted with DME containing penicillin/streptomycin and 10% fetal calf serum and 5–10 ml is dispensed into T-75 flasks using a large mouth pipet. The tissue pieces are dispersed on the bottom of the T flask by gentle swirling, after which the flask is tipped on its side so that the explants remain on the bottom of the flask (the medium is away from the explants). The flask cap is loosened and the flask left on its side for 1–2 hr in 5% CO_2–95% air at 37°. This allows the tissue to "dry" to the bottom of the flask so that when the flask is gently laid down and the medium covers the explants, the explants do not float. The flask is left untouched in the incubator for 1 week before changing the medium. The fibroblasts will start to grow out of the explants in 10–14 days. First passage is done at about 21 days when the cell layer is almost confluent. For chemotaxis, the cells should be used in low passages, three or less.

Preparation of Fibroblasts for Assay. The tissue culture dish is washed with 40 ml of phosphate-buffered saline (PBS) to remove the serum-containing medium. To detach the fibroblasts from the dish, 1.7 ml 0.025% (w/v) trypsin, 0.01% (w/v) EDTA, in PBS, is added to the culture dish. The cells are put back in the CO_2 incubator for 3 min, after which 10 ml of DME containing 20% (v/v) FCS, 1× L-glutamine, and 1× nonessential amino acids (NEAA) are added. At this point, the cells are detached and transferred from the tissue culture dish into one 50-ml centrifuge tube. The tube is spun at room temperature for 10 min at 1000 rpm. The supernatant is decanted and the cells resuspended in 10 ml of DME containing 1 mg/ml HSA with 1× L-glutamine and 1× NEAA. The washing procedure is repeated twice. The cells are finally suspended in DME containing 1 mg/ml HSA with 1× L-glutamine and 1× NEAA and the cell concentration is adjusted to 1.5×10^5 cells/ml.

Membrane preparation: Nucleopore membranes as received from the manufacturer are not suitable for use with fibroblasts because fibroblasts will not adhere or spread on them and therefore chemotaxis cannot occur. To make the membranes suitable they must be coated with gelatin. This is done as follows. The membranes are immersed in a glass beaker containing 0.5% (v/v) acetic acid solution. The beaker is heated for 20 min at 50°. After this, the membranes are washed extensively with distilled water and then immersed in a gelatin solution, 5 mg/ml in water, and incubated for 1 hr at 100°. They are then dried with a hair dryer, put in a glass Petri dish, and heated again for 1 hr at 100°. The membranes can be stored at room temperature in a closed container.

Chemotaxis Assay. Substances to be tested for chemotactic activity are diluted to the proper concentration with DME containing 1 mg/ml HSA with 1× L-glutamine and 1× NEAA and added to the lower compartment. Membranes are placed over each lower compartment in the correct order (0.45-μm Millipore first, then 8.0-μm Nucleopore), after which the upper compartment is filled with 0.34 ml of the cell suspension. The chemotactic plate is then put in a 5% CO_2–air incubator at 37° for 6 hr. At the end of the incubation the multiwell plates are separated, making sure not to separate the two filters, and then the membrane pairs are stained (Table I).

Counting Cells on the Membranes

Cell counts on chemotaxis membranes are done using high dry magnification (×400), five different fields per membrane. For monocyte and neutrophil chemotaxis the microscope eyepiece has a grid, and cell counts are done within the grid. With fibroblast chemotaxis the number of cells is relatively few so that cell counting is done over the entire microscopic field. Specifically, cells on the surface of the lower membrane, the 0.45-μm Millipore (cells that have migrated through the upper membrane), are the ones counted (Fig. 1). The positive and negative controls are counted first to confirm that the cells and the manipulations of the experiment were satisfactory. The 15 fields for each set of triplicates are averaged. The value thus obtained is corrected for the average value obtained from the negative controls also calculated as the average of five fields on each of three membranes. Excluding the controls, assays are read "blinded" to eliminate observer bias.

"Checkerboard" Analysis to Confirm Chemotactic Activity

The definitive test for chemotactic activity is a "checkerboard analysis," wherein a series of concentration gradients of the test substance are examined for their effects upon cell migration (Table II). This experimental design permits differentiating chemotaxis from stimulation of random motion, "chemokinesis." In contrast to chemotaxis, in which cell migration occurs only with a gradient of chemotactic agent, with chemokinesis cell migration is stimulated by the test substance irrespective of the presence of a gradient for the substance. A checkerboard study is done only after establishing that the test substance has a stimulatory effect upon cell migration.

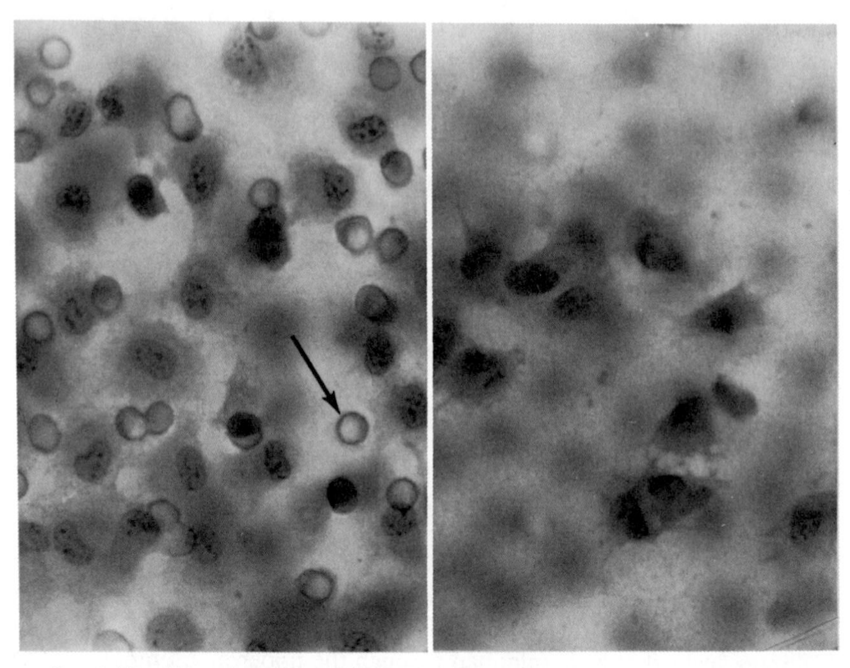

FIG. 1. Fibroblast migration to PDGF. The upper compartment contained a suspension of ligament fibroblasts. The lower compartment, separated from the upper compartment by an 8-μm nucleopore membrane overlaying a 0.45-μm Millipore Membrane, contained PDGF at 30 ng/ml in DME medium containing 1 mg/ml HSA. *Left:* After a 6-hr incubation, the upper surface of the 8-μm micropore membrane shows both fibroblasts and pores (arrow) of the membrane in focus at high-dry magnification, $\times 372$. *Right:* Focusing through the same membrane to the interface between the upper and lower membranes, some fibroblasts are seen clearly, but the pores of the upper membrane are no longer in focus, $\times 400$. These are cells that have migrated through the upper membrane in response to PDGF. In the absence of PDGF, few cells traverse the upper membrane (not shown).

To obtain a satisfactory "checkerboard," cell migration relative to the concentration of the putative chemotactic factor must be known so that it will be possible to select a range of concentrations for the gradients that are likely to show a range of responses of cell migration.

Additional details may be found in Ref. 51–55.

[51] S. H. Zigmond and J. G. Hirsch, *J. Exp. Med.* **137,** 387 (1973).

[52] J. I. Gallin and P. G. Quie (eds.), "Leukocyte Chemotaxis: Methods, Physiology, and Clinical Implications," Raven, New York, 1978.

[53] K. M. Lohr and R. Snyderman, *in* "Manual of Macrophage Methodology" (H. B. Herscowitz, H. T. Holden, J. A. Bellanti, and A. Ghaffar, eds.), Chap. 34. Dekker, New York, 1981.

[54] R. Snyderman, *in* "Methods for Studying Mononuclear Phagocytes" (D. O. Adams, P. J. Edelson, and H. Koren, eds.), Chap. 56. Academic Press, New York, 1981.

TABLE II
"CHECKERBOARD" ANALYSIS OF THE CHEMOTACTIC ACTIVITY OF PLATELET-DERIVED GROWTH FACTOR (PDGF) FOR HUMAN MONOCYTES[a]

		PDGF, upper compartment (ng/ml)			
		0	1	5	20
PDGF, lower compartment (ng/ml)	1	35 ± 2.6[b]	5 ± 2.0	3 ± 3.1	-1 ± 2.8
	5	54 ± 2.8	7 ± 2.3	2 ± 3.1	1 ± 2.8
	20	86 ± 2.8	34 ± 3.1	25 ± 2.6	2 ± 3.1

[a] From Deuel *et al.*[30]
[b] Cells per grid, SEM, $n = 15$.

General Comments

Samples for Chemotaxis. With many proteins potentially present in tested samples, care in handling is crucial. Pure test proteins should not be stored at highly diluted concentrations; storage should be in containers that minimize protein sticking to the container walls. For example, it is important to keep PDGF or PF_4 stock solutions at a concentration of at least 500 μg/ml in polypropylene tubes. When an aliquot of the stock solution is thawed it can be diluted to 0.5 μg/ml provided there is protein in the diluting solution, such as 1 mg/ml HSA. The diluted solutions of PDGF or PF_4 can be stored at $-20°$ and can be thawed two or three times without losing their activity.

Experimental Design. In the design of experiments and in the assessment of data, two principles are observed as follow: (1) Assays are run in triplicate and (2) all experiments include triplicate positive and negative controls. For experiments with neutrophils and monocytes a suitable positive control is formylmethionylleucylphenylalanine (Sigma) at 10^{-8} *M*, which yields between 80 and 150 cells above background per high power grid. The negative controls should be less than 50 cells per high power grid. For fibroblast chemotaxis, a suitable positive control is human fibrinopep-

[55] J. J. Castellot, Jr., M. J. Karnovsky, and B. M. Spiegelman, *Proc. Natl. Acad. Sci. U.S.A.* **79**, 5597 (1982).

tide B (Bachem, Inc., Torrance, CA) at 10^{-9} *M*.[56] It will yield approximately 50 cells per high power field above background. Medium blanks are less than 10 cells per high power field.

Acknowledgment

Supported in part by grants from USPHS NIH (HL29594, HL31102, HL14147) and the Monsanto Company.

[56] R. M. Senior, W. F. Skogen, G. L. Griffin, and G. D. Wilner, *J. Clin. Invest.* **77**, 1014 (1986).

[21] Secretion of von Willebrand Factor from Platelets

By MARIA FERNANDA LOPEZ-FERNANDEZ, JAVIER BATLLE, ZAVERIO M. RUGGERI, and THEODORE S. ZIMMERMAN

Introduction

Platelets contain approximately 15% of circulating von Willebrand factor (vWF). It is mostly stored in the α granules, but a small portion can be recovered from the membrane fraction.[1] Normal platelet vWF is not identical to circulating vWF because it is not complexed with the factor VIII procoagulant protein[2] and, when studied in an SDS–agarose gel discontinuous buffer system,[3] it contains larger multimeric forms than those normally present in the circulation. Structural abnormalities of platelet vWF are also detected in some patients with von Willebrand disease.[4]

[1] R. L. Nachman and E. A. Jaffe, *J. Exp. Med.* **141,** 1101 (1975).

[2] A. L. Bloom, *in* "Clinics in Haematology" (C. R. Rizza, ed.), Vol. 8, p. 53. Saunders, London, 1979.

[3] M. F. Lopez-Fernandez, M. H. Ginsberg, Z. M. Ruggeri, F. J. Batlle, and T. S. Zimmerman, *Blood* **60,** 1132 (1982).

[4] T. S. Zimmerman, Z. M. Ruggeri, and C. A. Fulcher, *in* "Progress in Hematology" (E. B. Brown, ed.), Vol. 8, p. 279. Grune & Stratton, New York, 1983.

Normal platelet vWF can be mobilized from the secretory pool to varying degrees by ADP, collagen, and thrombin.[3,5,6] The concentration of these agonists required for vWF release is the same or lower than those used for release of serotonin, lysosomal enzymes, or fibrinogen.[5] This process has the features of an energy-dependent secretory response because it is blocked by metabolic inhibitors.[5] When secreted vWF is analyzed in SDS–agarose gels, all platelet forms are released from collagen and ADP-stimulated platelets. In contrast, in the thrombin releasates, and in the presence of divalent cation, there is an absence of high-molecular-weight multimers. This suggests that thrombin induces reassociation of these larger forms with the stimulated platelets.[3] Evidence for this reassociation is severalfold. First, thrombin releases all the larger multimers, but they disappear from releasates in a time-dependent and thrombin dose-dependent manner. This agonist by itself does not produce loss of these multimers.[3] Second, following platelet activation by thrombin, there is a rapid increase in the amount of vWF expressed on the platelet surface measured by increased binding of specific anti-vWF antibodies.[6,7] Thus, platelets are not only enriched in potentially more hemostatically effective forms of vWF, but they also have the ability to reassociate endogenously secreted vWF to their membrane surfaces. It is conceivable that a dysfunction of this mechanism due to structurally abnormal platelet vWF or other platelet disorders may be responsible for a decrease of adhesion of platelet(s) to the subendothelium, platelet–platelet interaction, and consequently for a bleeding diathesis.

In this chapter, procedures for the isolation and analysis of platelet vWF, as well as its release from the secretory pool, are described in detail.

Platelet Isolation

Investigation of the platelet release reaction and of other platelet functions and parameters often requires a system from which the plasma components have been removed. There are different methods to wash platelets free of the plasma environment (see Chapters [1] and [2] in this volume). The first of these involves repeated centrifugation and resuspension of the platelets, usually using various physiological washing solutions containing inhibitors of aggregation[7,8] and often a layer of albumin or other

[5] J. Koutts, P. N. Walsh, E. F. Plow, J. W. Fenton, B. N. Bouma, and T. S. Zimmerman, *J. Clin. Invest,* **62,** 1255 (1978).

[6] Z. M. Ruggeri, R. Bader, F. Pareti, P. Mannucci, and T. S. Zimmerman, *Thromb. Haemostasis* **50,** 35 (Abstr.) (1983).

[7] J. N. George and A. R. Onofre, *Blood* **59,** 194 (1982).

[8] M. B. Zucker, M. J. Broekman, and K. L. Kaplan, *J. Lab. Clin. Med.* **94,** 675 (1979).

high-density substances to protect the platelets during centrifugation.[9,10] The second, pioneered by Tangen *et al.*,[11] depends on gel filtration for the removal of plasma components. Since this method does not completely separate platelets from plasma vWF, there is an alternative procedure that employs a combination of albumin density or gradient centrifugation and gel filtration of platelet-rich plasma (PRP).[12]

We are currently using two washing procedures. The first is designed for studying the structure of vWF in platelet lysates and includes centrifugation and resuspension of the cells in buffers containing inhibitors to prevent platelet aggregation and proteolysis of the platelet vWF. The other is used for release experiments and involves platelet separation by one step: washing and gel filtration in Sepharose 2B. This system provides functionally intact platelets and is quick, simple, and reproducible. Less than 0.5% of total starting vWF is present in the platelet-free supernatant, indicating little carry over of soluble plasma vWF.

Washed Platelets by Centrifugation for Obtaining Platelet vWF

ACD anticoagulant: Sodium citrate (25 g/liter), citric acid (13.64 g/liter), acid–dextrose (20 g/liter), pH 4.5

Buffer A: Sodium citrate·$2H_2O$ (0.013 *M*), glucose (0.028 *M*), sodium chloride (0.123 *M*), disodium EDTA (4 m*M*), pH 6.5

Buffer B: Trizma base (0.010 *M*), sodium chloride (0.14 *M*), disodium EDTA (5.5 m*M*), pH 7.5

Inhibitors: Phe-Phe-Ack (D-Phe-L-Phe-L-Arg-CH_2Cl), 29 m*M*, PPACK (D-Phe-L-Pro-L-Arg-CH_2Cl), 0.33 m*M*, and dansyl-GGACK (dansyl-L-Gly-L-Arg-CH_2Cl), 30 m*M* (Calbiochem-Behring Corp., La Jolla, CA), leupeptin 46 m*M* (Chemicon Corp., Los Angeles, CA). Use at 1 : 100 dilution in buffers A and B, and 1 : 30 dilution in ACD anticoagulant. These inhibitors may be stored at $-20°$ when in solution

Procedure

Five volumes of venous blood are collected by aseptic venipuncture into 50-ml polypropylene conical centrifuge tubes with 1 vol of acid–citrate–dextrose (ACD) anticoagulant plus inhibitors. PRP is obtained by three differential centrifugation steps at 1100 *g* for 1 min and 15 sec. The

[9] P. N. Walsh, *Br. J. Haematol.* **22,** 205 (1972).

[10] Z. M. Ruggeri, R. Bader, and L. de Marco, *Proc. Natl. Acad. Sci. U.S.A.* **79,** 6038 (1982).

[11] O. Tangen, M. H. Berman, and R. Marfey, *Thromb. Diath. Haemorrh.* **25,** 268 (1971).

[12] S. Timmons and J. Hawiger, *Thromb. Res.* **12,** 297 (1978).

PRP from each centrifugation is pooled in a 50-ml tube, taking care to avoid removing the buffy coat. After adding an equal volume of buffer Damon/IEC A plus inhibitors the tubes are centrifuged at 2200 rpm in a Damon/IEC centrifuge using a 269 rotor for 25 min at 4°. Before two repeat centrifugation steps at 2200 rpm for 15 min at 4°, the supernatant is poured off and the pelleted platelets resuspended, one in buffer A and the other in buffer B, both containing inhibitors. The resulting last pellet is suspended in no more than 2 ml of buffer B plus inhibitors. Count and adjust the cells with the same buffer to give an approximate concentration of 2×10^9 platelets/ml. After lysis by two cycles of freezing and thawing, the platelet membranes are pelleted by centrifugation at 7000 *g* for 5 min and the supernatants stored at −70° until used.

Washed Platelets by Gel Filtration for vWF Release Studies

Buffers

Modified Tyrode's solution: NaCl (8 g/liter), KCl (0.195 g/liter), $NaHCO_3$ (1.02 g/liter), glucose (1 g/liter), $MgCl_2 \cdot 6H_2O$ (0.123 g/liter), bovine serum albumin (BSA), 1 g/liter, pH 6.5

Lysis buffer: Trizma base (0.010 *M*), NaCl (0.140 *M*), disodium EDTA (5.5 m*M*), urea (8 *M*), PMSF (phenylmethylsulfonyl fluoride), 1 m*M*, SDS (2%), Triton X-100 (1%)

Procedure

Fifty milliliters of blood from drug-free volunteers is mixed with 10 ml ACD anticoagulant in 50-ml polypropylene conical centrifuge tubes (Corning Glass, Ithaca, NY). PRP is prepared by centrifuging at 120 *g* for 15 min at room temperature. The cells are pelleted by centrifugation at 1200 *g* for 20 min at 10° and the supernatant plasma (with the inhibitors described above) stored at −70° until used. The platelets are resuspended in 1–2 ml of calcium-free Tyrode's solution, pH 6.5, and passed over a 35×1.6 cm column containing Sepharose 2B (Pharmacia Fine Chemicals, Piscataway, NJ) that had been equilibrated with modified Tyrode's solution, pH 7.4. The flow rate of filtration is in the range of 15 to 25 ml/hr. The washed platelets are recovered in 4 to 5 ml by visual observation of the change in opacity of the column effluent. Platelets are counted and used at a final concentration of 10^9 cells/ml.

Release Experiments

Three hundred microliters of platelet suspension obtained by centrifugation and gel filtration is added to 12×75 mm polypropylene tubes followed by 4 μl of 100 mM $CaCl_2$ solution and 100 μl of stimulus or modified Tyrode's solution, pH 7.4. In some tubes, 5 mM EDTA (final concentration) is added before the addition of the agonist. The mixtures are incubated at 37° for the indicated time, without stirring. Macroscopic aggregation is not observed. The reaction is terminated by placing the tubes into ice–methanol for 20 sec followed by centrifugation at 7000 g for 30 min at 4°. Supernatants are removed and stored at −70°. In each experiment a separate aliquot of cells is subjected to five cycles of freezing by immersion in liquid N_2 followed by thawing in a 37° water bath, and another one is centrifuged and resuspended with lysis buffer.

Dose–Response Studies. Collagen (Millipore Corp, Freehold, NJ) is used at a final concentration of 1:250 μg/ml, ADP (grade I, Sigma Chemical Corp., St. Louis, MO) at 2–5 μM final concentration and α-thrombin at 0.01 to 2 U/ml. The reactions are stopped 30 min after the addition of the incubating agent.

Time Course Studies. The reactions are terminated at 2, 5, 10, 20, 30, and 60 min. The zero time is obtained by adding the appropriate stimulus when the platelet suspension is placed in ice–methanol. Thrombin is used at final concentration of 1 U/ml, collagen at 50 μg/ml, and ADP at 0.1 mM.

The quantification of the amount of vWF release is performed as described previously,[13] using an ^{125}I-labeled affinity-purified rabbit anti-vWF. The immunoprecipitates are analyzed with autoradiography.

Quantification of von Willebrand Factor by Radioquantitative Immunoelectrophoresis

Precoat $3\frac{1}{4} \times 4$ in. glass slides by spreading 2 ml of 1.5% medium electroendosmosis agarose (FMC, Rockland, ME) on warmed slides and allowing to air dry. Gels are prepared with 0.9% agarose in 0.025 M barbital, sodium salt, 0.005 M barbital, pH 8.6, at 54° to which sufficient rabbit anti-von Willebrand factor antibody has been added to result in a rocket height of 25 mm for an undiluted human plasma standard. Also added to the agarose is sufficient ^{125}I-labeled immunopurified rabbit anti-von Willebrand factor to allow a readable autoradiographic signal at 16 hr of exposure. Agarose (15 ml) is poured on the precoated slides.

Prior to running the gel 13 0.45-cm-diameter holes are punched along the length of the slide. Twenty-milliliter samples are placed in the wells and

[13] T. S. Zimmerman and J. R. Roberts, *in* "Immunoassays: Clinical Laboratory Techniques for the 1980's" (R. M. Nakamura, W. R. Dito, and E. S. Tucker III, eds.), p. 339. Liss, New York, 1980.

bibulous paper wicks are applied to the slides. Wicks are soaked in running buffer (0.075 *M* barbital, sodium salt; 0.015 *M* barbital, pH 8.6). Slides are electrophoresed at 6.0 mA per slide, constant current (approximately 17.5 V/cm) for 16 hr at room temperature. Slides are washed for 8 hr in normal saline, covered with wet bibulous paper, dried in a drying oven at 80°, the paper is removed, the slides placed in an X-ray exposure holder (Kodak) with XR-1 X-ray film (Kodak), and exposed for 16 hr.

Multimeric Structure of Secreted vWF

SDS-Agarose Electrophoresis

Our technique for resolving vWF multimers utilizes an SDS–agarose discontinuous electrophoresis system.[14,15] In this technique, the use of a stacking gel with multiphasic buffer systems concentrates proteins into very thin starting zones of high local protein concentrations, resulting in improved resolution. The samples are treated with SDS (sodium dodecyl sulfate) and then run in SDS–agarose gels. SDS, an ionic detergent, denatures proteins by binding avidly to them at about 1.4 mg of SDS/mg protein, imparting a uniform negative charge to all of the multimers and dispersing aggregates. The proteins are then subjected to an electric current at pH 8.35, causing them to migrate toward the anode. The pore size of agarose allows separation of the multimers of vWF by molecular weight. At the end of the electrophoresis the gel is fixed, reacted with ^{125}I-labeled anti-vWF antibody, and the vWF band detected by autoradiography.

Specific Details of Procedures

Buffers

Running gel buffer: 0.375 *M* Tris chloride, pH 8.8
Stacking gel buffer: 0.125 *M* Tris chloride, pH 6.8
Sample buffer (10 × concentrate): 10 m*M* Tris chloride and 1 m*M* disodium EDTA, pH 8.0
Electrophoresis buffer: 0.0495 *M* Trizma base, 0.384 *M* glycine, 0.1% SDS, pH 8.35
Phosphate-buffered saline (PBS): 150 m*M* NaCl and 10 m*M* Na_2HPO_4, pH 7.2

[14] Z. M. Ruggeri and T. S. Zimmerman, *Blood* **57,** 1140 (1981).
[15] T. S. Zimmerman, F. J. Batlle, Z. M. Ruggeri, and J. R. Roberts, *in* "Clinical Laboratory Assays: New Technology and Future Directions" (R. M. Nakamura, W. R. Dito, and E. S. Tucker III, eds.), p. 249. Masson, New York, 1983.

Pouring the Gel. A sandwich set is built with two glass plates; in one of them a plastic support (Gel-Bond film) of 12.7 × 28 cm has been applied previously and 1.5-mm spacers are set in between the edge of the Gel-Bond and the other plate. Clamp the whole apparatus together down both sides and across the bottom using large metal binder clips.

Dissolve Sigma type VI low-gelling-temperature (LGT) agarose (1.4% final concentration) and SDS (0.1% final concentration) in 0.375 *M* running gel by heating to 100°. Pour the hot agarose into the mold up to the top edge and store it at 4° for 2 to 3 hr. Then unclamp the assembly, remove tape across the top, and slide the upper glass plate down until the upper edge is 2.5 cm from the top. Cut the agarose along the edge of the glass plate and remove the gel above this line. Slide the glass plate up again and reclamp. Pour the stacking gel in the space so created. The stacking gel is prepared by dissolving Seakem high-gelling-temperature [HGT(P)] agarose (0.8% final concentration) in stacking gel buffer and SDS (0.1% final concentration). Leave at 4° for 1 hr (keep unused agarose at 56°), open the sandwich again, and remove gel and Gel-Bond. Punch into the stacking gel 10 sample wells (11 × 2 cm) at 10 mm from the interface between running and stacking gels.

Samples. Samples are diluted according to the concentration of vWF (usually plasma 1:20, lysates 1:10, and secreted vWF 1:4) with 1 × sample buffer to which 2% SDS and tracking dye (bromphenol blue, Sigma Chemical, St. Louis, MO) have been added. The samples are then heated at 56° for 30 min.

Electrophoresis. The gels should be electrophoresed while lying on a cooling platform adjusted to 16°, and using fresh electrophoresis buffer. Run 25 μl of samples into the agarose at 25 mA/gel, constant current, until the wells are just dry. Then fill the wells with stacking gel agarose and continue running at 9 mA/gel overnight (17–18 hr) until the dye front is 1 mm anodal wick. To minimize condensation or evaporation, a glass or plastic lid is suspended 3 mm above the gel.

Detection of vWF Multimers. Fix in 10% acetic acid and 25% α-propanol for 2 to 3 hr, wash in three changes of deionized H_2O for 1 hr each, and dry the gel. Equilibrate in PBS with 0.1% BSA and 0.02% azide for 30–60 min. Then the gel is soaked overnight in 100 ml of PBS, pH 7.2, 0.1% BSA, 0.02% NaN_3, and 0.2 mCi of ^{125}I-labeled affinity-purified rabbit anti-vWF antibody, radiolabeled by the chloramine-T method (at a specific activity of 200 mCi/mg of protein).[13] Finally, wash in 0.5 *M* saline (three changes), normal saline (overnight), and deionized water for 15 min, dry, expose to XRP-1 film (Kodak) at −70°, and develop.

Acknowledgments

We thank Ms. Claire Jackson for help in the preparation of this manuscript.

This work was supported in part by the U.S.-Spain Joint Committee for Scientific and Technological Cooperation Grant CA 83/014 and Grants HL 15491 and HL 31950 from the National Institutes of Health.

This is publication 3726 BCR from the Scripps Clinic and Research Foundation, La Jolla, California 92037.

[22] Secretion of Thrombospondin from Human Blood Platelets

By HENRY S. SLAYTER[1]

Among the components of the intracellular granules of human blood platelets, secreted through the surface-connected canalicular system, is a glycoprotein (with a molecular weight of 420,000) designated "thrombospondin,"[2] (TSN) which is composed of three disulfide-linked polypeptide chains of equal molecular weight.[2,3] Thrombospondin appears to arise in the α granules, and binds heparin rather firmly.[2,4-6] It has been reported that thrombospondin is the site of the endogenous lectin-like activity of platelets,[7] and thus may mediate their aggregation by binding to specific receptors, possibly fibrinogen, platelets,[8-10] or type V collagen.[11] A biologi-

[1] Research in the author's laboratory was supported by Grant HL33014PVI from the National Heart, Lung and Blood Institute, U.S. Public Health Service.

[2] J. W. Lawler, H. S. Slayter, and J. E. Coligan, *J. Biol. Chem.* **253,** 8609 (1978).

[3] N. L. Baenziger, G. N. Brodie, and P. W. Majerus, *J. Biol. Chem.* **247,** 2723 (1972).

[4] I. Hagen, *Biochim. Biophys. Acta* **392,** 243 (1975).

[5] R. Kaser-Glanzmann, M. Jakabora, and E. F. Luscher, *Chimia* **30,** 96 (1976).

[6] J. M. Gerrard, D. R. Phillips, G. M. R. Rao, E. F. Plow, D. A. Walz, R. Ross, L. A. Harker, and J. G. White, *J. Clin. Invest.* **66,** 102 (1980).

[7] E. A. Jaffe, L. L. R. Leung, R. L. Nachman, R. I. Levin, and D. F. Mosher, *Nature (London)* **295,** 246 (1982).

[8] T. K. Gartner, J. M Gerrard, J. G. White, and D. C. Williams, *Nature (London)* **289,** 688 (1981).

[9] L. L. R. Leung and R. L. Nachman, *J. Clin. Invest.* **70,** 542 (1982).

[10] R. L. Silverstein, L. L. R. Leung, P. C. Harpel, and R. L. Nachman, *J. Clin. Invest.* **74,** 1625 (1984).

[11] S. M. Mumby, G. J. Raugi, and P. Bornstein, *J. Cell Biol.* **98,** 646 (1984).

cal assay for thrombospondin based on agglutination of fixed activated platelets has been described elsewhere in this series.[12]

It has also been reported that thrombospondin is produced in other cell types, including endothelial cells,[13] fibroblasts,[14] and glial cells.[15] Thus, it may well eventually be shown to be a ubiquitous constituent of the extracellular matrix in a variety of tissues.

Thrombospondin has been characterized both physically[16] and chemically,[2] and electron microscopy has suggested that its structure is composed of several globular regions linked by connecting strands.[2,16,17] Known chemical and physical parameters of thrombospondin are summarized in Tables I and II. The N-terminal sequence of thrombospondin and that of the heparin-binding peptide (HBP) produced by plasmin digestion are homologous,[18] the cDNA sequence is published,[19] and the structure of thrombospondin is that of a three-stranded "bola" with the principal heads connected by 30-nm strands to a central disulfide knot region.[18]

It has also been shown that thrombospondin is present in normal human plasma at levels of 25–100 ng/ml and in normal serum at much higher levels ($15-30 \times 10^3$ μg/ml), implying that its bulk is derived from platelets during blood clotting.[15,16]

Thrombospondin is a rather peculiar protein in several respects. We have estimated its solubility to be at least 2 mg/ml in saline, but losses on surfaces of all kinds make quantitative work very demanding. Thus, chromatography, concentration steps, assays, binding experiments, and peptide purification and sequencing all require particular precautions to maintain undenatured thrombospondin in solution without significant losses.

The absorption coefficient of thrombospondin is 0.91 A_{280}/mg/ml, the isoelectric point is in the vicinity of 5, and, of particular interest, the cysteine content is 205 residues/molecule, a large fraction of which (ca. 80%) is concentrated in a disulfide-rich protease-resistant core.[18,20]

Since thrombospondin is now the subject of intense investigation in many laboratories interested in platelets and in the coagulation mechanism, it may be useful to describe in some detail the methods developed for

[12] S. A. Santoro and W. A. Frazier, this series, Vol. 144, p. 438.

[13] D. F. Mosher, M. J. Doyle, and E. A. Jaffe, *J. Cell Biol.* **93,** 343 (1982).

[14] S. M. Mumby, D. Abbott-Brown, G. J. Raugi, and P. Bornstein, *J. Cell. Physiol.* **120,** 280 (1984).

[15] A. S. Asch, L. L. K. Leung, J. Shapiro, and R. L. Nachman, *Proc. Natl. Acad. Sci. U.S.A.* **83,** 2904 (1986).

[16] S. Margossian, J. W. Lawler, and H. S. Slayter, *J. Biol. Chem.* **256,** 7495 (1981).

[17] J. Lawler, F. C. Chao, and C. M. Cohen, *J. Biol. Chem.* **257,** 12257 (1982).

[18] J. E. Coligan and H. S. Slayter, *J. Biol. Chem.* **259,** 3944 (1984).

[19] J. Lawler and R. D. Hynes, *J. Cell Biol.* **103,** 1635 (1986).

[20] J. W. Lawler and H. S. Slayter, *Thromb. Res.* **22,** 267 (1981).

TABLE I
CHEMICAL COMPOSITION OF THROMBOSPONDIN

Amino acid composition			Residues/10^5 daltons			Carbohydrate composition (wt %)	
mino cid[a]	Thrombo-spondin[b]	Thrombin-sensitive protein[c]	Thrombo-spondin[d]	HBP_p[d]	Sugar[e]	Thrombo-spondin[b]	Thrombin-sensitive protein[c]
ys	43	49	55	48	Sialic acid	0.7	3.9
is	22	17	22	12	GalNAc	None	None
rg	51	36	50	61	GlcNac	1.4	1.6
sp	160	142	177 (Asx)	117 (Asx)	Fuc	0.2	
hr	42	48	45	48	Man	0.6	
er	57	97	54	72	Gal	0.6	
ilu	95	121	101(Glx)	112 (Glx)	Glc	0.5	
ro	54	53	57	41	Hexuronic acid	None	
ily	89	114	81	86			
la	52	61	40	61			
ys	50	20	39[f]	ND[g]			
let	9	8	12	9			
al	51	53	66	91			
e	32	29	33	40			
eu	49	48	50	92			
he	26	22	26	37			
yr	25	17	25	9			

[a] Amino acids were analyzed on a Durrum D500 analyzer. Cysteine was determind as cysteic acid.
[b] From Lawler *et al.*[2]
[c] From Baenziger *et al.*[3]
[d] From Coligan and Slayter.[18]
[e] Amino sugars were analyzed on a Durrum D500 analyzer. Values for monosaccharides were obtained with the gas chromatographic procedure of Saglio and Slayter.[25] Hexuronic acid determination was obtained by the procedure of Dawes *et al.*[38]
[f] Determined as *S*-carboxymethylcysteine.
[g] ND, Not determined.

the secretion, purification, assay, and characterization of thrombospondin from platelets by various methods.

Secretion of Thrombospondin from Platelets

The secretion of thrombospondin from fresh platelets by various agents may be measured by radioimmunoassay, and compared with amount released in the serum of a freshly clotted blood control from the same individual, and with the amount released from an equivalent number of

TABLE II
PHYSIOCHEMICAL PROPERTIES OF THROMBOSPONDIN[a]

Property	Value
Molecular weight (M)	
Sedimentation equilibrium in TSE[b]	420,000
Sedimentation equilibrium in TSE and guanidine–HCl	410,000
SDS–PAGE	450,000
Molecular weight of reduced thrombospondin	
Sedimentation equilibrium in TSE and guanidine–HCl	133,000
SDS–PAGE	142,000
Partial specific volume *(v)*	0.714 ml/g
Sedimentation coefficient ($s^0_{20,w}$)	8.6S
Intrinsic viscosity (η)	40 ml/g_0
Stokes radius, calculated[c]	123 nm
Percentage α helix	3%
$M_i/M_{dithiothreitol}$[d]	3.05
Electron microscope shape[e]	
Thrombospondin	Bola
Maximum dimension	74 nm
cDNA-derived amino acid sequence[f]	

[a] From Margossian *et al.*[16]
[b] TSE: 0.02 *M* Tris–HCl (pH 7.6), 0.15 *M* NaCl, 1 m*M* EDTA.
[c] $r_s = [M(1 - v\rho)/6\pi\eta_o NS]$.
[d] From sedimentation equilibrium performed in guanidine–hydrochloride using the expression $[M(1 - v\rho)_{intact}]/[M(1 - v\rho)_{dithiothreitol}]$.
[e] From Coligan and Slayter.[18]
[f] From Lawler and Hynes.[19]

platelets by treatment with 1% Triton X-100, as follows (Table III): Blood is drawn in two lots: one anticoagulated with 5 m*M* EDTA, and one nonanticoagulated. The latter is allowed to clot and the serum removed for analysis. The former is used to prepare platelet-rich plasma (PRP), which is counted to determine the number of platelets/ml, using a hemocytometer (1 : 100 dilution), then treated at 25° with a nonionic detergent such as 1% Triton X-100 (to obtain optimum release data) or with agonists such as thrombin (0.5–1.0 units/ml PRP) and vortexed. Supernatants of a 5-min spin at 25°, at 7000 *g* are removed for analysis after 2 hr. Data from this and other release experiments using platelet-rich-plasma (PRP) are shown in Table III.

For purposes of preparation, thrombin is the most frequently used stimulant of the platelet release associated with aggregation of the platelet at a level of approximately 2 units/ml. Complete data are not available on the effectiveness of addition agents such as ADP, epinephrine, collagen, the ionophore A23187, or PAF-Acether. However, stimulation by the iono-

TABLE III[a]
RELEASE OF THROMBOSPONDIN FROM PLATELETS

Experiment	Platelets (per ml PRP)	Serum[b]	PRP[c]	Released by thrombin	Released by Triton X-100
A	2.57×10^8 [d]	0.83×10^{-4} ng/platelet (21 μg/ml serum)	0.04×10^{-4} ng platelet[c] (1.1 μg/ml PRP[d])	0.20×10^{-4} ng/platelet[e] (5.1 μg/ml PRP[d])	0.97×10^{-4} ng/platelet (25 μg/ml PRP[d])
B	4.6×10^8 [f]	—	—	—	1.5×10^{-4} ng/platelet (89 μg/ml PRP)
C	1.59×10^9 [f]	—	—	0.68×10^{-4} ng/platelet[g]	0.73×10^{-4} ng/platelet (116 μg/ml PRP[f])
D	1.59×10^9 [f]	—	—	0.55×10^{-4} ng/platelet[h]	—

[a] Unpublished experiments (Slayter, Miller, and Bates).
[b] Twenty-four-hour serum.
[c] Unreleased supernatant.
[d] PRP prepared from fresh anticoagulated blood. Expect ca. 3×10^8 platelets/ml in blood.
[e] 0.5 U/ml; Release 2 hr from fresh drawing from individual A.
[f] Red Cross platelet-rich plasma.
[g] 1.0 U/ml; 24-hr platelets from individual B.
[h] Same PRP as *d* but 48 hr later, released with 1.0 U/ml thrombin.

phore A23187 apparently does cause the secretion of thrombospondin.[21,22] ADP (10 μM) appears to activate fresh platelets to secrete thrombospondin but precise yields are again not available.[23]

Prevention of Losses at Surfaces

Due to the "sticky" nature of thrombospondin, some specific general precautions should be taken in working with it. Thus, 1% stock Siliclad (SCM Specialty Chemicals, Gainesville, FL) is used for coating surfaces used to contain the protein; that is, columns, tubes, beakers, pipets, etc. Column materials, until they have been used a number of times, remove significant amounts of thrombospondin from solutions passing through them, so that seasoned columns are useful. Concentrator membranes and dialysis membranes absorb significant quantities. Losses are often 25–100%, depending on the initial amount and concentration present. Precoating with a thrombospondin solution at 100 μg/ml or more if possible for overnight is helpful. Sometimes when a concentrator membrane is soaked in buffer for 10 hr the yield improves by 25%, but then reconcentration is still a problem. Column materials such as Sepharose 4B, when used fresh for thrombospondin, often bind 60% of the protein irreversibly! After liganding proteins such as fibrinogen to CNBr–Sepharose, typically 40–50% of thrombospondin subsequently passaged at 500 μg protein to 5 ml of column bed binds. Of this only a few percentage more may be removed with 2 M NaCl or 0.5% Triton $\times$100. Thus, binding studies involving thrombospondin must be approached with caution, and appropriate controls applied.

Nalgene tubes are soaked in Siliclad as well and, in addition, plastic tubes precoated with bovine serum albumin are used for collection of radiolabeled fractions. However, losses must be expected, planned for, and factored into analysis of results.

Preparation and Purification of Thrombospondin

Platelet concentrate from eight or more pheresis units is obtained, ideally within 2–4 hr after donation. Depending on the method of collection and storage the yield from platelets 24 to 48 hr old is found to diminish to about one-half to nil, respectively. (Release is largely inhibited

[21] T. F. Gartner and M. E. Doctor, *Thromb. Res.* **33,** 19 (1984).

[22] I. Fuse, A. Hattori, M. Higashihara, S. Takazawa, T. Takeshige, M. Hanano, R. Nagayama, T. Koike, H. Takahashi, and A. Shibata, *Scand. J. Hematol.* **36,** 44 (1986).

[23] G. Agam, O. Shohat, and A. Livne, *Proc. Soc. Exp. Biol. Med.* **177,** 482 (1984).

even in fresh platelets by anticoagulants containing procaine–HCl.) A 0.25 vol of anticoagulant acid–citrate–dextrose (ACD, National Institute of Health formula A) is added, and the platelets are pelleted by centrifugation at 900 *g* for 15 min. The platelets are resuspended in 0.02 *M* Tris–HCl buffer (pH 7.6) containing 0.15 *M* NaCl, 5 m*M* glucose, and 10% ACD (TSG–ACD). (Phosphate at pH 6.5 substituted for Tris at this stage seems to better protect platelet activity during washing.) The platelet suspension is centrifuged 200 *g* for 6 min to remove residual cells. The platelets are then washed two times in the above buffer and resuspended in 60 ml/10 units, or less, of the Tris buffer excluding ACD (TSG). The pH is checked and adjusted to pH 7.6, if necessary, prior to the addition of 3 units/ml of purified human thrombin to the stirring platelet suspension at 37°. After 2 min, hirudin (Pentapharm, Switzerland) is added to 6 units/ml, along with 0.1 m*M* leupeptin (Sigma, St. Louis, MO), 1 m*M* phenylmethylsulfonyl fluoride (PMSF) (including 1 m*M* EDTA only if the primary consideration is for very little proteolysis), and the aggregates allowed to settle in an ice bath. The supernatant is removed and centrifuged at 21,000 *g* for 35 min. The supernatant is applied to a column (5 × 90 cm) of Sepharose 4B at 4° equilibrated to 0.02 *M* Tris–HCl (pH 7.6), 0.15 *M* NaCl, [including also either 1 m*M* EDTA (TSE) or 1 m*M* Ca^{2+}, depending on whether the experiments to be conducted with the purified material require preservation of Ca^{2+}-dependent structures, or on the other hand prevention of proteolysis is the prime consideration] and 1 m*M* 2-mercaptoethanol (2-ME).The elution profile is shown in Fig. 1. The high-molecular-weight fractions (region A of Fig. 1) are pooled and applied to a column (0.9 × 8 cm) of immobilized heparin (coupled to Sepharose 4B by cyanogen bromide). Stepwise elution is carried out with 0.02 *M* Tris–HCl (pH 7.6), and 1 m*M* 2-ME containing 0.15, 0.25, 0.45, and 2.0 *M* NaCl (Fig. 1B). Fractions are analyzed for protein by the method of Lowry *et al.*,[24] and by fluorescence spectroscopy using excitation at 280 nm and emission at 350 nm, or by absorption spectrophotometry at 280 nm. Thrombospondin is eluted in 0.45 *M* NaCl and used immediately or stored at −80° in 25% sucrose. Multiple passages through a heparin affinity column resulted in virtually total removal of protein impurities from the preparation.[25]

When isolated by the methods described above, which are not identical to that given in Ref. 14, 1 to 2 mg of thrombospondin is obtained per unit, depending on the age and quality of the platelet concentrate. For this material, good agreement has been found between absorption coefficients

[24] O. H. Lowry, N. J. Rosebrough, A. L. Farr, and R. J. Randall, *J. Biol. Chem.* **258,** 785 (1951).

[25] S. D. Saglio and H. S. Slayter, *Blood* **59,** 162 (1982).

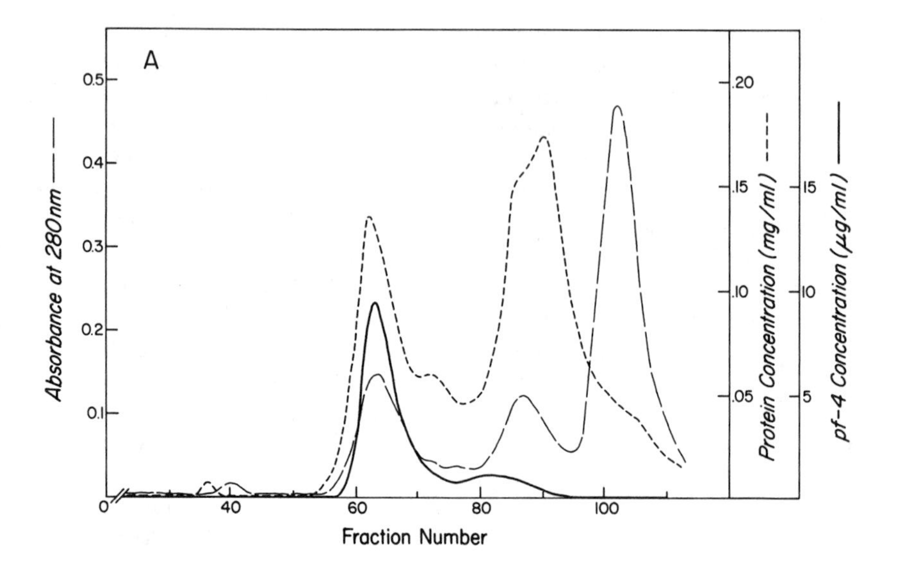

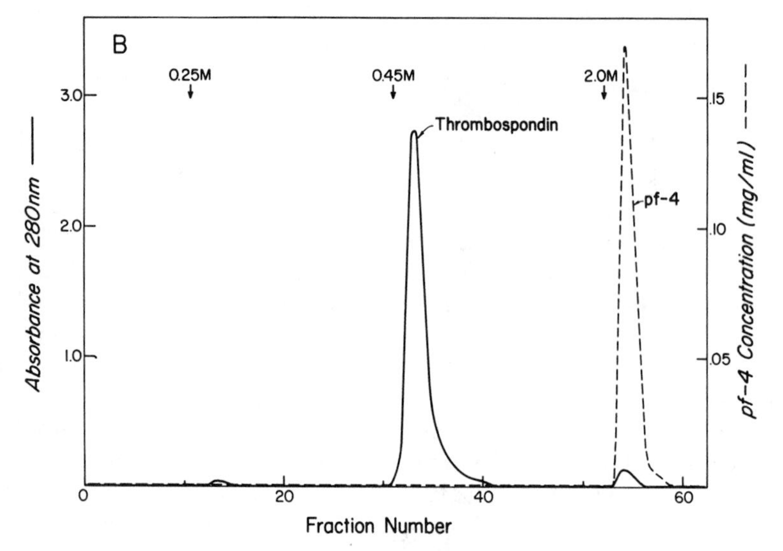

FIG. 1. (A) Profile of elution of crude supernatant from thrombin-treated platelets on Sepharose 4B column (5 × 90 cm). The absorbance at 280 nm, protein concentration, and platelet factor 4 concentration as a function of fraction number are indicated. Thrombospondin and platelet factor 4 were eluted in the high-molecular-weight peak at about fraction 65. (B) Elution of thrombospondin from a heparin–Sepharose 4B affinity column (0.9 × 8 cm). Thrombospondin is eluted at 0.45 *M* NaCl. Platelet factor 4 is eluted at 2.0 *M*. Commercial heparin–agarose is available from Sigma Chemical Company (St. Louis, MO).

obtained by relating special measurements to both Lowry and Kjeldahl analyses (0.95 and 0.885 A_{280}/mg/ml, respectively). The unreduced protein migrates as a single band on 3.3% SDS–PAGE with apparent molecular weight of 450,000 (Fig. 2). The reduced protein also migrates as a single band with an apparent molecular weight of 160,000 on 3.3% SDS–PAGE. On higher percentage acrylamide gels the unreduced protein does not penetrate far enough to permit accurate calibration; however, the reduced protein electrophoresed on 5 and 7.5% gels gives values of 150,000 and 142,000 D, respectively, but a few peculiarities deserve mention. Unreduced, a small residual thrombospondin aggregate band is often formed on top of the running or stacker gel. Reduction largely eliminates this material, which appears to consist of thrombospondin polymers (see Fig. 2). A small, variable amount of degradation of thrombospondin is often apparent, on SDS–PAGE, as a few very light bands in the molecular weight range 50,000–120,000. This is often overlooked if the gel is loaded with less than about 30–50 μg of protein, using Coomassie staining. Using autoradiographic detection, and loads of 2000–10,000 cpm ^{125}I-labeled protein, optimal exposures for native bands usually show the same proteolyzed bands as found with Coomassie staining, or Western blotting with anti-thrombospondin.

Thrombospondin usually remains soluble and undegraded at 4° at a concentration of 1.0 mg/ml in buffered saline for 1 week. Approximately 90% of the material is recovered if it is stored frozen at −20° for 1 week. This period may be extended greatly by the addition of 25% sucrose and storage at −80° Lyophilized samples can more readily be dissolved in aqueous solution when 25% sucrose is present. (Sucrose should be high-quality enzyme-free material.) Otherwise, they are soluble (but denatured) in pyridine or 50% acetic acid.

Heparin–Sepharose affinity chromatography removes high-molecular-weight proteins, which do not bind heparin, and also eliminates platelet factor 4, which is eluted from Sepharose 4B in the high-molecular-weight range, presumably as a complex with proteoglycan, but is found in significant quantity, virtually pure, in the 2.0 *M* eluate from the heparin affinity step. Heparin–Sepharose chromatography also concentrates the protein and eliminates the necessity for pressure dialysis concentration, which often results in significant losses. Use of 1% 2-ME is recommended for reduction along with brief treatment at 100°, to assure complete reduction of this highly disulfide-crosslinked protein. Residual bands at gel interfaces, sometimes persisting after this treatment, are thought to be due to cross-linkage by factor XIII.[26]

[26] G. W. Lynch, H. S. Slayter, B. E. Miller, and J. McDonagh, *J. Biol. Chem.* **262,** 1772 (1987).

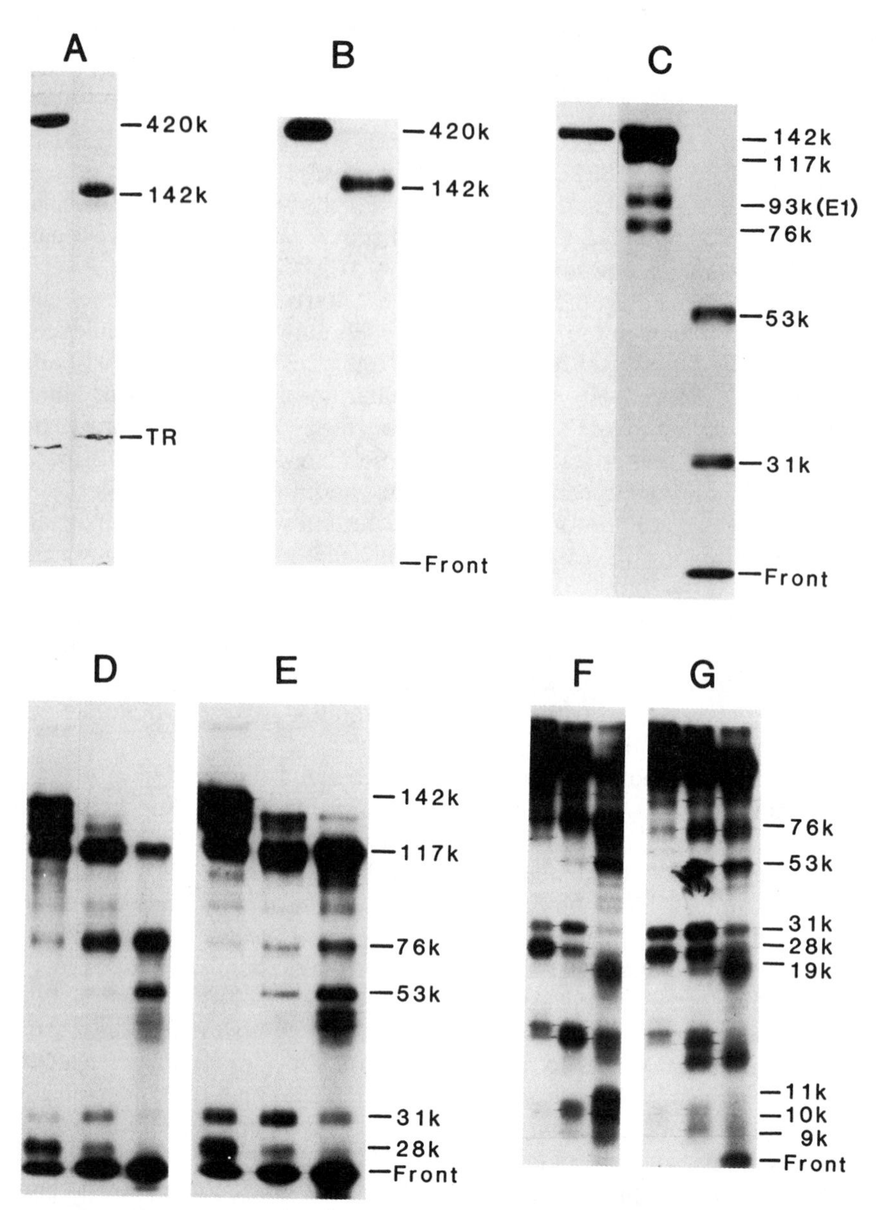

FIG. 2. SDS–PAGE analysis of thrombospondin using 5% gels stained with Coomassie brilliant blue staining (A), 10% gels and autoradiography (B), 10% gels and ^{125}I-labeled lectin blotting (C), 10% gels and simple blotting and autoradiography (D, E), and 15% gels, simple blotting, and autoradiography (F, G). A and B are shown unreduced (left lane) and reduced (right lane). C–F are shown only reduced. Apparent molecular weights of representative bands are indicated on the right-hand side. In C, left lane is native, reduced thrombospondin, middle lane is an early plasmin digest; right lane is a later plasmin digest. In D through G, early, middle, and late digests are shown in three-lane sequences, left to right. In D and F, thrombospondin was prepared in the presence of Ca^{2+}. Higher molecular weight peptides are shown more clearly on 10% gels (D, E) and smaller peptides are resolved on 15% gels (F, G).

Alternative methods of purification producing thrombospondin as identified by SDS–PAGE have appeared in the literature, but documentation of purity by radioimmunoassay for potential contaminants has not yet been reported.[12,27–30]

Micropreparative and Analytical Isoelectric Focusing

Preparative isoelectric focusing of thrombospondin can be performed on a 30 to 40% sucrose gradient in a liquid column (0.6 × 10 cm) supported by a plug of 5% polyacrylamide, as described elsewhere.[18]

Analytical isoelectric focusing is performed in slab gels obtained from LKB (ampholine PAGE plates, pH plates, pH range 3.5 to 9.5) and prepared and loaded as recommended by the manufacturer. A short slot is cut in the gel and filled with submicrogram quantities of ^{125}I-labeled thrombospondin. Use of larger loads tends to result in artifactual multiple bands of precipitated protein in the vicinity of the isoelectric point.

Preparation of Anti-Thrombospondin IgG

Thrombospondin antigen, carefully prepared either by multiple passage through a heparin affinity column, or by isoelectric focusing at 0.25 mg/0.5 ml, mixed 1 : 1 in 0.5 ml Freund's complete adjuvant, is injected subcutaneously and intramuscularly into New Zealand White rabbits with boosters administered bimonthly. Serum is 50% saturated with $(NH_4)_2SO_4$ and centrifuged at 15,000 *g* for 30 min. The pellet is washed once in 50% $(NH_4)_2SO_4$, then dialyzed into 0.025 *M* $NaPO_4$, pH 6.9, at 4° before passage through a DEAE (DE-52, Whatman, Springfield Mill, Maidstone, Kent, England) column, equilibrated in this buffer. At this point, the yield is 6.4 mg IgG/ml serum.[14] Unbound material is collected, dialyzed into 0.15 *M* NaCl, 0.01 *M* $NaPO_4$, pH 7.5, at 4°, and applied to an affinity column (Affi-Gel, Bio-Rad, Rockville Center, NY) to which 6 mg purified thrombospondin had been liganded. This is washed with 0.15 *M* NaCl, 0.01 *M* $NaPO_4$ pH 7.5, and anti-thrombospondin IgG eluted with 0.1 *M* glycine–HCl, 0.1 *M* NaCl, pH 2.5. Collection is directly into a dialysis sack suspended in 0.15 *M* NaCl, 0.01 *M* $NaPO_4$, pH 6.5, at

[27] J. Lawler, L. H. Derick, J. E. Connolly, J. H. Chen, and F. C. Chao, *J. Biol. Chem.* **260,** 3762 (1985).

[28] K. Takahashi, M. Aiken, J. W. Fenton II, and D. Walz, *Biochem. J.* **224,** 673 (1984).

[29] P. Clezardin, J. L. McGregor, M. Mandeh, F. Robert, M. Dechaudune, and K. J. Clemetson, *J. Chromatogr.* **296,** 249 (1984).

[30] G. P. Tuszynski, S. Srivastava, H. I. Switalska, J. C. Holt, C. S. Cierniewski, and M. J. Niewiarowski, *J. Biol. Chem.* **260,** 12240 (1985).

4°. Following dialysis, the IgG is stored frozen at 80°. The yield is usually 2.5% of the IgG precipitated by $(NH_4)_2SO_4$. Monoclonal antibodies were prepared by the method of Kohler and Milstein[31] and specificity assessed by Western blotting against proteolysed thrombospondin SDS–PAGE profiles.

Iodination of Thrombospondin

Thrombospondin (5 mg) can be iodinated using 1 μCi diiodinated Bolton–Hunter[32] reagent (New England Nuclear, Boston, MA), essentially according to the manufacturer's procedures,[28] or according to the Hunter procedure.[33] Specific activity of the labeled proteins is 0.7–1.2 μCi/μg.

The Bolton–Hunter method in our experience generally results in retention of 75–90% of the heparin-binding capacity, and results in higher specific activities than the less expensive chloramine-T method which, when carefully applied, results in retention of 60–80% of the heparin binding capacity, at high specific activity.

Blood Sample Collection for Radioimmunoassay

Blood is drawn through a 19- or 20-gauge needle, using "atraumatic" venipuncture technique, into two 10-ml vacutainer collection tubes, the second containing 0.6 ml 2.5% EDTA, 0.025% 2-chloroadenosine, 7.0% procaine–HCl (Thrombotect, Abbott Laboratories, N. Chicago, IL). The latter anticoagulant has been shown to be superior to several other common anticoagulants such as 0.1 ml 15% EDTA or ACD (acid–citrate–dextrose, NIH formula A, Becton-Dickinson, Rutherford, NJ).[25] The tubes with additives are inverted several times and placed in a melting ice bath within 10 sec after collection. Thirty to 120 min after collection, the samples are centrifuged at 2400 *g* for 20 min at 4° and 0.5-ml aliquots withdrawn from each tube, leaving 1 ml above the buffy coat. Typically, 1–2% of the original level of platelets is found in these aliquots by count. Subsequent removal of these residual platelets by ultrafiltration or ultracentrifugation has no appreciable effect on the level of thrombospondin determined by radioimmunoassay. One aliquot is placed in a plastic tube and another into a silicone-treated (Siliclad, Clay-Adams, Parsippany, NJ) plastic tube, capped, and stored at 4° until assay. The remaining 2.0 ml of plasma is passed through a 2.0-ml heparin–Sepharose column, equili-

[31] G. Kohler and C. Milstein, *Nature (London)* **256,** 495 (1975).
[32] A. E. Bolton and W. M. Hunter, *Biochem. J.* **133, 529** (1973).
[33] W. M. Hunter, *Nature (London)* **194,** 495 (1962).

brated to 0.15 *M* NaCl, 0.02 *M* Tris–HCl, 1 m*M* EDTA, pH 7.4, at 4°, and when eluted and collected, is found to contain about 5% of the original level of thrombospondin. This fraction has been designated "depleted plasma."

Radioimmunoassay of Thrombospondin

Thrombospondin standards and IgG are centrifuged at 10,000 *g* for 5 min to remove aggregates. Protein concentration is determined either by Lowry analysis[24] or by measurement of absorbance at 280 nm. All dilutions are performed with 0.15 *M* NaCl, 0.02 *M* $NaPO_4$, 5 m*M* EDTA, 3% BSA, 0.5% Triton X-100, pH 7.4, within 5 min of their addition into the assay tubes. EDTA is used in the assay buffer to avoid possible interactions with complement,[34] and the assay is conducted as previously described.[25] In actual assays, and in the absence of competing antigen, 40–80% of ^{125}I-labeled thrombospondin is precipitated. In addition to buffer standard curves, others obtained using thrombospondin-depleted plasma as a diluent are used to establish the usefulness of buffer diluted standards to determine levels in plasma samples.

As little as 35 ng/ml of thrombospondin in plasma can be measured on a typical standard curve, which showed linearity over a significant range when plotted on a log-logit basis.[25,35–37] The possibility that competition due to plasma components was affecting the assay results was eliminated by assaying normal and thrombospondin-depleted plasma to which standard aliquots of purified thrombospondin had been added. Results indicate that added thrombospondin was detected in addition to normal plasma levels. Also, standard curves using buffer and thrombospondin-depleted plasma as diluents, respectively, coincided within 95% confidence limits, suggesting that plasma factors do not interfere significantly in the assay.

Inhibition studies showed that fibrinogen, plasma fibronectin, platelet factor 4, β-thromboglobulin, and von Willebrand protein do not interact

[34] M. L. Egan, J. T. Lautenschlager, J. E. Coligan, and C. W. Todd, *Immunochemistry* **9,** 289 (1972).

[35] D. Rodbard, *Clin. Chem.* **20,** 1255 (1974).

[36] D. Rodbard, R. H. Lenox, A. L. Wray, and D. Ranseth, *Clin. Chem.* **22,** 350 (1976).

[37] Rodbard parameters for a typical saline and plasma standard curve pair are as follow: asymptotic bound, upper (saline) 5973 ± 130 (SD) cpm versus 4515 ± 126 (plasma); asymptotic bound, lower 228 ± 118 cpm (saline) versus 154 ± 123 (plasma); slope 1.06 ± 0.888 (saline) versus 0.943 ± 0.96 (plasma); 50% inhibition point 0.470 ± 0.038 μg/ml (saline) versus 0.514 ± 0.56 (plasma). A small amount of residual thrombospondin antigen remaining in the "depleted" plasma probably accounts for the slight observed deviation between the saline and standard plasma curves.

significantly with anti-thrombospondin IgG (i.e., less than 0.03, 0.03, 0.01, 0.004%, and none, respectively).[25]

The range of thrombospondin levels found in plasma of normal individuals is 24–280 ng/ml[25] or less.[38] Higher values could be due to variable trauma during venipuncture and storage, which is a continuing problem in the assay of platelet proteins. There was no correlation with sex, age,[25] or several pathological states, including surgery and a third trimester of pregnancy (unpublished observations).

It should be noted that thrombospondin may arise not only from platelet pools but from other cells such as the endothelial cells, fibroblasts, and macrophages. These other sources could impact significantly on measured levels in plasma or tissue related to specific pathological states.

Identification of Thrombospondin by Peptide Fingerprinting

The substructure of thrombospondin may be fingerprinted and further investigated by mapping glycosylation sites, interchain disulfide bond(s), and major plasmin-cleavable sites. Plasmin proteolysis is used to produce a series of fragments. Then through the use of SDS–PAGE fingerprinting and Western blotting techniques, coupled with reduction and lectin binding, heparin-affinity and N-terminal peptide sequencing data, these may be identified individually.

Purified thrombospondin is dialyzed against TSE. Digests are prepared by incubating ^{125}I-labeled thrombospondin ($0.5-1.0 \times 10^8$ cpm, final concentration 0.5 mg/ml) with plasmin (Sigma, human plasminogen activated with streptokinase, final concentration 5–25 μg/ml) for 5 min to 4 hr at 37° [or on occasion with TPCK-trypsin (Worthington, final concentration 5–20 μg/ml) for 5 to 30 min at 23°]. Soybean trypsin inhibitor (STI) (Worthington, final concentration 10–100 μg/ml) is added to the digests to terminate the reaction. Digests are stored in BSA-precoated siliconized glass tubes at 4° until use, up to 2 weeks.

A comparison of plasmin digest patterns prepared in the presence or absence of Ca^{2+} indicates subtle but observable differences in three bands —117K, 76K, and 53K. In Fig. 2D–G, the early digest patterns, are nearly identical, whereas the middle and late patterns reveal changes in the ratios of apparent density between both the 117K/76K and 53K, and the 76K/53K. The former ratio is greater than 1 in the presence of Ca^{2+} in the late digest, although not in the middle digest. The latter ratio is greater than 1 in the absence of Ca^{2+} in both the middle and late digests. The latter ratio in the late digest in the presence of Ca^{2+} is less than 1.

[38] J. Dawes, K. J. Clemetson, G. O. Gagstad, J. McGregor, P. Clezardin, C. U. Prowse, and D. S. Pepper, *Thromb. Res.* **29**, 569 (1983).

Selected Subjects on SDS–Polyacrylamide Gel Electrophoresis

It is clear that thrombospondin must be reduced thoroughly prior to electrophoresis.[2,39] In spite of the usual reduction steps, preparations often show bands corresponding to polymer—possibly due to crosslinking by factor XIII.[24] The use of thrombospondin itself as a standard in the high-molecular-weight range assures that high-molecular-weight thrombospondin peptides will be calibrated against the thermodynamically determined molecular weight of the native molecule, which can be expected to possess similar properties particularly with regard to disulfide content.

Electrophoretic Transfer of Proteins (Western Blotting)

This procedure is carried out essentially as described by Burnette.[40] Briefly, immediately after SDS–PAGE the peptide pattern in a slab gel is transferred in a sandwich to a glass or nitrocellulose membrane (pore size 0.223 μm, Schleicher and Schuell) by transverse electrophoresis using Whatman 3MM filter paper, porous polyethylene sheets (Scotchbrite), and plastic holders for making a gel–membrane sandwich. The transfer is done in the buffer of 20 m*M* Tris, 150 m*M* glycine, and 20% methanol (pH 8.3) at 200 mA for 15–20 hr. After the transfer, the blot of radioactive samples is dried and directly autoradiographed. Blots of unlabeled samples are used for peptide analysis as described elsewhere.[41]

Autoradiography of blots, as well as of the gels from which the blots were made (i.e., after electrophoretic transfer), indicates that there is good transfer yield of peptides smaller than 100K, but poor transfer for the larger fragments. A typical peptide profile is shown in Fig. 2D.

Heparin-Binding Peptides (HBP)

Sulfated polysaccharides bind to certain plasma proteins with significant interaction coefficients. We have developed data showing that there is a high-affinity site on purified thrombospondin for a fraction of heparin glycosaminoglycan[42] of M_r 8500, that the K_d of this interaction is about 2 n*M*, and that the number of binding sites for heparin is most probably three.[42] Also, we evaluated the ability of other sulfated polysaccharides to bind to thrombospondin based on their ability to inhibit the binding of

[39] U. K. Laemmli, *Nature (London)* **227**, 680 (1970).

[40] W. N. Burnette, *Anal. Biochem.* **112**, 155 (1981).

[41] R. M. Aebersold, D. B. Teplow, L. E. Hood, and S. B. H. Kent, *J. Biol. Chem.* **261**, 4229 (1986).

[42] H. S. Slayter, G. Karp, B. E. Miller, and R. D. Rosenberg, *Semin. Thromb. Hemost.* **13**, 369 (1987).

heparin. On the basis of the available data, we hypothesize that heparin-like molecules may regulate the activity of thrombospondin via interactions with the heparin-binding region.

HBP may be purified as described elsewhere.[18] On 15% gels the HBP-p is resolved into two bands of about 28K and 30K. Isoelectric focusing of the HBP-p on slab gels produces a split peak in the vicinity of pH 5.9, with secondary peaks at 5.7 and 6.1.[18] When experiments are performed with particular care (loading on both sides of the isoelectric point), the secondary peaks still occur. The p*I* of native thrombospondin is 4.7.[2]

The amino acid compositions of thrombospondin and of the HBP-p derived from it are similar, but HBP-p contains proportionately twice as much leucine, one-third as much tyrosine, half as much histidine, and a third less aspartic acid. The amino acid sequences of the first 20 NH_2-terminal residues of thrombospondin and of HPB-p are identical, indicating that HPB-p is derived from the NH_2-terminus of thrombospondin chains.

Identification of Molecular Thrombospondin by Electron Microscopy

Using a thin metal coating as described elsewhere,[18,43] the molecular dimensions are much less exaggerated than when platinum is used, so that there is a correspondingly greater potential for detection of detailed shape, but contrast becomes limiting. Therefore, darkfield electron microscopy is carried out using an annular condenser aperture, matched with a 40-μm objective aperture, and an exposure time on the order of 1 sec at a magnification of $\times$40,000. Figure 3 presents several electron micrographs of intact thrombospondin molecules, prepared as extremely thin rotary tungsten replicas. Corrections for metal cap build-up should be minimal.[18,43] Upon close examination of individual images, a pattern consistent with known physical and chemical data emerges. Figure 3 shows selected high-resolution micrographs of thrombospondin in which 7-nm heads appear clearly connected by strands to the centrum, which consists of an axis around which are distributed three small tabs of protein. On the basis of our experience with barely visible particles prepared by the same method, the molecular weight of the tabs is estimated to be of the order of 20,000. The peripheral heads are substantially larger with an estimated molecular weight of the order of 50,000, based on their size in comparison to other proteins of known molecular weight.[18,44] The 30-nm connecting

[43] H. S. Slayter, *Ann. N.Y. Acad. Sci.* **408,** 131 (1983).

[44] H. S. Slayter, *in* "Advances in Cell Biology" (K. Miller, ed.), Vol. 2. JAI Press, Greenwich, Connecticut, in press, 1988.

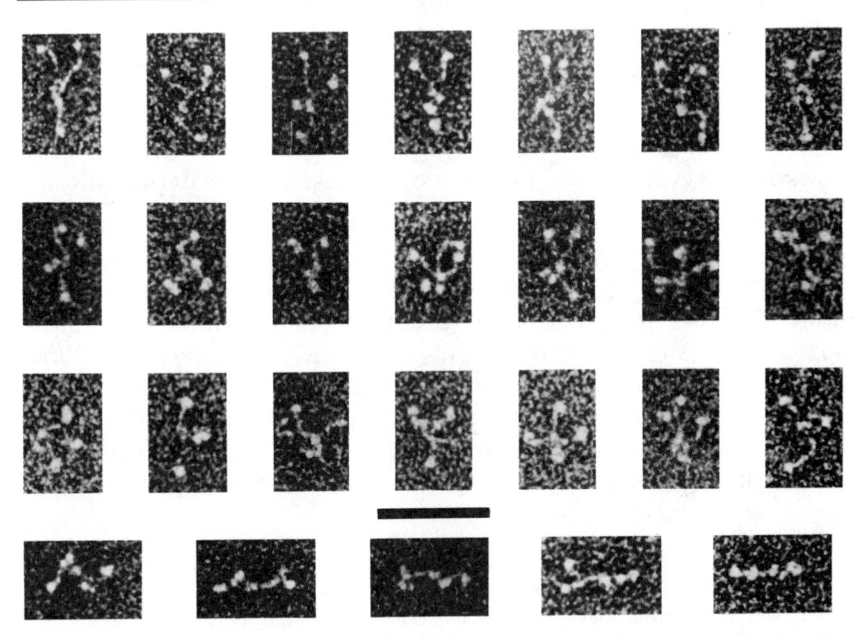

FIG. 3. Electron micrographs for thrombospondin molecules. Human thrombospondin, rotary shadowed with 10^{-7} g/cm^2 of tungsten, with darkfield imaging. $\times 154,000$. Peripheral 6.5-nm heads are attached to central regions of variable conformation by strands about 20 nm long. Bottom row is of molecules collapsed linearly. Minimal exaggeration of dimensions is expected, due to the small amount of metal used (9×10^{-8} g/cm^2). The intensity of Bragg reflections, which would prevent the use of the darkfield mode in thicker metal films, is at a minimum due to the small number of atoms per crystallite found in tungsten films of mass thickness 9.3×10^{-8} g/cm^2. (The negative contrast method using stains like uranyl acetate has to date not proved useful in facilitating visualization of thrombospondin.)

link would thus represent about 60,000 as determined by difference. The structure of the approximately spherical heads is somewhat variable, a fact which probably indicates that a loose tertiary structure is capable of assuming various conformations.

The problem of locating the position of peptide regions, such as the heparin-binding peptide, might be expected to be approached utilizing antibodies in electron microscopic mapping experiments similar to those we have previously executed with other molecules.[43,44] However, given HBP, loosely attached to the tertiary structure, with a molecular weight of 20K–30K, a specific monoclonal antibody (even if it possessed a high binding affinity) could be a marker for *any* one epitopic position on that peptide, which when fully extended could be as much as 95 nm long. An epitopic locus so identified could be anywhere throughout an entire 30-nm-long arm of thrombospondin! Thus, immunoelectron microscopic mapping experiments using monoclonal reagents are potentially ambiguous, and carefully prepared polyclonal antibodies may prove more useful for mapping.

Summary of Approaches to Assay of Released Thrombospondin

Thrombospondin may be assayed in unknown systems by the various procedures outlined herein, including radioimmunoassay, cysteine content, heparin binding, N-terminal sequence, proteolytic peptide finger print on SDS–PAGE, direct SDS–PAGE data on unreduced and reduced apparent molecular weights, and molecular electron microscopy. While any one method of assay is subject to uncertainty, several taken together provide rather secure evidence for the presence of thrombospondin. It may be that thrombospondin radioimmunoassay will be found useful clinically in view of recent reported increases in titer in induced thrombotic states in dogs,[45] and in epithelial tumors.[46]

[45] P. French, J. L. McGregor, M. Berruyer, J. Belleville, P. Touboul, J. Dawes, and M. Decharenne, *Thromb. Res.* **39,** 619 (1985).

[46] J. Dawes, P. Clezardin, and D. Pratt, *Semin. Thromb. Hemostasis* **13,** 378 (1987).

[23] Platelet Histidine-Rich Glycoprotein

By LAWRENCE L. K. LEUNG, PETER C. HARPEL, and RALPH L. NACHMAN

Introduction

Recent studies show that various plasma proteins are found in human platelets.[1-3] The characterization of a platelet constituent generally requires a sensitive and specific assay to quantify the platelet protein and involves studies to determine (1) the potential plasma contamination, (2) the subcellular localization and release of the protein following platelet stimulation, (3) the immunochemical relationship of the platelet protein

[1] R. L. Nachman and P. C. Harpel, *J. Biol. Chem.* **251,** 4514 (1976).

[2] E. F. Plow and D. Collen, *Blood* **58,** 1069 (1981).

[3] C. M. Chesney, D. Pifer, and R. W. Colman, *Proc. Natl. Acad. Sci. U.S.A.* **78,** 5180 (1981).

with its plasma counterpart, and (4) the specific functions of the platelet protein. Histidine-rich glycoprotein (HRGP), an α_2-glycoprotein in human plasma, apparently has diverse biological functions and interacts with divalent metals, plasminogen, heparin, thrombospondin, and lymphocytes.[4-9] Recently we presented evidence demonstrating the presence of HRGP in human platelets and its release following thrombin stimulation.[10]

Isolation and Purification of Plasma HRGP

HRGP was purified with minor changes in the method of Haupt and Heimburger,[4] as modified by Lijnen *et al.*[6] Soybean trypsin inhibitor (100 mg/liter final concentration) was added to acid–citrate–dextrose (ACD) plasma. Fifty percent polyethylene glycol, M_r 4000, was added to a final concentration of 6% to remove the fibrinogen and the precipitate removed by centrifugation. The supernatant was processed, as detailed by Lijnen *et al.*,[6] by adsorption to CM-cellulose (CM-52), elution with NH_4HCO_3, and affinity chromatography of the eluate on a column containing the high-affinity lysine-binding site of plasminogen.[6] Traces of IgG, fibrinogen, and plasminogen were detected in the HRGP preparation by enzyme-linked immunosorbent assays. These were removed by affinity chromatography utilizing the insolubilized rabbit IgG antisera directed against these contaminants. Purified plasma HRGP (kindly supplied by Dr. Norbert Heimburger) served as a standard. When analyzed by SDS–PAGE, purified HRGP showed a single band with an apparent molecular weight of 60,000 unreduced, and two bands with apparent molecular weights of 74,000 and 67,000 when reduced with 2% dithiothreitol. This is in agreement with previously published results.[4-6] Recent studies suggest that HRGP purified by this procedure, in the absence of aprotinin, may represent a proteolytic derivative of a native molecule with an apparent molecular weight of 75,000.[9]

Enzyme-Linked Immunosorbent Assay (ELISA) for HRGP

The final HRGP preparation was used to produce antisera by intradermal injection of the protein in Freund's complete adjuvant into New

[4] H. Haupt and N. Heimburger, *Hoppe-Seyler's Z. Physiol. Chem.* **353,** 1125 (1972).
[5] W. T. Morgan, *Biochemistry* **20,** 1054 (1981).
[6] H. R. Lijnen, M. Hoylaerts, and D. Collen, *J. Biol. Chem.* **255,** 10214 (1980).
[7] H. R. Lijnen, M. Hoylaerts, and D. Collen, *J. Biol. Chem.* **258,** 3803 (1983).
[8] L. L. K. Leung, R. L. Nachman, and P. C. Harpel, *J. Clin. Invest.* **73,** 5 (1984).
[9] H. R. Lijnen, D. B. Rylatt, and D. Collen, *Biochim. Biophys. Acta* **742,** 109 (1983).
[10] L. L. K. Leung, P. C. Harpel, R. L. Nachman, and E. M. Rabellino, *Blood* **62,** 1016 (1983).

Zealand White rabbits. The anti-HRGP was absorbed with insolubilized HRGP-depleted plasma, plasminogen, and human IgG to remove traces of contaminating antibodies from the HRGP IgG. The absorbed anti-HRGP produced a single precipitin arc when diffused against human serum and yielded a reaction of immunologic identity between serum and the purified HRGP. It did not react by ELISA with purified human fibrinogen, VIIIR:Ag, IgG, albumin, plasminogen, fibronectin, or thrombospondin. The methods used to develop the ELISA are essentially those detailed by Voller *et al.*[11] Microtitration plates were coated with rabbit anti-HRGP IgG. Portions (0.2 ml) of anti-HRGP IgG (5 μg/ml) in bicarbonate coating buffer (15 m*M* Na_2CO_3, 35 m*M* $NaHCO_3$, pH 9.6, with 0.02% NaN_3) were incubated in a humid chamber overnight at 4°. After washing the wells three times for 3 min each with phosphate-buffered saline containing 0.05% Tween 20 (PBS–Tween), serial dilutions of HRGP (10–120 ng/ml) in PBS–Tween containing bovine albumin (1 mg/ml) were added in duplicate to the coated wells. The plates were incubated for 3 hr at room temperature, the wells rewashed three times, and alkaline phosphatase-conjugated immunoaffinity-purified rabbit anti-HRGP IgG was added for 3 hr at 37°. The wash step was repeated, and the substrate (0.2 ml) *p*-nitrophenyl phosphate (1 mg/ml in 10% diethanolamine buffer, pH 9.8) added. Color development was followed by repeated readings at 405 nm in a Titertek multiscan photometer. The change in absorbance per minute (ΔA_{405} min^{-1}) was calculated by subtracting the 5-min value from the 25-min reading and dividing by 20. The binding of HRGP was linear with the concentration of HRGP added to the well up to 120 ng/ml (Fig. 1). Control studies using nonimmune IgG-coated plates showed no binding of HRGP. This HRGP standard curve was reproduced with each experiment in this study. Using this ELISA, the plasma concentration of HRGP was determined to be 104.5 ± 7 μg/ml ($n = 6$), which is in good agreement with the published value.[4-6]

Detection and Quantification of HRGP in Platelets

Venous blood was obtained in plastic syringes using 0.11 *M* sodium citrate as anticoagulant from normal volunteers who had no medication during the previous 2 weeks. Platelets were collected and washed once with 120 m*M* KCl, 75 m*M* Tris–HCl, 12 m*M* sodium citrate, pH 6.3, and twice with 150 m*M* NaCl, 10 m*M* Tris–HCl, 1 m*M* EDTA, pH 7.4, as

[11] A. Voller, D. Bidwell, and A. Bartlett, *in* "Manual of Clinical Immunology" (N. R. Rose and H. Friedman, eds.), p. 506. American Association for Microbiology, Washington, D. C., 1976.

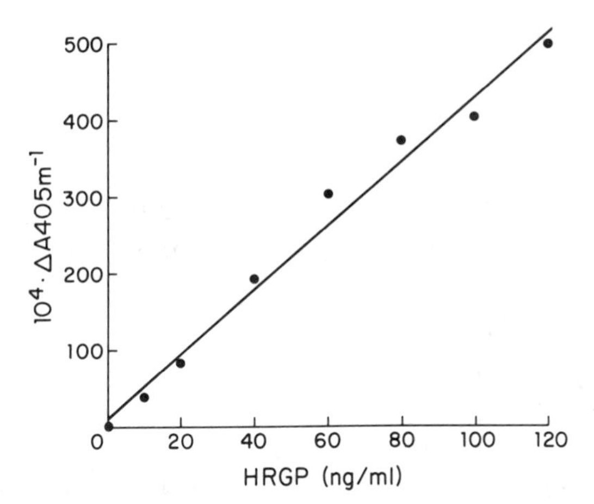

FIG. 1. ELISA standard curve for HRGP.

previously described.[12] The washed platelets were resuspended in this EDTA–Tris–saline buffer, with the final EDTA concentration adjusted to 5 m*M*. More than 99.9% of the cells in these samples were platelets, as determined by phase-contrast microscopy. For determination of HRGP content in washed platelets, the platelet pellet was lysed by incubation with 1% Triton X-100 for 30 min at 22°. Complete lysis was established by microscopic examination, and cell debris was removed by centrifugation in a Beckman microfuge B for 2 min. HRGP was detected in all seven extracts, and the range of HRGP was 231–486 ng/10^9 platelets, with a mean level of 371 ng HRGP/10^9 platelets.

Three approaches were used to determine the potential contribution of contaminating plasma HRGP to the level of HRGP measured in the platelet detergent extracts. First, after centrifugation to remove the washed platelets for HRGP determination, the supernatant was assayed for the presence of HRGP by ELISA. No HRGP was detected in the supernatant. Second, [^{125}I]HRPG was added to the platelet-rich plasma, and the recovery of radioactivity in the platelet lysates was measured. Based on this analysis, plasma HRGP accounted for only 11.5 ng/10^9 platelets. To assess the presence of tightly bound plasma HRGP on the platelet surface, washed platelets were incubated with monospecific, affinity-purified ^{125}I-labeled anti-HRGP IgG and the platelet-associated IgG was measured. Affinity-purified, monospecific anti-HRGP IgG and nonimmune rabbit

[12] L. L. K. Leung, T. Kinoshita, and R. L. Nachman, *J. Biol. Chem.* **256,** 1994 (1981).

IgG were radiolabeled with ^{125}I by the modified chloramine-T method.[13] When ^{125}I-labeled anti-HRGP was incubated with increasing amounts of adsorbed HRGP on a microtitration plate, there was a direct correlation between the amount of ^{125}I-labeled anti-HRGP bound and the amount of HRGP on the plate, indicating the functional integrity of anti-HRGP following radiolabeling. Control experiments using nonimmune rabbit ^{125}I-labeled IgG did not show binding. To assess the binding of these ligands to platelets, duplicate samples of 0.3 ml washed platelets at 0.4×10^9/ml in EDTA–Tris–saline buffer were incubated with the labeled IgG at 2 μg/ml (precentrifuged at high speed to remove aggregates) for 30 min at 22°. Aliquots of 0.2 ml were then removed and centrifuged for 2 min through 0.5 ml silicone oil in a Beckman microfuge B. This system effectively separated platelet-bound from -unbound IgG with $\geq$95% platelet recovery in the tip of the centrifuge tube. The tips were amputated and the nanograms IgG bound were determined from the specific radioactivity for the IgG; 37.3 ng of the monospecific anti-HRGP IgG was bound per 10^9 platelets compared to 5.9 ng nonimmune IgG/10^9 platelets. Assuming that one monospecific anti-HRGP molecule bound one HRGP molecule, 12.6 ng HRGP resided on the surface of 10^9 platelets and was accessible to the antibody. Thus, plasma HRGP, either in the platelet-suspending medium or on the surface of the platelets, accounted for less than 3.4% of the detectable platelet HRGP.

Secretion of HRGP from Platelets

The release of HRGP from platelets following thrombin stimulation was assessed. Platelets were labeled with [^{14}C]serotonin (22 μCi/μmol) in platelet-rich plasma by incubation at 37° for 30 min and then isolated by centrifugation. The labeled cells were suspended in 1 ml of the EDTA–Tris–saline buffer, incubated with 2 μM imipramine at 37° for 5 min, then stimulated with thrombin for 3 min at 22°. [^{14}C]Serotonin and HRGP present in the supernatant after centrifugation at 11,750 rpm in a Beckman microfuge for 2 min were quantified. Percentage secretion was calculated relative to constituents solubilized by addition of 1% Triton X-100 to unstimulated platelets (Table I). The extent of HRGP release from platelets was directly related to the dose of thrombin added and correlated with the extent of serotonin release. Thrombin stimulation at the concentrations employed in this study did not cause platelet lysis.[2] Alternatively, secretion experiments can be done with human platelets suspended in either Tyrode's or HEPES buffer (see Chapters [1] and [2] in this volume).

[13] P. McConahey and F. Dixon, *Int. Arch. Allergy Appl. Immunol.* **29**, 185 (1966).

TABLE I
RELEASE OF HRGP AND SEROTONIN FROM THROMBIN-STIMULATED PLATELETS[a]

Thrombin (U/ml)	Percentage release	
	HRGP	[^{14}C]Serotonin
0	0	0
0.01	0	3.6 ± 0.8
0.1	36 ± 12	33.4 ± 2.6
1.0	71.9 ± 7	70.4 ± 6.5
3.0	79.7 ± 13	75.7 ± 5.6

[a] Results are the mean (± SD) of three separate experiments, with duplicate determinations in each experiment.

Analysis of Plasma HRGP and Thrombin-Stimulated Platelet Releasate by Monospecific Anti-HRGP on Immunoblot

To investigate the immunochemical characteristics of the platelet HRGP, purified plasma HRGP and thrombin-stimulated platelet releasate were analyzed using the monospecific anti-HRGP on electrophoretic blots. Anti-HRGP reacted with two major bands in the reduced plasma HRGP sample with molecular weights of 74,000 and 67,000, in agreement with previously published results.[4-6] Similarly, two major bands were identified in the reduced platelet releasate sample (Fig. 2). It is of note that the relative intensity of the two bands was different in the two samples. The reason for these differences remains to be determined. Other minor bands of lower molecular weight range were also identified in both reduced samples, possibly representing fragments produced by proteolysis. Control studies using the preimmune rabbit sera were negative.

Detection of HRGP in Marrow Megakaryocytes

Human megakaryocytes isolated from normal marrow tissue were studied for the presence of HRGP using the immunofluorescence technique. Using the monospecific anti-HRGP, HRGP was detected in virtually all morphologically recognizable megakaryocytes. Homogeneous staining of variable intensity was observed in all cells areas, except the nucleus (Fig. 3). The pattern of fluorescence staining was similar to that of

[14] E. M. Rabellino, R. B. Levene, L. L. K. Leung, and R. L. Nachman, *J. Exp. Med.* **154**, 88 (1981).

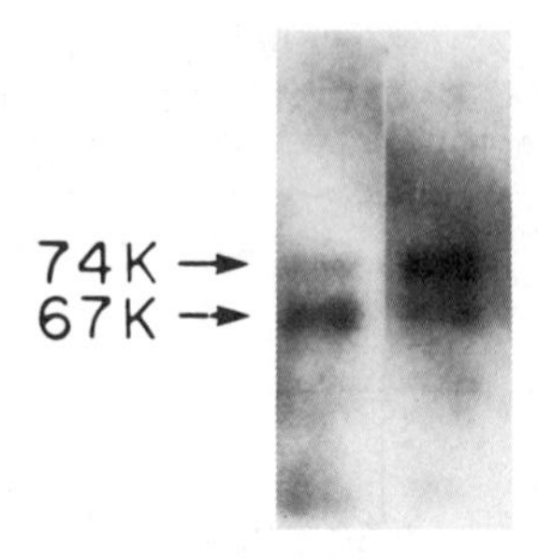

FIG. 2. Electrophoretic blot analysis of purified plasma HRGP and thrombin-stimulated platelet releasate with monospecific anti-HRGP. (A) Plasma HRGP; (B) platelet releasate. The proteins were reduced with 2% dithiothreitol and separated by SDS–PAGE (Laemmli system, 3.9% stacking, 7.5% separation gel), then electrophoretically transferred onto nitrocellulose paper. After incubation with monospecific anti-HRGP sera (1 : 50 dilution), HRGP was identified by the addition of ^{125}I-labeled protein A.

intracellular von Willebrand factor.[14] The megakaryocytic morphology of the fluorescent stained cells was confirmed by concomitant examination of the cells under phase-contrast microscopy. There was no staining of erythroid and myeloid precursor cells with anti-HRGP. The specificity of the megakaryocyte staining was demonstrated by complete inhibition of cell staining after absorption of anti-HRGP with purified HRGP but not with bovine serum albumin. Control studies using preimmune sera did not show any cell staining.

Platelet Aggregation Studies

Anti-HRGP IgG at 0.5 mg/ml did not inhibit aggregation of gel-filtered platelets induced by thrombin, arachidonic acid, or collagen. Anti-whole-platelet membrane IgG inhibited platelet aggregation under similar conditions.

Comments

These studies demonstrate that HRGP is present in human platelets. The platelet HRGP does not appear to represent a trapped plasma contaminant in the washed platelet pellet, as suggested by three separate lines of evidence: (1) platelets isolated from plasma containing trace-labeled HRGP did not contain significant amounts of plasma protein; (2) binding studies using radiolabeled anti-HRGP revealed only minor traces of

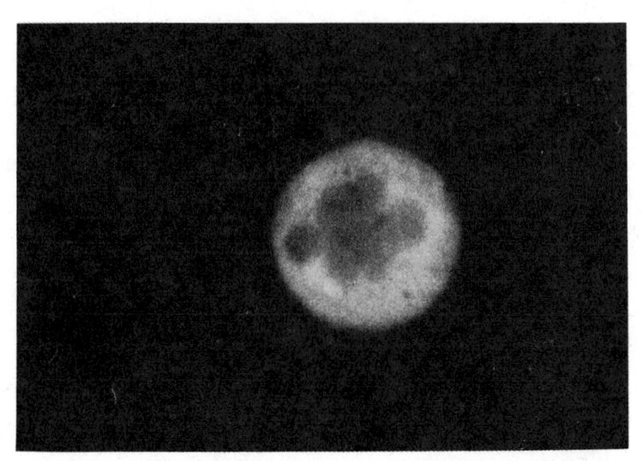

FIG. 3. Immunofluorescent staining of isolated human marrow megakaryocyte by anti-HRGP. ×970.

HRGP on the washed platelet surface; (3) supernatants of washed platelets contained no detectable HRGP prior to thrombin-induced release of significant amounts of the protein. The observation that bone marrow megakaryocytes stained for intracellular HRGP (Fig. 3) is further evidence that HRGP is an intrinsic platelet protein. The extent of HRGP release was dependent on the thrombin dose and correlated directly with the extent of serotonin release (Table I). Preliminary studies demonstrate that the distribution of HRGP in platelet subcellular fractions is similar to that of thrombospondin, a known α-granule protein.[15]

By immunoblot analysis, the platelet releasate sample showed a similar pattern with the plasma HRGP sample (Fig. 2), indicating that the platelet and plasma HRGP are similar by immunochemical analysis. It is of interest that the relative intensity of the two bands in the platelet releasate sample was different from that of the plasma HRGP sample, possibly related to different extent of proteolysis during the isolation procedures.

Compared to a plasma level of 100 μg/ml, 4×10^8 platelets in 1 ml of blood contain 148 ng HRGP (based on a mean level of 371 ng/10^9 platelets). Thus, platelet HRGP constitutes 0.14% of the blood level on a volume basis. Platelets from 1 ml of platelet-rich plasma contain approximately 330 μg of total protein,[16] thus, platelet HRGP accounts for 0.045% platelet proteins. In plasma with 70 mg of protein/ml, plasma HRGP constitutes 0.15% of plasma proteins. Therefore, there is no relative concentration of HRGP in platelets on a protein basis. Similar relationships

[15] J. W. Lawler, H. S. Slayter, and J. E. Coligan, *J. Biol. Chem.* **253,** 8609 (1978).

[16] R. L. Nachman, A. J. Marcus, and D. Zucker-Franklin, *J. Lab. Clin. Med.* **69,** 651 (1967).

have been described for platelet α_2-macroglobulin, α_1-antitrypsin,[1] and α_2-plasmin inhibitor,[2] as contrasted to other proteins, such as factor VIIIR: Ag and fibrinogen, which are concentrated in the platelets relative to plasma. Since platelets are concentrated in the fibrin clot, HRGP released by platelets following thrombin stimulation may achieve a high local concentration and thus play a significant role in modulating fibrinolysis in the microenvironment of the platelet plug. The biologic importance of plasma as well as platelet HRGP remains to be fully clarified.

Acknowledgment

We thank Barbara Ferris and T. S. Chang for expert technical assistance. This work was supported by Grant HL 18828 (Specialized Center of Research in Thrombosis) from The National Institutes of Health.

[24] Platelet High-Molecular-Weight Kininogen

By ALVIN H. SCHMAIER and ROBERT W. COLMAN

The plasma kininogens (high and low molecular weight) are substrates from which the vasoactive peptide bradykinin is released. High-molecular-weight kininogen (HMWK) is the procofactor for activation as well as a substrate of the plasma enzymes factor XIIa, kallikrein, and factor XIa. Plasma kallikrein cleaves HMWK in a three-step sequence pattern (Fig. 1).[1] The first cleavage yields a "nicked" kininogen composed of two disulfide linked chains of M_r 62,000 and 56,000 on reduced SDS gels. The second cleavage yields bradykinin and an intermediate kinin-free protein of equal molecular weight to "nicked" HMWK. A third cleavage results in a stable kinin-free protein composed of two disulfide-linked chains of M_r 62,000 and 46,000. Since one gene directs the synthesis of both plasma kininogens, the molecules apparently differ due to gene splicing. The distinguishing feature between HMWK and low-molecular-weight kininogen (LMWK) is the presence of the M_r 56,000 light chain on the carboxy-terminal end of HMWK compared to the M_r 5000 light chain of LMWK. The light chain of HMWK contains unique antigenic sites and possesses

[1] K. Mori and S. Nagasawa, *J. Biochem. (Tokyo)* **89,** 1465 (1981).

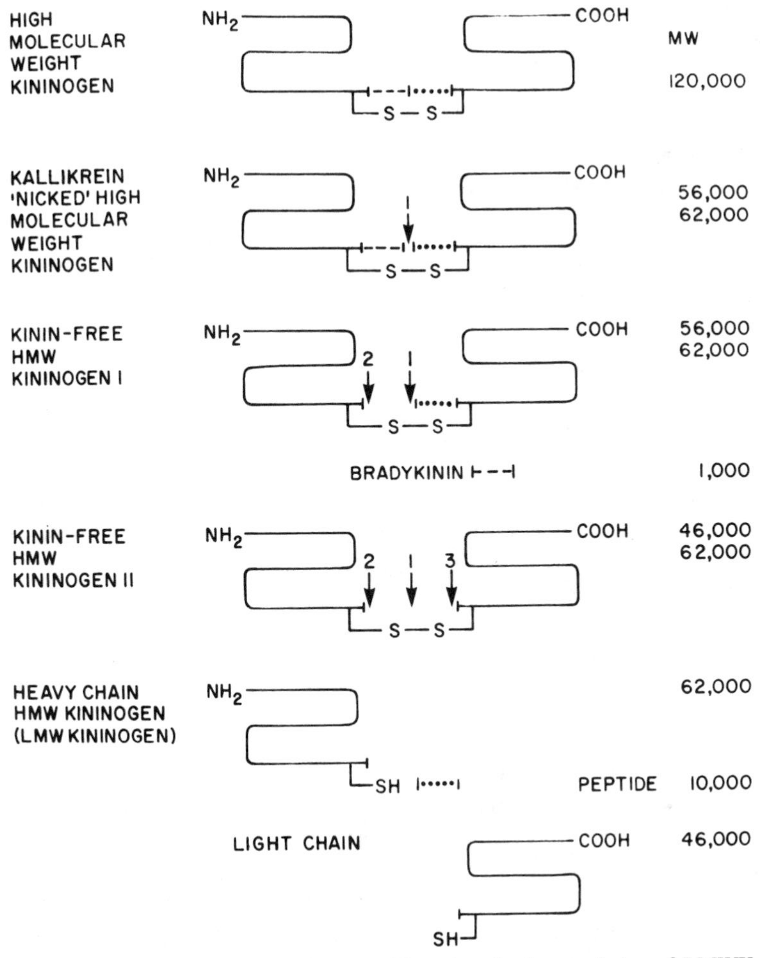

FIG. 1. Structure of plasma kininogens. Note that the heavy chains of LMWK and HMWK are identical and the light chain of LMWK is not shown.

coagulant activity due to its ability to bind to surfaces as well as prekallikrein and factor XII. Both these properties serve to distinguish it from LMWK.

Platelets have been proposed as an alternate pathway of intrinsic coagulation activation in the absence of factor XII.[2] Platelets have associated factor XI-like activity and factor XI antigen.[3] Factor XIa specifically binds

[2] P. N. Walsh, *Blood* **43,** 597 (1974).

[3] G. P. Tuszynski, S. J. Bevacqua, A. H. Schmaier, R. W. Colman, and P. N. Walsh, *Blood* **59,** 1148 (1982).

to activated platelets in the presence of HMWK.[4] The identification of a platelet form of HMWK[5,6] indicates that the platelet could participate in factor XI activation, as well as serve as a negatively charged surface for contact-phase zymogen activation *in vivo.*

Coagulant Assay of HMWK

The coagulant assay of HMWK is based on the kaolin-activated partial thromboplastin time of Proctor and Rapaport[7] with the modifications of Hardisty and MacPherson.[8] The general principle of the assay is a one-stage, phospholipid-dependent, kaolin-activated assay with factor-deficient plasma.

Reagents

Kaolin (Fisher Scientific Co., Pittsburg, PA): 10 mg/ml
Tris–saline buffer (TBS): 0.01 *M* Tris–HCl, pH 7.4, 0.15 *M* NaCl
Inosithin (Associated Concentrates, Woodside, NY): Stored in a stock solution of 1% in TBS at −70° and freshly thawed and utilized in the assay in a 0.1% solution in TBS
$CaCl_2$: 0.025 *M* stock solution
10 × 75 mm glass tubes
Total kininogen-deficient plasma (Williams-trait plasma)

Procedure

In these assays, 100 μl kaolin (10 mg/ml), 100 μl of 0.1% inosithin, 100 μl of HMWK-deficient plasma, and 100 μl of test sample, in this sequence, are introduced into a glass tube, mixed, and incubated for 5 min at 37° before addition of 100 μl 0.025 *M* $CaCl_2$. The time from the introduction of calcium to the observation of a clot detected by tube tilting in a 37° water bath is measured. Samples to be assayed are compared against a daily standard curve constructed from pooled normal plasma diluted in TBS from 1/10 to 1/1000. The data are plotted on log–log paper and the relationship between coagulation time and dilution is linear over this range. One coagulant unit is defined as that amount which is presented in 1 ml of pooled normal plasma.

[4] D. Sinha, F. S. Seaman, A. Koshy, and P. N. Walsh, *J. Clin. Invest.* **73,** 1550 (1984).
[5] A. H. Schmaier, A. Zuckerberg, C. Silverman, J. Kuchibhotla, G. P. Tuszynski, and R. W. Colman, *J. Clin. Invest.* **71,** 1477 (1983).
[6] D. M. Kerbiriou-Nabias, F. O. Garcia, and M.-J. Larrieu, *Br. J. Haematol.* **56,** 273 (1984).
[7] R. R. Proctor and S. I. Rapaport, *Am. J. Clin. Pathol.* **36,** 212 (9161).
[8] R. M. Hardisty and J. C. MacPherson, *Throm. Diath. Haemorrh.* **7,** 215 (1962).

Purification of HMWK

HMWK is purified by a modified technique of Kerbiriou and Griffin.[9] Four hundred and fifty milliliters of blood from two donors is freshly collected into 50 ml of 3.8 g% sodium citrate containing 37.5 mg SBTI/50 ml, 150 mg Polybrene/50 ml, 10 m*M* EDTA, 10 m*M* benzamidine, and 0.02% NaN_3 and platelet-poor plasma is prepared by centrifugation at 1,000 *g* for 30 min at 23°. All reagents for the purification are purchased from Sigma Chemical Company. In this modified procedure, 0.1 *M* EACA is added to all buffers and DFP (2 m*M* final concentration) is added to the platelet-poor plasma before the purification and to the pooled material containing the HMWK after the ammonium sulfate precipitation and prior to chromatography on SP Sephadex. Once the plasma is incubated for 30 min with DFP, it is then diluted 2:3 with 0.03 *M* Tris–Cl, 3 m*M* benzamidine, 3 m*M* EDTA, 0.06% NaN_3, and 0.3 *M* EACA, pH 8.0. The initial procedure is batch adsorption of QAE-Sephadex. The gel is equilibrated with 0.1 *M* Tris, 1 m*M* benzamidine, 1 m*M* EDTA, 0.1 *M* EACA, pH 8.0, containing 50 μg/ml Polybrene (4–6 mΩ). After the plasma is applied, it is washed with 0.121 *M* Tris, 0.044 *M* succinic acid, 0.103 *M* NaCl, 1 m*M* EDTA, 1 m*M* benzamidine, 0.02% NaN_3, and 0.1 *M* EACA, pH 7.7 (14–15 mΩ) until the absorbance at 280 nm is less than 0.1. The HMWK is eluted from the anion exchange with 0.199 *M* Tris, 0.075 *M* succinic acid, 0.182 *M* NaCl, 1 m*M* EDTA, 1 m*M* benzamidine, 0.02% NaN_3, and 0.1 *M* EACA, pH 7.4 (23–25 mΩ). After 50% ammonium sulfate precipitation and dialysis of the resuspended precipitate against 0.05 *M* sodium acetate, 0.075 *M* NaCl, 1 m*M* EDTA, 1 m*M* benzamidine, 0.02% NaN_3, and 0.1 *M* EACA, pH 5.3 (7–9 mΩ), the HMWK is applied to a column of SP-Sephadex equilibrated in the same buffer. After washing the column, a linear gradient is applied with 0.1 *M* sodium acetate, 0.15 *M* NaCl, 1 m*M* EDTA, 1 m*M* benzamidine, 0.02% NaN_3, and 0.1 *M* EACA, pH 5.3 (15–18 mΩ) in the proximal chamber and the same buffer containing 0.5 *M* NaCl (32–38 mΩ) in the distal chamber. The HMWK elutes after $1\frac{1}{2}$ to 2 column volumes has passed from the gradient.

Radiolabeling of HMWK

Purified HMWK is radiolabeled with $Na^{125}I$ using chloramide, 1,3,4,6-tetrachloro-3a,6a-diphenylglycouril (Iodogen, Pierce Chemical Co., Rockford, IL) by a modified method of Fraker and Speck.[10] Purified HMWK

[9] D. M. Kerbiriou and J. H. Griffin, *J. Biol. Chem.* **254,** 12020 (1979).
[10] P. J. Fraker and S. C. Speck, Jr., *Biochem. Biophys. Res. Commun.* **80,** 849 (1978).

(50–200 μg) in acetate buffer is made pH 8.0 by adding 2.5 *M* Tris and incubated with $Na^{125}I$ in a plastic vial precoated with Iodogen (2–4 μg) for 15–35 min on ice. The iodination reaction is stopped by the addition of sodium metabisulfite (60 μg/ml final concentration) and the free ^{125}I is separated from protein-bound ^{125}I by gel filtration on a 0.8 × 10 cm column of Sephadex G-50 equilibrated in 0.01 *M* Tris–Cl, 1.0 *M* NaCl, pH 8.0 containing 0.25% gelatin. The specific radioactivity of the protein usually varies from 1 to 8 μCi/μg.

Preparation and Characterization of Antisera Antibodies to HMWK

HMWK is an α-globulin with an isoelectric point of 4.7.[11] Since the heavy chain of HMWK is identical to the heavy chain of LMWK, injection of purified HMWK into animals will usually result in antisera that recognizes both HMWK and LMWK. Polyclonal antisera to plasma kininogens is produced in goats by intramuscular injection of 500 μg purified HMWK in Freund's complete adjuvant followed by a 250-μg injection of HMWK in Freund's complete adjuvant 4 weeks later.[12,13] The specificity of the antisera to HMWK is tested by its ability to neutralize the coagulant activity of plasma HMWK. In order to perform a neutralization experiment, the crude antisera is precipitated with 50% ammonium sulfate followed by two successive kaolin adsorptions (40 mg/ml) for 30 min at 37° so that the final crude antibody preparation has negligible HMWK, prekallikrein, factor XII, or factor XI coagulant activity.

Monospecific antisera to total plasma kininogen can most simply be produced by adsorption with total kininogen-deficient plasma. Anti-total kininogen antisera results in a double precipitin arc against normal human plasma on immunoelectrophoresis (Fig. 2).[13] Further adsorption of the antisera with Fitzgerald plasma (plasma deficient in HMWK but not LMWK), purified LMWK, or the heavy chain of HMWK will result in monospecific antisera to HMWK (Fig. 2).[5] The most direct means to produce antisera that uniquely recognizes HMWK is to immunize with the purified light chain of HMWK.

Immunochemical Assays

HMWK in plasma can be quantified and characterized by radial immunodiffusion, electroimmunodiffusion, and crossed immunoelectrophoresis. However, studies on platelets required the development of a more sensitive immunochemical assay.

[11] H. Kato, S. Iwanga, and S. Nagasawa, this series, Vol. 80, p. 172.

[12] B. A. L. Hurn and S. M. Chantler, this series, Vol. 70, p. 104.

[13] A. H. Schmaier, L. Silver, A. L. Adams, G. C. Fischer, P. C. Munoz, L. Vroman, and R. W. Colman, *Thromb. Res.* **33,** 51 (1984).

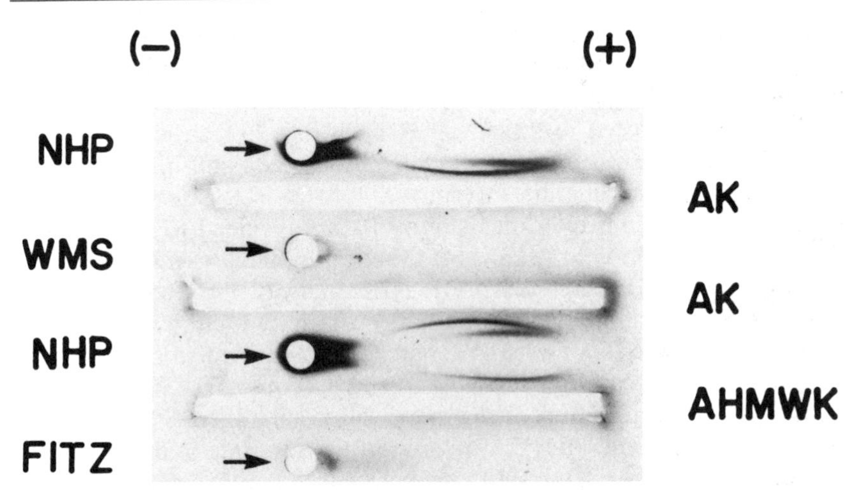

FIG. 2. Immunoelectrophoresis of antisera to plasma kininogens. Antisera to total kininogen (AK) (both HMWK and LMWK); NHP, normal human plasma; WMS, Williams plasma, i.e., plasma deficient in total kininogen; FITZ, Fitzgerald plasma, i.e., plasma deficient in only HMWK. AHMWK, Antisera to HMWK.

Competitive Enzyme-Linked Immunosorbent Assay (CELISA) for High-Molecular-Weight Kininogen (HMWK)

Reagents and Supplies

Polystyrene cuvettes (Gilford Instr. Lab., Inc., Oberlin, OH)
Specific goat antisera or antibody to the light chain of HMWK
Rabbit anti-goat whole immunoglobulin conjugated with alkaline phosphatase (Sigma Chemical Corp., St. Louis, MO)
Substrate: *p*-Nitrophenyl phosphate disodium (Sigma Chemical Corp., St. Louis, MO)
Radioimmunoassay-grade bovine serum albumin (Sigma Chemical Corp., St. Louis, MO)
Coupling buffer: 0.1 M Na_2CO_3, pH 9.6
Substrate buffer: 0.05 M Na_2CO_3, 1 mM MgCl, pH 9.8
PBS–Tween: 0.01 M sodium phosphate, pH 7.4, 0.15 M NaCl containing 0.05% Tween 20

Procedure. This assay was based on a modification of the procedure of Engvall.[14] On day 1, 100–500 ng of purified HMWK diluted in 0.1 M Na_2CO_3, pH 9.6, is linked to the surface of polystyrene cuvette wells by overnight incubation at 37°. On the same day, incubation mixtures in 1.5-ml conical polypropylene tubes precoated with 0.2% bovine serum

[14] E. Engvall, this series, Vol. 70, p. 419.

albumin are made containing the following: 0.15 ml of antigen (standards or test samples) diluted in PBS–Tween and 0.15 ml of a previously optimal titered specific goat anti-human HMWK antisera or antibody. These samples are incubated overnight at 37°. On day 2, each antigen-linked cuvette well is washed three times with PBS–Tween and incubated with 0.2% bovine serum albumin in water for 1 hr at 37°. After washing the cuvettes, 0.2 ml from each incubation mixture is added to each cuvette well and incubated for 2 hr at 37°. At the conclusion of this incubation, the rewashed wells are exposed to a previous optimally titered second antibody (rabbit anti-goat whole immunoglobulin) conjugated with alkaline phosphatase diluted in PBS–Tween. After another 2.5-hr incubation at 37°, the washed cuvettes receive sequentially timed additions of 0.4 ml of *p*-nitrophenyl phosphate disodium (1 mg/ml) in substrate buffer. At precise time intervals (20–30 min) after the addition of the substrate to each well, the amount of hydrolysis of the substrate in each well is either stopped in each well with sequential timed additions of 0.4 ml of 2 *M* NaOH (final concentration 1 *M* NaOH) or is sequentially measured spectrophotometrically in a PR 50 EIA Processor-Reader (Gilford Instr. Lab., Inc.) at 405 nm. Since this is a competitive assay, the amount of optical absorbance is inversely proportional to the amount of antigen.

The development of an assay such as this is dependent on the optimal titrations of linked antigen, specific antibody, and enzyme-conjugated second antibody. The determinations of these optimal titrations can only be done empirically. In general, to obtain maximal sensitivity, the amount of antigen used for coating is decreased as far as practicable and the amount of antibody added should be limited. One approach is a checkerboard titration, as described by Engvall,[14] of dilutions of various amounts of linking antigen with one titer of specific antibody versus various titers of specific antibody with one concentration of linked antigen. This titration approach presupposes that there is one optimal titer of enzyme-conjugated second antibody which is best for all amounts of specific antibody. Using commercial second antibody conjugates and different batches of specific antisera, this is often not the case. Another approach, which we use, is to decide in advance the range of sensitivity of the assay desired and then optimize the titration of the reagents at hand. For example, for design of an assay with a linear portion of the standard curve between 5 and 100 ng, antigen in 100- to 500-ng amounts will be linked to the cuvette. Specific antisera in multiple dilutions, followed by the enzyme-conjugated secondary antibody at various concentrations, will be sequentially introduced (Fig. 3). The optical absorbance measured represents the total value (or the amount of absorbance if no antigen is preincubated with the antisera) to be obtained in the final CELISA. After comparison of the total absorbance of

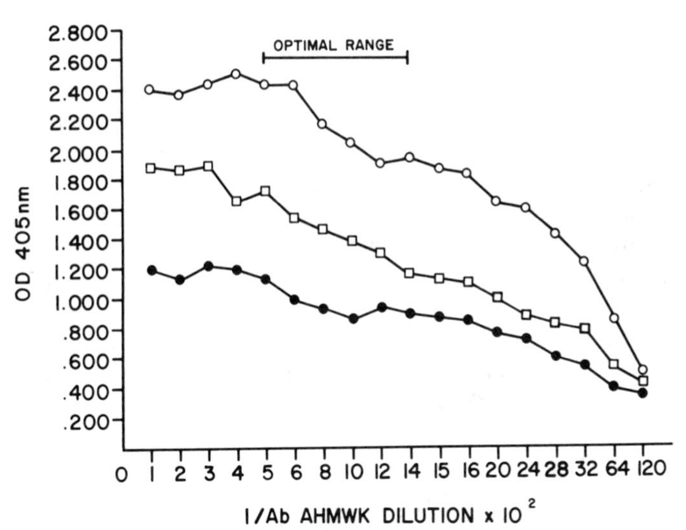

FIG. 3. Titration of primary and secondary antibody for CELISA. The dilution of primary antibody (abscissa) is plotted against the optical density (OD) at 405 nm using three titers, 1/500 (○), 1/750 (□), 1/1000 (●) of second antibody conjugated with alkaline phosphatase. The optimal range of both primary and secondary antibody for use in the CELISA is given by the bar.

multiple specific antisera dilutions using a few different conjugated antibody titers, the titers of the two antibodies used, in sequence, that give the highest absorbance over the range of optical linearity of the spectrophotometer for a fixed reaction time (usually 20–30 min) are chosen. In Fig. 3, using these principles we chose an initial dilution of the primary antibody at 1/500 with a dilution of the secondary antibody conjugated with alkaline phosphatase at 1/500. The final primary antibody dilution, if none was consumed in the overnight incubation, would be 1/1000. For this specific batch of primary antibody and conjugate, the total absorbance would be 2.0—the upper limit of linearity on our instrument and a value that would give the steepest slope on our competitive assay.

Although we use automated equipment, the entire assay can be performed in polystyrene test tubes and read manually in a spectrophotometer. A crucial feature for the success of this type of assay is that each reaction after the addition of substrate is read or stopped with NaOH at precisely the same time interval after beginning the enzyme hydrolysis. This manipulation requires some automated instrumentation. We have found it best to perform standard curves and test samples in triplicate at 10–15 different dilutions. In this laboratory, data for the standard curve and test sample determination are analyzed by a computer.[15] Antigen values are determined in absolute amounts and original concentration are calculated considering the dilutions of the standards and test samples.

Using the CELISA, HMWK antigen as a purified protein reconstituted into total kininogen immunodeficient plasma, and as antigen in normal plasma gives superimposable, parallel competition inhibition curves (Fig. 4). On 20 individually obtained normal plasma, HMWK antigen assayed as a concentration of 105 μg/ml. The interassay coefficient of variation of a single plasma sample assayed four times over a 1-month period is 3.0%. These values compare favorably as to sensitivity and precision with a previously reported radioimmunoassay for HMWK.[16] Using this radioimmunoassay, the mean value of plasma HMWK in normal plasma by radioimmunoassay was 90 μg/ml.

Preparation of Platelets

Albumin Density Gradient Centrifugation and Gel Filtration (AGGF). Platelets are washed by a modified combined technique of albumin density gradient centrifugation and gel filtration[17] (see Chapter [2] in this volume). Eight milliliters of platelet-rich plasma is layered on a 2.5-ml discontinuous increasing albumin density gradient (10, 15, 20, 30, 40%) prepared according to the method of Walsh *et al.*[18] Forty percent stock solution of albumin (Sigma, bovine serum albumin #4503) is prepared by the procedure of Walsh *et al.*[18] In preparation of the albumin gradient, it is diluted in HEPES-buffered Tyrode's solution (3.5 g/liter HEPES, 8 g/liter NaCl, 0.2 g/liter KCl, 1 g/liter $NaCO_3$, 0.05 g/liter NaH_2PO_4, 0.2 g/liter $MgCl_2$, 3.5 g/liter BSA, 1 g/liter dextrose, pH 7.35). After centrifugation for 20 min at 900 *g* at 23°, the platelet layer above the 40% albumin cushion is removed and 5 ml of resuspended platelets is applied to a 60-ml column of Sepharose 2B in Hepes-buffered Tyrode's solution. Void volume fractions are pooled.

Identification of Platelet HMWK

Four issues were considered in the identification of platelet HMWK which exists in trace quantities to the amount of HMWK in the surrounding plasma: (1) What was the degree of plasma contamination in the washed platelet aliquots that were studied for the total platelet HMWK

[15] P. F. Cannellas and A. E. Karu, *J. Immunol. Methods* **47,** 375 (1981).
[16] D. Proud, J. V. Pierce, and J. J. Pisano, *J. Lab. Clin. Med.* **95,** 563 (1980).
[17] S. Timmons and J. Hawiger, *Thromb. Res.* **12,** 297 (1978).
[18] P. N. Walsh, D. C. B. Mills, and J. G. White, *Br. J. Haematol.* **36,** 281 (1977).

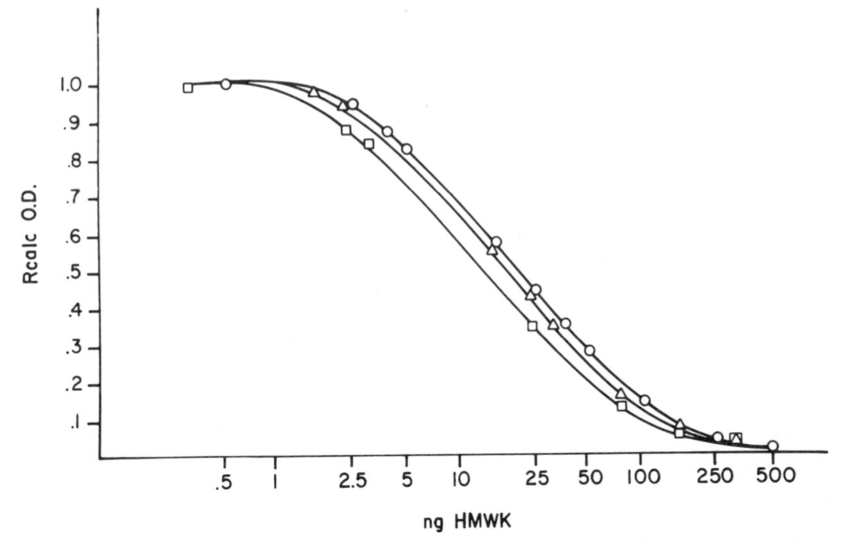

FIG. 4. Competitive inhibition curves of CELISA for HMWK. The ordinate is the relative absorbance (Rcalc OD) and the abscissa is the absolute amount of antigen incubated with antisera. Inhibition curve produced by purified HMWK (□); inhibition curve produced by pooled normal plasma (○); inhibition curve produced by total kininogen-immunodeficient plasma reconstituted with purified HMWK (△). Curves calculated by computer program.[15] From *J. Clin. Invest.* **71,** 1477 (1983) by copyright permission of the American Society for Clinical Investigation.

content; (2) was the HMWK antigen found with platelets really platelet-associated plasma HMWK: (3) was plasma HMWK taken up by platelets; and (4) what contribution did the suspension medium of washed platelets make to the total amount of platelet HMWK measured in platelet lysates.

[^{125}I]HMWK is introduced in platelet-rich plasma (PRP) to be used as a tracer to determine the amount of the radiolabel recovered in the final washed platelets. This technique is used to estimate the amount of plasma HMWK that might contaminate the washed platelet aliquot. Its use presupposes that the radiolabeled protein is in equilibrium with the unlabeled plasma HMWK. Whether the [^{125}I]HMWK was incubated 5 min or 3 hr with the PRP, the percentage of the tracer remaining with the washed platelets was about 0.03%. At best, this technique is a lowest estimate of plasma contamination in washed platelets. Plasma antigen can be tightly bound and nonexchangeable with the platelet surface. In order to determine whether we are measuring platelet-associated plasma HMWK, a semiquantitative indirect antibody consumption assay using the CELISA was developed to estimate the amount of HMWK antigen on the unstimulated platelet surface. In our studies on platelet HMWK, unactivated platelets gave a superimposable competitive inhibition curve with the suspension medium of these platelets indicating that little (≤ 3 ng/10^8 platelets) platelet-associated plasma HMWK was tightly bound and nonexchangeable with the platelet surface.[5] These experiments were performed using fresh platelets pretreated with 1 μM PGE_1 (Sigma) and separated by albumin gradient centrifugation and gel filtration. Alternatively, platelets were prepared by the technique of Mustard *et al.*[19] without the addition of PGE_1 (see Chapter 1 in this volume).

Measurement of Total Platelet HMWK

To determine the total amount of HMWK in lysed platelets, AGGF platelets are used. PEG_1 (1 μM) is included in the collection anticoagulant, resuspension buffer for the platelets from the albumin gradient, and the gel filtration buffer. Without PGE_1, direct measurement of the platelet suspension buffer for HMWK antigen shows a 2- to 3-fold increase in antigen levels, suggesting that platelets were activated during the washing procedure. Platelets are then solubilized at 22° for 30 min with 0.5% Triton X-100 and assayed for total HMWK antigen levels. In 15 normal donors, the mean platelet HMWK levels was 55 ng $\pm$ 22/10^8 (mean $\pm$ SD) platelets (Fig. 5). Direct measurement on the suspension buffer of each aliquot of platelets revealed a mean value of 2.8 ng $\pm$ 2.1/10^8 platelets or 5% of the total. One individual with a congenital absence of platelet α granules (the gray platelet syndrome) had a total platelet HMWK levels of 16 ng/10^8 platelets. Another individual with a plasma deficiency of both HMWK and LMWK was found to have a platelet HMWK level of less than 5 ng/10^9 platelets. The ability of platelets to take up plasma HMWK was studied by incubating washed platelets from a patient with total kininogen deficiency in the presence of normal plasma for 1 hr at 37°. After separating these platelets from plasma by AGGF, the level of platelet HMWK was still lower than 5 ng/10^9 platelets. This result indicated that the platelet and plasma pools of HMWK are separate. Last, platelet HMWK antigen from normal platelets was immunochemically identical to plasma HMWK as evidenced by the platelet antigen producing parallel competitive inhibition to plasma antigen (Fig. 6).

Secretion of Platelet HMWK

Secretion studies for platelet HMWK are performed on AGGF platelets. Values for the total content of platelet HMWK are obtained by lysing platelets by freezing and thawing four times on dry ice and at 37° after the

[19] J. F. Mustard, D. W. Perry, N. H. Ardlie, and M. A. Packham, *Br. J. Haematol.* **22,** 193 (1972).

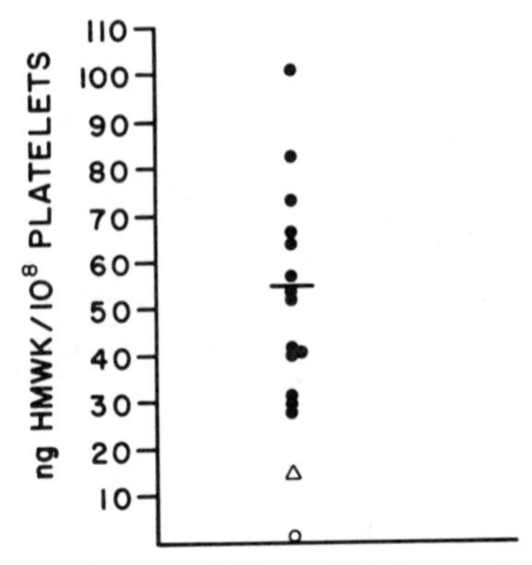

FIG. 5. Total platelet HMWK in normal (●), total kininogen-deficient platelets (○), and gray platelets (△).

platelets are diluted 1:3 with deionized water or by adding 0.2–0.5% Triton X-100 for 30 min at 22°. By either technique, no difference in the amount of total platelet HMWK is measured. A platelet dense granule marker for secretion studies was obtained by incubated platelets with 5-hydroxy[^{14}C]tryptamine (New England Nuclear) and its secretion is

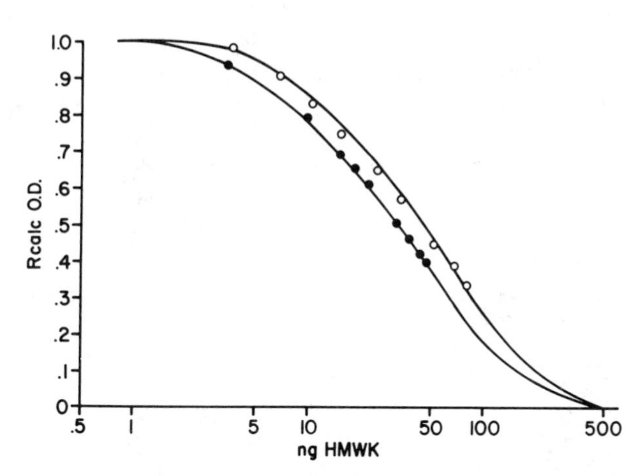

FIG. 6. Competition inhibition curve showing that platelet HMWK antigen is immunochemically indistinguishable from plasma HMWK antigen. Inhibition curve produced by pooled normal plasma (○); inhibition curve produced by solubilized platelets (●). From *J. Clin. Invest.* **71,** 1477 (1983) by copyright permission of the American Society of Clinical Investigation.

assessed by the method of Jerushalmy and Zucker.[20] Platelet lysis during agonist stimulation of platelets is determined by lactate dehydrogenase loss.[21] Platelet α granule secretion is assessed by the measurement of low-affinity platelet factor 4 secretion.[22]

Platelets for secretion studies were also incubated with the metabolic inhibitors antimycin A (15 μg/ml), 2-deoxy-D-glucose (30 mM), and D-gluconic acid δ-lactone (10 mM) (all purchased from Sigma) to inhibit platelet aerobic and anerobic glycolysis[23] as well as glycogenolysis.[24] The addition of D-gluconic acid δ-lactone is essential to block 95% of the platelets' ability to respond to platelet agonists.

For concentration-dependent reactions or fixed dose secretion experiments, platelets in glass or plastic cuvettes are incubated in an aluminum block or water bath at 37° positioned over a magnetic stirrer (1000 to 1200 rpm). At precisely 10 min from the introduction of the stimulus, each cuvette is placed on ice. Aliquots of activated platelets used for 5-hydroxy-[^{14}C]tryptamine determination are centrifuged in a microcentrifuge tube at 12,000 g containing cold 135 μM formaldehyde, 5 mM EDTA (four parts platelets/one part formaldehyde–EDTA). The addition of this mixture prevents artifactual loss of 5-hydroxy[^{14}C]tryptamine during the centrifugation procedure[24] (see also chapter [17] in this volume). Other platelet supernatants of activated platelets and controls for other studies are directly obtained after a 12,000 g centrifugation in a microcentrifuge. All samples are then immediately frozen at −70° until the time of assay. All secretion studies are performed with a nonstimulated control. Percentage secretion (or loss) is determined by the ratio of the supernatant of the agonist-treated specimen to the supernatant of the platelet lysates after the value of the control supernatant is subtracted from both (see Chapter 17 in this volume).

Platelet HMWK is secreted in a concentration-dependent manner using different agonists (Fig. 7). With both A23187 (Calbiochem-Behring) and collagen (Worthington) at lower doses, secretion of platelet HMWK paralleled that of the α-granule marker, low-affinity platelet factor and differed from secretion of the dense granule marker 5-hydroxy[^{14}C]tryptamine. These functional data are consistent with an α-granule location for platelet HMWK.[25] The α-granule localization of platelet HMWK has been

[20] Z. Jerushalmy and M. B. Zucker, *Thromb. Diath. Haemorrh.* **15,** 413 (1966).
[21] F. Wroblewski and J. S. Ladue, *Proc. Soc. Exp. Biol. Med.* **90,** 210 (1955).
[22] B. Rucinski, S. Niewiarowski, P. James, D. A. Walz, and A. Budzynski, *Blood* **53,** 47 (1979).
[23] A. H. Schmaier and R. W. Colman, *Blood* **56,** 1020 (1980).
[24] C. A. Dangelmaier and H. Holmsen, *Method Haematol.* **8,** 92 (1983).
[25] A. H. Schmaier, P. M. Smith, A. D. Purdon, J. G. White, and R. W. Colman, *Blood* **67,** 119 (1986).

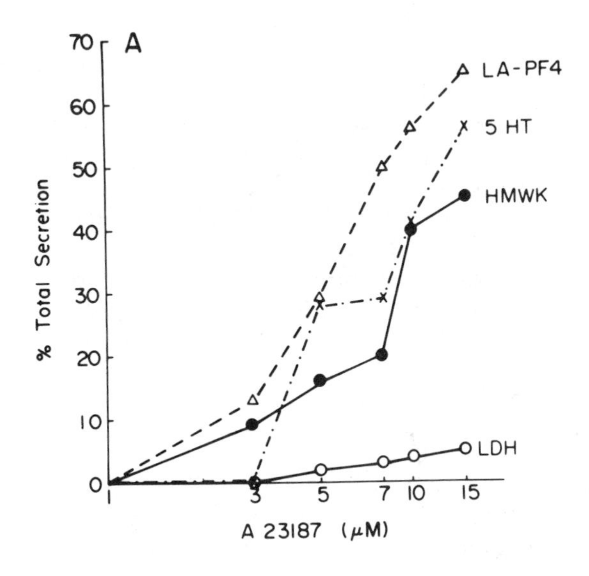

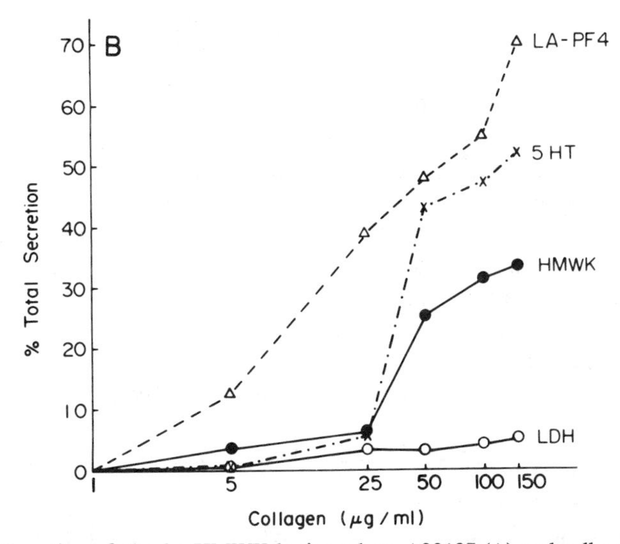

FIG. 7. Secretion of platelet HMWK by ionophore A23187 (A) and collagen (B). Data plotted from one representative experiment with agonist. Final concentration is plotted on the abscissa of the dose–response of agonist to percentage of total secretion, plotted on the ordinate. LA-PF_4, low-affinity platelet factor 4; 5-HT, 5-hydroxy[^{14}C]tryptamine; LDH, lactate dehydrogenase. From *J. Clin. Invest.* **71,** 1477 (1983) by copyright permission of the American Society for Clinical Investigation.

confirmed by studies on a patient with the gray platelet syndrome who had levels of total platelet HMWK 29% of normal (Fig. 5) and by platelet subcellular localization studies[26] using the technique of Fukami *et al.*[27] Using maximal doses of A23187 and collagen, 46 and 32%, respectively, of the total platelet HMWK was secreted. The extent of the total amount of secretion of HMWK with these agonists at maximal doses was less than that seen with the α-granule marker, low-affinity platelet factor 4 ($\geq$66% secretion). The secretion of only 25 to 40% of the total platelet content of a high-molecular-weight kininogen contained within platelet α granules, in contrast to low-affinity platelet factor 4,[22] has been noted previously for platelet fibronectin,[28] von Willebrand factor,[29] and C1 inhibitor.[30] This finding suggests that platelet α granules may differentially secrete their granule contents or that the proteins remain tightly bound to the external platelet membrane after secretion.

Expression of Platelet HMWK on the Activated Platelet Surface

Since platelets only secrete 40% of their total platelet HMWK a study was performed to determine if some of the remaining platelet HMWK, not secreted, became localized on the external membrane of the activated platelets using the CELISA in a quantitative indirect antibody consumption assay.[25] Four hundred and fifty milliliters of blood from two donors is collected (1 : 10) into 73 m*M* citric acid, 3 m*M* trisodium citrate, and 2% dextrose and, after adjusting the pH to 6.5 with the anticoagulant, the platelets are washed by the technique of Mustard *et al.*[19] Apyrase is prepared from potatoes by the method of Molnar and Lorand[31] and is titered so that the amount used is the minimal amount necessary to prevent second wave platelet aggregation with a threshold dose of ADP. The final washed platelets are resuspended in HEPES-buffered Tyrode's solution to a concentration from 3 to 9×10^9 platelets/ml. These washed platelets are used to prepare activated platelets.

A scheme for the preparation of activated platelets is shown in Fig. 8.

The washed platelets are divided into three aliquots. Two aliquots are treated with PGE_1 (1 μM) and incubated 40 min in a 37° water bath. A

[26] K. L. Kaplan, M. J. Broekman, A. Chernoff, G. R. Lesznik, and M. Drilling, *Blood* **53**, 604 (1979).

[27] M. H. Fukami, J. S. Bauer, G. J. Stewart, and L. Salganicoff, *J. Cell Biol.* **77**, 389 (1978).

[28] M. H. Ginsberg, R. G. Painter, C. Birdwell, and E. F. Plow, *J. Supramol. Struct.* **11**, 167 (1979).

[29] J. Koutts, P. N. Walsh, E. F. Plow, J. W. Fenton, B. N. Bouma, and T. S. Zimmerman, *J. Clin. Invest.* **62**, 1255 (1978).

[30] A. H. Schmaier, P. M. Smith, and R. W. Colman, *J. Clin. Invest.* **75**, 242 (1985).

[31] J. Molnar and L. Lorand, *Arch. Biochem. Biophys.* **93**, 353 (1961).

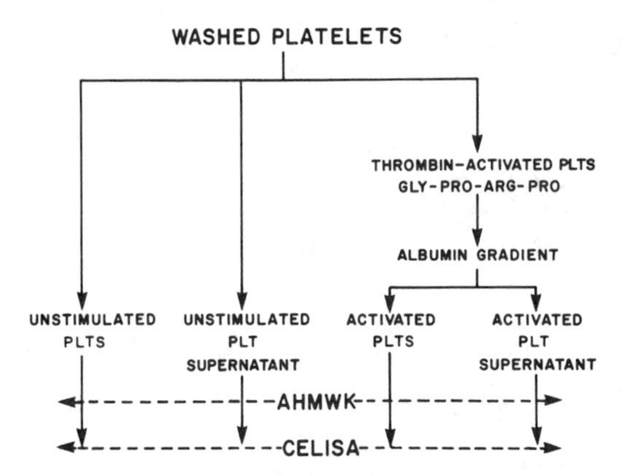

FIG. 8. Scheme for indirect antibody consumption assay to determine the expression of platelet HMWK antigen on the external platelet membrane. See text for method.

third aliquot receives 0.4 m*M* Gly-Pro-Arg-Pro (Sigma). The Gly-Pro-Arg-Pro is to prevent fibrin polymerization and platelet aggregation of activated platelets.[32]

After treating the platelets with Gly-Pro-Arg-Pro, thrombin (0.5 U/ml) is introduced and incubated for 10 min at 37° without stirring. Following thrombin stimulation, the activated platelets are then layered on a 2.5-ml discontinuous albumin density gradient and centrifuged at 900 *g* for 20 min. The activated platelets form a layer over the 40% albumin cushion and the activated platelet supernatant—a platelet releasate—remains above the gradient. The activated platelets are then resuspended in HEPES-buffered Tyrode's solution containing 0.4 m*M* Gly-Pro-Arg-Pro. A second platelet aliquot is centrifuged at 12,000 *g* and the supernatant collected. Thus, four platelet specimens are collected—an aliquot of intact, unactivated platelets, supernatant from unactivated platelets, activated platelet supernatant, and activated platelets, themselves.

The platelet samples are prepared in order to be employed in a modified use of the CELISA for HMWK as an indirect antibody consumption assay. Since membrane-expressed platelet HMWK would not be soluble to allow for a direct determination of its presence and solubilization of activated platelets would liberate intracellular platelet HMWK, an experimental design was developed to determine if surface-expressed platelet HMWK could be detected when whole platelets were incubated with the antisera directed to HMWK. Conditions for this antibody consumption assay for surface-expressed platelet HMWK were developed so that intraplatelet HMWK (that which was neither secreted nor expressed on the external

[32] E. J. Harfenist, M. A. Guccione, M. A. Packham, and J. F. Mustard, *Blood* **59,** 952 (1982).

membrane) would not be interfering. The four platelet-derived specimens are then incubated with anti-HMWK antisera for 1 hr at 37° in the indirect antibody consumption assay using the CELISA for HMWK as previously described. The objective of the assay is to determine the extent by which each of the four platelet specimens could reduce the titer of the starting anti-HMWK antibodies. After incubation with the four platelet specimens, four adsorbed anti-HMWK antibody aliquots are obtained for further analysis.

Four antibody samples per experiment are then compared by the CELISA assay for HMWK using known amounts of purified HMWK antigen to determine the slope of the competitive inhibition curve produced by each batch of platelet-adsorbed or platelet supernatant-adsorbed antisera. The competitive inhibition curves generated on the CELISA by the four aliquots of adsorbed antisera are produced by the method of incubation of samples indicated previously. However, since the aim of the assay is to determine whether the titer of the antisera would be decreased (consumed) by incubation with platelets or their supernatants, the final competitive inhibition curves are analyzed by nonlinear regression to determine the slope of the competitive inhibition curve produced by antisera adsorbed with each of the aliquots of platelet material. In all experiments the differently adsorbed aliquots of antisera are reacted with the same amounts of purified HMWK (1 to 125 ng). The data are plotted as the concentration of the purified HMWK used to determine the slope of the competitive inhibition curve of the adsorbed antisera on the abscissa versus the optical density at 405 nm on the ordinate. The measured value of this assay is the slope of the competitive inhibition curve.

Quantification of the amount of platelet HMWK expressed on the surface of the platelet is obtained by comparing the values of the slopes produced by each sample of platelet- or platelet supernatant-adsorbed antisera with a standard curve of the anti-HMWK antisera adsorbed with known concentrations of purified HMWK (Fig. 9). In the generation of this curve, the anti-HMWK antisera is adsorbed by purified HMWK at various concentrations (20 to 2000 ng/ml) for 1 hr at 37°. After incubation, the adsorbed antisera is then interacted with purified HMWK (1 to 125 ng) to determine the adsorbed antisera's competitive inhibition curve. These slopes are also calculated by nonlinear regression. The data from these latter experiments are plotted as concentration of purified HMWK used to adsorb the antisera on the abscissa versus the slope of the resultant competitive inhibition curve on the ordinate. As can be seen in Fig. 9, anti-HMWK antisera initially diluted 1 : 500 produced a competitive inhibition curve with a calculated slope by nonlinear regression of 0.9. Incubating the anti-HMWK antisera initially diluted 1 : 500 in separate aliquots

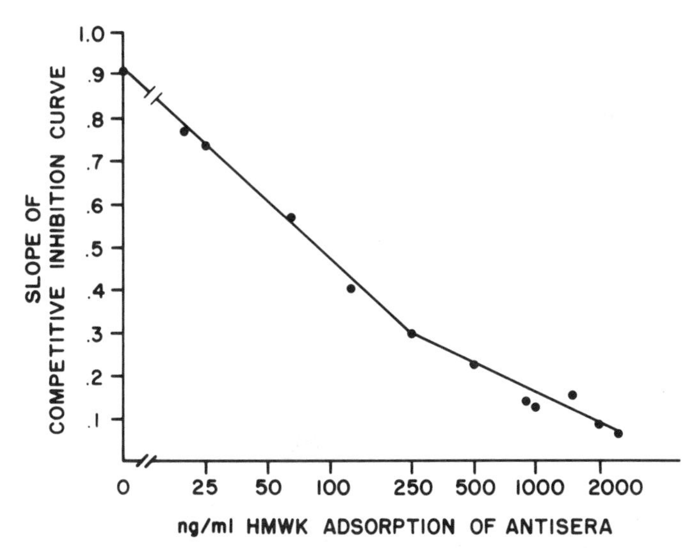

FIG. 9. The relationship of the slope of the competitive inhibition curve produced by anti-HMWK antisera versus the concentration of HMWK used to adsorb the antisera. Anti-HMWK antisera was prediluted 1:500 in PBS–Tween and incubated for 1 hr at 37° in the absence or presence of purified HMWK at a concentration from 20 to 2000 ng/ml in the suspending buffer. At the conclusion of the 1-hr incubation, the adsorbed antisera was incubated overnight at 37° with an equal volume of purified HMWK in an absolute amount from 1 to 125 ng. A CELISA assay was performed as indicated in text. The competitive inhibition curves produced by the unadsorbed and adsorbed antisera were calculated by nonlinear regression. In the figure the calculated slope of the adsorbed antisera was plotted on the ordinate and the concentration of purified HMWK used to adsorb the antisera for 1 hr was plotted on the abscissa. From *Blood* **67**, 119 (1986) by copyright permission of Grune & Stratton.

with increasing concentrations of purified HMWK (20 to 2000 ng/ml) for 1 hr at 37° resulted in competitive inhibition curves with reduced calculated slopes (0.77 to 0.06).

In one representative experiment (Fig. 10), the competitive inhibition curves produced by antisera adsorbed with the supernatant of unstimulated platelets (slope 0.34) and unstimulated platelets (slope 0.33) themselves gave parallel and almost superimposable curves. The competitive inhibition curve produced by antisera adsorbed with the activated platelet supernatant was flattened with a decreased slope (0.20) when compared to that produced by the supernatant of unstimulated plates (Fig. 10). Since platelet HMWK is secreted by thrombin-activated platelets,[5] this finding indicated that secreted platelet HMWK adsorbed and decreased the titer of the anti-HMWK antibody. The competitive inhibition curve produced by activated platelets (slope 0.26) was similar to the curve which characterized

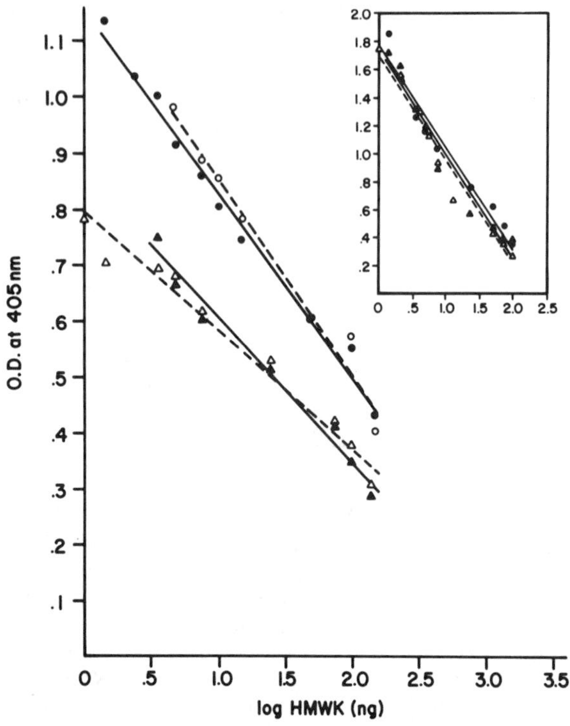

FIG. 10. Expression of platelet HMWK antigen on the activated platelet surface. The competitive inhibition curves produced by anti-HMWK antisera adsorbed with unstimulated platelet supernatant (○), unstimulated platelets (●), activated platelet supernatant (△), and activated platelets (▲) are plotted. See text for the preparation of each of the platelet samples. The ordinate is the observed optical density (OD) at 405 nm and the abscissa is the absolute amount (log HMWK in nanograms) of purified HMWK incubated with the prepared antisera. *Inset:* The competitive inhibition curves produced by anti-HMWK antibody consumed by total kininogen-deficient platelets. The competition inhibition curves produced by anti-HMWK antisera adsorbed with unstimulated (●) and activated (▲) total kininogen-deficient platelets as well as the activated platelet supernatant of total kininogen deficient platelets (△) are plotted. All data in the inset are plotted as above. In both plots, the competitive inhibition curves were calculated by nonlinear regression. From *Blood* **67**, 119 (1986) by copyright permission of Grune & Stratton.

the material released by platelets (Fig. 10). Antisera adsorbed with activated platelets showed a competitive inhibition curve which had a reduced slope and left shifted when compared to unstimulated platelets. Knowing the number of platelets in each experiment, the amount of HMWK (plasma or platelet) associated with the surface of the platelet or in its suspending medium is estimated by comparing the slopes of the competitive inhibition curve produced by the antisera adsorbed with the platelet material with the slopes of the competitive inhibition curves produced by known concentrations of purified HMWK. In the four experiments, unstimulated platelets and their supernatant had a mean of 4.9 ng HMWK/10^8 platelets and 4.2 ng HMWK/10^8 platelets, respectively, associated with the material. Alternatively, activated platelets and their supernatant had a mean of 17.3 ng HMWK/10^8 platelets and 17.2 ng HMWK/10^8 platelets, respectively, associated with the aliquots. The consumption of anti-HMWK antibody by activated platelet and its supernatant is specific for expressed platelet HMWK antigen because adsorption of the anti-HMWK antisera by unstimulated total kininogen-deficient platelets, thrombin-activated total kininogen-deficient platelets, and activated deficient platelet supernatant produced competitive inhibition curves with calculated slopes of 0.71, 0.78, and 0.77, respectively (Fig. 10, inset). These slopes were similar to the competitive inhibition curve produced by unadsorbed antisera and when corrected for the number of platelets in the experiment gave values for available HMWK less than 1 ng HMWK/10^8 platelets.

These combined results indicate that some of the platelet HMWK, in addition to being secreted, is also expressed on the external membrane of the activated platelet. The expression of platelet HMWK on the activated platelet surface suggests that the platelet membrane has a receptor for this protein. This notion is supported by the recent findings that exogenous HMWK specifically binds to saturable sites on unstimulated[33] and activated[33,34] platelets. The possible presence of a receptor for HMWK on the platelet would suggest involvement of platelets in contact activation.

Function of Platelet HMWK

Since both prekallikrein and factor XI circulate in plasma complexed to HMWK,[35-37] activated profactor forms of HMWK[38] which have increased surface-binding activity[39,40] may bring these zymogens to the platelet surface for activation. Platelet HMWK, which binds to the unactivated platelet surface with higher affinity (K_d 1.0 nM) and a larger number of molecules per platelet (1800/platelet)[33] than either factor XI[41] or factor

[33] E. Gustafson, D. Schutsky, L. C. Knight, and A. H. Schmaier, *J. Clin. Invest.* **78,** 310 (1986).

[34] J. Greengard and J. H. Griffin, *Biochemistry* **23,** 6863 (1984).

[35] R. E. Thompson, R. Mandle, Jr., and A. P. Kaplan, *J. Clin. Invest.* **60,** 1376 (1977).

[36] C. F. Scott and R. W. Colman, *J. Clin. Invest.* **65,** 413 (1980).

[37] R. Mandle, Jr., R. W. Colman, and A. P. Kaplan, *Proc. Natl. Acad. Sci. U.S.A.* **73,** 4179 (1976).

[38] C. F. Scott, L. D. Silver, M. Schapira, and R. W. Colman, *J. Clin. Invest.* **73,** 954 (1984).

[39] T. Sugo, H. Kato, S. Iwanaga, and S. Fujic, *Thromb. Res.* **24,** 329 (1981).

[40] T. Sugo, N. Ikari, H. Kato, S. Iwanaga, and S. Fujic, *Biochemistry* **19,** 3215 (1980).

[41] J. S. Greengard, P. N. Walsh, and J. Griffin, *Blood* **58** (Suppl.), 194 (1981).

XIa,[4] may be the platelet receptor for these proteins. This hypothesis suggests that platelet HMWK may have a role in the initiation of intrinsic coagulation.

Platelet calpains [calcium-activated cysteine protease(s)] cleave HMWK and concomitant with cleavage increase its coagulant activity.[25] α-Cysteine protease inhibitor, the major plasma inhibitor of tissue calpains, is identical to plasma LMWK.[42] HMWK is a potent inhibitor of platelet calpain.[43] Platelet HMWK, as well as being a substrate for platelet calpains, may also function as an intraplatelet inhibitor of these enzymes.

Last, platelet HMWK may serve as a source of peptides reactive with endothelial cells. Bradykinin, which is derived from the kininogens, is a potent stimulator of endothelial cell prostaglandin synthesis.[44,45]

[42] I. Ohkubo, K. Kwachi, J. Takasawa, H. Shiokawa, and M. Sasaki, *Biochemistry* **23,** 5691 (1984).

[43] A. H. Schmaier, H. Bradford, L. D. Silver, S. Farbee, C. F. Scott, D. Schutsky, and R. W. Colman, *J. Clin. Invest.* **77,** 1565 (1986).

[44] S. L. Hong, *Thromb. Res.* **18,** 787 (1980).

[45] D. J. Crutchley, J. W. Ryan, J. S. Ryan, and G. H. Fisher, *Biochim. Biophys. Acta* **751,** 99 (1983).

[25] Platelet α_2-Antiplasmin

By EDWARD F. PLOW, LINDSEY A. MILES, and DESIRE COLLEN

Introduction

Platelets contain α_2-antiplasmin, the primary inhibitor of plasmin in blood.[1,2] Also referred to in the literature as α_2-plasmin inhibitor, α_2-proteinase inhibitor, antiplasmin, and the primary plasmin or fibrinolysis inhibitor (as summarized in Ref. 3), this proteinase inhibitor is a single-chain glycoprotein of M_r 70,000. It rapidly forms a 1:1 stoichiometric complex with plasmin, and the enzyme:inhibitor complex is stable in sodium dodecyl sulfate. This interaction involves two distinct steps: (1) a noncovalent interaction of α_2-antiplasmin with the high-affinity lysine-

[1] E. F. Plow and D. Collen, *Blood* **58,** 1069 (1981).

[2] G. O. Gogstad, H. Stormorken, and N. O. Solum, *Thromb. Res.* **31,** 387 (1983).

[3] B. Wiman, this series, Vol. 80, p. 395.

binding site of plasmin(ogen),[4,5] and (2) formation of a tetrahedral intermediate or ester bond between the reactive site of the inhibitor and the active site of the enzyme.[6] In addition to α_2-antiplasmin, several other components of the fibrinolytic system are present in platelets, including the substrate, fibrinogen, 7.1 μg/10^8 platelets[7]; the zymogen, plasminogen, 21 ng/10^8 cells[8,9]; a rapidly acting plasminogen activator inhibitor (PAI-1)[10]; and the modulator of plasminogen activation, histidine-rich glycoprotein, 37.1 ng/10^8 platelets.[11] Less well defined in molecular terms are the plasminogen activator and plasminogen proactivator activities within platelets.[12] Other proteinase inhibitors are also present in platelets, including α_1-antitrypsin and α_2-macroglobulin.[13] These molecules are capable of inhibiting plasmin in a purified system, but are unlikely to function in this capacity in plasma to a significant extent unless α_2-antiplasmin is depleted.[14] All of these components, including α_2-antiplasmin, are present in low concentrations in platelets relative to their plasma levels. While they may contribute to the regulation of fibrinolysis at sites of platelet deposition and secretion, direct evidence for their role in fibrinolysis remains to be demonstrated. The role of platelet inhibitors of plasmin which may not circulate in plasma also remains unspecified.[15-20]

α_2-Antiplasmin antigen is present in platelets at a level of 62 $\pm$ 24 ng/10^9 platelets (range = 33–114 ng, $n = 10$). Of the various peripheral blood cells tested, α_2-antiplasmin antigen was detected only in platelets. It is an intrinsic component of platelets based on the following criteria: (1) radiolabeled α_2-antiplasmin, added to plasma, is not recovered at significant levels within washed platelet preparations; (2) antibodies to α_2-antiplasmin do not bind to platelets, indicating that the plasma antigen does not tightly

[4] B. Wiman and D. Collen, *Eur. J. Biochem.* **84,** 573 (1978).
[5] U. Christensen and I. Clemmensen, *Biochem. J.* **163,** 389 (1977).
[6] B. Wiman and D. Collen, *J. Biol. Chem.* **254,** 9291 (1979).
[7] J. M. Gerrard, D. R. Phillips, G. H. R. Rao, E. F. Plow, D. A. Walz, R. Russ, L. A. Harker, and J. G. White, *J. Clin. Invest.* **66,** 102 (1980).
[8] J. C. Holt and S. Niewiarowski, *Circulation* **62** (Suppl. 3), 342 (1980).
[9] L. A. Miles and E. F. Plow, *J. Biol. Chem.* **260,** 4303 (1985).
[10] L. A. Erickson, M. H. Ginsberg, and D. J. Loskutoff, *J. Clin. Invest.* **74,** 1465 (1984).
[11] L. L. K. Leung, P. C. Harpel, R. L. Nachman, and E. M. Rabellino, *Blood* **62,** 1016 (1983).
[12] S. Thorsen, P. Brakman, and T. Astrup, *Hematol. Rev.* **3,** 123 (1972).
[13] R. L. Nachman and P. C. Harpel, *J. Biol. Chem.* **251,** 4514 (1976).
[14] N. Aoki, M. Moroi, M. Matsuda, and K. Tachiya, *J. Clin. Invest.* **60,** 361 (1977).
[15] K. Wakabayashi, K. Fujikawa, and T. Abe, *Thromb. Diath. Haemorrh.* **24,** 76 (1970).
[16] P. Ganguly, *Clin. Chim. Acta* **39,** 466 (1972).
[17] P. T. K. Mui, H. L. James, and P. Ganguly, *Br. J. Haematol.* **29,** 627 (1975).
[18] S. Moore, D. S. Pepper, and J. D. Cash, *Biochim. Biophys. Acta* **279,** 160 (1975).
[19] J. H. Joist, S. Niewiarowski, N. Nath, and J. F. Mustard, *J. Lab. Clin. Med.* **87,** 659 (1976).
[20] M. S. Hansen and I. Clemmensen, *Biochem. J.* **187,** 173 (1980).

associate with the cell surface; and (3) the antigen is secreted from stimulated platelets. The release of α_2-antiplasmin from platelets is virtually complete at high doses of a platelet stimulus such as thrombin in the absence of significant cell lysis. Secretion of platelet α_2-antiplasmin parallels that of platelet factor 4.[1] This concordance suggests the presence of α_2-antiplasmin within α granules, and subcellular localization studies support this conclusion.[2] In a patient congenitally deficient in plasma α_2-antiplasmin, the platelet levels of the antigen were also decreased but the extent of the deficiency in the platelets and in plasma was not identical (platelet antigen = 31% of normal; plasma antigen = <1.3% of normal).[21] Of the described α_2-antiplasmin-deficient kindreds, positive bleeding histories have been reported for some heterozygotes.[21,22] Levels of platelet α_2-antiplasmin could potentially influence the expression of bleeding problems in these individuals.

The method originally used for detection and quantitation of platelet α_2-antiplasmin antigen was a competitive inhibition radioimmunoassay of the double-antibody type. Any immunoassay system of sufficient sensitivity should be suitable for this purpose. The purification and characterization of α_2-antiplasmin has been previously considered in another volume of this series, which provides greater detail on these aspects.[3]

Radioimmunoassay for α_2-Antiplasmin

Purification of α_2-Antiplasmin. Currently used purification procedures of α_2-antiplasmin take advantage of its affinity for the high-affinity lysine-binding site of plasminogen. Fresh plasma or fresh frozen plasma is usually used as the source of the inhibitor. These sources are first depleted of plasminogen on lysine–Sepharose. Plasminogen binds efficiently to this affinity matrix and is eluted with 6-aminohexanoic acid. The recovered plasminogen can then be used to construct a plasminogen–Sepharose column or to prepare a plasminogen fragment containing the high-affinity lysine-binding site by elastase digestion. This fragment, LBS1, or plasminogen, is coupled to cyanogen bromide-activated Sepharose. The plasminogen-depleted plasma may be precipitated with ethanol or polyethylene glycol, primarily for the purpose of removing fibrinogen, prior to application to the affinity columns, and α_2-antiplasmin is eluted with 6-aminohexanoic acid. Frequent contaminants of the α_2-antiplasmin preparation at this point are histidine-rich glycoprotein and fibrinogen, and steps that

[21] L. A. Miles, E. F. Plow, K. J. Donnelly, C. Hougie, and J. H. Griffin, *Blood* **59,** 1246 (1982).

[22] C. Kluft, E. Vellenga, E. J. P. Brommer, and G. Wijngaards, *Blood* **59,** 1169 (1982).

have been used to remove these contaminants have included chromatography on DEAE-Sephadex A-50, concanavalin A-Sepharose, and/or Ultrogel AcA 44. The purified α_2-antiplasmin is assessed for homogeneity by polyacrylamide gel electrophoresis in sodium dodecyl sulfate and should yield a single protein band under reducing and nonreducing conditions.

Antisera to α_2-Antiplasmin. Antisera to α_2-antiplasmin is prepared in rabbits using injection doses of 50–200 μg.[23] The antigen may be administered initially in Freund's complete adjuvant followed by a weekly or biweekly injection in incomplete adjuvant. Antisera obtained after the third or fourth boost are generally of sufficient titer to develop a radioimmunoassay. Specificity of the antiserum may be assessed by immunoelectrophoresis. When diffused against the purified inhibitor or plasma, a single immunoprecipitin arc in the α_2 region should be formed.

Development of the Radioimmunoassay. α_2-Antiplasmin is radioiodinated by a modified chloramine-T procedure to obtain the ligand for use in radioimmunoassays.[24] In a typical radioiodination procedure, 100–200 μg α_2-antiplasmin in 0.2 *M* sodium phosphate buffer, pH 7.3, is used. ^{125}I (1 mCi) and 20 μg chloramine-T are added to the α_2-antiplasmin, and the sample is mixed and incubated at 22° for 5 min. The radioiodination is stopped by addition of 20 μg sodium metabisulfite and 0.1% KI. Bovine serum albumin (0.1%) is added as a carrier. Free iodine may be removed by dialysis or by gel filtration. The radioiodinated α_2-antiplasmin is examined by polyacrylamide gel electrophoresis and the precipitability of radioactivity in 15% trichloroacetic acid should exceed 85%. Specific activities of 1–3 μCi/μg are routinely obtained. α_2-Antiplasmin may also be radioiodinated using lactoperoxidase.[25]

A quantitative radioimmunoassay using ^{125}I-labeled α_2-antiplasmin and its antiserum follows the general protocols described for development of double-antibody radioimmunoassays.[26] Goat anti-rabbit Ig of sufficient titer may be used for immunoprecipitation. Serial dilutions of purified α_2-antiplasmin are used to develop the standard curve for quantitation of the antigen. With ^{125}I-labeled α_2-antiplasmin at 0.1 n*M* and a dilution of the antiserum to bind 30–50% of the radiolabeled ligand, the assay sensitivity is approximately 10 ng/ml.[1] As a control for specificity, normal plasma should yield antigen levels of approximately 70 μg/ml. The potential unique complexity of such assays is the possibility that free α_2-antiplas-

[23] E. F. Plow, B. Wiman, and D. Collen, *J. Biol. Chem.* **255,** 2902 (1980).

[24] T. S. Edgington, E. F. Plow, C. I. Chavkin, D. H. De Heer, and R. M. Nakamura, *Bull. Cancer* **63,** 673 (1976).

[25] N. Aoki, H. Saito, T. Kamiya, K. Koie, Y. Sakata, and M. Kobakura, *J. Clin. Invest.* **63,** 877 (1979).

[26] A. R. Midgley, Jr., and M. R. Hepburn, this series, Vol. 70, p. 266.

min and the plasmin : α_2-antiplasmin complex may not be detected with equal sensitivity in the radioimmunoassay.[27] This can be assessed by determining the capacity of the assay to quantitate purified α_2-antiplasmin versus the plasmin : α_2-antiplasmin complex or normal plasma compared to plasma activated with streptokinase or urokinase.

Measurement of Platelet α_2-Antiplasmin. As α_2-antiplasmin is an α-granule constituent, its secretion can be measured following stimulation of platelets with typical secretagogues such as thrombin, collagen, or calcium ionophore. The release of α_2-antiplasmin parallels that of other platelet α-granule constituents, and the simultaneous measurement of a second released protein, such as β-thromboglobulin, can serve as an excellent control. As there is no evidence for retention or reassociation of secreted α_2-antiplasmin with platelets, the total level of α_2-antiplasmin antigen measured within detergent lysates of platelets should be attained in releasates if a full platelet secretory response is elicited. The time course, stimulus dependence, and susceptibility to metabolic inhibitors of α_2-antiplasmin secretion will parallel the parameters observed for other platelet α-granule constituents. A radioimmunoassay, such as that described above, can then be used to quantitate platelet α_2-antiplasmin in detergent extracts of the cells or in the supernatant fluid of platelets stimulated to undergo a secretory response. It has not yet been demonstrated whether an activity assay can be used to accurately quantitate α_2-antiplasmin levels in platelet releasates or extracts.

Potential problems in the quantitation of platelet α_2-antiplasmin can arise from (1) carryover of plasma α_2-antiplasmin into the platelet preparation, or (2) inadvertent platelet activation and secretion during cell isolation. Either of these pitfalls can be successfully avoided by a platelet isolation procedure such as described in Ref. 28, which involves differential centrifugation followed by gel filtration of the platelets on Sepharose 2B, or protocols such as those described in Section I of this volume.

[27] E. F. Plow, F. De Cock, and D. Collen, *J. Lab. Clin. Med.* **93,** 199 (1979).

[28] G. A. Marguerie, E. F. Plow, and T. S. Edgington, *J. Biol. Chem.* **254,** 5357 (1979).

[26] Immunocytochemical Localization of Platelet Granule Proteins

By JAN J. SIXMA, JAN-WILLEM SLOT, and HANS J. GEUZE

Secretion of small molecules, proteins, and lysosomal enzymes from storage organelles is an important function of blood platelets. Small molecules such as ATP, ADP, serotonin, Ca^{2+}, and pyrophosphate are stored in the dense granules.[1] These granules are absent in the more severe forms of storage pool deficiency where there is a lack of a secretable pool of ATP, ADP, and serotonin.[2,3] Fractionation studies have confirmed these observations. Dense granules are quite sensitive to disruption, but with the use of gentle techniques for tissue homogenization and of differential centrifugation, dense granule fractions of high purity have been obtained.[4] Morphologically, dense granules are characterized by their high electron density which makes them visible even in whole-platelet mounts. Electron density is due to the high local concentration of divalent cations, Ca^{2+} in the human species and Mg^{2+} in many animal species. Dense granules are usually spherical but occasionally they project a tail which gives them a comet-like appearance. Dense granules are not easily identified in cryosections.

Blood platelets also possess a different storage organelle, α granules, containing a variety of proteins.[5] These can be subdivided into several overlapping groups. The first group comprises proteins that are involved in the adhesion of blood platelets to the vessel wall, or to one another. Proteins that belong to this group are fibrinogen, von Willebrand factor, fibronectin, and thrombospondin. These four proteins appear to bind to the membrane of activated platelets. A second group of proteins includes those that have a direct action on the vessel wall. They include β-thromboglobulin and platelet factor 4, which have chemotactic effects on leukocytes, and platelet-derived growth factor (PDGF), which is chemotactic

[1] M. H. Fukami and L. Salganicoff, *Thromb. Haemostasis* **38,** 963 (1977).
[2] H. Holmsen and H. J. Weiss, *Br. J. Haematol.* **191,** 643 (1970).
[3] H. Holmsen and H. J. Weiss, *Annu. Rev. Med.* **30,** 119 (1979).
[4] M. N. Fukami, J. S. Baner, G. J. Stewart, and L. Salganicoff, *J. Cell Biol.* **77,** 389 (1978).
[5] J. G. White, *Blood* **33,** 598 (1968).

and growth stimulating for smooth muscle cells (see Chapter [20] in this volume). A third group of proteins includes those involved in coagulation and fibrinolysis. They encompass coagulation factor V, plasminogen, α_2-antiplasmin, C$\bar{1}$ inhibitor, antithrombin III, histidine-rich glycoprotein, and tissue plasminogen activator inhibitor. High-molecular-weight kininogen has also been encountered in secretory granules but only at very low concentration.[6] A fourth group of proteins is without a defined function in regard to the vessel wall, blood platelets, or the coagulation and fibrinolytic systems. Albumin is the most prominent member of this group.

There are approximately 80 α granules per platelet. The α granules are spherical or ellipsoidal with a mean diameter of 200 nm and often contain a dark eccentric nucleoid which has been attributed to local accumulation of glycosaminoglycans.[7] The localization of the α-granule-specific proteins has been confirmed by observations in a rare bleeding disorder, the so-called "gray platelet syndrome," which is characterized by the absence of granules.[8,9] With the exception of albumin, all stored proteins studied were indeed absent in the platelets of these patients. Albumin was decreased.[10] Its continuous presence was attributed to localization in the surface-connected canalicular system which is prominent in this disorder.[9]

Whether there are one, or more, subset(s) of α granules is unclear because of their nonparallel pattern of protein secretion. For instance, all low-molecular-weight substances stored in dense granules are released simultaneously and in a parallel fashion. In contrast, the α-granule protein β-thromboglobulin is secreted almost completely, whereas von Willebrand factor is partially released with the same stimuli.[11,12] This has led to the hypothesis of granule heterogeneity according to which distinct subclasses of these granules may contain specific proteins.

A third type of secretory granule in platelets is formed by lysosomes. The existence of at least two types of lysosomes in platelets has been postulated.[13] On the basis of the secretion pattern and studies on the gray platelet syndrome, it has become evident that lysosomes differ from α granules. Direct electron microscopic observations have not yielded an obvious candidate for lysosomes. Whereas histochemical studies at the

[6] A. H. Schmaier, P. M. Smith, A. D. Purdon, J. G. White, and R. W. Colman, *Blood* **67,** 119 (1986).

[7] S. S. Spicer, W. B. Green, and J. H. Hardin, *J. Histochem. Cytochem.* **17,** 781 (1969).

[8] J. G. Breton-Gorius, W. Vainchenker, A. Nurden, S. Levy-Toledano, J. Caen, and J. G. White, *Am. J. Pathol.* **102,** 10 (1981).

[9] J. G. White, *Am. J. Pathol.* **95,** 445 (1979).

[10] A. T. Nurden, J. J. Kunicki, D. Dupuis, C. Soria, and J. P. Caen, *Blood* **59,** 709 (1982).

[11] J. Koutts, P. N. Walsh, E. F. Plow, J. W. Fenton, B. N. Bouma, and T. S. Zimmerman, *J. Clin. Invest.* **68,** 1255 (1978).

[12] M. B. Zucker, M. J. Broekman, and K. L. Kaplan, *J. Lab. Clin. Med.* **941,** 675 (1979).

[13] J. W. N. Akkerman, S. Niewiarowski, and H. Holmsen, *Thromb. Res.* **17,** 249 (1979).

ultrastructural level have suggested that they may be vesicular in nature, this was hard to ascertain because a product of the histochemical reaction obscured the actual morphology.[14,15] Nevertheless, this group of organelles contains a heparitinase which is able to degrade the proteoglycans of the vessel wall,[16] and elastase,[17] among others.

Immunocytochemical methods shed new light on these problems. It is the purpose of this chapter to review these methods with the main emphasis on the potential of postembedding techniques for immunocytochemistry and the problems encountered with their use.

Methodology

Immunocytochemical methodology has been reviewed extensively.[18,19] The following description highlights the key points. Immunocytochemical techniques can be classified according to the accessibility of the antigen. *Preembedding techniques* are most suitable for antigens that are directly accessible to the antibodies. These techniques can also be used for the localization of intracellular antigens after the cellular membranes have been made permeable for antibodies by chemical or mechanical treatment. In *postembedding techniques* the immunoreaction is performed on the surface of ultrathin sections of embedded tissue. An important advance in recent years has been the introduction of ultracryomicrotomy in which ultrathin cryosections are immunolabeled. These sections may be embedded after the immunoreaction in order to protect the structures against shrinking during drying. Most of our data on the immunocytochemical localization of platelet granule proteins have been obtained with ultrathin cryosections; hence we will focus our discussion on this technique.

Fixation. Because antigenic markers in a given tissue sample can be destroyed by a fixative, the method of fixation must be selected accordingly. One of the most popular procedures, Karnovsky's fixative method, comprises a mixture of 1% paraformaldehyde and 1% glutaraldehyde, but 2% paraformaldehyde with added glutaraldehyde varying between 0.05 and 0.5% is generally more effective in retaining antigenicity. Glutaraldehyde has the advantage that it produces better crosslinking thus preventing local diffusion and improving the ultrastructure, but it may impair or destroy antigenicity. This is occasionally observed with concentrations of glutaraldehyde as low as 0.1%. Paraformaldehyde alone has also been used

[14] M. E. Bentfeld and D. F. Bainton, *J. Clin. Invest.* **56,** 1635 (1975).
[15] M. E. Bentfeld-Barker and D. F. Bainton, *Blood* **59,** 472 (1982).
[16] J. Yahalom, A. Eldor, Z. Fuks, and I. Vlodavsky, *J. Clin. Invest.* **74,** 1842 (1984).
[17] Y. Legrand, G. Pigneaud, and J. P. Caen, *Haemostasis* **6,** 180 (1977).
[18] K. T. Tokuyasu, *Histochem. J.* **12,** 381 (1980).
[19] G. Griffiths, K. Simons, G. Warren, and K. T. Tokuyasu, this series, Vol. 96, p. 435.

but 2% paraformaldehyde gives a poor morphology. Higher concentrations up to 8% paraformaldehyde provide a reasonable morphology. Blood platelets should not be fixed directly in 8% paraformaldehyde because this solution is very hypertonic. A practical way to avoid this has been the use of increasing concentrations of paraformaldehyde. The best procedure in our hands has been the following: 15 ml of blood is collected into 35 ml of 2% paraformaldehyde in 0.1 *M* sodium phosphate buffer, pH 7.4, and, after gentle mixing, fixed for 30 min at room temperature. Thereafter, platelet-rich plasma is prepared at a somewhat lower centrifugation speed than normal (100 *g*, 15 min, room temperature) and a platelet pellet is then made (100 *g*, 10 min, room temperature). Fixation is then continued by incubation of the pellet for 30 min at room temperature in 2% paraformaldehyde in 0.1 *M* phosphate buffer. This is then replaced for 30 min by 4% paraformaldehyde in the same buffer and this is then replaced by 8% paraformaldehyde in 0.05 *M* phosphate buffer for 1 hr. The pellet is then incubated twice for 15 min in 0.05 *M* phosphate buffer.

Cryosectioning. For cryosectioning the blood platelet pellets are first soaked with 10% gelatin in 0.1 *M* phosphate buffer at 37°. The platelet pellet containing gelatin is then solidified by pressing it between two Petri dishes on ice. The formed gelatin–platelet block is fixed with either 2 or 8% paraformaldehyde for 1 hr and stored in 2% paraformaldehyde at 4°. Before cryosectioning, the gelatin blocks are trimmed to the right size and infiltrated with 2.3 *M* sucrose to prevent ice crystal formation during freezing. The gelatin–platelet blocks are then mounted on the specimen stage, frozen in liquid nitrogen, and transferred to the ultramicrotome. Ultrathin (50–100 nm) sections are cut with a glass knife at a 45° angle at −90 to −100°. The sections are picked up from the knife by touching them with a drop of 2.3 *M* sucrose in 0.1 *M* phosphate buffer in a platinum wire loop. Outside the cryochamber the sections are allowed to stretch by thawing on the sucrose drop. Stretching takes no more than 10 sec and can be enhanced by warm breath. The sucrose drop with sections on them is then transferred to carbon-coated Formvar grids with sections downward, so that the sections are pressed against the grids with the sucrose on top. The sucrose is then allowed to diffuse away by placing the grids upside down on top of 3-mm-thick layers of 2% gelatin in 0.1 *M* phosphate buffer solidified in melting ice and kept wet by the nearby presence of moistened filter paper. When cutting is completed, the gelatin is placed at room temperature and the grids are left on it for 15 min after liquefaction of the gelatin so that they are covered with gelatin, which prevents nonspecific adherence of immune reagents.

Immunolabeling. Grids with sections are taken from the gelatin and washed for a few minutes by putting them on a drop of phosphate-buffered saline containing 0.02 *M* glycine in the case of glutaraldehyde-fixed tissue in order to quench aldehyde groups in the tissue. This and all subsequent steps are performed at room temperature. The grids are then placed on 5- to 10-μl droplets of the appropriate antibody for 30 min to 1 hr. In most instances affinity-purified IgG fractions are used in concentrations varying between 10 and 50 μg/ml. After the incubation, excess antibody is washed away by putting the grids several times on drops of phosphate-buffered saline. Subsequently the grids are placed for 20–30 min on 5- to 10-μl drops of a protein A–colloidal gold complex, prepared by adding protein A to colloidal gold solution as described by Roth.[20] Colloidal gold solutions containing spherical particles of any desired diameter in the range of 3–15 nm can be prepared by reducing a gold chloride solution with a tannic acid–citrate mixture according to Slot and Geuze.[21] Larger particles can be prepared by the method of Frans.[22] Reactive protein A–gold preparation without aggregates and free of unbound protein A can then be obtained by gradient centrifugation.[23] Gold particles varying in size between 5 and 10 nm are most commonly used. Before use, the protein A–gold solution was diluted in PBS containing 1% bovine serum albumin to prevent nonspecific staining.

Colloidal gold particles can also be bound to antibodies directed against the first antibody.[24] These IgG–colloidal gold complexes are suitable markers for ultrathin cryosections,[25] and have been used for localization studies in platelets. After incubation with the protein A–gold complex, the grids are washed thoroughly on four successive drops of phosphate-buffered saline for at least 5 min in each drop. More than one antigen can be localized on a single tissue section by a simple extension of the immunolabeling procedure. The section is first incubated for 5 min with free protein A (0.1 mg/ml) immediately after the first protein A–gold marker.[26] The sections are then washed, incubated with the second antibody, washed again, and then treated with the second protein A–gold complex, which is of a different size.

Processing for EM Observation. When cryosections are first passed over various drops of distilled water and then air dried, cell morphology is

[20] J. Roth, M. Bendayan, and L. Orci, *J. Histochem. Cytochem.* **28,** 55 (1978).

[21] J. W. Slot and H. J. Geuze, *Eur. J. Biol.* **38,** 87 (1985).

[22] G. Frans, *Nature (London), Phys. Sci.* **241,** 20 (1973).

[23] J. W. Slot and H. J. Geuze, *J. Cell Biol.* **90,** 533 (1981).

[24] J. de Mey, M. Moeremans, G. Geuens, R. Muydens, and M. de Brabander, *Cell Biol. Int. Rep.* **5,** 889 (1981).

[25] J. W. Slot and H. J. Geuze, *in* "Immunolabelling for Electron Microscopy" (J. M. Pollak and I. M. Vandell, eds.), p. 129. Elsevier, Amsterdam, The Netherlands, 1984.

[26] H. J. Geuze, J. W. Slot, P. van der Ley, and R. J. C. Scheffer, *J. Cell Biol.* **89,** 653 (1981).

difficult to see. Much better results can be obtained by a combination of staining with uranyl acetate and embedding in methylcellulose.[18] These grids are first stained for 10 min with 2% neutral uranyl solution (obtained by mixing equal volumes of 4% uranyl acetate and 0.3 *M* potassium oxalate), then washed on three drops of distilled water, and subsequently stained for 5 min with 2–4% uranyl acetate. The grids are then transferred without washing to three drops of 1.5% methylcellulose for 30 sec each. They are then picked up with a wire loop; the excess methylcellulose is removed and the grids are left to dry.

The uranyl is easily extracted from the section, even though the grids are transferred from the uranyl to the methylcellulose solution without washing. Therefore a useful variant, introduced by Griffiths *et al.*,[19] is a protective step in which the uranyl acetate staining is done while the grids are on the methylcellulose drops. Uranyl acetate is then added to 2% methylcellulose at a final concentration of 0.2–0.4%.

Immunofluorescence. Semithick sections (500–1000 nm) are prepared at a higher temperature (−60 to −70°), lifted on a drop of 2.3 *M* sucrose, and transferred to glass slides covered with 1% gelatin in 0.1% $KCr(SO_4)$. The sucrose is allowed to diffuse away in droplets of 2% gelatin in phosphate buffer (0.1 *M*, pH 7.4). The protein A–gold complex is replaced by a fluorescein- or rhodamine-labeled second antibody. The sections are finally embedded in 50% glycerol and covered with a glass coverslip which is sealed off with nail polish. Such specimens can be stored for weeks in the refrigerator.

Localization of Platelet Granule Proteins

Immunofluorescent Studies on Permeabilized Platelets. The initial studies on the localization of platelet granule proteins made use of immunofluorescent staining of mildly fixed (usually with acetone or methanol) air-dried platelets. A punctate staining pattern was observed for coagulation factor V, von Willebrand factor, fibronectin, platelet factor 4, and β-thromboglobulin.[27–30] Platelet factor 4 was also demonstrated in the cytoplasm with this technique.[31] Recently formaldehyde-fixed platelets permeabilized with Triton X-100 were used for the demonstration of the codistribution of von Willebrand factor, thrombospondin, fibrinogen, fi-

[27] K. Breederveld, J. C. Gidings, J. W. Ten Cate, and A. L. Bloom, *Br. J. Haematol.* **29,** 405 (1985).

[28] J. C. Giddings, L. R. Brookes, F. Piovella, and A. L. Bloom, *Br. J. Haematol.* **52,** 79 (1982).

[29] M. H. Ginsberg, R. G. Painter, C. Birdwell, and E. F. Plow, *J. Supramol. Struct. Cell Biochem.* **11,** 167 (1979).

[30] H. H. Ginsberg, L. Taylor, and R. G. Painter, *Blood* **55,** 661 (1980).

[31] R. Ryo, R. F. Proffito, and T. F. Deuel, *Thromb. Res.* **17,** 629 (1980).

bronectin, and β-thromboglobulin.[32] Thrombospondin, fibronectin, and β-thromboglobulin were found to have the same distribution as fibrinogen, in double-label immunofluorescence. The staining for von Willebrand factor was finer but in general similar, suggesting a possible differential intragranular localization.

Immunofluorescent Studies on Frozen Sections. Direct staining on frozen sections has the advantage of better structural integrity, easier accessibility, and higher resolution due to the thickness of the section. von Willebrand factor, albumin, fibrinogen, platelet factor 4, β-thromboglobulin, and factor XIII were studied with this technique.[33] Factor XIII was localized in the cytoplasm[34]; all of the other proteins were present in α granules. Codistribution was found for von Willebrand factor with fibrinogen, and of β-thromboglobulin and platelet factor 4 with fibrinogen.[35]

Ultrastructural Studies Based on Immunoelectron Microscopy

In one study formaldehyde-fixed platelets permeabilized with saponin were used for immunoperoxidase staining at the ultrastructural level.[36,37] Not all granules were reactive and saponin interfered with localization of antigens on membranes. These studies were further pursued with staining of ultrathin sections. This technique was also used in the other studies regarding the ultrastructural localization. Fibrinogen, platelet factor 4, and β-thromboglobulin were found in α granules and not in the cytoplasm.[38] We found platelet factor 4 concentrated at the dark eccentric nucleoid in the α granules and fibrinogen around it, as illustrated in Fig. 1. β-Thromboglobulin had no preferential distribution.[38] The even distribution of β-thromboglobulin was confirmed in other studies and the differential distribution of platelet factor 4 and fibrinogen was also observed but not as a consistent pattern.[36,37,39] Albumin was found to codistribute with β-thromboglobulin, whereas the cytoplasmic distribution of factor XIII was confirmed at the ultrastructural level.[34] All studies showed that some granules which were morphologically similar to α granules did not stain (see further below).

[32] J. D. Wencel-Drake, E. F. Plow, T. S. Zimmerman, R. G. Painter, and H. H. Ginsberg, *Am. J. Pathol.* **115,** 156 (1984).

[33] J. W. Slot, B. N. Bouma, R. Montgomery, and T. S. Zimmerman, *Thromb. Res.* **13,** 871 (1978).

[34] J. J. Sixma, A. van den Berg, M. J. Schiphorst, N. J. Geuze, and J. McDonagh, *Thromb. Haemostas.* **51,** 388 (1984).

[35] H. J. Sander, J. W. Slot, B. N. Bouma, H. J. Geuze, and J. J. Sixma, *Thromb. Haemostasis* **42,** 408 (1979).

[36] P. A. Stenberg, M. A. Schuman, S. P. Levine, and D. F. Bainton, *J. Cell Biol.* **98,** 748 (1984).

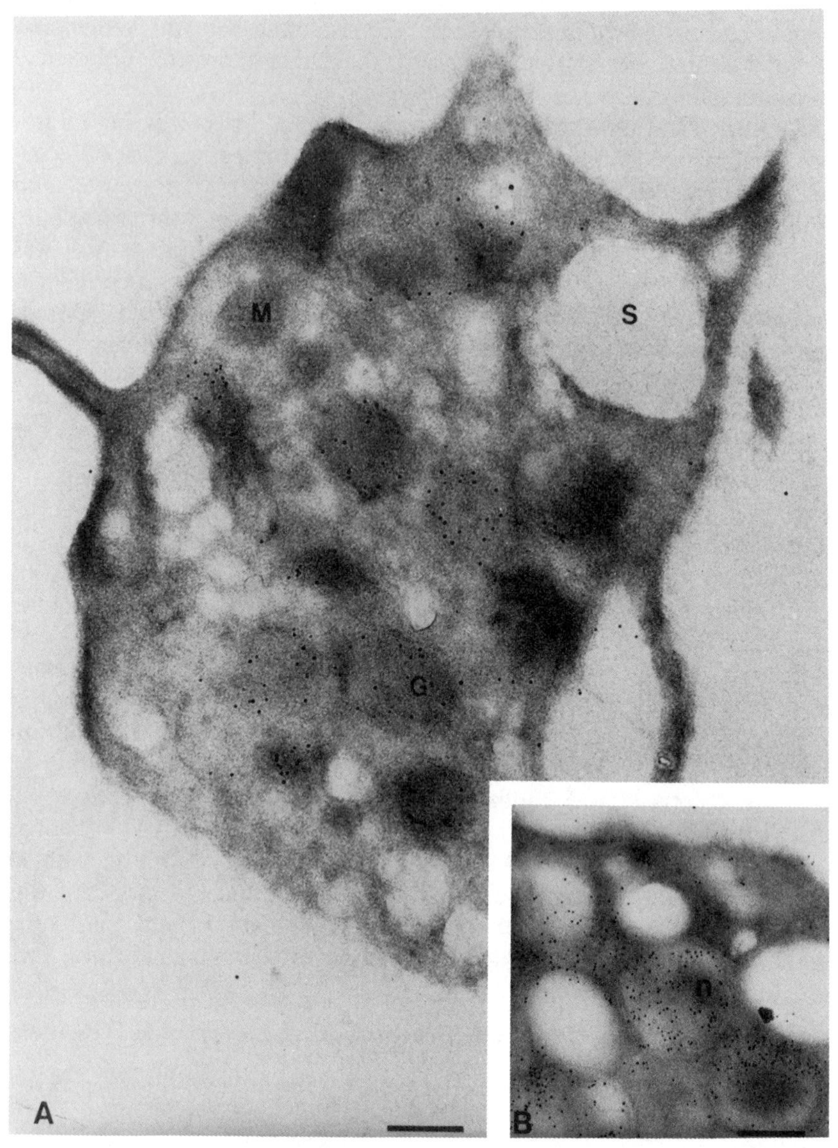

FIG. 1. (A) Localization of fibrinogen on cross-section of washed platelet. Gold label (8 nm in size) is present on profiles of granules (G) but not on a mitochondrion (M) or on surface-connected canalicular system (S). (B) More intense staining with 5-nm gold reveals that the nucleoid (N) remains unlabeled. The bar corresponds to 0.2 μm. Reproduced with permission from Sander *et al.*[38]

Lysosomes were localized with an antibody to cathepsin D.[40] This antibody gave good and reproducible staining in megakaryocytes in which primary as well as secondary lysosomes could be demonstrated (Fig. 2). Staining of lysosomes in platelets was only occasionally found and only in platelets which were recently fixed in 2% paraformaldehyde. These lysosomes tended to be smaller than α granules. Immunofluorescent studies showed the presence of at least one lysosome in almost every platelet cross section.[41] This was confirmed in a preliminary study with a monoclonal antibody specific for lysosomes, which also indicated that platelet lysosomes may be of the same size as α granules.[42]

Immunocytochemical Studies of Platelet Secretion

Immunocytochemical studies have been performed with indirect immunofluorescence on permeabilized platelets as well as by immunoelectron microscopy on cryosections.[30,36] Both studies gave the same results. Released product was found in dilated sacs of the surface-connected system with a similar distribution of all proteins studied. Secondary binding to the platelet membrane was demonstrated for fibrinogen and platelet factor 4. Some binding was also observed for β-thromboglobulin but this tended to be less.[36] Secretion of lysosomal enzymes occurs probably also via the surface-connected canalicular system as was suggested by the positive staining for cathepsin D in the dilated profiles of this system after thrombin treatment of platelets.[40]

Concluding Remarks

Immunocytochemical studies have established without doubt that the secreted platelet proteins are present in α granules. Almost all α granules were positive for all antigens studied and in every investigation codistribution of the relevant antigens was seen. This suggests that there is no heterogeneity in α granules but that all proteins are present in the same granule. Some caution is needed, however, because all ultrastructural studies are unanimous in that they find occasional granules with a morphology similar to α granules, which do not stain. The most likely explanation is

[37] P. A. Stenberg, M. A. Shuman, and S. P. Levine, *Histochem. J.* **16,** 983 (1983).

[38] H. J. Sander, J. W. Slot, B. N. Bouma, P. A. Bolhuis, D. S. Pepper, and J. J. Sixma, *J. Clin. Invest.* **72,** 1277 (1983).

[39] T. D. Pham, K. L. Kaplan, and V. B. Butler, *J. Histochem. Cytochem.* **31,** 905 (1983).

[40] J. J. Sixma, A. van den Berg, A. Hasilik, K. von Figura, and H. J. Geuze, *Blood* **65,** 1287 (1985).

[41] J. J. Sixma, unpublished observations.

[42] H. K. Nieuwenhuis and J. J. Sixma, unpublished observations.

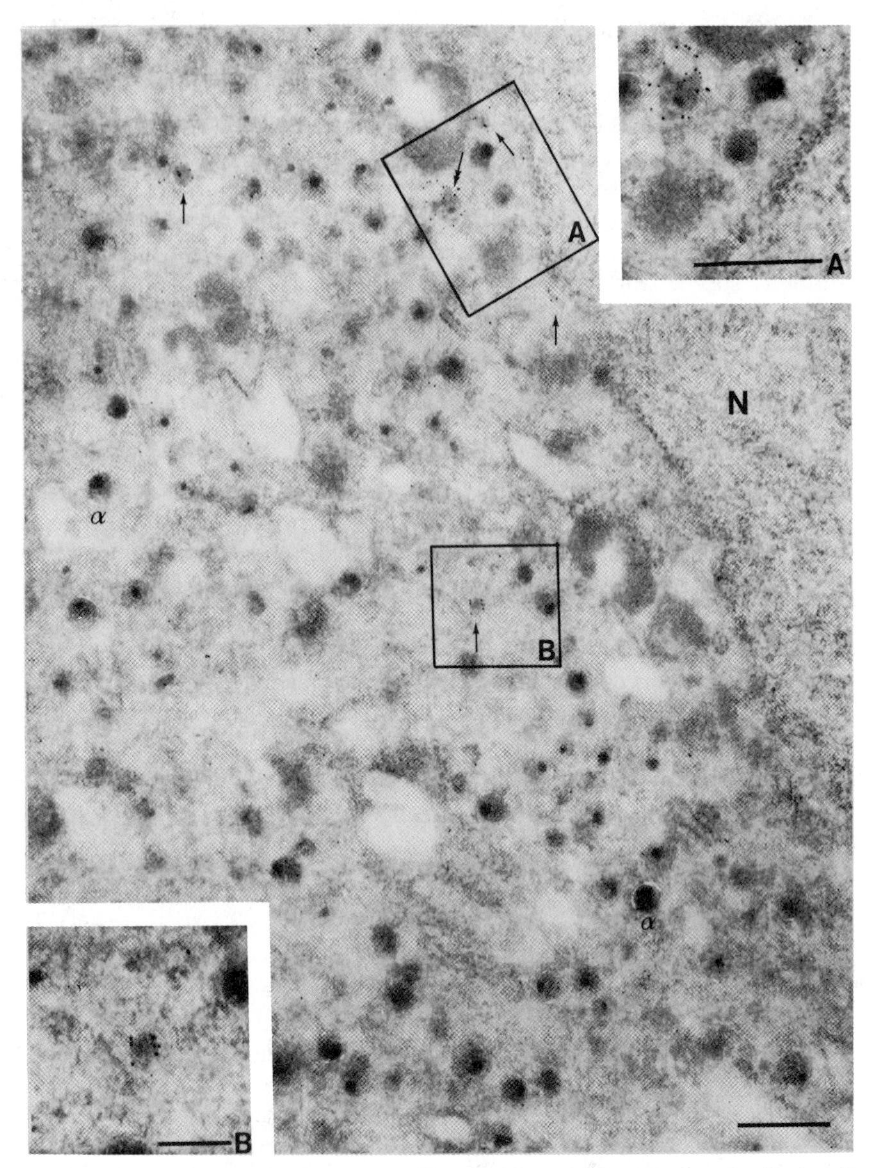

FIG. 2. Megakaryocyte stained with antibody to cathepsin D to show the localization of lysosomes (arrows). Inset A shows a secondary lysosome in a higher magnification. Inset B shows a primary lysosome in a higher magnification. α, α granules. The bars in the overview and in inset A correspond to 0.5 μm; in inset B the bar corresponds to 0.25 μm.

that they represent lysosomes. Recent unpublished observations[42] suggest that lysosomes may be more frequent than was surmised from previous studies with values of 10–20% of the granules representing lysosomes. Morphologically they tend to be less dense and to lose their contents during sectioning but this difference may not always be apparent. Another explanation for the lack of staining of all α-granule profiles in cross sections may be that this is random and shows a lack of accessibility of the relevant antigens. The number of granules that is not stained is rather low. It appears unlikely, therefore, that a specific subclass of α granules is present. The immunocytochemical studies have also confirmed previous observations on the route of secretion.[43–45] Both lysosomal enzymes and α-granule proteins are apparently secreted via the surface connected canalicular system. Further studies are still required to establish the precise morphology of lysosomes and to investigate whether the two classes of lysosomes which were postulated on the basis of selective permeabilization[13] can indeed be identified morphologically.

[43] J. G. White, *Am. J. Pathol.* **58,** 31 (1970).
[44] R. Holme, J. J. Sixma, E. H. Murer, and T. Hovig, *Thromb. Res.* **3,** 347 (1973).
[45] M. J. Droller, *Scand. J. Haematol.* **11,** 35 (1973).

[27] PADGEM Protein

By CINDY L. BERMAN, ERIK L. YEO, BARBARA C. FURIE, and BRUCE FURIE

Introduction

Platelets play a fundamental role in hemostasis and thrombosis. Upon activation of resting platelets the surface membrane is structurally and functionally altered, resulting in the expression of receptors for ligands such as fibrinogen, fibronectin, factor Va, factor XIIIa, and thrombospondin. These platelet-bound ligands participate in platelet–platelet and platelet–endothelial cell interactions and in the propagation of coagulation. To identify, isolate, and characterize platelet surface constituents unique to the activation process, we have utilized an immunologic approach. Monoclonal anti-platelet antibodies were developed and selected for their ability to recognize only the surface of activated but not resting platelets. A monoclonal antibody with this property, KC4, was characterized. KC4 was employed to identify and purify by affinity chromatography the antigen against which it is directed.[1] Studies of this antigen, a newly identified platelet protein, have shown that the KC4 antigen is an integral membrane glycoprotein which is internal in resting platelets but expressed on the cell surface only after platelet activation and granule secretion. This observation has lead to the hypothesis that the protein is a component of an internal granule membrane which is fused with the external membrane during platelet activation. Therefore, this glycoprotein has been descriptively named *p*latelet *a*ctivation-*d*ependent *g*ranule to *e*xternal *m*embrane (PADGEM) glycoprotein. Further studies have shown that PADGEM protein is an integral component of the α granule in resting platelets,[2] and is translocated to the external surface membrane upon platelet activation. As PADGEM protein is an α-granule membrane protein, it is a marker of the movement of α granules to the membrane surface during platelet secretion and of platelet activation. The function for PADGEM protein, currently unknown, may involve a role as an internal receptor during α-granule coalescence, movement, or release or may relate to the unique receptor and procoagulant activities of the activated platelet membrane.

Immunochemical Reagents

Preparation of Anti-PADGEM Monoclonal Antibodies

BALB/c mice are immunized intraperitoneally with aggregated thrombin-activated platelets ($1-5 \times 10^9$) in 250 μl of HEPES buffer (3.8 m*M* HEPES, 0.14 *M* NaCl, 3 m*M* KCl, 1 m*M* $MgCl_2$, 3.8 m*M* NaH_2PO_4, 0.1% dextrose, and 0.35% BSA, pH 7.35) biweekly for 2 months. The mice are rested for 3 months and a final injection is performed before their splenocytes are hybridized[3] with sp2/0 cells to form monoclonal antibody-producing hybridomas. Using an ELISA system (described below), clones making anti-platelet antibodies specific for activated platelets are selected; these antibodies do not bind to resting platelets. Large quantities of monoclonal antibody are prepared by intraperitoneal injection of the hybridoma

[1] S.-C. Hsu-Lin, C. L. Berman, B. C. Furie, D. August, and B. Furie, *J. Biol. Chem.* **259,** 9121 (1984).

[2] C. L. Berman, E. L. Yeo, J. D. Wencel-Drake, M. H. Ginsberg, B. C. Furie, and B. Furie, *J. Clin. Invest.* **78,** 130 (1986).

[3] G. Kohler and C. Milstein, *Nature (London)* **256,** 495 (1975).

cells into pristane-treated mice. The ascites that develops is applied to a protein A–Sepharose column (Pharmacia) for the purification of the monoclonal antibody KC4, a murine monoclonal anti-PADGEM antibody (subclass IgG_1 κ) isolated in our laboratory.[1] The protein A–Sepharose column is washed extensively with 0.1 *M* sodium phosphate, pH 8.0, and bound antibodies are eluted with 0.1 *M* sodium citrate, pH 6.0.[4] Elution conditions of other antibodies may differ since the affinity of IgG interaction with protein A varies depending on subclass. The KC4 antibody is dialyzed against Tris-buffered saline (TBS: 20 m*M* Tris–HCl, 0.15 *M* NaCl, pH 7.5) and stored at $-70°$ until use.

Preparation of Polyclonal Anti-PADGEM Antibodies

Polyclonal anti-PADGEM protein antiserum is produced by immunization of a New Zealand White rabbit with purified PADGEM protein. An initial injection of PADGEM protein (50 μg in 200 μl saline) emulsified in Freund's complete adjuvant (200 μl) is given subcutaneously; this is followed 1 month later by an injection of PADGEM protein emulsified similarly in Freund's incomplete adjuvant. Subsequent injections of 50 μg of protein in saline are made at monthly intervals. Antibodies specific to the PADGEM protein are affinity purified from the antiserum on a PADGEM protein–Sepharose 4B column. The column is prepared by covalently linking purified PADGEM protein to cyanogen bromide-activated Sepharose 4B at a ratio of 3 mg of protein/ml of Sepharose 4B.[5] Rabbit anti-PADGEM protein antiserum (12 ml) is applied to a 5-ml PADGEM protein–Sepharose column. Nonspecifically bound proteins are removed by washing the column with 50 ml of 0.5 *M* NaCl until the optical density at 280 nm of the column effluent is less than 0.01. Anti-PADGEM antibodies are eluted with 4 *M* guanidine–HCl. The elution is monitored by absorption at 280 nm and the fractions of maximum absorption pooled and dialyzed three times against 1 liter of TBS. The isolated antibody migrates as a single band on SDS gels under nonreducing conditions. The monospecificity of the affinity-purified polyclonal anti-PADGEM protein antibodies is established by the Western blot technique. The anti-PADGEM protein antibodies react only with a single platelet protein corresponding in molecular weight to the PADGEM protein.

[4] P. L. Ey, S. J. Prowse, and C. R. Jenkin, *Immunochemistry* **15**, 429 (1978).

[5] P. Cuatrecasas, M. Wilchek, and C. B. Anfinsen, *Proc. Natl. Acad. Sci. U.S.A.* **61**, 636 (1969).

Measurement of Anti-PADGEM Antibodies

Enzyme-Linked Immunosorbent Assay (ELISA)

The ELISA is primarily used for identifying murine anti-platelet antibodies with the desired properties. Parallel ELISAs are performed using either fixed gel-filtered thrombin-activated platelets or fixed resting platelets bound to microtiter wells to select antibodies specific only for the surface of activated platelets.

Reagents

ELISA solution: 0.05 *M* Tris–HCl, 0.14 *M* NaCl, 1.5 m*M* $MgCl_2$, 2 m*M* 2-mercaptoethanol, 0.05% Tween 20, 0.05% NaN_3, pH 7.2

0.05 *M* Sodium phosphate, 1.5 m*M* $MgCl_2$, 100 m*M* 2-mercaptoethanol, pH 7.2

0.5 *M* Na_2CO_3

Procedure. To prepare the ELISA plates, gel-filtered platelets[6] are prepared in HEPES buffer in the absence of bovine serum albumin (BSA) and resuspended at 10^3 cells/ml. Thrombin-activated platelets are prepared by the addition of 0.15 U thrombin/ml of gel-filtered platelets suspended in buffer containing 2.5 m*M* EDTA. The platelets are stirred for 15 min after the addition of thrombin. The platelet suspension (100 μl) is added to 100 μl of 2.5% glutaraldehyde contained in each well of a 96-well microtiter plate (Immunolon II, Dynatech). The plates are centrifuged at 1000 *g* for 10 min, covered, and incubated at 4° for 1 hr. The platelets fixed to the microtiter wells are washed three times with TBS to remove the glutaraldehyde. Special care must be taken during all washes as the platelets are easily washed off the microtiter wells. Dipping the plates gently into the rinsing solution is recommended. TBS with 1% BSA is added to the wells to block nonspecific binding of antibody and the plates incubated at 4° for several hours. After the wells are washed with TBS as above, they may be filled with 60% (v/v) glycerol and the coated plates stored for several months at −70°. Prior to use, the glycerol is removed by washing the plates three times with TBS.

The assay is performed as follows: murine antibody solution (100 μl; cell culture supernatant or antiserum dilution) is added to the coated wells and the plates incubated at 37° for 1 hr. The wells are washed three times with ELISA solution. Sheep anti-mouse immunoglobulin conjugated with β-galactosidase (50 μl in ELISA solution, BRL) is added and the plates incubated at 22° for 2 hr with constant shaking. After washing three times with ELISA solution, color is developed by the addition of 50 μl of *p*-nitro-

[6] S. Timmons and J. Hawiger, *Thromb. Res.* **12,** 297 (1978).

phenyl-β-D-galactoside (1 mg/ml) in 0.05 *M* sodium phosphate, 1.5 m*M* $MgCl_2$, 100 m*M* 2-mercaptoethanol, pH 7.2. The plates are incubated for 2 hr at room temperature or overnight at 4°, and the reaction is stopped by the addition of 25 μl of 0.5 *M* Na_2CO_3. The release of *p*-nitrophenol is monitored at 405 nm on a Dynatech MR580 MICROELISA Auto-Reader.

Double-Antibody Suspension/Fluid Phase Radioimmunoassay (RIA)

Reagents

TBS, 1% BSA, human IgG (50 μg/ml)
^{125}I-Labeled F(ab')$_2$ fragments of sheep anti-mouse IgG
TBS, 1% BSA, 0.05% Tween 20

Procedure. A solution-phase immunoassay, albeit more cumbersome than the solid-phase assays, is associated with fewer binding artifacts and is preferred for quantitative evaluation of antibody–antigen binding. For this reason, the following assay was devised. Resting gel-filtered platelets or platelets activated with 0.15 U thrombin/ml are fixed by the addition of glutaraldehyde to a final concentration of 3%. The platelets are stirred slowly for 30 min, and washed twice with TBS. Fixed resting or thrombin-activated platelets (0.5 ml at 10^3 platelets/ml) are suspended in TBS containing 1% BSA and 50 μg/ml of human IgG in a 1.5-ml microfuge tube. After incubation at 22° for 15 min, the platelets are sedimented by centrifugation in a Beckman microfuge for 30 sec, and the supernatant aspirated. The culture fluid (0.5 ml) from a cloned hybridoma cell line or ascites fluid (0.5 ml) containing murine antibody is added and the platelets resuspended and incubated at 37° for 30 min. After sedimentation by centrifugation, the platelets are washed with 0.5 ml of TBS, 1% BSA, and human IgG (50 μg/ml). ^{125}I-Labeled F(ab')$_2$ fragments of sheep anti-mouse IgG (2–10 μCi/μg estimated specific radioactivity, New England Nuclear) are diluted in TBS, BSA, human IgG buffer such that 100 μl contains 40,000 cpm. After the addition of 100 μl, the platelets are resuspended and incubated at 37° for 5 min. After washing three times with TBS, 1% BSA containing 0.05% Tween 20, the platelets are sedimented by centrifugation and the supernatant aspirated. The bottom of the microfuge tube, containing the platelet pellet, is cut off with tin snips and assayed for ^{125}I in a Beckman Gamma 8000 spectrometer.

Purification of PADGEM Protein

The PADGEM protein is purified from crude platelet extracts by affinity chromatography using monoclonal antibody bound to a solid support.

The KC4 antibody is purified from ascites fluid and then covalently linked to Sepharose 4B. Sepharose (7 ml) is washed with cold phosphate-buffered saline, pH 7.4 (PBS), and activated with 200 mg of cyanogen bromide per milliliter of Sepharose.[5] The activated Sepharose is mixed with 25 mg of purified KC4 antibody in PBS and stirred overnight at 4°. After washing sequentially with 1 liter each of 1 *M* ethanolamine, pH 8.0, PBS, 1 *M* NaCl, and 0.15 *M* NaCl, the KC4 antibody–Sepharose is packed into a column, equilibrated in PBS, 0.02% NaN_3, pH 8.0, and stored at 4° for use in the affinity purification of PADGEM protein.

Reagents

TBS
TBS, 2.5 m*M* EDTA
TBS, 1% Triton X-100
1 *M* NaCl, 1% Triton X-100
0.05 *M* Diethylamine, pH 11.5

Procedure

Fresh or outdated platelets from one unit of blood are washed in TBS, 2.5 m*M* EDTA and resuspended in 3 ml of TBS, 1% Triton X-100. After sonication three times at 100 W for 20 sec at 4° using a Heat Systems-Ultrasonics, Inc. cell disrupter, the platelet extract is centrifuged at 100,000 *g* for 30 min at 4° in a Beckman L3-50 ultracentrifuge. The supernatant may be stored at −20°. For PADGEM protein purification, the platelet preparation is filtered through a 0.8-μm membrane (Millipore), and 3 ml of the filtrate is applied to the KC4–Sepharose 4B column. With the KC4 monoclonal antibody, the highest yield of PADGEM protein is obtained when the sample is allowed to incubate on the column for 30–60 min at 22°. The column is washed with 50 ml of TBS, 1% Triton X-100 and then 50 ml of 1 *M* NaCl, 1% Triton X-100 to remove nonspecifically bound proteins. Triton is removed by extensive washing with TBS until the optical density at 280 nm of the column effluent is below 0.01. Bound antigen is then eluted with 0.05 *M* diethylamine, pH 11.5, as described by McEver *et al.*[7] Elution is monitored by the optical density at 280 nm, and the fractions of highest optical density pooled and neutralized with 1.0 *N* HCl. The purified PADGEM protein is dialyzed three times against 1 liter of TBS, concentrated against sucrose, and redialyzed against TBS. Approximately 2 mg of pure PADGEM protein can be obtained by this method.

The purified protein migrates as a major diffuse band on SDS gels upon electrophoresis under nonreducing conditions (Fig. 1). This is a pattern

[7] R. P. McEver, N. L. Baenziger, and P. W. Majerus, *J. Clin. Invest.* **66,** 1311 (1980).

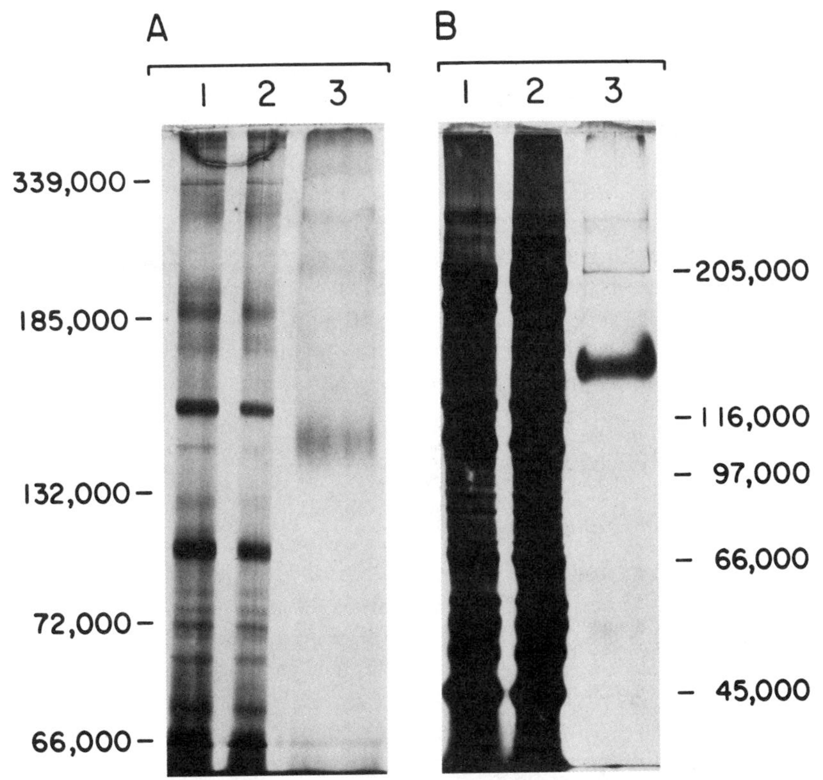

FIG. 1. SDS gel electrophoresis of immunoaffinity-purified KC4 antigen. Platelet membrane proteins were extracted into Triton X-100, and the proteins applied to a KC4 antibody–agarose column. The bound protein was eluted with diethylamine and analyzed by SDS gel electrophoresis in 6% gels. Lane 1, solubilized platelet membrane proteins; lane 2, solubilized platelet membrane proteins which did not bind to KC4 antibody–agarose; lane 3, bound fraction of solubilized platelet membrane proteins eluted with diethylamine. (A) Nonreduced; (B) reduced. The migration of proteins of known molecular weight is indicated. Proteins were visualized with silver stain.

characteristic of heavily glycosylated proteins. The protein can be stained with Coomassie brilliant blue or periodic acid–Schiff stain. The dominant protein band corresponds to an apparent molecular weight of 140,000. The character of this band is unchanged in the presence of Ca^{2+} or EDTA. In SDS gels run under reducing conditions, the purified KC4 antigen migrates as a single narrow band also with a molecular weight of 140,000. Based on these characteristics, the protein is most easily visualized in a gel run under reducing conditions and stained with silver stain.

Properties of PADGEM Protein

This immunologic approach has enabled us to identify and purify a unique platelet membrane protein which is expressed on the surface membrane of activated platelets.[1] PADGEM protein is a single-chain glycoprotein with a molecular weight of 140,000 under both reducing and nonreducing conditions. PADGEM protein is an integral component of the α granule in resting platelets and is translocated to the external surface membrane upon platelet activation.[2] These properties suggest that PADGEM protein is a constituent of the α-granule membrane. Its expression on the cell surface is directly proportional to the extent of α-granule secretion. Upon activation of platelets with thrombin, 13,000 molecules of PADGEM protein are expressed per platelet.

Assays of PADGEM Protein

Direct Suspension/Fluid Phase RIA

A direct suspension/fluid phase radioimmunoassay was developed to quantitate the interactions of antibody with thrombin-activated and resting platelets. This assay is typically used to quantitate the expression of PADGEM protein on the platelet surface. It is generally performed using gel-filtered platelets (fresh or fixed). However, since normal human plasma does not contain PADGEM protein,[1] platelet-rich plasma can be used. Although the difference between resting and activated platelets can be observed when the assay is performed using platelet-rich plasma, it should be noted that the results of the assay are more difficult to interpret due to the complications of clot formation.

Reagents

TBS, 3% BSA, ^{125}I-labeled KC4 antibody
Density centrifugation medium (7:93, Apiezon oil C:dibutyl phthalate)

Procedure. The purified KC4 monoclonal antibody is labeled with ^{125}I using the chloramine-T method.[8] Glycyl tyrosine is added to a final concentration of 20 m*M*, and the protein-bound iodine separated from free iodine on a Sephadex G-25 column. The labeled antibody has a final specific radioactivity of approximately 3 μCi/μg. Gel-filtered platelets (400 μl) at a concentration of 10^8 platelets/ml in HEPES buffer or platelet-rich plasma (400 μl) are mixed in a 12×75 mm polypropylene test tube with 100 μl of TBS, 3% BSA containing 20,000 cpm of ^{125}I-labeled KC4

[8] W. M. Hunter and F. C. Greenwood, *Nature (London)* **194,** 495 (1962).

monoclonal antibody. After adding 10 μl of either thrombin (final concentration 0.15 U/ml) or TBS, the platelet suspension is incubated at 22° for 15 min without stirring. A 450-μl aliquot of the reaction mixture is then gently layered onto 0.4 ml of density centrifugation medium in a 1.5-ml microfuge tube. The free and platelet-bound antibodies are separated by centrifugation in a Beckman microfuge for 3 min.[9] Separation of the platelet pellet from the fluid phase may be accomplished by either of two methods. The oil and supernatant are both aspirated or the tube frozen in dry ice. The tip of the tube containing the platelet pellet is then cut off and the ^{125}I-labeled KC4 antibody bound to the platelet pellet is quantitated in a Beckman Gamma 8000 spectrometer. After correcting for nonspecific binding, the percentage bound is determined by comparison to the total cpm used per tube.

The interactions of the KC4 antibody with unfixed activated and resting platelets have been studied using this suspension/fluid phase radioimmunoassay. As shown in Fig. 2, the monoclonal antibody displayed preferential and saturable binding to thrombin-activated platelets, whereas there was minimal binding to resting platelets.

Western Blot Analysis

The presence of PADGEM protein can be qualitatively determined using the Western blot method.[10] This analysis is not quantitative but can be used for comparative purposes and has the advantage that the platelet sample need not be fresh. Furthermore, this assay procedure can detect PADGEM protein which is not expressed on the cell surface (i.e., intracellular), and may also be used to detect fragments of the protein.

Reagents

25 m*M* Tris, 192 m*M* glycine, 20% (v/v) methanol, pH 8.3
TBS, 3% BSA, 50 μg/ml mouse immunoglobulin
^{125}I-Labeled KC4 antibody
TBS, 0.01% Tween 20, 0.02% NaN_3

Procedure. Samples containing washed platelets or purified protein are solubilized in 3% SDS for analysis. Solubilized proteins are separated by SDS–polyacrylamide gel electrophoresis (6% acrylamide, 30 : 0.8 acrylamide : bisacrylamide) under nonreducing conditions, and transferred electrophoretically to nitrocellulose paper in 25 m*M* Tris, 192 m*M* glycine, 20% methanol, pH 8.3, for 18 hr at 150 mA. After transfer, the nitrocellulose blots are incubated with TBS containing 3% BSA and nonimmune

[9] H. Feinberg, H. Michel, and G. V. R. Born, *J. Lab. Clin. Med.* **84,** 926 (1974).
[10] H. Towbin, T. Staehelin, and J. Gordon, *Proc. Natl. Acad. Sci. U.S.A.* **76,** 4350 (1979).

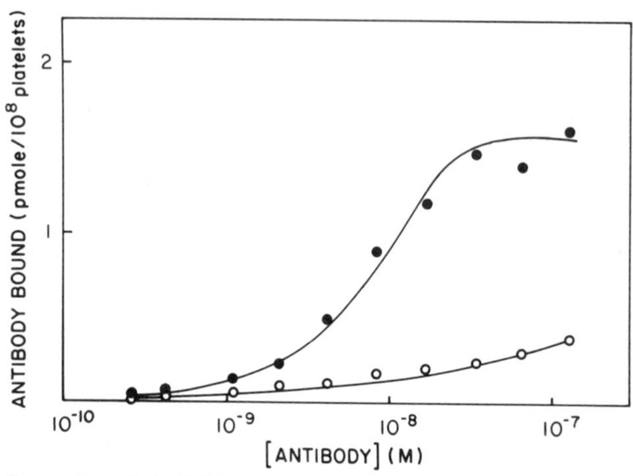

FIG. 2. Interaction of the KC4 monoclonal antibody with unfixed thrombin-activated platelets and unfixed resting platelets. Platelets (100 μl; 1×10^9/ml) in HEPES buffer and 100 μl of ^{125}I-labeled KC4 antibody (concentrations as indicated) were incubated at 22° for 15 min. The free and bound antibodies were separated by centrifugation in an oil mixture. The antibody bound to platelets is expressed in picomoles per 10^8 platelets. ●, thrombin-activated platelets; ○, resting platelets.

mouse IgG (50 μg/ml) for 1 hr at 22° with constant shaking. ^{125}I-Labeled KC4 antibody ($1-5 \times 10^6$ cpm) is added to the incubation mixture and shaken for an additional hour at 22°. The blots are washed four times with 500 ml of TBS, 0.01% Tween 20, 0.02% NaN_3 for a total of 1 hr, dried thoroughly, and autoradiographed with Kodak X-Omat AR film for 18 hr. The KC4 antibody binds to a single band present in both the solubilized thrombin-activated and resting platelets. This band migrates with an apparent molecular weight of about 140,000 (Fig. 3). Platelets, surface-labeled with ^{125}I using the lactoperoxidase method[11] and run for comparison, show the characteristic band pattern of GPIIb, GPIIa, and GPIII. The protein antigen of the KC4 antibody migrated between glycoproteins IIb and IIa.

In some instances, the rabbit anti-PADGEM antiserum is used to identify PADGEM protein. The samples are prepared and subjected to electrophoresis as described above. After transfer of the proteins to the nitrocellulose paper, the blots are incubated for 1 hr with TBS and 3% BSA; 0.1 vol of rabbit anti-PADGEM protein antiserum is added and the blots incubated for an additional 2 hr. Excess primary antibody is removed by washing the blots five times with 200 ml of TBS over a 30-min period. The blots are then exposed to goat anti-rabbit ^{125}I-labeled $F(ab')_2$ (3×10^6

[11] D. R. Phillips and P. P. Agin, *J. Biol. Chem.* **252,** 2121 (1977).

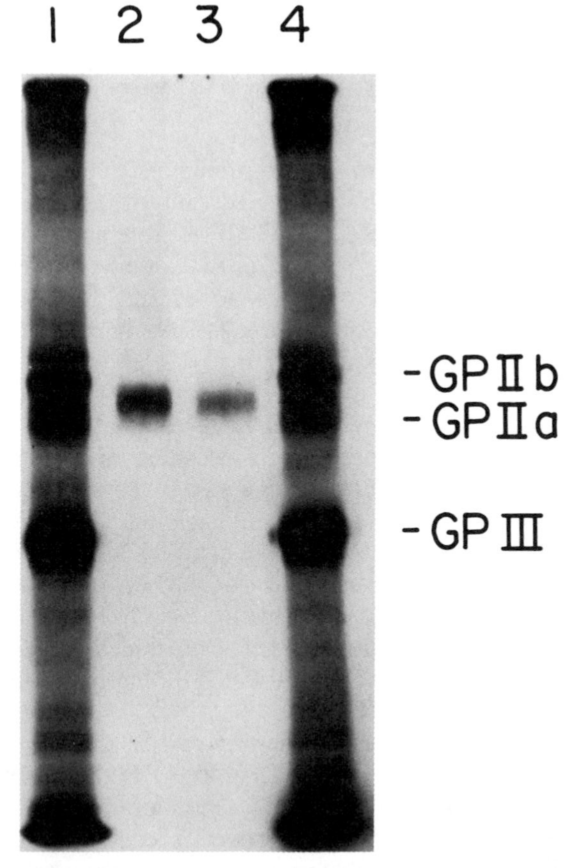

FIG. 3. Immunoblotting of platelet proteins with KC4 monoclonal antibody. SDS extracts of resting and thrombin-activated platelets were prepared and subjected to electrophoresis in nonreduced SDS gels. ^{125}I-Surface-labeled platelet proteins were run as standards. After electrophoretic transfer and blotting with ^{125}I-labeled KC4 antibody, the blot was developed by autoradiography. Lanes 1 and 4, ^{125}I-labeled platelet protein; lane 2, solubilized unlabeled resting platelets; lane 3, solubilized unlabeled thrombin-activated platelets. The proteins were blotted with ^{125}I-labeled KC4 antibody. Glycoproteins IIb, IIa, and III are identified, for comparison.

cpm, 2–10 μCi/μg, New England Nuclear) in TBS, 3% BSA for 3 hr before being washed, dried, and autoradiographed as above.

Competition RIA

To quantitate PADGEM protein, a competition solution-phase RIA is employed.

Reagents

TBS, 0.1% BSA

3% (v/v) goat anti-rabbit antiserum in TBS, 2.5% PEG-6000, 0.02% NaN_3

Procedure. In 12 × 75 mm polypropylene test tubes, 100-μl samples containing PADGEM protein were mixed with 100 μl of a solution containing 20,000 cpm of ^{125}I-labeled PADGEM protein and 0.1 mg of normal rabbit immunoglobulin. Rabbit anti-PADGEM protein antiserum (100 μl), diluted in TBS, 0.1% BSA, is added and the reaction mixture incubated for 2 hr at room temperature. The resultant antibody–antigen complexes are precipitated by the addition of 1 ml of a solution which is 3% (v/v) goat anti-rabbit IgG antiserum in TBS, 2.5% polyethylene glycol (PEG-6000), 0.02% NaN_3. The precipitate is sedimented by centrifugation at 1000 *g* for 20 min, the supernatant is decanted, and the pellet assayed for ^{125}I. The percentage binding is calculated by comparison to the total counts used per tube, and nonspecific binding is determined in the absence of competitor and primary antibody. The quantity of PADGEM protein in the sample can then be estimated by comparison to a standard curve. Figure 4 shows the results of a competition assay in which samples of purified PADGEM or sonicated platelets were added to the system. The assay was performed with rabbit anti-PADGEM protein antiserum at a 100-fold dilution in TBS, 0.1% BSA.

In our experience, nonspecific binding can be as high as 15% while maximal specific binding may only reach 40%. For this assay, PADGEM protein has been labeled with ^{125}I using the chloramine-T method.[8] The incorporation of radioactive iodine appears to be quite low. It is possible that the protein also loses some of its antigenicity upon labeling. However, attempts to repurify PADGEM protein on the KC4 antibody column did not improve the result of the RIA. Alternative labeling procedures, specifically those for labeling glycoproteins, can be applied here.

Detection of PADGEM Protein

Immunofluorescence

PADGEM protein can be detected in fixed intact and permeabilized platelets both before and after activation with thrombin.[2] The procedure which follows has been adapted from a method published by Wencel-Drake *et al.*[12]

[12] J. D. Wencel-Drake, E. F. Plow, T. S. Zimmerman, R. G. Painter, and M. H. Ginsberg, *Am. J. Pathol.* **115,** 156 (1984).

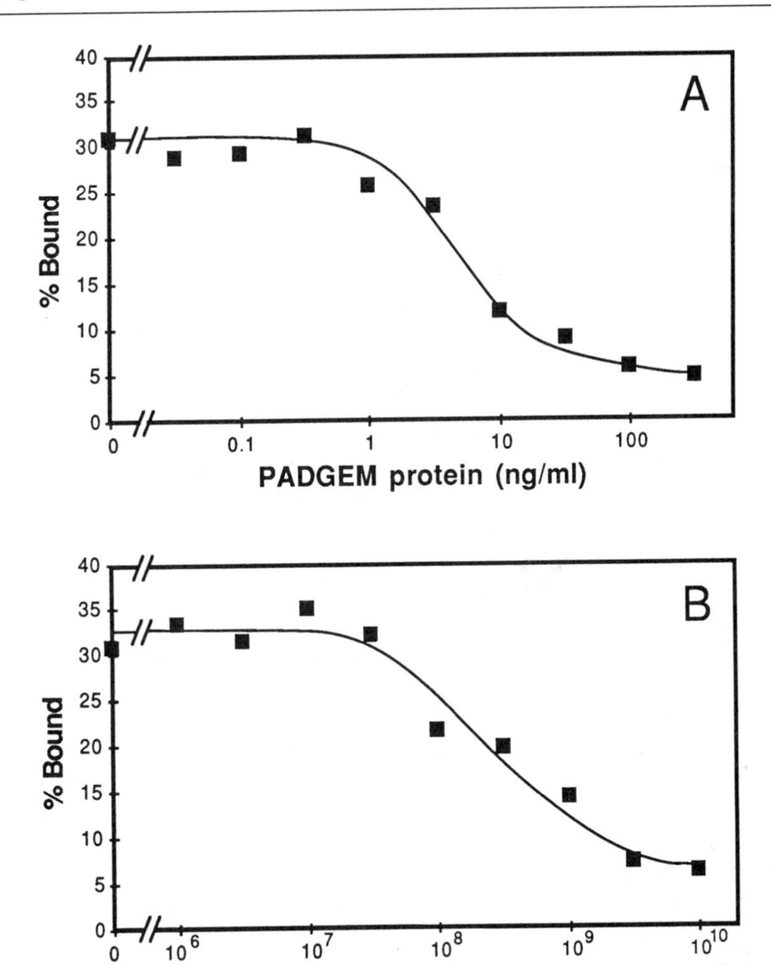

FIG. 4. Competition radioimmunoassay of PADGEM protein. Rabbit anti-PADGEM protein antiserum diluted 1:100 in TBS, 0.1% BSA was mixed with ^{125}I-labeled PADGEM protein (20,000 cpm) and 0.1 mg normal rabbit γ-globulin. Purified PADGEM protein (A) or sonicated washed platelets (B) diluted in TBS, 0.1% BSA were added and the reaction mixture incubated overnight at 4°. The immune complexes, precipitated by goat anti-rabbit antiserum, were assayed for ^{125}I. The nonspecifically bound protein has been subtracted.

Reagents

4% Paraformaldehyde
40 m*M* NH_4Cl in TBS, pH 7.4
TBS, 0.1% BSA, pH 8.0
0.1% Triton X-100

Poly(L-lysine) (200 μg/ml) in PBS
KC4 monoclonal antibody (100 μg/ml) in TBS, 0.1% BSA, pH 8.0
Fluorescein-conjugated goat $F(ab')_2$ anti-mouse IgG (100 μg/ml) in TBS, 0.1% BSA, pH 8.0
90% Glycerol in TBS, pH 9.6

Procedure. Gel-filtered platelets are prepared at a concentration of 5×10^8 cells/ml in Tyrode's buffer and 1-ml aliquots placed into 12×75 mm plastic test tubes. An equal volume of fresh 4% paraformaldehyde which has been warmed to 37° is added to each tube, the tubes covered, and placed on ice for 1 hr. To each tube, 2 ml of 40 m*M* NH_4Cl in TBS, pH 7.4, is added and centrifuged at 350 *g* for 20 min at 4°. The supernatant is decanted, the fixed platelets are resuspended in 1 ml of NH_4Cl–TBS, and centrifuged again. The platelet pellet is resuspended in 200 μl of TBS, 0.1% BSA, pH 8.0. Glass coverslips are prepared by wiping clean, dipping in methanol, and flaming dry. The dry coverslips are then soaked in poly(L-lysine) (200 μg/ml in PBS) for at least 30 min, and set on Parafilm after the excess liquid is removed by blotting the edge of the coverslip on a Kimwipe. The platelet suspension (100 μl) is loaded onto each coverslip and allowed to settle for 30 min. Cells which have not adhered to the glass are removed by touching the edge of the coverslip to a Kimwipe. Coverslips are rinsed by dipping into TBS, 0.1% BSA and blotting. When Triton permeabilization of the cells is desired, the slip is placed coated face down on 100 μl of 0.1% Triton X-100, and incubated at 22° for 3 min. The coverslips are again washed with TBS, 0.1% BSA and incubated face down on a drop of 1% BSA. After rinsing through three drops TBS, 0.1% BSA, the slips were blotted, loaded with 100 μl of KC4 monoclonal antibody (at a concentration of 100 μg/ml) diluted in TBS, 0.1% BSA, and incubated for 20 min at 22°. Unbound antibody is removed by dipping the slips in TBS, 0.1% BSA and then rinsing through four drops of TBS, 0.1% BSA and blotting. The cells are counterstained by incubation for 20 min at 22° with 100 μl of fluorescein-conjugated goat $F(ab')_2$ anti-mouse IgG (at a concentration of 100 μg/ml). Excess stain is removed by dipping in TBS, 0.1% BSA, rinsing through four drops of TBS, 0.1% BSA, and blotting. The slips are then placed face down on 75 μl H_2O, blotted, and placed face down on a small drop of 90% glycerol in TBS, pH 9.6. The coverslips can be stored at −20° until use. The platelets are viewed with a Zeiss Universal microscope equipped with an HBO 50-W mercury lamp and an IVFI epifluorescence condenser with a BP 546 excitation filter, a KT 580 chromatic splitter, and are photographed with Tri-X Panchromatic film.

Electron Immunocytochemistry

Immunocytochemistry is performed as previously described.[13] In brief, platelets are fixed using a mixture of 2% paraformaldehyde and 0.1% glutaraldehyde. Unreacted aldehyde is blocked by treatment with NH_4Cl–TBS, pH 7.4. The cells are embedded in a low-gelling-temperature agarose. Agarose blocks are infused with 1.3 *M* sucrose, mounted on copper bullets, and stored in liquid N_2 until sectioning. Sections are prepared according to the method of Tokuyasu and Singer.[14,15] Silver–gold ultrathin sections are prepared using a Sorvall MT 2-B ultramicrotome with an LTC-2 cryokit attachment. Rabbit anti-PADGEM protein antibody is used as primary antibody in an indirect immunolabeling procedure. Staining specificity is assessed with the use of preimmune rabbit serum. Affinity purified and biotinylated goat anti-rabbit IgG is used as the second antibody. Antibody–antigen interactions are visualized with 5-nm colloidal gold–avidin conjugate. Thawed ultrathin sections are conditioned with 2% gelatin and 0.1% BSA. After immunolabeling, the sections are washed briefly with distilled water and negatively stained with 0.2% phosphotungstic acid and 1% dextran. Sections are viewed on a Hitachi 12-A transmission electron microscope with an accelerating voltage of 75 kV and are photographed with Kodak electron microscope film N. 4489.

Comments

We have discovered an integral membrane protein specific to the surface of activated platelets.[1] Further characterization has shown that this protein is a glycoprotein of the α granule in resting platelets and is translocated to the cell surface membrane upon platelet activation.[2] PADGEM protein has been shown to be distinct from the major surface glycoproteins of platelets.[2] Using a similar approach with an activated platelet-specific monoclonal antibody S12, McEver and Martin[16] have purified a protein antigen with essentially identical characteristics to PADGEM protein. Recent results of Western blot analysis and crossed immunoelectrophoresis show that the S12 monoclonal antibody and the KC4 monoclonal antibody are directed against the same protein (unpublished observations)

[13] J. D. Wencel-Drake, R. G. Painter, T. S. Zimmerman, and M. H. Ginsberg, *Blood* **65,** 929 (1985).

[14] K. T. Tokuyasu, *J. Cell Biol.* **57,** 551 (1973).

[15] K. T. Tokuyasu and S. J. Singer, *J. Cell Biol.* **71,** 894 (1976).

[16] R. P. McEver and M. N. Martin, *J. Biol. Chem.* **259,** 9799 (1984).

indicating the identity of PADGEM and GMR 140.[17] This protein may function specifically in coagulation as a receptor for an adhesive protein or a soluble coagulation factor thus imparting upon the surface of the activated platelet its unique properties. It may also be possible that PADGEM protein plays a role in secretion in general, interacting with the cytoskeleton in the movement, coalescence, and secretion of the α granules. These and other aspects of PADGEM protein warrant further investigation.

[17] P. E. Stenberg, R. P. McEver, M. A. Shuman, Y. V. Jacques, and D. F. Bainton, *J. Cell Biol.* **101,** 880 (1985).

[28] Dark-Field Electron Microscopy of Platelet Adhesive Macromolecules

By HENRY S. SLAYTER

Introduction

The possibility of routinely exploiting the dark-field mode in high-resolution electron microscopy of coagulation and adhesive proteins, present in plasma and released from platelet α granules, has seemed particularly attractive. Such biological macromolecules are inherently low-contrast specimens in which substructural dimensions approach the level of noise contributed by contrasting agents. Hitherto, details of macromolecular structure have, almost without exception, been resolved by bright-field operation using metal coating of the order of 10^{-6} g/cm^2. Such coating, however, contributes 2.5 nm or more to the size of macromolecular features and therefore obscures fine detail. At the current level of 0.3- to 0.4-nm point-to-point instrumental resolution in bright field, contrast becomes the limiting factor, so that a reevaluation of prospects for high-resolution dark-field operation is in order.

Dubochet[1] has discussed the relevant theory and the practical and theoretical limitations of dark-field electron microscopy (EM). Attempts to exploit dark-field EM have encountered inherent difficulties related to intensity, specimen stability, chromatic aberration, and highly demanding specimen preparation methods. In particular, much higher incident electron intensities are required than for bright-field operation. Consequently,

[1] J. Dubochet, *in* "Principles and Techniques of Electron Microscopy" (M. A. Hayat, ed.), p. 115. van Nostrand–Reinhold, New York, 1973.

naked macromolecules are "fried" within a time interval which is of the order (1 sec) of that required for recording the relatively low-intensity dark-field image. Even more than in bright-field operation, potential signal contrast is lost as a consequence of background scattering from the specimen support. The thickness, average atomic number, and therefore the scattering power, of support films are comparable to those of the biological macromolecules themselves, a condition which tends to reduce effective contrast. Finally chromatic aberration imposes a fundamental limitation: In dark field, inelastically scattered electrons, differing in wavelength from the incident beam, create circles of confusion around the image points formed by elastically scattered electrons in dark field. In particular, the light elements of which specimen supports are composed tend to scatter inelastically.

The possibility of minimizing these difficulties using metal coating in conjunction with dark field has not been extensively explored until recently. Metal coating of replicas, which minimizes specimen damage by high incident intensities, also produces preferential accumulation of material of high atomic number in areas of interest. Thus, not only is specimen contrast enhanced, but chromatic aberration limitations could also be minimized since the proportion of inelastically scattered electrons decreases with increasing atomic number of the scatterer. Unfortunately, however, Bragg reflections arise from atomic planes within metal crystallites in the size range normally produced by currently applied coating procedures. Oriented at random, and highly susceptible to translocation as a function of focal level, these reflections easily create unacceptable confusion of image detail.

Tilted illumination, until recently the only form available for dark-field operation, poses additional problems of a technical nature. Apart from the fact that it is difficult to obtain tilted beams of adequate intensity, asymmetric contamination rapidly creates varying and therefore almost uncontrollable astigmatism and other image defects.

Given the limitations just discussed, it is not surprising that dark-field imaging has never been applied to the electron microscopy of macromolecules on a routine basis. Most applications of the technique on uncontrolled specimens have revealed additional specimen detail only after iterative processing of images,[2,3] and even then tend to leave much to the imagination.[4]

Here we detail the use of refractory metal coatings of minimal thickness

[2] A. M. Fiskin, D. V. Cohn, and G. S. J. Peterson, *J. Biol. Chem.* **252,** 8261 (1977).
[3] F. P. Ottensmeyer, J. H. Bazett-Jones, and G. B. Price, *Ultramicroscopy* **3,** 303 (1978).
[4] A. Klug, *Chem. Scr.* **14,** 291 (1978–79).

such that Bragg reflections do not confuse the image, and demonstrate the application of the currently available annular dark-field illumination system to routine dark-field imaging of platelet adhesive macromolecules.

Several shortcomings of dark-field electron microscopy and of thin metal replication of macromolecular structures are shown to be overcome by combination of these methods. In the elucidation of macromolecular conformation, image contrast, specimen stability, ease of specimen preparation, and ratio of elastically to inelastically scattered electrons are all improved, while beam damage is minimized.

Preparation of Proteins

Procedures for preparation of purified proteins have been described elsewhere as follows: Fibrinogen,[5] von Willebrand factor,[6] thrombospondin,[7,8] except that 1 mM Ca^{2+} was present in the preparation procedure used here instead of EDTA.

Metal Replication with Tungsten

Metal replication (coating) was carried out by the mica-replica method, in general, as previously described in detail[9] and amended for thin coating by tungsten.[10-12] For dark-field use, amounts of coating metal have been diminished to the point at which fine macromolecular features begin to disappear in bright-field micrographs due to contrast limitations. However, in dark field the same fine features reappear. Macromolecular solutions in 0.15 M ammonium acetate were diluted 1 : 1 with glycerol and sprayed at freshly cleaved mica as previously described.[9-12] After a 20-hr outgassing at 10^{-8} torr, preparations were coated with a very thin layer of tungsten by electron beam evaporation. During this process, specimens were rotated in a plane oriented at 5° to the vapor stream. Mass thickness of the background film of tungsten, as determined by the piezoelectric monitor, was 9.3×10^{-8} g/cm². (The range of mass thicknesses used for bright field is

[5] P. A. Norton and H. S. Slayter, *Proc. Natl. Acad. Sci. U.S.A.* **78,** 1661 (1981).

[6] H. S. Slayter, J. Loscalzo, P. Bockenstadt, and R. I. Handin, *J. Biol. Chem.* **260,** 8559 (1985).

[7] S. S. Margossian, J. W. Lawler, and H. S. Slayter, *J. Biol. Chem.* **256,** 7495 (1981).

[8] J. E. Coligan and H. S. Slayter, *J. Biol. Chem.* **259,** 3944 (1984).

[9] H. S. Slayter, *Ultramicroscopy* **1,** 341 (1976).

[10] H. S. Slayter, *Ann N.Y. Acad. Sci.* **408,** 131 (1983).

[11] H. S. Slayter, *in* "Principles and Techniques of Electron Microscopy" (M. A. Hayat, ed.), Vol. 9, p. 174. Litton, New York, 1978.

[12] M. Ohtsuki, M. S. Isaacson, and A. V. Crewe, *Proc. Natl. Acad. Sci. U.S.A.* **76,** 1228 (1979).

10^{-6} to 10^{-7} g/cm^2.) This metal layer was subsequently coated by 2.5 nm of carbon followed by a coating of nitrocellulose.

Using a thin metal coating as described above, contrast is much less than with a thicker coating of, for example, platinum. Molecular dimensions are thus much less exaggerated, so that there is correspondingly greater potential for accurate measurement of molecular dimensions or detection of differences in size, and of molecular features, even though contrast becomes limiting. Dark-field electron microscopy was carried out on these specimens, using an annular condenser aperture matched with a 40-μm objective aperture. The intensity of Bragg reflections, which would prevent the use of the dark-field mode in conjunction with thicker metal films, was at a minimum due to the presence of an extremely small number of atoms per crystallite.

Electron Microscopy

Dark-field operation was carried out in a JEM 100cx transmission electron microscope, using a top entry stage. Annular dark-field condenser illumination was used in place of the standard tilted beam dark-field mode. Compensation of astigmatism was carried out in bright field, with a 40-μm objective aperture inserted, centered, and the astigmatism correction readjusted as necessary in dark field. Images were recorded very close to focus at $\times$20,000–53,000 using 0.7-sec exposures.

Comparison of Dark-Field versus Bright-Field Image

Figure 1 illustrates the difficulties encountered in attempting to obtain dark-field images of relatively heavily metal-coated macromolecules. Fibrinogen particles appear clearly in dark field when contrasted with tungsten using only 1×10^{-7} g/cm^2 (Fig. 1A), but they are obscured by the substantial background intensity, and in addition, that arising from Bragg reflections, when thicker coatings (ca. 10^{-6} g/cm^2) of either tungsten or platinum are used (Fig. 1B); farther under focus (Fig. 1C), the Bragg images shift position significantly, further complicating the image. When the thickness of metal coating is reduced, however, Bragg reflections do not detract substantially from contrast within the target object. As demonstrated by comparing Fig. 1A with Fig. 2, coating thicknesses of 1×10^{-7} g/cm^2 barely reveal the connecting links between D and E nodules in bright-field imaging of a field of fibrinogen molecules. In observing fibrinogen and other macromolecules, features which are difficult to see in bright field are consistently more clearly delineated in the comparable dark-field images. Features of different size or complex shape are preferentially en-

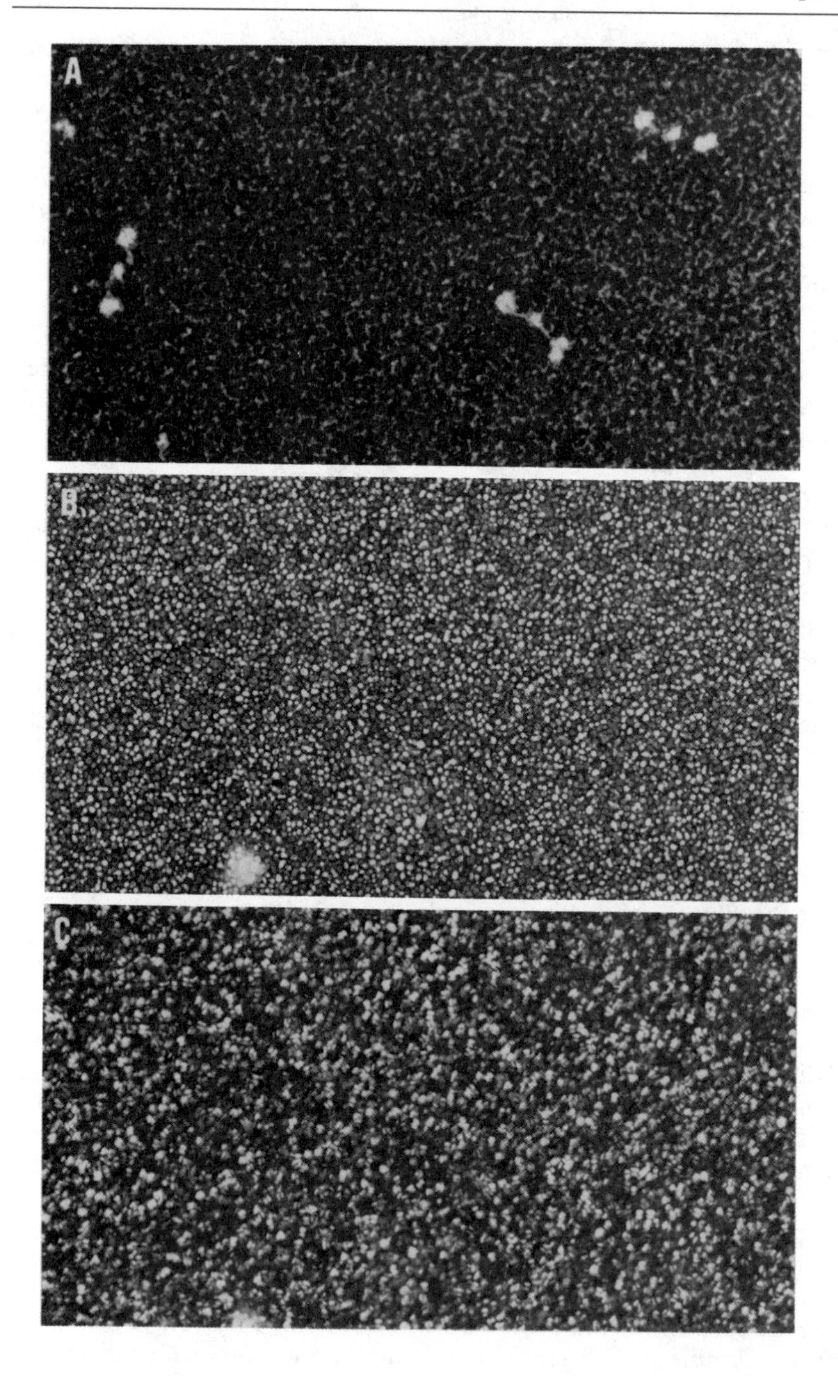
A
B
C

hanced at different critical coating thickness. It is hoped that improvements in the coating technique will provide more continuous films and will further enhance this approach.

Dark-Field Electron Microscopy of Adhesive Proteins

Figure 3A–D shows dark-field images of a number of tungsten-coated multimers of von Willebrand factor. Protomer units within the multimer are indicated in the image by bars. It can be seen that each unit contains a mirror plane of symmetry, located at a central "knot," and that paired heads exist at each end of the protomer. These molecular features are distinguishable only with the thin metal coatings, and disappear when metal-coating mass thicknesses exceed ca. 1×10^{-7} g/cm^2 as required for bright-field contrast.

Figure 4A shows a dark-field image of a tungsten-coated preparation of human thrombospondin, which may be compared with the identical field in bright field shown in Fig. 4B. Thrombospondin is shaped generally as a flexible "bola-like" structure, with 3-fold symmetry, each third of which consists of a peripheral 7-nm lobe attached to a centrum by a strand (about 30×2 nm). The increase in overall contrast and in the clarity of definition of low-contrast features, such as the connecting linkage and the central tabs of polypeptide associated in variable positions relative to the centrum, is immediately apparent. The negative contrast method using stains such as uranyl acetate has to date not proved useful in facilitating visualization of either thrombospondin or von Willebrand protein. The electron micrographs of intact thrombospondin molecules shown in Fig. 4 are prepared from extremely thin tungsten replicas. In these rotary-shadowed replicas, some metal would be expected to accumulate on the sides of particles, but correction for metal cap build-up should be minimal,[10] and is thought to be about 0.7 nm here. Even though contrast is marginal under these circumstances, fine detail can be visualized. Thin strands clearly delineated, in dark field, to be connecting the globular domains are often overshadowed by the latter. Thus, as described elsewhere,[9] they are difficult to detect

FIG. 1. (A) Fields of human fibrinogen, comparing the results obtained with relatively small as opposed to large mass thicknesses of metal in dark field. Rotary metal coated with ca. $<10^{-7}$ g/cm^2 of tungsten, showing the triad structure in dark field. $\times$288,000. (B) Dark field, in focus, using approximately 10^{-6} g/cm^2 of tungsten. Platinum fibrinogen is not distinguishable from background. $\times$153,000. (C) Dark field, underfocus, showing strong Bragg reflections which are offset from their respective reflecting crystallites [same field as (B)]. $\times$153,000.

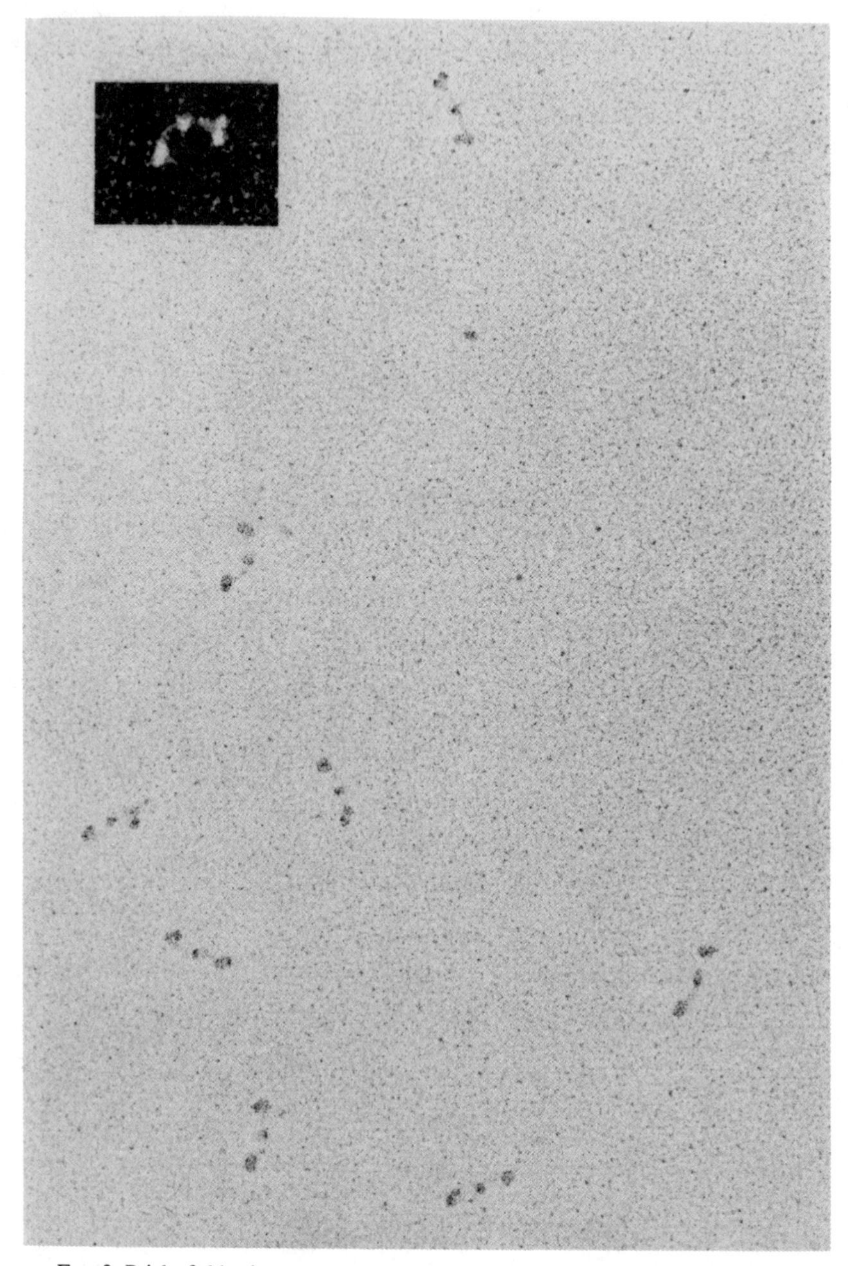

FIG. 2. Bright-field micrograph of human fibrinogen, rotary metal coated with ca. 10^{-7} g/cm^2 of tungsten, showing the triad structure invariably. Particular attention is called to the difficult-to-resolve interconnecting strands. ×220,000. The latter, along with other details such as the pear-shaped neck region, are more consistently well delineated in dark field: See inset for dark-field comparison ×320,000.

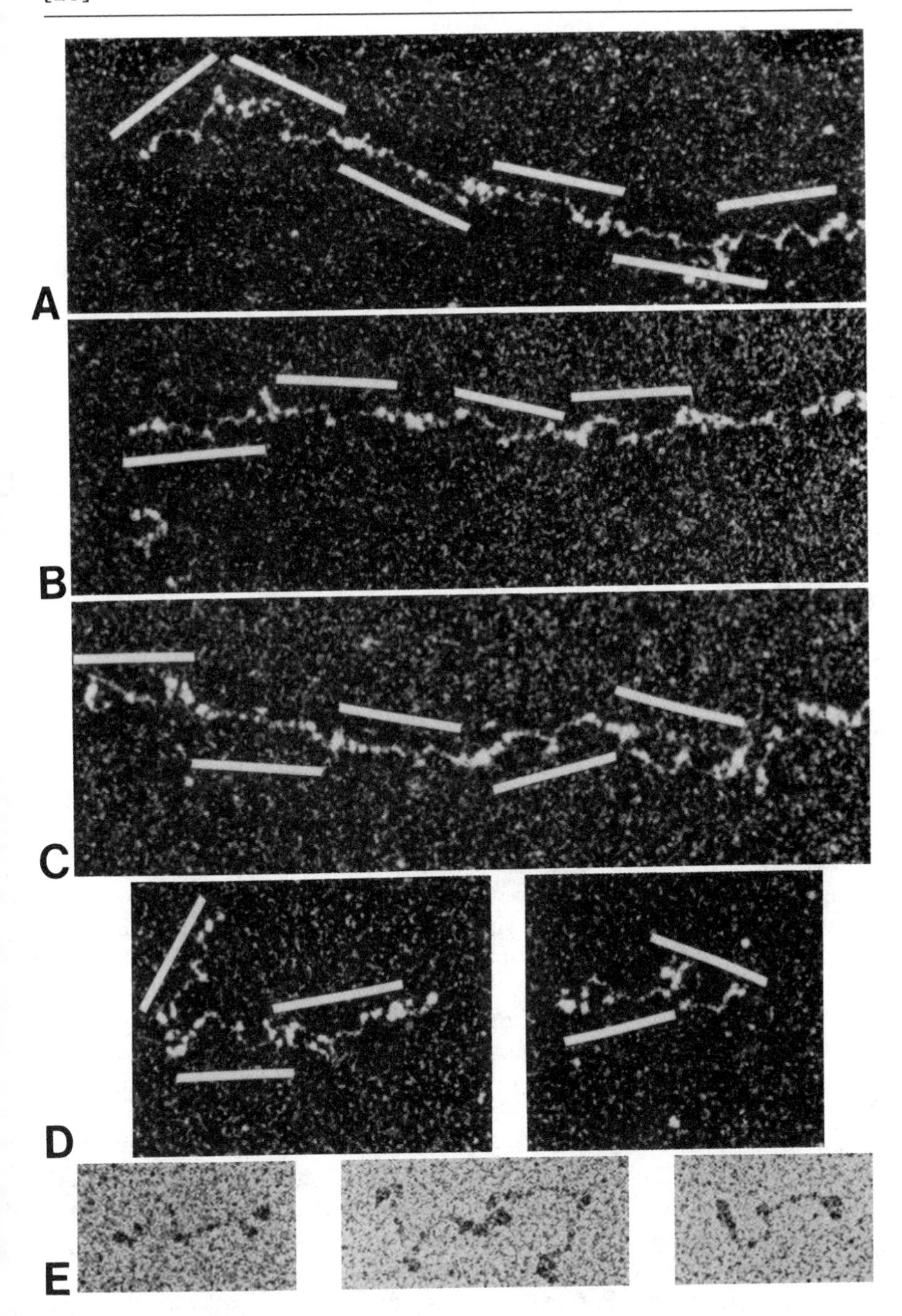

FIG. 3. von Willebrand protein multimer, coated with $<1 \times 10^{-7}$ g/cm^2 of tungsten. (A–D) Dark field (×128,000). (E) Bright field (×171,000).

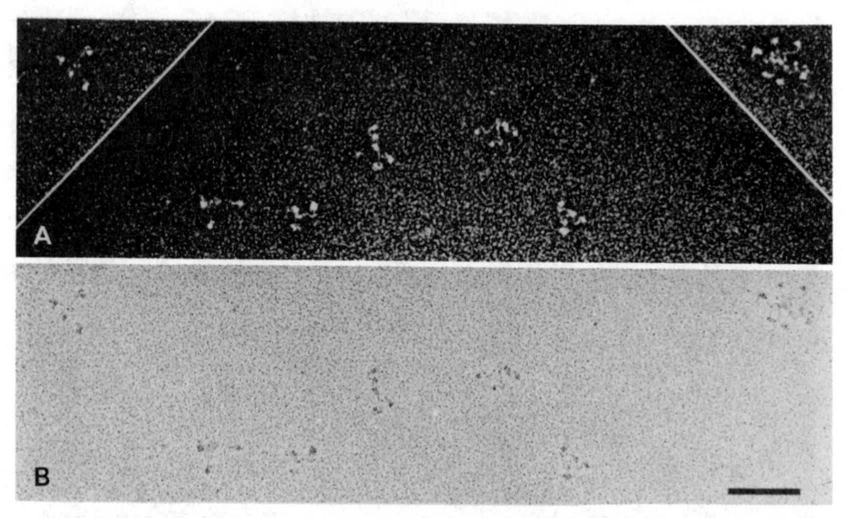

FIG. 4. Thrombospondin, tungsten coated with 10^{-7} g/cm^2. (A) Dark field (×98,000). Note: Thrombospondin complex in upper right. (B) Bright-field image of same field as in (A) (×98,000).

using the bright-field mode, given the shadowing angle and amount of metal used here. Upon close examination of individual images, a pattern consistent with known physical and chemical data emerges.

Summary

Application of these methods to a number of macromolecular species shows, in general, that features which are only occasionally or marginally detected in bright field may be visualized clearly and consistently through the use of dark-field illumination. Given the minimal average thicknesses of contrasting metal required for dark-field examination (ca. $< 10^{-7}$ g/cm^2, or less than 0.4 nm thickness), distortions of molecular features by accumulation of coating metal, although not avoided altogether, are certainly minimized. Thus, application of minimal metal coating in conjunction with high-resolution dark-field electron microscopy should facilitate solution of structural problems in molecular biochemistry generally. The method is of particular current interest in the specific area of coagulation and adhesive protein conformation. The advantages of metal coating macromolecules for dark-field observation are summarized in Table I.

In the case of macromolecules or macromolecular complexes which are of limited width (5 nm or less) even though extended lengthwise, contrast available for visibility tends to be limiting. It is for such structures that dark

TABLE I
DARK-FIELD ELECTRON MICROSCOPY: ADVANTAGES AND DISADVANTAGES

Advantages in conjunction with very thin metal replicas of macromolecules
Specimen preparation reproducible and not arduous
Very high contrast using minimal coatings
Tungsten coating minimizes effects of specimen damage
Greater proportion of elastic imaging electrons give lower chromatic aberration
Thin coatings minimize distortion of macromolcular features
Charging and heating minimized by metal film
Disadvantages of naked macromolecules
High beam intensity required can produce specimen beam damage
Requires thin substrate
Nonspecific heavy-metal contaminants must be minimized
Relatively long photographic exposure may be required
Resolution limited by significant fraction of inelastic imaging electrons

field offers a productive alternative. The dark-field method seems promising for study of certain macromolecular preparations of adhesive proteins vis-à-vis isolation of receptor loci. These include the complex between fibrinogen and GPIIb/IIIa; the complex between GPIb and von Willebrand protein, and the proposed complex between thrombospondin and fibrinogen. The use, for relating epitopic sites to topographical position, of Fab fragments directed against specific epitopes on coagulation macromolecules (and/or their complexes) of otherwise well-defined structure may also be extended by means of the dark-field approach.[13]

Acknowledgments

Research in the author's laboratory was supported by Grants GM 14237 from the National Institute of General Medical Sciences and HL22270 from the National Heart, Lung and Blood Institute, U.S. Public Health Service. The author is grateful also to E. M. Slayter, Cindy Bates, and Barbara Miller for helpful discussions.

[13] H. S. Slayter, *in* "Advances in Cell Biology" (K. Miller, ed.), Vol. 2. JAI Press, Connecticut, in press, 1988.

[29] Measurement of Secretion of Lysosomal Acid Glycosidases

By HOLM HOLMSEN and CAROL A. DANGELMAIER

Human platelets are rich in acid hydrolases and the most prominent are shown in Table I.[1-6] For all the acid (ecto) glycosidases it has been ascertained that they are present in lysosome-like organelles and secreted from the platelets during stimulation with "strong" agonists such as thrombin. This is also the case for arylsulfatase (EC 3.1.6.1), "aliphatic" phosphatase, elastase, endoglycosidase, but not for acid arylphosphatase (i.e., the phosphatase that splits the ester bond between phosphoric acid and a phenolic OH) which is definitely not secreted from platelets.

Although the significance of both the released and retained acid hydrolases is not clear, their secretion constitutes a distinct platelet response that is monitored frequently. The kinetic properties of the enzymes found in platelet lysates[1] have been studied and the two methods for determination of two major acid glycosidases described below are specifically developed for gel-filtered platelets (GFP).

In contrast to the contents of the dense granules and the α granules, which are completely secreted upon maximal stimulation, platelet acid hydrolases are only partially secreted. Two groups of acid glycosidases may be distinguished according to their degree of secretion: 60% of some (e.g., β-hexosaminidase) and 30% of others (e.g., β-glucuronidase, β-galactosidase) are secreted with maximal stimulation.[2] The reason for this partial secretion is not clear, but it has been speculated that the secretable fraction is soluble within the lysosome while the retained enzyme is bound to the inner leaflet of the lysosomal membrane. This view is supported by the findings that the secretable form of β-hexosaminidase is of a type that does not bind to endocytic receptors while the retained form is of a type that

[1] C. A. Dangelmaier and H. Holmsen, *Anal. Biochem.* **104,** 182 (1980).
[2] H. Holmsen and H. J. Day, *J. Lab. Clin. Med.* **75,** 840 (1970).
[3] J. Besse, W. Farr, and E. Gruner, *Klin. Wochenschr.* **44,** 1049 (1966).
[4] R. L. Nachman and B. Ferris, *J. Clin. Invest.* **47,** 2530 (1968).
[5] H. P. Ehrlich and J. L. Gordon, *in* "Platelets in Biology and Pathology" (J. L. Gordon, ed.), Vol. 1, p. 353. North-Holland, Amsterdam, The Netherlands, 1976.
[6] A. Oldberg, C.-H. Heldin, A. Wasteson, C. Busch, and M. Hook, *Biochemistry* **19,** 5755 (1980).

TABLE I
ACID HYDROLASES FOUND IN HUMAN PLATELETS[a]

Proteases	Phosphatases	Sulfatases	(Ecto) Glycosidases (D-configuration)	Endoglycosidases
Cathepsin D and E	Arylphosphatase	Arylsulfatase	β-Hexosaminidase	Heparitinase
Carboxypeptidase A (cathepsin A)	"Aliphatic" phosphatase (β-glycerophosphatase)		β-Glucuronidase	
Proline carboxypeptidase			β-Galactosidase	
Elastase			α-Mannosidase	
Collagenase			α-Arabinosidase	
			α-Galactosidase	
			α- and β-Fucosidase	

[a] From Refs. 1–6.

does bind, and that preincubation of platelets with primary amines (which do not induce secretion of acid hydrolases) causes 100% of β-hexosaminidase to be secreted upon thrombin stimulation.[7,8]

Platelet β-galactosidase is rapidly inactivated by the formaldehyde fixation step, whereas β-hexosaminidase (=β-*N*-acetylglucosaminidase and β-*N*-acetylgalactosaminidase) and β-glucuronidase are not so quickly affected by this treatment. We have therefore limited the methodological description below to measurement of one representative enzyme for each of the two secretion groups relatively insensitive to formaldehyde treatment.

The lysosomal secretion measurement is performed exactly as described earlier.[9] Acid hydrolase secretion is considerably slower than secretion of dense granules[10,11] and α-granule markers[9] and is not evoked by weak platelet agonists.[12] The samples referred to below represent the supernatants or noncentrifuged controls of the formaldehyde–EDTA-treated incubation mixtures of gel-filtered platelets.

Colorimetric Method

Principle

A glycosidase cleaves the *p*-nitrophenolglyceride and liberates free *p*-nitrophenol; after alkalinization the free *p*-nitrophenol can be determined by its absorbance at 410 nm (A_{410}).

Reagents: Preparation and Stability of Solutions. *p*-Nitrophenyl-*N*-acetyl-β-D-glucosaminide (substrate for β-hexosaminidase) and *p*-nitrophenyl-β-D-galactopyranoside (substrate for β-galactosidase) are made up to 20 m*M* in 0.15 *M* NaCl and may be stored in aliquots at −20° for at least 3 months. The substrates are not entirely soluble at this concentration and undissolved solids must be removed before use. *p*-Nitrophenol standard (10 m*M*, Sigma) and the 0.1 *M* citrate–phosphate buffers (pH 4.5 and 5.0) are stable at 4°.

[7] G. D. Vladutiu, C. A. Dangelmaier, V. Amigone, B. A. van Oost, and H. Holmsen, *Biochim. Biophys. Acta* **802,** 435 (1984).

[8] B. A. van Oost, J. B. Smith, H. Holmsen, and G. D. Vladutiu, *Proc. Natl. Acad. Sci. U.S.A.* **82,** 2375 (1985).

[9] H. Holmsen and C. W. Dangelmaier, this volume [16].

[10] H. Holmsen, C. A. Setkowsky, B. Lages, H. J. Day, H. J. Weiss, and M. C. Scrutton, *Blood* **46,** 131 (1975).

[11] See also Fig. 1, this volume [16], p. 200.

[12] H. Holmsen and S. Karpatkin, *in* "Hematology" (W. J. Williams, E. Beutler, A. J. Erslev, and M. A. Litchmanm, eds.), 3rd Ed., p. 1149. McGraw-Hill, New York, 1983.

Procedure

Treatment and Stability of Sample. When the platelets are gel filtered in an albumin-containing medium, the albumin should be checked for "enzyme activity." It is advisable to use crystalline human albumin which contains very little β-hexosaminidase and virtually no β-galactosidase.

After the secretion experiment has been performed, Triton X-100 is added to aliquots of supernatants and total (control) samples in a final concentration of 0.1%. The enzyme activities should be measured immediately after the solubilization step, as prolonged storage at $-20°$ of the formaldehyde-containing sample causes some decay of the enzyme activities.

Assay. Mix and prewarm at 37°: 100 μl of citrate–phosphate buffer (pH 5.0 for β-galactosidase and pH 4.5 for β-hexosaminidase) and 50 μl of 20 m*M* substrate. Start incubation by adding 150 μl of sample. Incubate at 37° for 150 min for β-galactosidase or 30 min for β-hexosaminidase. Stop the reaction by adding 250 μl of 0.08 *N* NaOH.

Read absorbance at 410 nm against a reagent blank with a composition identical to the incubation mixture, but "sample" is substituted with the Tyrode's buffer; the latter is incubated for the same time as the test sample. This will also represent the enzyme activity in the commerical albumin preparation.

Standard Curve. Mix together:

100 μl of suspension medium
50 μl of *p*-nitrophenol (0, 0.15, 0.3, 0.45, 0.60 m*M* in the Tyrode's buffer used for gel filtration)
100 μl of buffer (pH corresponding to enzyme activity)
50 μl substrate (corresponding to enzyme activity)
250 μl of 0.08 *N* NaOH

Read at 410 nm against a blank for *p*-nitrophenol containing Tyrode's buffer.

Calculations of Enzyme Activity. Calculate from standard curves $F = x/y$, where x is the millimolar concentration of the sample and y is the absorbance at 410 nm (A_{410}). In order to obtain enzyme activity (μmol/min) use factor (f):

$$f = F/\text{incubation time (minutes)}$$

$$\text{Total activity} = A_{410} \times f = \mu\text{mol/min/ml sample}$$

$$\text{Activity per } 10^{11} \text{ cells} = (A_{410} \times f)/[(\text{platelet count}/\mu\text{l})(10^{-5})](10^{3})$$

where "platelet count" refers to the number of platelets in 1 ml of sample.

Normal Values

When measured as described above the normal levels of β-hexosaminidase in human platelets is 4 ± 0.63 μmol/min/10^{11} cells and those of β-galactosidase is 0.37 ± 0.12 μmol/min/10^{11} cells.[1]

Calculation of Percentage Secretion

In the formula for percentage secretion given above[7] the unit terms for enzyme activity disappear, and the percentage secretion of acid glycosidase can simply be computed as

$$\text{Secretion (\%)} = (A_{410} \text{ in } S_{\text{test}} - A_{410} \text{ in } S_{\text{control}})/(A_{410} \text{ in } T_{\text{control}} - A_{410} \text{ in } S_{\text{control}})(100)$$

Fluorimetric Method

Principle

A glycosidase cleaves the 4-methylumbelliferone glycoside and free 4-methylumbelliferone is liberated; after alkalinization the freed 4-methylumbelliferone fluoresces at 450 nm when excited at 365 nm.

Preparation of Reagents and Stability of Solutions

In general, 4-methylumbelliferone glycosides are sparingly soluble in the acid buffers used. Saturated solutions of 4-methylumbelliferone *N*-acetylglucosaminide and 4-methylumbelliferone glucuronide are made in 0.1 *M* phosphate–citrate buffer at pH 4.0 and 5.0, respectively. These are kept at 4° overnight or at 37° for 30 min, and undissolved solids are removed by centrifugation. The clear supernatants are distributed in small aliquots and can be kept for at least a month at −20° without detectable decomposition (hydrolysis). However, repeated freezing and thawing of these solutions causes marked hydrolysis, and it is advisable to thaw one aliquot for each run and discard any unused, thawed substrate.

A standard of 4-methylumbelliferone (1 m*M* in water) may be stored for several months (usually, a few drops of 0.1 *N* NaOH must be added to the dissolved reagent).

Procedure

The formaldehyde-containing samples from the secretion experiment[7,8] are solubilized with 0.1% Triton X-100 (final concentration) and can be frozen overnight before assaying. However, enzyme activity decays slowly in the presence of formaldehyde (even while frozen) and it is advisable to measure the enzymatic activities immediately after Triton X-100 addition.

Assay. Incubation starts by addition of sample. Mix 75 μl of substrate in buffer prewarmed at 37° with 50 μl of sample. Incubate at 37° for 6 min (β-hexosaminidase) or 30 min (β-glucuronidase). Incubate 75 μl of substrate with 50 μl of H_2O for the respective times. Both mixtures are used as blanks and also serve to monitor the free 4-methylumbelliferone that is slowly formed during prolonged storage, as well as measures of nonenzymatic hydrolysis of the substrate. Stop the reactions by adding 1.0 ml of 0.2 *M* glycine–NaOH, pH 10.7. The fluorescence at 450 nm and excitation at 365 nm are determined.

Standard Curve. Standards include 0.2 ml of H_2O and aqueous solutions of 5, 10, 15, and 25 μ*M* 4-methylumbelliferone. They are mixed with 1.8 ml of glycine–NaOH, pH 10.7, and read as above.

Calculations of Total Enzyme Activity. Calculate from the standard curve $F = x/y$, where x is the micromolar concentration of the standard in the cuvette and y is the fluorescence reading.

To calculate the total activity (μmol/min/ml of the sample):

$$\text{Total activity} = F[(R_{\text{corr}} \times 1.125)/(0.05t \times 1000)]$$
$$= (22.75)(F/t)(R_{\text{corr}})(1/1000)$$

where R_{corr} is the fluorescence reading of the sample minus the fluorescence reading of the blank and t is the incubation time (in minutes).

To calculate the specific activity expressed as

(1) (μmol/min)/mg protein:

$$\text{Specific activity} = \text{total activity}/P$$

where P is protein (μg/ml).

(2) (μmol/min)/10^{11} platelets:

$$\text{Specific activity} = \text{total activity}/C \times 1000$$

where C is the platelet count/μl $\times 10^{-5}$.

Normal Values

When measured as described above, the normal levels in human platelets are 1.1 and 4.3 μmol/min/10^{11} platelets for β-glucuronidase and β-hexosaminidase, respectively.[1]

Calculation of Percentage Secretion

In the formula for percentage secretion given above[7] the unit terms for enzyme activity (but not the correction for the fluorescence of the blank) disappear. The percentage secretion can thus be computed:

$$\text{Secretion (\%)} = (R_{corr} \text{ in } S_{test} - R_{corr} \text{ in } S_{control}) / (R_{corr} \text{ in } T_{control} - R_{corr} \text{ in } S_{control})(100)$$

Accuracy and Precision. The precision of the measurement of secreted acid hydrolases depends on the platelet count and the background activities in the suspending medium, i.e., the albumin used. With crystalline human albumin and a platelet count of $1.5 \times 10^5/\mu l$ or more the precision of the measurement is $\pm 10\%$ or less, both for β-hexosaminidase and β-glucuronidase.

Adaptation for PRP. Acid hydrolases cannot be easily measured in PRP. The high background activities of the acid hydrolases in the plasma (which appear in the $S_{control}$ term in the formula) make the measurement grossly insensitive.

[30] Platelet Heparitinase

By ROBERT D. ROSENBERG

A series of specific exoglycosidases are responsible for the degradation of heparin and heparan sulfate within mammalian species.[1] These enzymes sequentially release inorganic sulfates and monosaccharide residues from the nonreducing ends of the above mucopolysaccharides.[2] This process is facilitated by the action of endoglycosidases that cleave internal glycosidic bonds and thereby increase the number of nonreducing terminals available to the exoglycosidases.[3] One of these endoglycosidases is present within

[1] E. Buddecke and H. Kresse, *in* "Connective Tissues: Biochemistry and Pathophysiology" (R. Fricke and R. Hartman, eds.), p. 131. Springer-Verlag, Berlin, Federal Republic of Germany, 1974.

[2] E. F. Neufeld and M. Cantz, *in* "Lysosomes and Storage Diseases" (H. G. Hers and F. Van Hoof, eds.), p. 262. Academic Press, London, 1973.

[3] B. Arbogast, J. Hopwood, and A. Dorfman, *Biochem. Biophys. Res. Commun.* **75,** 610 (1977).

human platelets and has been termed platelet heparitinase.[4] Platelet heparitinase has recently been purified to homogeneity.[5] This enzyme cleaves glucuronsyl–glucosamine bonds and thereby releases heparan sulfate chains from their attachments to core proteins which are integral components of the cell membrane. The action of the platelet heparitinase within the cardiovascular system may alter the nonthrombogenic properties of endothelial cells,[6] modulate the growth of the underlying smooth muscle cells,[7] and facilitate the extravasation of blood cells.[8] Thus, the above endoglycosidase could play a critical role in initiating thrombotic and atherosclerotic lesions within the circulatory tree. A related enzyme is also found within leukocytes, macrophages, activated T lymphocytes, and highly malignant tumor cells.[9-12] It has been suggested that the above endoglycosidase may be involved in the early events of extravasation and diapedesis of leukocytes and macrophages in response to inflammation at an extravascular locale,[9] permit activated T lymphocytes to penetrate the vascular basal lamina and to reach antigen-bearing sites within the body,[10] and determine the extent to which tumor cells may distribute to specific organs during metastatic spread.[11,12]

Assay Method

Principle

Platelet heparitinase catalyzes the cleavage of certain glucuronsyl–glucosamine bonds within heparin and heparan sulfate, but does not scission other glycosaminoglycan chains. Platelet heparitinase is routinely assayed by quantitating the liberation of ^{125}I-labeled heparin fragments from ^{125}I-labeled heparin–Sepharose 4B.[5] Alternately, the activity of the above enzyme can be measured by determining the extent of release of

[4] A. Oldberg, C. Helden, A. Wasteson, C. Busch, and M. Hook, *Biochemistry* **19,** 5755 (1980).

[5] G. M. Oosta, L. V. Favreau, D. L. Beeler, and R. D. Rosenberg, *J. Biol. Chem.* **257,** 11249 (1982).

[6] J. A. Marcum, J. B. McKenney, and R. D. Rosenberg, *J. Clin. Invest.* **74,** 1 (1984).

[7] J. J. Castellot, Jr., L. V. Favreau, R. D. Rosenberg, and M. J. Karnovsky, *J. Biol. Chem.* **257,** 11256 (1982).

[8] J. Yahalom, A. Eldor, Z. Fuks, and I. Vlodavsky, *J. Clin. Invest.* **74,** 1842 (1984).

[9] Y. Matzner, M. Bar-Ner, J. Yahalom, R. Ishai-Michaeli, and I. Vlodavsky, *J. Clin. Invest.* **76,** 1306 (1985).

[10] Y. Naparstek, I. R. Cohen, Z. Fuks, and I. Vlodavsky, *Nature (London)* **310,** 241 (1984).

[11] I. Vlodavsky, Z. Fuks, M. Bar-Ner, Y. Ariav, and V. Schirrmacher, *Cancer Res.* **43,** 2704 (1983).

[12] M. Nakajima, D. R. Welch, T. Irimura, and G. L. Nicolson, *in* "Cancer Metastasis: Experimental and Clinical Strategies" (M. Nakajima, ed.), p. 113. Liss, New York, 1986.

small oligosaccharides from radiolabeled heparan sulfate which cannot be precipitated with cetylpyridinium chloride (see alternate assay method).

Reagents

Sodium acetate, 0.1 *M*, pH 5.0

^{125}I-Labeled heparin–Sepharose 4B: Crude heparin containing the carbohydrate linkage region and the covalently attached polypeptide was purchased from Diosynth, Chicago, Illinois and purified by three successive precipitations at 1.2 *M* NaCl with cetylpyridinium chloride.[13] The anticoagulant potency of this product (CPC heparin) averaged 183 USP U/mg. To label CPC heparin with ^{125}I, *N*-succinimidyl-3-(4-hydroxyphenyl)propionate was initially coupled to the mucopolysaccharide by mixing 150 μl of the derivatizing agent at 6.67 mg/ml dissolved in 66% ethanol with 200 μl of heparin (15 mg/ml) dissolved in 0.05 *M* sodium borate, pH 9.5. After maintaining the above reaction mixture at 0° for 3 hr, visible precipitate was removed by centrifugation at 7000 *g* for 5 min and excess derivatizing agent was eliminated by exhaustive dialysis versus distilled water. The derivatized heparin was labeled with ^{125}I (100 μCi) mixed with 40 μl of the mucopolysaccharide in 50 μl of 0.05 *M* sodium phosphate, pH 7.5. After 1–2 min of incubation, 300 μg of sodium metabisulfate in 50 μl of 0.05 *M* sodium phosphate, pH 7.5, was introduced with subsequent addition of 50 μl of 0.21 *M* potassium iodide. The entire mixture was then diluted to approximately 10 ml with distilled water and applied to a DEAE-cellulose column (1.3×2 cm) equilibrated with 0.01 *M* Tris–HCl, pH 7.5. The chromatographic matrix was washed with 50 ml of distilled water and ^{125}I-labeled heparin was eluted with 3 *M* NaCl in 0.01 *M* Tris–HCl, pH 7.5. The radiolabeled product was dialyzed exhaustively against distilled water, concentrated to about 1 ml by rotary evaporation, and then dialyzed against 0.1 *M* $NaHCO_3$, pH 8.5. ^{125}I-labeled heparin–Sepharose 4B was prepared by mixing 40 μg of ^{125}I-labeled heparin (see above) dissolved in 0.1 *M* $NaHCO_3$, pH 8.5, with 5 ml of "packed" Sepharose 4B that had been activated with cyanogen bromide according to the method of Porath *et al.*[14] After maintaining the above slurry at 0° for 16–18 hr, unreacted sites were subsequently blocked by adding 1 ml of ethanolamine at 0° for 4 hr. The final product was washed exhaustively with 3 *M* NaCl in 0.01 *M* Tris–HCl, pH 7.5, and then equilibrated with 0.1 *M* NaCl. This matrix typically contained about 1×10^7 cpm of ^{125}I-labeled heparin/ml of "packed" gel.

[13] U. Lindahl, J. A. Cifonelli, B. Lindahl, and L. Roden, *J. Biol. Chem.* **240**, 2817 (1965).

[14] J. Porath, R. Axen, and S. Ernback, *Nature (London)* **215**, 1491 (1967).

Procedure

The assay of platelet heparitinase was initiated by adding 100 μl of a given concentration of enzyme to 400 μl of a solution containing 20,000 cpm of substrate. The buffer utilized in the dilution of platelet endoglycosidase and suspension of ^{125}I-labeled heparin–Sepharose was 0.1 *M* sodium acetate, pH 5.0. The reaction mixture was stirred by continuous rotation at 24° for 30–60 min, and then centrifuged at 3300 *g* for 5 min at 40°. The supernatant was carefully drawn off and 150 μl of this fluid was counted for ^{125}I. Each assay measurement was conducted in duplicate and each assay sequence included a series of standard dilutions of purified platelet heparitinase (step VIII, see below Purification Procedure). Plots of enzyme concentration versus liberated ^{125}I-labeled heparin were linear with a correlation coefficient of 0.993 until about 25% of the mucopolysaccharide had been released. All determinations should be conducted within the linear region of the titration curve. One unit of platelet endoglycosidase is arbitrarily defined as that amount of enzyme which releases 3×10^6 cpm of ^{125}I-labeled heparin/ml/hr.

Alternate Assay Method

The second assay that has been utilized to quantitate the activity of platelet heparitinase is based upon the decreased precipitability with cetylpyridinium chloride of radiolabeled heparan sulfate as compared to undegraded polysaccharide.[4] The standard incubation mixture containing 25 μl of enzyme, 125 μl of 75 m*M* 2-(*N*-morpholino)ethanesulfonic acid (MES), pH 6.0, and 5000 cpm of [^{3}H]heparan sulfate was incubated at 37° for 2 hr. The radiolabeled polysaccharide was prepared by chemical ^{3}H-acetylation of free amino groups either directly or immediately upon partial deacetylation.[15] After addition of 50 μg of heparin in 50 μl of water as carrier, the samples were digested with papain; 200 μl of 0.8 *M* NaCl and 0.1 *M* acetate buffer, pH 5.5, containing 0.75 mg of papain was added and the mixtures were incubated at 60° for 6 hr. Polysaccharide within the papain digest were precipitated by adding 100 μl of 5% cetylpyridinium chloride and incubating the samples at 37° for 30 min. The final concentration of NaCl of 0.32 *M* was chosen because prior experiments indicated that a major portion of radiolabeled heparan sulfate can be precipitated at this ionic strength with cetylpyridinium chloride. It may be necessary to slightly modify the concentration of NaCl depending on the polysaccharide preparation employed as substrate. After centrifugation at 2000 *g* for

[15] M. Hook, J. Resenfeld, and U. Lindahl, *Biochem. Biophys. Res. Commun.* **120,** 365 (1986).

10 min at 37°, the ^3H radioactivity in the supernatant was determined. The amount of nonprecipitated and therefore degraded radiolabeled oligosaccharide should show a linear increase with the amount of enzyme added in the range of 35–85% of ^3H radioactivity recovered in the supernatant.[4,5]

Purification Procedure

Purification of Platelet Heparitinase

Platelet heparitinase was isolated from outdated human platelets by freeze–thaw solubilization of the enzyme (step I), heparin–Sepharose chromatography (step II), DEAE-cellulose chromatography (step III), hydroxylapatite chromatography (step IV), octyl-agarose chromatography (step V), concanavalin A-Sepharose chromatography (step VI), initial Sephacryl S-200 gel filtration (step VII), and final Sephacryl S-200 gel filtration (step VIII).

The individual steps have been performed 10 to 20 times and have not deviated markedly from the average specific activity and recovery estimates provided in Table I. All buffers contained 1 m*M* sodium azide as a bacteriocidal agent.

Step I: Solubilization of Platelet Heparitinase. Approximately 150 units of outdated human platelets stored at −20° were rapidly thawed at 37°, acidified to pH 5.0 with glacial acetic acid, and centrifuged at 8000 *g* for 30 min at 4°. Direct measurement of the resultant supernatant solution revealed that significant amounts of heparin-cleaving activity had been liberated from the above cellular elements. Homogenization of the residual platelet pellet with a Waring blender for 1 min in the presence or absence of 0.1% (w/v) Triton X-100 detergent released less than 10% additional heparin-cleaving activity.

Step II: Heparin–Sepharose Chromatography. Approximately 800 ml of the step I product containing about 411,000 A_{280} units of protein were filtered through a heparin–Sepharose column (5.1 × 33 cm) that had been equilibrated with 0.15 *M* NaCl in 0.01 *M* sodium acetate–acetic acid, pH 5.0. The chromatographic matrix was then washed with 650 ml of 0.45 *M* NaCl in 0.01 *M* sodium acetate–acetic acid, pH 5.0. Subsequently, the platelet heparitinase was eluted with a linear salt gradient that employed a reservoir containing 2 liters of 1.1 *M* NaCl in 0.01 *M* sodium acetate–acetic acid, pH 5.0, and a mixing chamber containing 2 liters of 0.45 *M* NaCl in 0.01 *M* sodium acetate–acetic acid, pH 5.0. Chromatography was conducted at 4° with flow rates maintained at 100 ml/hr. During a typical fractionation, the platelet endoglycosidase eluted over a range of 0.4 *M* NaCl with the peak of activity centered at an added NaCl concentration of

TABLE I

PURIFICATION OF HUMAN PLATELET HEPARITINASE[a]

Step and procedure	Protein (mg)	Activity (milliunits)	Specific activity (milliunits/mg)	Recovery (%)	Purification (-fold)
I. Freeze–thaw	2,469,990	112,947	0.046	100	
II. Heparin–Sepharose	5,321	84,710	15.9	75	346
III. DEAE-cellulose	2,640	59,129	22.4	52.4	487
IV. Hydroxylapatite	427	46,236	108.3	40.9	2,354
V. Octyl-agarose	126	36,545	290	32.4	6,304
VI. Concanavalin	3.96	22,484	5,678	19.9	123,435
VII. First Sephacryl S-200	1.64	16,539	10,085	14.6	219,239
VIII. Second Sephacryl S-200	0.570	6,285	11,026	5.6	239,696

[a] These data represent average results obtained employing 900 units of outdated human platelets.

0.73 *M* NaCl. Samples possessing specific potencies of 10 milliunits/mg or greater were pooled. When stored at 4°, step II products were stable for 3 to 24 weeks. Prior to step III, step II preparations were concentrated by ultrafiltration from about 1150 to 200 ml using Amicon PM30 membranes, and dialyzed extensively against 0.15 *M* NaCl in 0.01 *M* Tris–HCl, pH 7.2.

Step III: DEAE-Cellulose Chromatography. The step II preparation containing about 887 A_{280} units of protein was filtered at 20 ml/hr through a DEAE-cellulose column (2.8 × 30 cm) equilibrated at 4° with 0.15 *M* NaCl in 0.01 *M* Tris–HCl, pH 7.2. The chromatographic matrix was then washed with an additional 190 ml of the above solution at the same flow rate and temperature. Under these conditions, the platelet heparitinase does not bind to DEAE-cellulose and can be recovered in the initial column effluent. This preparation was concentrated by ultrafiltration from about 200 to 10 ml using PM30 membranes and then stored at −80° prior to use. In this state, products were stable for at least 4 weeks.

Step IV: Hydroxylapatite Chromatography. Three step III preparations were pooled and then dialyzed extensively against 0.15 *M* NaCl in 0.01 *M* sodium phosphate, pH 6.0. The resultant product obtained from 450 units of outdated platelets was subdivided into two equivalent portions. Each half of the preparation consisting of about 660 A_{280} units of protein was individually filtered at 10 ml/hr through a hydroxylapatite column (2.5 × 28 cm) that had been equilibrated at 4° with 0.15 *M* NaCl in 0.01 *M* sodium phosphate, pH 6.0. The platelet heparitinase was eluted with a linear salt gradient that utilized a reservoir containing 1000 ml of 0.15 *M* NaCl in 0.60 *M* potassium phosphate, pH 6.0, and a mixing chamber containing 1000 ml of 0.15 *M* NaCl in 0.01 *M* potassium phosphate, pH 6.0. Chromatography was conducted at 4° and flow rates were maintained at 40 ml/hr. The platelet enzyme eluted over a range of 0.15 *M* potassium phosphate with the peak of mucopolysaccharide-cleaving activity centered at an added potassium phosphate concentration of 0.31 *M*. In all instances, fractions exhibiting specific activities of 70 milliunits/mg or greater were pooled. The two separate hydroxylapatite products were pooled, concentrated by ultrafiltration from about 600 to about 50 ml utilizing PM30 membranes, and stored at −80° prior to use. Under these conditions, preparations of platelet endoglycosidase were stable for at least 12 weeks.

Step V: Octyl-Agarose Chromatography. The step IV preparation obtained from 450 units of outdated platelets was subdivided into three equivalent portions. Each of these products containing about 71 A_{280} units of protein was individually filtered at 10 ml/hr through an octyl-agarose column (1.4 × 11 cm) that had been equilibrated at 24° with 0.15 *M* NaCl in 0.60 *M* potassium phosphate, pH 6.0. The columns were washed with the above buffer at the same flow rate and temperature until the absorbance readings reached baseline levels. Under these conditions, the platelet heparitinase does not bind to the octyl-agarose and can be recovered in the initial column wash. The three separate octyl-agarose preparations were pooled, concentrated by ultrafiltration from about 120 to about 10 ml utilizing PM30 membranes, and stored at −80° prior to use. In this state, products were stable for at least 8 weeks.

Step VI: Concanavalin A-Sepharose Chromatography. The step V preparation obtained from 450 units of outdated platelets and consisting of about 63.0 A_{280} units of protein was filtered at 2 ml/hr through a concanavalin A-Sepharose column (1.25 × 2.5 cm) that had been equilibrated at 4° with 0.15 *M* NaCl in 0.01 *M* sodium phosphate, pH 6.0. The lectin-Sepharose matrix was washed with an additional 10 to 15 ml of the above solution at the same flow rate as well as temperature, and bound platelet heparitinase was harvested with 6 ml of 1 *M* glucose and 0.5 *M* NaCl in 0.01 *M* sodium phosphate, pH 6.0. Subsequently, 10 ml of the above eluting solution was mixed with the chromatographic matrix and the resultant suspension was maintained at 4° for 12 hr. At the end of this period of time, the supernatant liquid was carefully drawn off and then examined for mucopolysaccharide-cleaving activity. The above fluid was usually found to contain significant amounts of platelet endoglycosidase and was therefore combined with enzyme initially eluted from the concanavalin A-Sepharose column. The step VI preparation was concentrated by ultrafiltration from 16 ml to ~700 μl with PM10 membranes and stored at 4° prior to use. Under these conditions, the above products were stable for at least 2 weeks.

Step VII: Initial Sephacryl S-200 Gel Filtration. The step VI preparation isolated from 450 units of outdated platelets and containing about 2.0 A_{280} units of protein was further purified by gel filtration. To this end, 700 μl of this material which represented an entire step VI product was chromatographed on a Sephacryl S-200 column (0.6 × 128 cm) that had been equilibrated at 4° with 0.5 *M* NaCl in 0.01 *M* Tris–HCl, pH 7.5. The filtration procedure was conducted at a flow rate of 3 ml/hr. Two discrete protein peaks were observed with the component of lower molecular size exhibiting the bulk of the mucopolysaccharide-cleaving activity. As indicated, fractions with specific activities of 9500 milliunits/mg or greater were pooled. These products were stored at −80° prior to use and were stable for at least 3 weeks.

Step VIII: Final Sephacryl S-200 Gel Filtration. In order to remove trace amounts of impurities, step VII preparations were rechromatographed on Sephacryl S-200. To this end, two of the step VII products obtained from 900 units of outdated platelets and containing about

1.6 A_{280} units of protein were concentrated by ultrafiltration from about 12 ml to about 700 μl with PM10 membranes. This material was chromatographed on Sephacryl S-200 exactly as previously described. During a typical separation, the peak of mucopolysaccharide-cleaving activity coincides with the peak of protein. Fractions with specific potencies of at least 10,500 milliunits/mg were pooled and stored at $-80°$ prior to use.

A summary of the purification procedures is given in Table I. Platelet heparitinase is purified about 240,000-fold with an approximate yield of 6%. The purified enzyme is homogeneous as judged by polyacrylamide gel electrophoresis at acidic pH.

Properties

Stability. Homogeneous preparations of platelet heparitinase are stable at 0° for several weeks and may be stored at $-80°$ indefinitely.

Physicochemical Properties. Platelet heparitinase exhibits a molecular weight of 134,000 as judged by gel filtration under nondissociating conditions. Furthermore, the enzyme is apparently composed of a single polypeptide chain with no subunit structure.[5]

Specific Enzymatic Properties of Platelet Heparitinase. The mucopolysaccharide-cleaving potency of platelet heparitinase has been analyzed as a function of pH with the final enzyme concentration set at 10 milliunits/ml diluted in 0.1 mg/ml of albumin and the final ionic strength maintained at 0.1. The results of this study indicate that the platelet endoglycosidase is maximally active over a broad pH ranging from 5.5 to 7.5. Furthermore, the above enzyme possesses minimal ability to cleave heparin at pH values less than 4.0 or greater than 9.0.[5] The proteolytic activity of the step VIII product has also been examined with the ^{125}I-labeled α-casein assay. At concentrations as high as 50 μg/ml, the platelet enzyme exhibited no capacity to cleave radiolabeled casein.

The substrate specificity of platelet heparitinase has been determined by identifying susceptible linkages within the heparin molecule that can be cleaved by the platelet component. Our studies demonstrate that the platelet heparitinase is only able to hydrolyze glucuronsyl–glucosamine linkages. Furthermore, our investigations of the structure of the disaccharide which lies immediately on the nonreducing end of the cleaved glucoronic acid residue indicate that N-sulfation of the glucosamine moiety or ester sulfation of the iduronic acid group are not essential for bond scission.[5] These results confirm as well as extend observations on substrate specificity made by Oldberg *et al.*[4] utilizing preparations that were 1000-fold less pure than the enzyme employed in our studies.

Subcellular Localization of Platelet Heparitinase. In order to establish the subcellular site of localization of the above enzyme, washed human platelets were homogenized using the nitrogen cavitation technique. Membranes, as well as organelles, were separated by sucrose density gradient ultracentrifugation. The various fractions obtained by this procedure were comparable to those previously described by Broekman *et al.*[16] It was apparent that the relative specific activities of β-glucuronidase and β-*N*-acetylglucosaminidase were maximal in fraction 4 (mitochondria and lysosomes), whereas that of platelet factor 4 was highest in fraction 7 (α granules). The distribution profile of the relative specific activity of platelet heparitinase was virtually identical with that of the acid hydrolases but distinct from that of platelet factor 4. The lysosomal localization of the above enzyme suggests that release of the platelet endoglycosidase will demand a higher concentration of platelet agonists than is required for secretion of proteins stored in the α granules.

This hypothesis is compatible with our observation that addition of small amounts of thrombin to platelet-rich plasma results in the discharge of only $\sim 10\%$ of the available endoglycosidase but permits virtually complete liberation of the α-granule component thrombospondin.[5] Furthermore, the above supposition is also strengthened by the data of Oldberg *et al.*,[4] which indicate that release of platelet endoglycosidase is more readily attained with thrombin or collagen than with epinephrine or ADP. It is of interest to note that human platelets possess relatively small amounts of this enzyme. If one takes into account the number of platelets employed as starting material as well as the final yield of the heparitinase obtained, it is possible to calculate that each platelet must contain about 1000 molecules of this endoglycosidase. This level of enzyme is considerably less than that noted for other platelet constituents such as β-thromboglobulin, platelet factor 4, or thrombospondin,[17-19] but is similar to the quantities of platelet-derived growth factor found within the α granules.[20]

[16] M. J. Broekman, N. P. Westmoreland, and P. Cohen, *J. Cell Biol.* **60,** 507 (1974).
[17] S. Moore, D. S. Pepper, and D. Cashet, *Biochim. Biophys. Acta* **379,** 360 (1975).
[18] R. I. Handin and H. J. Cohen, *J. Biol. Chem.* **251,** 4273 (1976).
[19] S. S. Morgossian, J. W. Lawler, and H. S. Slayter, *J. Biol. Chem.* **256,** 7495 (1981).
[20] T. F. Duel, J. S. Huang, R. T. Proffitt, J. U. Baenziger, D. Chang, and B. B. Kennedy, *J. Biol. Chem.* **256,** 8896 (1981).

Section IV

Regulation of Platelet Function

[31] Interaction of Extracellular Calcium with the Surface Membrane of Human Platelets

By LAWRENCE F. BRASS and SANFORD J. SHATTIL

Introduction

Extracellular Ca^{2+} is required for several aspects of human platelet function. For example, the binding of fibrinogen to the glycoprotein IIb–IIIa complex on the platelet surface requires Ca^{2+}, as does platelet secretion in response to agonists such as ADP and epinephrine. In theory, extracellular Ca^{2+} could be involved in these events either by interacting with extracellular proteins, such as fibrinogen, or by binding to and altering the function of components of the platelet surface membrane. Alternatively, extracellular Ca^{2+} could enter the platelet and influence intracellular, Ca^{2+}-dependent reactions. Both the binding of Ca^{2+} to fibrinogen and the role of cytoplasmic Ca^{2+} in the platelet secretory process have been studied in some detail. In contrast, relatively little is known about the role of Ca^{2+} bound to the platelet surface membrane. In this chapter we will describe methods for (1) quantitating surface-bound Ca^{2+} on unstimulated and stimulated platelets, (2) identifying some of the membrane components responsible for this binding, and (3) ascertaining the role of this bound Ca^{2+} in platelet function.

Platelets contain approximately 20 nmol of $Ca^{2+}/10^8$ cells, most of which is confined to the dense storage granules.[1] The remaining Ca^{2+} appears to be either sequestered within intracellular organelles, such as the dense tubular system and mitochondria, or bound to membranes and cytoplasmic proteins. Thus, despite the fact that platelets are normally bathed in a plasma milieu rich in Ca^{2+}, the Ca^{2+} that is associated with the platelet surface membrane represents only a small fraction of total platelet Ca^{2+}. In order to measure such a relatively small pool of Ca^{2+}, we have taken advantage of the sensitivity of $^{45}Ca^{2+}$ tracer methodology. The general approach is to separate platelets from their plasma milieu by gel filtration and then incubate the platelets under isotonic conditions with $^{45}CaCl_2$ over a range of extracellular free Ca^{2+} concentrations. The amount of tracer associated with the platelet is then determined. Since extracellular

[1] J. M. Gerrard, D. A. Peterson, and J. G. White, *Platelets Biol. Pathol.* **2,** 407 (1981).

$^{45}Ca^{2+}$ exchanges not only with surface-bound Ca^{2+} but also with several pools of intracellular Ca^{2+},[2] a distinction must then be made between surface-bound $^{45}Ca^{2+}$ and intracellular $^{45}Ca^{2+}$.

Measurement of Surface-Bound Ca^{2+} on Unstimulated Platelets

Preparation of Platelets

Since manipulation of platelets can damage them and result in platelet shape change, aggregation, and secretion, the method of transferring platelets from plasma to incubation buffer must be relatively gentle and the platelets should be monitored for evidence of damage. Although several methods of transfer are available, we have found that gel filtration produces the most satisfactory results.

1. Fresh blood from normal donors is anticoagulated with 13 m*M* sodium citrate. Platelet-rich plasma is obtained by differential centrifugation of the blood at 180 *g* for 15 min. The platelets are then incubated with 1 m*M* aspirin for 30 min at room temperature.[2] Aspirin is used to block cyclooxygenase and thereby help prevent the major Ca^{2+} fluxes that occur across the platelet surface during secretion. When measurement of platelet serotonin release is to be carried out, the platelet-rich plasma is also incubated with [^{14}C]serotonin, 0.2 μCi/ml (New England Nuclear, Boston, MA), and then imipramine (1 μM) is added just before gel filtration to inhibit serotonin reuptake.[3,4]

2. Platelets are separated from each 5–7 ml of plasma by gel filtration on a 40- to 50-ml Sepharose 2B column using an elution buffer containing 137 m*M* NaCl, 2.7 m*M* KCl, 1 m*M* $MgCl_2$, 5.6 m*M* glucose, 1 mg/ml bovine serum albumin, 3.3 m*M* NaH_2PO_4, and 20 m*M* HEPES, pH 7.4.[2,5] This and all other buffers are prepared using distilled water that has been passed over a Chelex 100 ion-exchange resin (Bio-Rad, Richmond, CA). This reduces the residual free Ca^{2+} in the final buffer to approximately 5 μM. Platelets obtained by gel filtration should retain the visible "swirl" characteristic of discoid platelets in suspension, and they should not form microscopic or macroscopic aggregates, either spontaneously or when stirred in an aggregometer cuvette. On the other hand, after the addition of 100 μg/ml of fibrinogen, these platelets should undergo primary aggregation when stirred in the presence of 0.2–10 μM ADP or epinephrine.

[2] L. F. Brass and S. J. Shattil, *J. Biol. Chem.* **257,** 14000 (1982).
[3] Z. Jerushalmy and M. B. Zucker, *Thromb. Diath. Haemorrh.* **15,** 413 (1966).
[4] P. N. Walsh and G. Gagnatelli, *Blood* **44,** 157 (1974).
[5] L. F. Brass and S. J. Shattil, *J. Clin. Invest.* **73,** 626 (1984).

Incubation Conditions for Ca^{2+}-Binding Assay

1. Incubations of the gel-filtered platelets are carried out at 22° in polypropylene tubes. To avoid the variable temperatures of "room temperature" incubation, the tubes can be incubated at 22° in a temperature-controlled water bath. As described in detail in the next section, EGTA (0.5 mM) and $^{45}CaCl_2$ (1 – 1500 μM) are added to the gel-filtered platelets to obtain final extracellular free Ca^{2+} concentrations ranging from 10^{-9} to 10^{-3} M. Previous $^{45}Ca^{2+}$ exchange studies have indicated that surface-bound Ca^{2+} exchanges with extracellular Ca^{2+} with a $t_{1/2}$ of < 1 min, which is at least 20 times faster than the most rapidly exchangeable of the intracellular Ca^{2+} pools.[2,6] Therefore, in studies of surface-bound Ca^{2+}, isotopic equilibrium can be assumed to be present after 5 to 10 min. Additional time serves only to increase the amount of tracer that enters the cells. In the standard assay, unstirred platelets are incubated with the EGTA and $^{45}CaCl_2$ for 10 min in a total reaction volume of 3.0 ml. After the 10 min, three 400-μl aliquots from each tube are removed and layered on top of 0.5 ml of silicone oil ["550 fluid," 80 parts; "200 fluid," 20 parts (v/v); William Nye, Inc., New Bedford, MA) in a conical bottom microcentrifuge tube. The binding reaction is stopped by sedimenting the platelets through the oil at 12,000 g for 4 min in a microcentrifuge. At the same time, excess EGTA (3 mM final concentration) is added to the platelets remaining in the reaction tube and an additional three 400-μl aliquots are removed and sedimented through silicone oil. Since multiple reaction tubes are processed simultaneously, 4 to 5 min typically elapse between the time the EGTA is added to the reaction tubes and the time the platelets are sedimented. This is sufficient time for $^{45}Ca^{2+}$ to be removed from the platelet surface, but not enough time for significant loss of intracellular $^{45}Ca^{2+}$($t_{1/2} > 20$ min). Aliquots of the remaining platelet suspension are used to measure the specific activity. As described in the section Definition of Surface-Bound Ca^{2+}, the amount of surface-bound Ca^{2+} is derived from the difference in the values for platelet-associated Ca^{2+} that are obtained when the samples are sedimented with and without excess EGTA.

2. Following centrifugation of the platelets through the silicone oil, the supernatant fluid is aspirated, each tube is placed upside down in a test tube rack for at least 5 min, and the inside of the tube is swabbed with a cotton-tipped applicator to remove traces of $^{45}Ca^{2+}$ not associated with the platelet pellet. The tip of the tube containing the pellet is separated from the main body of the tube with a dog nail cutter and placed in a scintillation vial. The platelet pellet is then digested by incubation overnight at 45°

[6] L. F. Brass, *J. Biol. Chem.* **259,** 12563 (1984).

with 0.75 ml Protosol (New England Nuclear). Afterward, 10 ml of nonaqueous scintillation cocktail is added and the $^{45}Ca^{2+}$ is quantitated in a scintillation counter.

Setting and Measuring the Free Ca^{2+} Concentration

Membrane proteins that bind Ca^{2+} in a specific and saturable manner may have dissociation constants less than 1 μM. To examine the binding of Ca^{2+} to such proteins, the extracellular free Ca^{2+} concentration must be reduced to levels that are less than the concentration of residual Ca^{2+} in the water used to prepare the gel filtration buffer. A number of Ca^{2+} buffers can be used to accomplish this task, including EGTA, EDTA, HEDTA, and nitrilotriacetic acid.[7] We chose EGTA because it is a good Ca^{2+} buffer at physiological pH in the range of free Ca^{2+} concentrations that are of particular interest in these studies (10^{-8} to 10^{-6} M) and because it is relatively insensitive to the presence of Mg^{2+} in the medium.

Extracellular free Ca^{2+} concentrations $> 10^{-5}$ M can be measured using any of the commercially available Ca^{2+}-specific electrodes that have a linear response in this range. Such electrodes can also be used to measure the dissociation constant for Ca^{2+}-EGTA under the actual experimental conditions. For Ca^{2+} concentrations $< 10^{-5}$ M, metallochromic indicators, such as arsenazo III, or fluorescent Ca^{2+} binders, such as Quin 2 or Fura 2 (in the free acid form), can be used to measure the free Ca^{2+} concentration. Both techniques, however, require calibration using solutions in which known low free Ca^{2+} concentrations are produced by combining $CaCl_2$ with a Ca^{2+} buffer such as EGTA.

Methods for calculating the free Ca^{2+} concentration in solutions containing EGTA are discussed in detail elsewhere[8,9] and will be described only briefly here. EGTA binds Ca^{2+} and, to a lesser extent, Mg^{2+}. Since EGTA also binds H^+, the absolute association constants for the binding of Ca^{2+} and Mg^{2+} to EGTA must be corrected for the pH (or, more accurately, the H^+ activity of the solution). As adapted from Fabiato and Fabiato,[8]

$$K'_{Ca} = (K_{Ca-EGTA} + K_{Ca2+EGTA}*K_1*H)/SUM \quad (1)$$

$$K'_{Mg} = (K_{Mg-EGTA} + K_{Mg2-EGTA}*K_1*H)/SUM \quad (2)$$

$$SUM = 1 + K_1*H + K_1K_2*H^2 + K_1K_2K_3*H^3 + K_1K_2K_3K_4*H^4 \quad (3)$$

[7] A. C. H. Durham, *Cell Calcium* **4,** 33 (1983).

[8] A. Fabiato and F. Fabiato, *J. Physiol. (Paris)* **75,** 463 (1979).

[9] H. Portzehl, P. C. Caldwell, and J. C. Ruegg, *Biochim. Biophys. Acta* **79,** 581 (1964).

where K'_{Ca} and K'_{Mg} are the apparent association constants for Ca^{2+} and Mg^{2+} binding to EGTA at a particular pH, H is the H^+ concentration, and $K_{Ca-EGTA}$, $K_{Ca2-EGTA}$, $K_{Mg-EGTA}$, and $K_{Mg2-EGTA}$ are the absolute association constants for the binding of Ca^{2+} and Mg^{2+} to EGTA. K_1, K_2, K_3, and K_4 are the association constants for H^+ binding to EGTA. Values for the absolute association constants have been obtained experimentally and have varied somewhat with the method used and the temperature and composition of the solution. We have used the values published by Sillen and Martell[10] and, more recently, the revised values of Fabiato.[11] Expressed as logs, these values are $K_{Ca-EGTA} = 10.95$, $K_{Ca2-EGTA} = 5.33$, $K_{Mg-EGTA} = 5.21$, $K_{Mg2-EGTA} = 3.37$, $K_1 = 9.58$, $K_2 = 8.97$, $K_3 = 2.80$, and $K_4 = 2.12$. The free Ca^{2+} and Mg^{2+} concentrations are calculated using Eqs. (4) and (5):

$$[Ca]_f = [Ca]_{tot}/(1 + K'_{Ca}[EGTA]_f) \quad (4)$$

$$[Mg]_f = [Mg]_{tot}/(1 + K'_{Mg}[EGTA]_f) \quad (5)$$

where $[EGTA]_f$, the free EGTA concentration, is calculated from Eq. (6):

$$[EGTA]_f = [EGTA]_{tot}/(1 + K'_{Ca}[Ca]_f + K'_{Mg}[Mg]_f) \quad (6)$$

For most purposes, two approximations can be used. The first is that the free Ca^{2+} concentration is unaffected by the presence in the buffer of PO_4^{3-}, which has a much lower affinity for Ca^{2+} than does EGTA ($10^{1.70}$). The second approximation is that the free Mg^{2+} concentration in the buffer is sufficiently close to the total Mg^{2+} concentration that the latter can be substituted for the former, permitting the free Ca^{2+} concentration to be determined from the positive root of a quadratic equation involving only three known species: the total Ca^{2+}, Mg^{2+}, and EGTA concentrations. Alternatively, Eqs. (4), (5), and (6) can be solved iteratively using a computer. The latter approach is required if additional Ca^{2+}-binding species, such as ATP, oxalate, or a second Ca^{2+} buffer, are present. In our studies, the free Ca^{2+} concentrations are calculated using a program called FREECAL, which has been written in BASIC for use on the Apple Macintosh microcomputer.[12] Examples of the final free Ca^{2+} concentrations resulting from the addition of various amounts of $CaCl_2$ to the incubation buffer are shown in Table I.

As a further refinement, the real association constant for Ca^{2+} binding to EGTA under the conditions of pH, temperature, and solution composi-

[10] L. G. Sillen and A. E. Martell, Special Publication No. 25. The Chemical Society, Burlington House, London, 1971.

[11] A. Fabiato, *J. Gen. Physiol.* **78,** 457 (1981).

[12] The SCATCH and FREECAL programs are available from Dr. Brass upon request.

TABLE I
CALCULATED FREE Ca^{2+} CONCENTRATIONS IN THE PLATELET INCUBATION BUFFER[a]

Total Ca^{2+} (μM)	Calculated free Ca^{2+} (M)
20	2.87×10^{-9}
60	9.40×10^{-9}
100	1.72×10^{-8}
140	2.68×10^{-8}
180	3.88×10^{-8}
220	5.42×10^{-8}
260	7.47×10^{-8}
300	1.03×10^{-7}
340	1.47×10^{-7}
380	2.18×10^{-7}
420	3.61×10^{-7}
460	7.77×10^{-7}
500	5.62×10^{-6}

[a] The calculated free Ca^{2+} concentrations were computed by the FREECAL computer program with the following inputs: EGTA, 0.5 mM; $MgCl_2$, 1 mM; NaH_2PO_4, 3.3 mM; and NaCl, 137 mM; 2.7 mM KCl, pH 7.4; $\log(K'_{Ca}) = 7.19$, $\log(K'_{Mg}) = 1.95$.

tion used for the binding studies can be measured. As a check on the value calculated for K'_{Ca} using Eq. (1), we have measured the binding of Ca^{2+} to EGTA in our incubation buffer using the method of Bers.[13] In this method, the free Ca^{2+} concentration obtained by combining Ca^{2+} and EGTA is measured across the range in which the Ca^{2+} electrode is linear (10^{-6} to 10^{-3} M). The electrode is calibrated using reference Ca^{2+} standards without EGTA. K'_{Ca} is derived from the linear plot of $[Ca - EGTA]/[Ca]_f$ vs $[Ca - EGTA]$. The value of $10^{7.04}$ that we obtained by this method for K'_{Ca} at pH 7.3 is in good agreement with the calculated value of $10^{7.00}$.

In the Ca^{2+}-binding studies, EGTA is added to the platelet suspensions after gel filtration. The desired free Ca^{2+} concentrations from 10^{-9} to 10^{-3} M are obtained by adding $CaCl_2$ from either commercially available reference solutions or from solutions prepared in the laboratory from dehydrated $CaCl_2$. Adding $CaCl_2$ to EGTA-containing solutions tends to decrease the pH by displacing H^+ from the EGTA. Although the presence of 20 mM HEPES in the incubation buffer minimizes these changes in pH, it is recommended that the actual pH be measured at each Ca^{2+} concentration.

[13] D. M. Bers, *Am. J. Physiol.* **242,** C404 (1982).

Definition of Surface-Bound Ca^{2+}

1. Platelet-Associated $^{45}Ca^{2+}$. After the platelets in the incubation mixture have been centrifuged through silicone oil, the amount of $^{45}Ca^{2+}$ contained in each platelet pellet represents the sum of platelet-associated $^{45}Ca^{2+}$ and extracellular $^{45}Ca^{2+}$ trapped within the pellet. The trapped extracellular volume, which averages 0.35 $\mu l/10^8$ platelets, can be measured directly using [^{3}H]sorbitol.[2,5] In our experience, neither the extracellular free Ca^{2+} concentration nor the addition of platelet agonists significantly affects this value. After subtracting trapped extracellular $^{45}Ca^{2+}$, the remaining $^{45}Ca^{2+}$ in the pellet is referred to as "platelet-associated $^{45}Ca^{2+}$."

2. Distinguishing Surface-Bound from Intracellular $^{45}Ca^{2+}$. Platelet-associated $^{45}Ca^{2+}$ represents the sum of the $^{45}Ca^{2+}$ that has become bound to the platelet surface and the $^{45}Ca^{2+}$ that has entered the cell. Surface-bound $^{45}Ca^{2+}$ is defined as the difference between the values for platelet-associated $^{45}Ca^{2+}$ obtained before and after the addition of 3 mM EGTA to the platelet incubation mixture just prior to sedimenting the platelets. This additional EGTA immediately decreases the extracellular free Ca^{2+} to $\sim 10^{-9}$ M, thereby causing the rapid dissociation of Ca^{2+} from the platelet surface. Thus, surface-bound $^{45}Ca^{2+}$ represents the Ca^{2+} that binds to the platelet during the initial 10-min incubation period which can be removed from the platelet by the excess EGTA. The amount of surface-bound $^{45}Ca^{2+}$ determined by this method agrees well with the amount obtained when lanthanides, such as La^{3+} or Gd^{3+}, are used instead of EGTA to displace surface-bound Ca^{2+}.[2,5] However, the EGTA method described here is preferable because of several technical problems associated with the use of lanthanides.[14]

Analysis of Surface-Bound Ca^{2+}

An analysis of surface-bound Ca^{2+} has to take into account the likely presence of more than one class of saturable Ca^{2+}-binding sites as well as the existence of nonsaturable Ca^{2+}-binding sites on membrane proteins and phospholipids. Figure 1 shows Ca^{2+} binding to normal platelets at free Ca^{2+} concentrations between 10^{-9} and 10^{-3} M. The inflection in the curve near 10^{-7} M suggests that at least two classes of saturable binding sites are present. The continued ascent of the curve at free Ca^{2+} concentrations $>10^{-4}$ M suggests the presence of either additional, nonsaturable binding sites or large numbers of saturable sites with $K_d >> 10^{-3}$ M. We have chosen to analyze these data as a sum of two or more independent saturable classes of binding sites plus nonsaturable binding. Rather than mea-

[14] C. G. Dos Remedios, *Cell Calcium* **2,** 29 (1981).

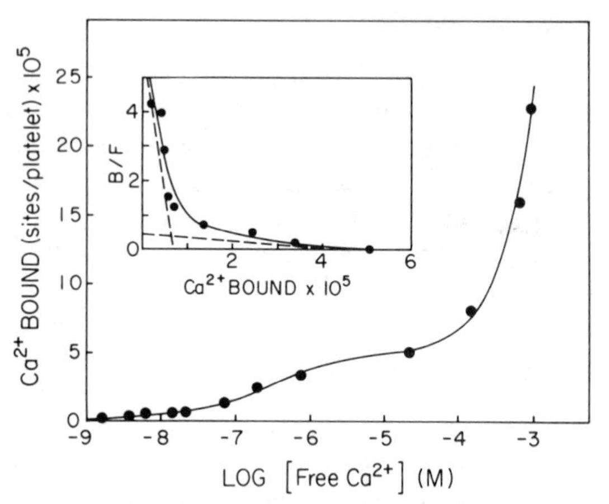

FIG. 1. Surface-bound Ca^{2+} in unstimulated platelets. $^{45}CaCl_2$ (0–1 mM) and EGTA (0.5 mM) were added to gel-filtered platelets to obtain extracellular free Ca^{2+} concentrations from 10^{-9} to 10^{-3} M. Ten minutes later, surface-bound Ca^{2+} was measured. The solid line through the data points is a best fit curve and reflects two classes of saturable binding sites as well as nonsaturable binding. In the inset, the same data are shown as a Scatchard plot with the results obtained at free Ca^{2+} concentrations $>10^{-4}$ M omitted for clarity. The broken lines are the estimates for each class of saturable sites. In this example, there were 67,000 sites/platelet with a high affinity for Ca^{2+} ($K_d = 5$ nM) and 420,000 sites/platelet with a lower affinity ($K_d = 0.35$ μM). Nonsaturable binding was 0.08% of the free Ca^{2+} concentration. Reproduced with permission from Brass and Shattil.[2]

suring and subtracting the amount of nonsaturable binding at each Ca^{2+} concentration, a term for nonsaturable binding is included in the data analysis. This approach, which was previously described by Munson and Rodbard for use in their binding program LIGAND,[15] is summarized in Eq. (7):

$$\text{Bound } Ca^{2+} = F\left(\frac{N_1}{F+K_1} + \frac{N_2}{F+K_2} + k\right) \tag{7}$$

where F is the free Ca^{2+} concentration, N_1 and N_2 are the number of saturable Ca^{2+}-binding sites in each of two classes, K_1 and K_2 are the dissociation constants for each class of sites, and k is the proportionality constant that characterizes nonsaturable Ca^{2+} binding. N, K, and F are expressed as moles/liter and k is expressed as a fraction without units. Although this type of analysis could be extended mathematically to include three or more classes of saturable binding sites, the useful limit is usually two.

[15] P. J. Munson and D. Rodbard, *Anal. Biochem.* **107,** 220 (1980).

For each set of binding data, a set of values for N_i, K_i, and k can be generated by computerized nonlinear regression analysis in which a curve generated by Eq. (7) is fit to the observed data points. Any of a number of published schemes can be used to accomplish this task, including LIGAND.[15] We have chosen to adapt the generalized nonlinear regression curve-fitting routine described by Koeppe and Hamann[16] because of its compactness and ease of translation of different forms of BASIC. Modifications were made in the method used to assign potential new values to N_i, K_i, and k in order to improve the speed at which the program arrives at the optimal values for these parameters. The resulting program, called SCATCH, was written in BASIC.[12]

Using SCATCH to generate a curve of the type shown in Fig. 1 requires an initial set of estimated values for N_1, N_2, K_1, K_2, and k. Estimates for N and K can be obtained graphically from a plot of the data in the form of bound/free vs bound (a Scatchard plot). An initial value for k can be obtained graphically from the higher concentrations of the bound vs free plot. Upon input of these estimates and the observed values for the amount of bound Ca^{2+} at each free Ca^{2+} concentration, the total least-squares error (E) is calculated using Eq. (8):

$$E = \sum \left(\frac{\text{Bound}_{\text{obs}}}{\text{Bound}_{\text{calc}}} - 1 \right)^2 \tag{8}$$

$\text{Bound}_{\text{calc}}$ at each free Ca^{2+} concentration is calculated using Eq. (7) and the current values of N_i, K_i, and k. A new value for E is then calculated after the computer assigns a new value to one of the five parameters selected at random. If the new value for E is smaller than the previous E, this value is adopted, ending the first cycle of iteration. Usually only 100–200 such cycles are required to obtain an optimal set of parameter values such as those used to generate the curve shown in Fig. 1. When data from multiple experiments are combined, a weighting factor can be used to take into account the degree of scatter in $\text{Bound}_{\text{obs}}$ at each free Ca^{2+} concentration. The final result is a set of values for N_1, N_2, K_1, K_2, and k, a plot of bound vs $\log[Ca^{2+}]_{\text{free}}$, and a Scatchard transformation of the data (Fig. 1). The precision of the final set of parameter values in each study is estimated using the method of support planes described by Duggleby[17] or the radex range as described by Koeppe and Hamann.[16] The results of a number of Ca^{2+}-binding studies performed with platelets obtained from normal donors are summarized in Table II. The platelet surface contains

[16] P. Koeppe and C. Hamann, *Comput. Programs Biomed.* **12,** 121 (1980).

[17] R. G. Duggleby, *Eur. J. Biochem.* **109,** 93 (1980).

TABLE II
Ca^{2+}-BINDING SITES ON UNSTIMULATED PLATELETS

Site	K_d (nM)	Sites/platelet
Higher affinity	9 ± 2	$86,000 \pm 11,000$
Lower affinity	400 ± 100	$389,000 \pm 35,000$

two classes of binding sites for Ca^{2+}, with K_d values for Ca^{2+} of 10 nM and 0.4 μM, respectively.

This approach to analyzing the data for surface-bound Ca^{2+} is attractive mathematically, but carries several important caveats. First, the view that only two classes of saturable sites are present is potentially misleading. Equation (7) generates a curve that is a good fit to the observed data, but tends to force potentially heterogeneous sites with similar dissociation constants to appear as two homogeneous classes of sites with averaged dissociation constants. Although additional terms can be added to Eq. (7), without major improvements in the fit of the curve to the observed data it becomes increasingly difficult to determine whether such sites actually exist.

The second major caveat concerns the range of free Ca^{2+} concentrations over which the data are collected. Equation (7) will attempt to force the data to fit the chosen model regardless of whether the experimental range of free Ca^{2+} concentrations extends sufficiently far above and below the K_d to encompass both classes of binding sites. Too narrow a range of free Ca^{2+} concentrations can lead to large errors in the calculated values for N and K. This type of problem can usually be avoided by inspection of the bound vs log(free) plot.

The final caveat is the most important. As recently discussed by Klotz and Hunston,[18] mathematically defined binding sites may not have a biological reality. Therefore, it is important to perform independent studies that can associate the theoretical binding sites with physically detectable entities, such as proteins within the membrane. As discussed in a later section, we have used platelets deficient in one or more membrane glycoproteins to confirm the existence of the Ca^{2+}-binding sites defined by our binding studies. In addition, we have used independent techniques to examine the interaction of Ca^{2+} with the solubilized membrane proteins. The results of these studies support the physical existence of the mathematically defined Ca^{2+}-binding sites on the platelet surface.

[18] I. M. Klotz and D. L. Hunston, *J. Biol. Chem.* **259,** 10060 (1984).

Measurement of Surface-Bound Ca^{2+} on Stimulated Platelets

As with unstimulated platelets, efforts must be made to prevent secretion when measuring surface-bound Ca^{2+} on stimulated platelets. Thus, incubations of aspirin-treated, gel-filtered platelets are performed at 22° and without stirring. To date, only ADP and epinephrine have been used as agonists because, unlike thrombin, these agonists do not cause platelet secretion when platelet prostaglandin synthesis is blocked by aspirin. Ca^{2+}-Binding studies are performed as for unstimulated platelets, except for the following additions and modifications:

ADP is dissolved in distilled water and epinephrine in 0.8 m*M* ascorbic acid. Typically, 1/100 vol of the agonist is added to the gel-filtered platelets simultaneously with the $^{45}CaCl_2$ and EGTA to obtain final agonist concentrations of 0.2–10 μM. Unstimulated platelets are assayed in parallel to allow direct comparison of agonist-stimulated Ca^{2+} binding with basal Ca^{2+} binding. To date, only Ca^{2+} binding to the lower affinity class of binding sites ($K_d = 0.4\ \mu M$) has been studied on stimulated platelets.[2,5] These studies show that normal platelets develop approximately 138,000 new lower affinity Ca^{2+}-binding sites per platelet after stimulation with 10 μM ADP and about 65,000 new sites after stimulation with 10 μM epinephrine (Fig. 2). To test the specificity of the agonist effect, receptor antagonists, such as 50–500 μM ATP (in the case of ADP) and 10 μM yohimbine or phentolamine (in the case of epinephrine), can be added to the platelet suspension 5 min before the addition of the agonist.[2]

Incubation of platelets at temperatures higher than 22° in the presence of free Ca^{2+} concentrations less than 1 μM results in a time-dependent loss of platelet aggregability that is due to a dissociation of the glycoprotein IIb–IIIa heterodimer complex within the surface membrane.[19,20] Since such a perturbation might also affect other surface membrane reactions such as Ca^{2+} binding, we have avoided performing Ca^{2+}-binding studies at temperatures above 22°.

Identification and Further Characterization of Ca^{2+}-Binding Proteins within the Platelet Surface Membrane

The relatively high affinities of the two classes of binding sites for Ca^{2+} suggests that these sites are associated with proteins exposed on the platelet surface rather than with other membrane constituents, such as phospholipids. We have used two basic approaches to identify and characterize

[19] S. J. Shattil, L. F. Brass, J. S. Bennett, and P. Pandhi, *Blood* **66**, 92 (1985).
[20] L. A. Fitzgerald and D. R. Phillips, *J. Biol. Chem.* **260**, 11366 (1985).

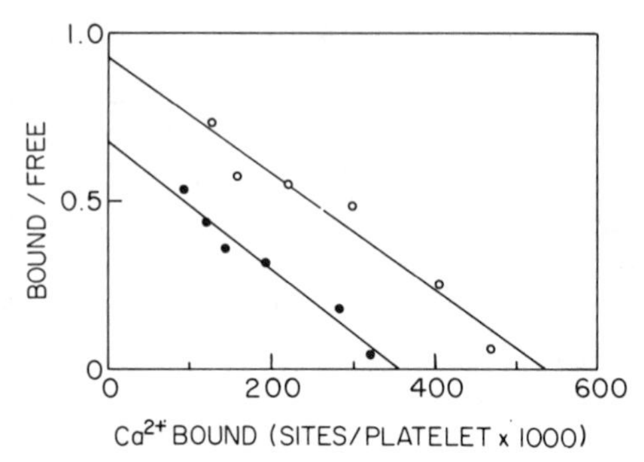

FIG. 2. The effect of ADP on Ca^{2+} binding to normal platelets. $CaCl_2$ and EGTA were added to gel-filtered platelets to obtain free Ca^{2+} concentrations from 30 nM to 10 μM. At the same time, ADP (10 μM) was added to half of the platelets. Ten minutes later, the amount of surface-bound Ca^{2+} on platelets stimulated by ADP (open circles) or on unstimulated platelets (filled circles) was measured. The results obtained are displayed as a Scatchard plot. This figure illustrates the data obtained on platelets from a single normal donor. ADP increased the number of Ca^{2+}-binding sites on these normal platelets from 354,000 sites/platelet ($K_d = 0.22\ \mu M$) to 534,000 sites/platelet ($K_d = 0.24\ \mu M$). Reproduced from Brass and Shattil, *J. Clin. Invest.* **73**, 626–632 (1984) by permission of The American Society for Clinical Investigation.

these sites further. The first applies the techniques developed for quantitating Ca^{2+}-binding sites on normal platelets to the study of platelets with well-characterized membrane glycoprotein deficiencies. These studies have identified the glycoproteins IIb and/or IIIa as the major Ca^{2+}-binding proteins on the platelet surface. The second approach uses monoclonal antibodies to examine Ca^{2+}-dependent conformational changes in both the solubilized IIb–IIIa complex and in the membrane-associated IIb–IIIa complex.

Heritable Deficiencies of Platelet Glycoproteins

Glanzmann's thrombasthenia is an autosomal recessive bleeding disorder in which platelets fail to aggregate because they are deficient in the membrane glycoproteins IIb and IIIa. Glycoproteins IIb and IIIa have been shown to form a Ca^{2+}-dependent heterodimer complex when extracted from the platelet membrane.[21,22] We measured Ca^{2+}-binding sites on the surface of platelets from three donors with homozygous thrombasthenia. All three had $<7\%$ of the normal amount of glycoprotein IIb and IIIa as determined by the direct binding to platelets of an anti-IIb–IIIa monoclonal antibody. Unstimulated thrombasthenic platelets contained an average

[21] T. J. Kunicki, D. Pidard, J.-P. Rosa, and A. T. Nurden, *Blood* **58**, 268 (1981).

[22] L. K. Jennings and D. R. Phillips, *J. Biol. Chem.* **257**, 10458 (1982).

of 92% fewer higher affinity Ca^{2+}-binding sites ($K_d = 9$ nM) and 63% fewer lower affinity Ca^{2+}-binding sites ($K_d = 0.4$ μM) than normal platelets (Fig. 3). The surface membrane of normal platelets contains about 45,000 copies of the IIb–IIIa complex per cell.[23] Assuming that the differences in Ca^{2+}-binding sites between normal and thrombasthenic platelets are due exclusively to the absence of glycoproteins IIb and IIIa, each IIb–IIIa complex would have approximately eight saturable binding sites for Ca^{2+}. However, these studies of surface-bound Ca^{2+} do not clarify which of these two glycoproteins binds extracellular Ca^{2+}. Studies with solubilized IIb and IIIa have suggested that glycoprotein IIb binds Ca^{2+} but IIIa does not.[24,25] This conclusion is supported by the recently published predicted amino acid sequence of GPIIb, which contains four regions homologous to the calcium-binding domains found in calmodulin and troponin C.[26]

Thrombasthenic platelets have also been studied after ADP stimulation. In contrast to the marked deficiency of Ca^{2+}-binding sites found on unstimulated thrombasthenic platelets, ADP-stimulated thrombasthenic platelets developed a normal number of new Ca^{2+}-binding sites (Fig. 4). This suggests that the newly detectable sites are not associated with glycoproteins IIb and IIIa.

In an effort to identify other Ca^{2+}-binding proteins on the platelet surface, studies were also performed with platelets from two donors with the Bernard–Soulier syndrome.[5] Bernard–Soulier platelets are larger than normal and lack several membrane glycoproteins, including glycoproteins Ib, V, and IX. Glycoprotein Ib serves as the receptor for the ristocetin-induced binding of von Willebrand factor to platelets. Unstimulated Bernard–Soulier platelets had twice the normal number of saturable Ca^{2+}-binding sites, an observation that may be related to the larger size and increased surface area of these platelets. After stimulation with ADP, Bernard–Soulier platelets developed an average of 216,000 new lower affinity Ca^{2+}-binding sites per platelet, an increase that is also greater than normal (Fig. 4). These observations suggest that the Ca^{2+}-binding sites present on normal platelets are not associated with any of the glycoproteins deficient in Bernard–Soulier platelets.

[23] J. S. Bennett, J. A. Hoxie, S. F. Leitman, G. Vilaire, and D. B. Cines, *Proc. Natl. Acad. Sci. U.S.A.* **80,** 2417 (1983).

[24] K. Fujimura and D. R. Phillips, *J. Biol. Chem.* **258,** 10247 (1983).

[25] S. Karpatkin, R. Ferziger, and D. Dorfman, *J. Biol. Chem.* **261,** 14266 (1986).

[26] M. Poucz, R. Eisman, R. Heidenreich, S. Silver, G. Vilaire, S. Surrey, E. Schwartz, and J. Bennett, *J. Biol. Chem.* **262,** 8476 (1987).

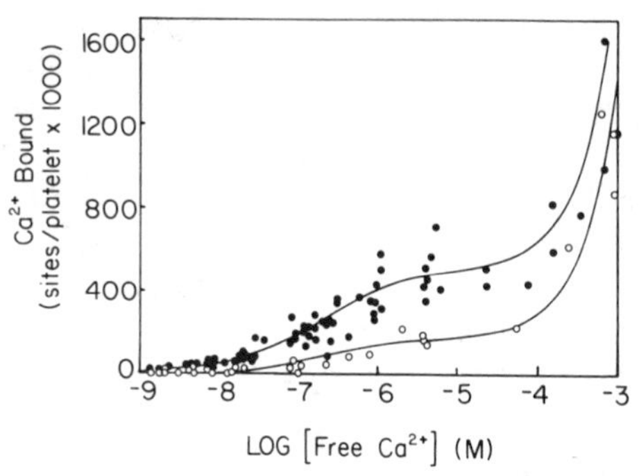

FIG. 3. Ca^{2+}-Binding sites on unstimulated thrombasthenic platelets. Gel-filtered platelets were prepared from three donors with thrombasthenia (open circles) and seven normal donors (filled circles). Surface-bound Ca^{2+} was measured at free Ca^{2+} concentrations from 1 n*M* to 1 m*M*. Platelets from all donors were studied across this entire range of free Ca^{2+} concentrations. Each data point represents the mean of triplicate determinations. The solid lines were generated by nonlinear regression analysis of the grouped data. Reproduced from Brass and Shattil, *J. Clin. Invest.* **73,** 626–632 (1984) by permission of The American Society for Clinical Investigation.

Use of Monoclonal Antibodies to Study Ca^{2+}-Binding Proteins on the Platelet Surface

In order to examine the relationship between the Ca^{2+} bound to IIb and IIIa and the Ca^{2+} required to maintain the integrity of the IIb–IIIa complex, a method was needed to determine whether IIb and IIIa are present in the membrane as the individual glycoproteins or as the intact complex. Although this problem has been approached with a variety of techniques, we have found anti-platelet monoclonal antibodies to be useful. A2A9 is a monoclonal antibody that binds to the intact IIb–IIIa complex, but not to either of the individual glycoproteins.[23] The antibody binds to approximately 45,000 sites on normal platelets, but does not bind to thrombasthenic platelets.

Studies published by Zucker and Grant[27] show that normal platelets lose their ability to bind fibrinogen after incubation at 37° with a Ca^{2+} chelator, such as EGTA or EDTA. This observation raised the possibility that the IIb–IIIa complex might dissociate under these conditions. At 37° (but not at 22°) the binding of A2A9 to intact platelets and platelet membranes is Ca^{2+} dependent.[28] Thus at 1 m*M* extracellular free Ca^{2+}, A2A9 binding is maximal. At 1 n*M* free Ca^{2+}, A2A9 binding is abolished (Fig. 5). A2A9 binding is half-maximal at 0.4 μM, which is the same as the K_d for the binding of Ca^{2+} to the lower affinity class of Ca^{2+}-binding sites

[27] M. B. Zucker and R. A. Grant, *Blood* **52,** 505 (1978).

[28] L. F. Brass, S. J. Shattil, T. J. Kunicki, and J. S. Bennett, *J. Biol. Chem.* **260,** 7875 (1985).

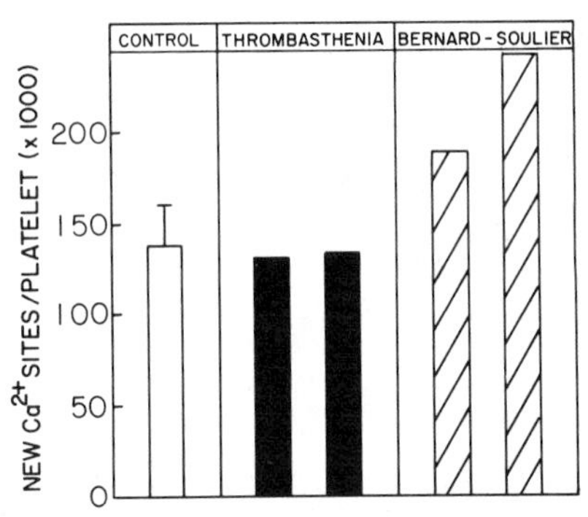

FIG. 4. ADP-induced Ca^{2+}-binding sites on normal, thrombasthenic, and Bernard–Soulier platelets. The values shown are the new Ca^{2+}-binding sites stimulated by 10 μM ADP and represent the difference between the number of Ca^{2+}-binding sites measured in the presence and absence of ADP. The results for the normal platelets are mean ± SEM for eight donors. The values shown for the thrombasthenic and Bernard–Soulier platelets are the results from each individual donor. Reproduced from Brass and Shattil, *J. Clin. Invest.* **73,** 626–632 (1984) by permission of The American Society for Clinical Investigation.

on IIb–IIIa. This observation suggests that some of the "surface-bound" Ca^{2+} associated with glycoprotein IIb and IIIa functions to hold the IIb–IIIa complex together. It also indicates that at physiological free Ca^{2+} concentrations (1 mM), these glycoproteins are normally present in the form of heterodimer complexes within the surface membrane of intact platelets. A similar conclusion has been drawn using nonimmunological techniques to quantitate the IIb–IIIa complex in platelet membranes.[19,20]

Surface-Bound Ca^{2+} and Platelet Function

Two different approaches can be taken to determine whether surface-bound Ca^{2+} is necessary for normal platelet function.

1. Gel-filtered platelets labeled with [^{14}C]serotonin are prepared as for the Ca^{2+}-binding studies, except aspirin is omitted to permit platelet secretion. Fibrinogen (Kabi), 0.1 mg/ml, is added to 0.5 ml of platelets in an aggregation cuvette. Then various amounts of $CaCl_2$ and 0.5 mM EGTA are added to obtain a range of free Ca^{2+} concentrations (Table I). The Mg^{2+} concentration in the buffer is 1 mM. The platelets are stirred at 1000 rpm in an aggregometer, and platelet aggregation and serotonin release are measured 3 min after the addition of ADP or epinephrine. Using this method, it has been demonstrated that at least 10^{-8} M Ca^{2+} is required for maximal aggregation and serotonin release in response to these two agonists.[5] This value is similar to the K_d of the higher affinity Ca^{2+}-binding sites on the platelet surface membrane.

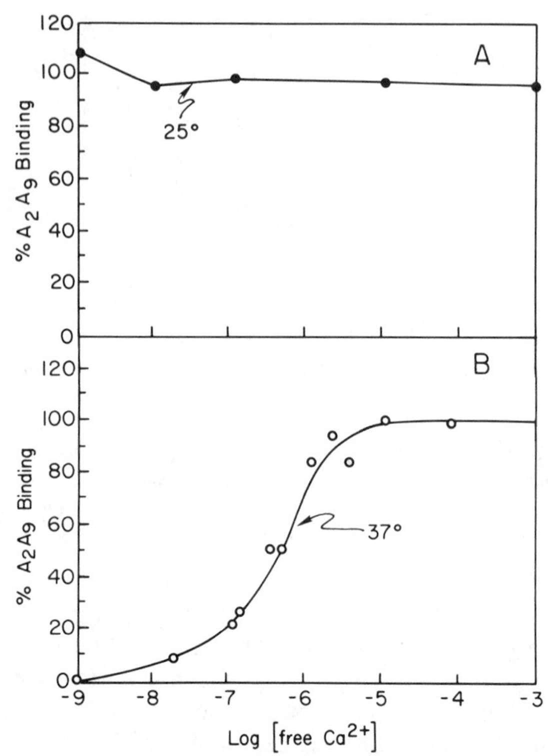

FIG. 5. Ca^{2+} Dependence of ^{125}I-labeled A2A9 binding to intact platelets at 22° and 37°. Gel-filtered platelets were incubated for 30 min at either 22° (A) or 37° (B) with 0.5 m*M* EGTA and sufficient $CaCl_2$ to give the final free Ca^{2+} concentrations that are shown. Afterward ^{125}I-labeled A2A9 was added. The extent of ^{125}I-labeled A2A9 binding was determined 15 min later by sedimenting the platelets through silicone oil and measuring platelet-associated ^{125}I. The results shown are the mean of two such studies and are expressed as a percentage of maximum binding.

2. In the second approach, a lanthanide, Gd^{3+}, is used instead of EGTA to displace Ca^{2+} from the platelet surface before performing the aggregation and serotonin release studies. The pH of the gel filtration buffer must be reduced to 6.9 or less and PO_4^{3-} must be omitted to avoid formation of insoluble gadolinium species. The gel-filtered platelets are then incubated for 10 min at room temperature in the presence of 10–100 μM $GdCl_2$. Increasing concentrations of Gd^{3+} displace increasing amounts of surface-bound Ca^{2+}. This results in decreased ADP- and epinephrine-induced platelet aggregation and serotonin release, suggesting that these platelet functions are inhibited by the displacement of surface-

bound Ca^{2+}.[5] It should be emphasized that experiments with lanthanides must be interpreted cautiously. $LaCl_3$ has been reported to stimulate both platelet secretion and cAMP production, effects that are not observed at the concentrations of $GdCl_3$ used in our studies.

[32] Calcium-45 Exchange Techniques to Study Calcium Transport in Intact Platelets

By LAWRENCE F. BRASS and ELIZABETH BELMONTE

Prior to stimulation by an agonist, human platelets maintain a low cytosolic free Ca concentration and a steep plasma membrane Ca concentration gradient. In other tissues, this is accomplished by the concerted action of processes that (1) limit Ca influx, (2) promote Ca efflux, and (3) sequester Ca within labile storage sites such as endoplasmic reticulum, sarcoplasmic reticulum, and mitochondria. In platelets the existence of such processes has been addressed by a variety of techniques. The results suggest that the platelet dense tubular system, which is thought to be derived from megakaryocyte endoplasmic reticulum, is able to function in a manner analogous to sarcoplasmic reticulum. That is, the dense tubular system is believed to be able to sequester Ca from the cytosol of resting platelets and release Ca in response to IP_3 during platelet activation. This conclusion is based on observations of ATP-dependent ^{45}Ca transport by platelet membrane vesicles isolated by differential centrifugation or flow electrophoresis. Less is known about Ca efflux from platelets. Although the process is presumably energy dependent, the transport mechanism has not yet been identified. Part of the explanation for this paucity of information lies in the small size of platelets, which limits the applicability of methods developed in larger cells.

Although classical cell fractionation techniques have not proved entirely adaptable to platelets, fluorescent probes such as Quin 2 and aequorin have been used to measure the steady-state cytosolic free Ca concentrations in resting platelets. The value obtained with Quin 2, approximately 0.1 μM, is similar to levels reported in other tissues.[1,2] Higher values are obtained with aequorin, which may reflect the respon-

[1] T. Rink, S. Smith, and R. Tsein, *FEBS Lett.* **148,** 21 (1982).
[2] P. Johnson, J. A. Ware, and E. W. Salzman, this volume [33].

siveness of this probe to local variations in the cytosolic free Ca concentration.[2] Although useful for information about the free Ca concentration in platelets, studies with these probes reveal little about the dynamics of Ca movements in intact platelets. In order to complement the data obtained by these methods, we have used ^{45}Ca to examine Ca transport and the distribution of exchangeable Ca in intact resting platelets.[3-6] The present chapter describes the methods used and briefly examines some of the adaptations that are required for studies in activated platelets. The general procedure can be summarized as follows: (1) Platelets are isolated from plasma and placed in an isotonic medium with defined ionic composition. (2) After reequilibration at a selected extracellular Ca concentration, ^{45}Ca influx or efflux is observed. (3) An exponential curve-fitting procedure is used to determine the ^{45}Ca content and half-time of exchange ($t_{1/2}$) of the Ca pools associated with the platelets. (4) Based on this information, a simple model of Ca exchange is used to calculate the complete set of parameters that define Ca movements into and within platelets. Each of these steps will be discussed separately.

Platelet Preparation

The same procedure as described in Chapter [31] of this volume is used.[7]

^{45}Ca Influx and Efflux

All steps are normally performed at room temperature. Immediately after gel filtration the platelets are equilibrated with 0.01 to 1.0 m*M* $CaCl_2$ for 30 min before adding a tracer quantity of ^{45}Ca (2 to 10 μCi/ml). The extracellular free Ca concentration is measured with a Ca-sensitive electrode. Thirty minutes was found to be sufficient for a return to steady-state conditions. Afterward, in the studies of Ca influx, triplicate or quadruplicate 0.2-ml aliquots of the platelet suspension are removed at intervals up to 4 hr after the addition of tracer. Each aliquot is diluted in 5 ml of wash buffer whose composition is the same as the gel filtration buffer with 5 m*M* EGTA added and glucose and albumin omitted. After 3 min the diluted platelets are filtered through 0.45-μm Millipore HAWP filters (Millipore Corp.) which are then rinsed with two additional 5-ml aliquots of wash

[3] L. F. Brass and S. J. Shattil, *J. Biol. Chem.* **257**, 14000 (1982).
[4] L. F. Brass, *J. Biol. Chem.* **259**, 12563 (1984).
[5] L. F. Brass, *J. Biol. Chem.* **259**, 12571 (1984).
[6] L. F. Brass, *J. Biol. Chem.* **260**, 2231 (1985).
[7] L. F. Brass and S. J. Shattil, this volume [31].

buffer. This washing procedure is required to remove ^{45}Ca bound to the platelet surface as well as any traces of high-specific-activity extracellular ^{45}Ca that might otherwise remain on the filters. Calcium-45 that has entered the platelets is then quantitated by scintillation counting using an aqueous cocktail.

Studies of ^{45}Ca efflux begin the same way, but in this case the platelets are gel filtered a second time after a 2-hr loading period. The purpose of the second gel filtration is to remove unincorporated tracer. In order to preserve steady-state conditions, $CaCl_2$ is added to the elution buffer at the same final concentration as was present during the loading period. This process takes approximately 10 min. Since ^{45}Ca bound to the platelet surface exchanges with a $t_{1/2} < 1$ min,[3] little, if any, surface-bound ^{45}Ca remains after the second gel filtration. Immediately afterward, the ^{45}Ca-loaded platelets are diluted 5-fold with column elution buffer and tracer efflux is followed for 3 hr. At any time, the amount of ^{45}Ca remaining inside the platelets is determined by filtering duplicate 1-ml aliquots of platelet suspension through the Millipore filters. Because the specific activity of the extracellular ^{45}Ca is low, it is not necessary to dilute the platelet suspension further before filtering. Each filter is then rinsed twice with 5-ml aliquots of wash buffer (Fig. 1).

Curve-Fitting Procedures

In both the influx and the efflux studies, the accumulated data for the amount of platelet-associated ^{45}Ca (R_T) observed at various times are analyzed as a sum of exponential terms using Eqs. (1) and (2):

$$R_T = \sum_i R_i(e^{-\lambda_i t}) \qquad \text{(efflux)} \tag{1}$$

$$R_T = \sum_i R_i(1 - e^{-\lambda_i t}) \qquad \text{(influx)} \tag{2}$$

In each case R_i (cpm/10^8 platelets) and λ_i (min^{-1}) are constants that reflect the size and turnover of i individual exchangeable Ca pools within the platelet. The half-time of exchange ($t_{1/2}$) of each pool is equal to $\ln(2)/\lambda_i$. In the influx studies, R_i is the amount of ^{45}Ca present in each pool at equilibrium (infinite time). In the efflux studies, R_i is the amount of ^{45}Ca present in each pool at the start of the efflux period. In theory, this should be at the point during the second gel filtration when the platelets leave the ^{45}Ca-containing medium after a loading period of infinite duration. In practice, the starting point on the efflux curves (t_0) is taken immediately after gel filtration and the values for R_i are corrected for the actual length of the loading period as described in the next section. In order to normalize the

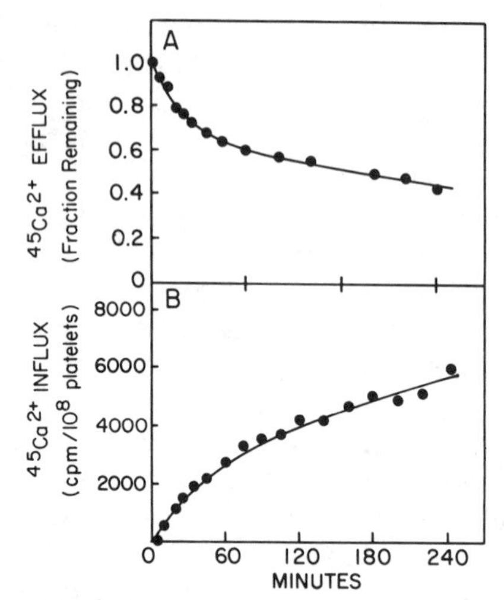

FIG. 1. Calcium-45 influx and efflux under steady-state conditions. Gel-filtered platelets were incubated for 30 min with 0.2 m*M* Ca after which a tracer amount of ^{45}Ca was added. (A) After a 2-hr loading period, the platelets were gel filtered a second time using buffer that contained 0.2 m*M* $CaCl_2$. The subsequent efflux of ^{45}Ca is expressed as a fraction of the amount of intracellular ^{45}Ca present at the start of the efflux period. The solid line was generated using Eq. (1). (B) Calcium-45 influx was observed for 4 hr. The solid line was generated using Eq. (2). Reproduced with permission from Brass.[4]

results from different studies, the efflux data can also be expressed as fractional efflux by dividing by the calculated value for $R_T(t_0) = \Sigma R_i$.

Although larger numbers of terms can be used in Eqs. (1) and (2), in general we have found that a good fit to the observed data is obtained using either two (efflux) or three (influx) exponential terms. The extra term in the influx studies is required because of the presence of a small amount of residual extracellular and/or surface-bound ^{45}Ca. The actual fitting of the curves defined by Eqs. (1) and (2) to the observed data is accomplished with a program called ECURVE written in BASIC for use on either a Hewlett-Packard series 80 microcomputer with plotter or an Apple Macintosh. This program uses a modified version of the nonlinear regression algorithm described by Koeppe and Hamann.[8] This algorithm determines the values for R_i and λ_i which minimize the residual sum of squares (E) shown in Eq. (3):

$$E = \sum_{j=1}^{n} \left(\frac{y(j)^{\circ}}{y(j)} - 1 \right)^2 \tag{3}$$

[8] P. Koeppe and C. Hamann, *Comput. Programs Biomed.* **12,** 121 (1980).

where $y(j)^o$ and $y(j)$ are, respectively, the observed and calculated values for the amount of platelet-associated ^{45}Ca at each of the n times of observation. Depending upon the type of study, $y(j)$ is calculated using Eq. (1) or (2). In the examples described in this chapter, E was typically <0.02. Although this suggests a tight fit of the curve to the data, it is also important to inspect the curve to be sure that the program has not been trapped in a local minimum value for E that is suboptimal (see below).

At the start of each analysis, the observed data and initial estimates for R_i and λ_i are entered. The initial estimates can be obtained by a quick sketch on graph paper of cpm vs time (influx) or log cpm vs time (efflux). In the latter case, the process can be automated by having the computer do linear regression on the points on the semilog plot with $t > 1$ hr to obtain estimates for R_2 and λ_2. Estimates for R_1 and λ_1 are obtained by subtraction. After input of the initial parameter estimates, multiple cycles of iteration are performed in which the current value of E is compared to a new value calculated using a slightly altered set of values for the parameters R_i and λ_i. This new set of values is obtained by selecting one of the parameters at random and varying its value by either a preselected percentage ("fixed percentage" cycle) or randomly within preselected limits ("boundary limits" cycle). If the new value for E is smaller than the old value, the new set of parameters is accepted and a new cycle begins. Otherwise, the new parameters are rejected and the old values are restored before beginning the next cycle. In general, we have found that 300–400 such cycles are sufficient to minimize E. Usually the first three-quarters of these are fixed percentage cycles in which the randomly selected parameter is either increased (even numbered cycles) or decreased (odd numbered cycles) by 2–4%. This approach, which is a modification of the published algorithm, makes the curve-fitting procedure more tolerant of poor choices for the original parameter estimates and results in a more rapid fit of the curve to the data. In the boundary limit cycles, new values for each parameter are selected randomly within boundaries initially set at 5–10% of the current value. The total calculation time depends upon the computer used for the analysis, the total number of data points entered, and the number of terms in Eq. (1) or (2). In the examples given in this chapter, this was typically no more than 10 min. At any time the cycling process can be temporarily interrupted to display the current values for E and for each parameter, the number of cycles that have elapsed since the last cycle in which the value of E improved and, if desired, a plot of the regression curve using the current parameter values. This information can be used to prematurely abort the cycling process, either because improvements in the value of E have become inconsequential or because the regression analysis is trapped in a local minimum. When the latter problem occurs, it is usually sufficient to restart the program with a different set of initial parameter estimates. The final output is a set of values for R_i and λ_i and a plot of the regression curve and the data. An example of a ^{45}Ca influx and efflux study is shown in Fig. 1.

The precision of the final values for each parameter can be estimated by methods such as the radex range as described by Koeppe and Hamann[8] or the method of support planes described by Duggleby.[9] Both methods yield numbers that reflect the amount each parameter can be varied without increasing the value of E by more than an arbitrary amount. By either criterion, we have generally found the "standard error" in our efflux studies to be $<10\%$. This method of curve fitting is not limited to the types of studies described in this chapter, but can also be applied whenever the equation for the desired curve is known and can be substituted for Eqs. (1) and (2). The use of this program to analyze ligand-binding data is described elsewhere.[7]

Analysis of the ^{45}Ca Exchange Data

The curve-fitting procedure produces a set of parameters that reflects the presence of two exchangeable Ca pools within platelets. Although the following discussion will focus upon the ^{45}Ca efflux data, similar results can be obtained by analyzing the two intracellular pools observed in the influx studies. In general, however, we have found the efflux data to be easier to work with. This is due in part to the greater variability in the influx data arising from the presence of residual high-specific-activity extracellular ^{45}Ca. As a result it is possible to achieve a better regression curve "fit" in the efflux studies. In either case the data are analyzed using an open series three-compartment model. In this model, extracellular Ca (the first of the three compartments) exchanges directly with the more rapidly exchangeable of two intracellular Ca compartments (pool 1), which in turn exchanges with the more slowly exchangeable intracellular Ca compartment (pool 2). Pool 2 does not directly exchange with the extracellular space. This model can be represented schematically as

$$\text{Extracellular Ca} \underset{k_{10}}{\overset{J_1 k_{01}}{\rightleftarrows}} \text{pool 1} \underset{k_{21}}{\overset{J_2 k_{12}}{\rightleftarrows}} \text{pool 2} \tag{4}$$

where J_1 and J_2 represent Ca flux between the pools and k_{10}, k_{12}, k_{01}, and k_{21} are rate constants. As used here, the term "Ca flux" refers to Ca movement. "Efflux" refers specifically to Ca movement out of the cell and "influx" refers to Ca movement into the cell across the plasma membrane. At steady state, by definition, Ca influx equals Ca efflux, Ca flux from pool

[9] R. Duggleby, *Eur. J. Biochem.* **109,** 93 (1980).

1 into pool 2 equals Ca flux from pool 2 into pool 1 and the sizes of both pools remain constant. Calcium "exchange" refers to the bidirectional movement of Ca under steady-state conditions. A series model was originally selected for these studies because of the experience in other tissues, subsequently confirmed in platelets, that pool 2 is located in an intracellular compartment, such as endoplasmic reticulum, which would not be expected to communicate directly with the extracellular space. Since the specific activity of the extracellular Ca remains negligible during the efflux studies, the platelets can be treated as an open system even though the medium is not changed during the period of observation.

Values of J_i and k_{ij} in the efflux studies are calculated using equations published by Uchikawa and Borle.[10] Since those authors have discussed their methods in detail (including the derivation of the equations) elsewhere, only the equations themselves and their application to ^{45}Ca efflux studies in platelets will be presented here. Using the nomenclature in Eqs. (1), (2), and (4):

$$J_1 = C_1\lambda_1 + C_2\lambda_2 \quad (\text{pmol}/10^8 \text{ platelets/min}) \tag{5}$$

$$J_2 = S_1S_2\lambda_1\lambda_2/J_1 \quad (\text{pmol}/10^8 \text{ platelets/min}) \tag{6}$$

$$S_1 = J_1^2/(C_1\lambda_1^2 + C_2\lambda_2^2) \quad (\text{pmol}/10^8 \text{ platelets}) \tag{7}$$

$$S_2 = C_1 + C_2 - S_1 \quad (\text{pmol}/10^8 \text{ platelets}) \tag{8}$$

$$k_{10} = J_1/S_1 \quad (\text{min}^{-1}) \tag{9}$$

$$k_{12} = J_2/S_1 \quad (\text{min}^{-1}) \tag{10}$$

$$k_{21} = J_2/S_2 \quad (\text{min}^{-1}) \tag{11}$$

S_1 and S_2 are the size of each of the intracellular pools. In an open system, k_{01} is indeterminate.[10] C_1 and C_2 are constants calculated by correcting R_1 and R_2 for the actual length of the tracer loading period in minutes (U) after using the specific activity of the medium (Φ) to convert cpm to pmol:

$$C_1 = R_1/[\Phi(1 - e^{-\lambda_1 U})] \quad (\text{pmol}/10^8 \text{ platelets}) \tag{12}$$

$$C_2 = R_2/[\Phi(1 - e^{-\lambda_2 U})] \quad (\text{pmol}/10^8 \text{ platelets}) \tag{13}$$

Strictly speaking, Φ (cpm/pmol) refers to the specific activity of the medium at the point when isotopic equilibrium has been reached. However, since the total Ca content of the intracellular exchangeable Ca pools is so much lower than the Ca content of the extracellular space, the specific activity of the medium at the start of the loading period can be used to

[10] T. Uchikawa and A. Borle, *Am. J. Physiol.* **234,** R29 (1978).

approximate Φ.[10,11] Equations (12) and (13) are based upon the premise that, by definition, the rate of turnover of the two Ca pools (λ_1 and λ_2) under steady-state conditions is the same during both the loading period and the period during which ^{45}Ca efflux is observed.[11]

In practice, the efflux studies are performed and analyzed as follows. Gel-filtered platelets are loaded with $^{45}CaCl_2$ for 2 hr and then gel filtered a second time. The specific activity of the extracellular medium (Φ) is determined prior to the second gel filtration by measuring total ^{45}Ca and total Ca in aliquots of the medium. Calcium-45 efflux is then observed for 3 hr and values for λ_1, λ_2, R_1, and R_2 are determined by exponential curve fitting [Eq. (1)]. These values are, in turn, used to calculate values for S_1, S_2, J_1, J_2, k_{10}, k_{12}, and k_{21} [Eqs. (5)–(13)]. The results obtained in studies using platelets from 34 normal donors are summarized in Table I.

The surface-bound exchangeable Ca pool has been excluded from this discussion. Although the method used in the ^{45}Ca efflux studies precludes observations of this pool, such observations can be included in the influx studies by modifying the method used to measure platelet-associated ^{45}Ca. Instead of washing the platelets with EGTA-containing medium before measuring platelet-associated ^{45}Ca, aliquots of platelet suspension are rapidly sedimented through silicone oil by centrifugation for 2 min at 12,000 *g* in an Eppendorf microcentrifuge. Calcium-45 in the platelet pellet is the sum of internalized ^{45}Ca, surface-bound ^{45}Ca, and extracellular ^{45}Ca trapped within the pellet. The amount of trapped ^{45}Ca can be calculated and then subtracted by incubating identical aliquots of platelet suspension with [^{3}H]sorbitol or [^{14}C]inulin instead of ^{45}Ca. The data for platelet-associated ^{45}Ca vs time are analyzed using three terms in Eq. (2) to obtain R_i and λ_i. Pool sizes and exchange rates are calculated using a mixed model in which surface-bound Ca exchanges in parallel with the intracellular exchangeable Ca pools.[3,11]

Locations of the Intracellular Ca Pools

Studies using ^{45}Ca are able to define exchangeable Ca pools within platelets mathematically, but reveal little about the physical locations of the pools. In other tissues examined with similar techniques, the more rapidly exchangeable Ca pool (pool 1) has proved to be located in the cytosol. The more slowly exchangeable Ca pool (pool 2) has been assigned to sites of Ca sequestration such as mitochondria and endoplasmic reticulum. In order to determine whether a similar distribution exists in platelets, the sizes of the two Ca pools were determined after equilibrating the platelets at various Ca concentrations. Over a 100-fold range of extracellu-

[11] A. Borle, this series, Vol. 39, p. 513.

TABLE I
KINETICS OF Ca EXCHANGE IN NORMAL UNSTIMULATED PLATELETS[a]

Parameter	Value	Unit
Pool 1		
$t_{1/2}$	17 ± 1	min
S_1	52 ± 3	pmol/10^8 platelets
J_1	1.53 ± 0.09	pmol/10^8 platelets/min
k_{10}	30.0 ± 1.2	min^{-1} (×1000)
Pool 2		
$t_{1/2}$	313 ± 13	min
S_2	183 ± 12	pmol/10^8 platelets
J_2	0.57 ± 0.04	pmol/10^8 platelets min
k_{12}	11.2 ± 0.5	min^{-1} (×1000)
k_{21}	3.20 ± 0.12	min^{-1} (×1000)

[a] Calcium-45 efflux was measured at steady state in buffer containing 0.2 mM $CaCl_2$. The results shown are the mean ± SE ($n = 34$). Adapted from Brass.[4]

lar free Ca concentrations, the size of the more rapidly exchangeable Ca pool increased only 5-fold. Most of this increase occurred below 0.1 mM Ca. Above 0.1 mM Ca, the size of the pool was relatively unaffected by changes in extracellular Ca. In contrast, the size of the more slowly exchangeable Ca pool continued to increase with the extracellular Ca concentration (Fig. 2).

This behavior suggests that the two Ca pools are physically, as well as mathematically, distinct and that the more slowly exchangeable pool may help to buffer the size of the rapidly exchangeable pool. To determine the locations of the pools, advantage can be taken of the ability of digitonin to differentially permeabilize cellular membranes in an order determined by their cholesterol content. In general, plasma membranes are affected at much lower digitonin concentrations than either mitochondria or endoplasmic reticulum.[12-14] Figure 3 shows the release of ^{45}Ca from gel-filtered platelets exposed to increasing concentrations of digitonin. Two increments in ^{45}Ca release are seen, the first beginning at approximately 10–30 μM digitonin. The second occurs at digitonin concentrations > 500 μM. Only 80% of the total ^{45}Ca within the platelets is released by digitonin. Further increases in the digitonin concentration have no effect.

In order to relate release of ^{45}Ca to loss of membrane integrity, release of marker enzymes from within the cells was also measured. Total platelet-

[12] E. Murphy, K. Coll, T. Rich, and J. Williamson, *J. Biol. Chem.* **255,** 6600 (1980).
[13] M. Meredith and D. Reed, *J. Biol. Chem.* **257,** 3747 (1982).
[14] J. Akkerman, R. Ebberink, J. Lips, and G. Christiaens, *Br. J. Haematol.* **44,** 291 (1980).

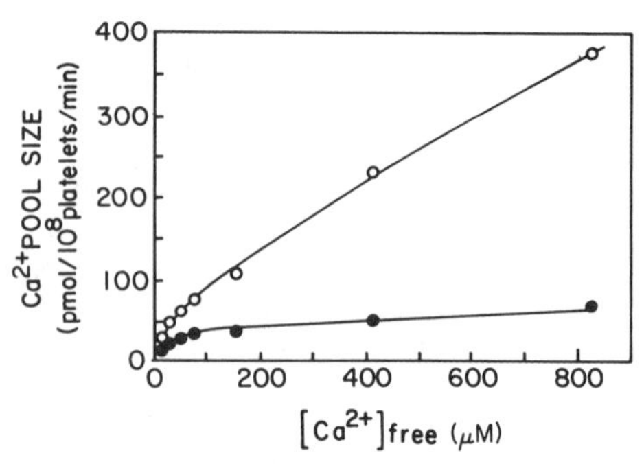

FIG. 2. The effect of $[Ca]_o$ on the sizes of the intracellular Ca pools. Steady-state ^{45}Ca efflux was observed using platelets equilibrated with 0.01 to 1.0 mM $CaCl_2$ before addition of the tracer. The extracellular free Ca concentration was measured with a Ca-sensitive electrode and remained constant throughout the loading period and the efflux period. S_1 (closed circles) and S_2 (open circles) were calculated from the observed data using Eqs. (7) and (8). Reproduced with permission from Brass.[4]

associated activity of each enzyme was determined by solubilizing the platelets with 0.2% Triton X-100. Lactate dehydrogenase (LDH), an enzyme present in the platelet cytosol, was released by the same digitonin concentrations that were associated with the first burst of ^{45}Ca release (Fig. 3). In nine such studies, 100 μM digitonin released 100% of platelet LDH and 50 ± 3% of the ^{45}Ca that had entered the platelets during a 2-hr loading period. For comparison, the fraction of intracellular ^{45}Ca contained located in each exchangeable Ca pool can be calculated using the data in Table I. Since at steady state these values apply to ^{45}Ca influx as well as to ^{45}Ca efflux, Eq. (2) can be used to calculate the fraction of total platelet ^{45}Ca that should be located in the rapidly exchangeable pool under these conditions. The result obtained, 55%, is in close agreement with the fraction of ^{45}Ca released by concentrations of digitonin that release lactate dehydrogenase from the platelet, supporting the concept that the rapidly exchangeable Ca pool is located in the cytosol.

The second burst of ^{45}Ca release occurs at the same concentration of digitonin that liberates glutamate dehydrogenase, a mitochondrial marker, and NADH dehydrogenase, a marker for endoplasmic reticulum.[15,16] Using the logic applied above, Ca pool 2 could be located in either organelle. Therefore, the size of each pool and the cytosolic free Ca concentration have been measured in the presence of either (1) trifluoperazine, an

[15] E. Schmidt, *in* "Methods in Enzymatic Analysis" (H. Bergmeyer, ed.), p. 650. Verlag Chemie–Academic Press, New York, 1974.

[16] J. Hochstadt, D. Quinlan, D. Radar, C. Lin, and D. Dowd, *Methods Membr. Biol.* **5,** 117 (1975).

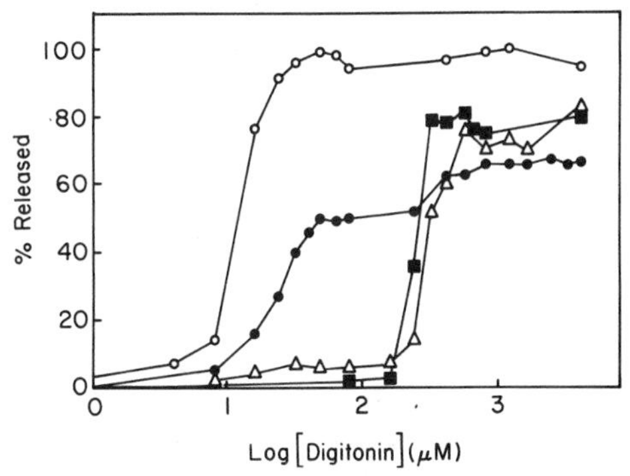

FIG. 3. Release of marker enzymes and intracellular ^{45}Ca by digitonin. Gel-filtered platelets were incubated for 2 hr with 0.2 m*M* $^{45}CaCl_2$ and then gel filtered a second time to remove unincorporated tracer. Immediately afterward, digitonin (1 μ*M* to 5 m*M*) and EGTA (1 m*M*) were added. After a 5-min incubation at 4°, the platelets were sedimented and ^{45}Ca (closed circles), lactate dehydrogenase (open circles), glutamate dehydrogenase (solid squares) and NADH dehydrogenase (triangles) were measured in the supernatant. Each result is the mean of at least three experiments. The values are expressed as a fraction of total platelet-associated activity determined by completely dissolving the platelets with 0.2% Triton X-100. Reproduced with permission from Brass.[4]

inhibitor of Ca uptake into the dense tubular system, (2) a combination of a mitochondrial uncoupler, carbonyl cyanide *m*-chlorophenylhydrazone (CCCP),[17] and KCN or (3) a combination of CCCP, KCN, and 2-deoxyglucose, which, by inhibiting both glycolysis and oxidative phosphorylation, decreases platelet ATP levels.[18] Adding trifluoperazine or CCCP + KCN + deoxyglucose decreased the size of the slowly exchangeable Ca pool and increased the cytosolic free Ca concentration. Inhibition of mitochondrial function did neither (Table II). These data suggest that pool 2 is located in the platelet dense tubular system, but not in mitochondria. Additional support for this conclusion has been obtained in studies of ^{45}Ca uptake into intact platelets permeabilized with low concentrations of digitonin or saponin. The K_M for the ruthenium red-inhibitable (i.e., mitochondrial) component of ^{45}Ca uptake in this system is approximately 1 μ*M*.[19] This value is similar to that obtained for mitochondrial Ca uptake in other tissues and is probably too high for platelet mitochondria to be solely responsible for the 0.1 μ*M* cytosolic free Ca concentration found in unstimulated platelets. In contrast, the platelet dense tubular system has a K_M of 0.1 μ*M* for ATP-dependent Ca uptake. It is possible, however, that

[17] P. Heytler and W. Pritchard, *Biochem. Biophys. Res. Commun.* **7,** 272 (1962).
[18] H. Holmsen and L. Robkin, *Thromb. Haemostasis* **42,** 1460 (1979).
[19] L. F. Brass, unpublished observations (1984).

TABLE II
EFFECT OF INHIBITORS ON Ca HOMEOSTASIS IN INTACT PLATELETS[a]

Inhibitor	Parameter	Control	Inhibitor	n	p
TFP (20 μM)	S_1	35 ± 3	45 ± 4	5	<0.025
	S_2	153 ± 21	97 ± 13	5	<0.025
	Free	113 ± 14	304 ± 16	6	<0.001
CCCP (10 μM) + KCN (2 mM)	S_1	32 ± 2	45 ± 4	6	<0.01
	S_2	152 ± 21	141 ± 31	6	NS
	Free	113 ± 14	123 ± 16	6	NS
CCCP (10μM) + KCN (2 mM) + deoxyglucose (10 mM)	S_1	59 ± 4	73 ± 6	6	<0.05
	S_2	260 ± 43	196 ± 26	6	<0.05
	Free	113 ± 14	271 ± 34	6	<0.005

[a] Calcium-45 efflux was measured at 0.2 mM extracellular Ca with each inhibitor present during both the ^{45}Ca loading period and efflux. The values for S_1 and S_2 are expressed as pmol/10^8 platelets/min. "Free" refers to the cytosolic free Ca concentration measured with Quin 2 1 hr after the addition of the inhibitors. The p values were calculated by paired t test. NS, Not significant. TFP, trifluoperzaine. Adapted from Brass.[4]

mitochondria contribute to the recovery of basal cytosolic free Ca concentrations after platelet stimulation.

Problems of Application and Interpretation

This approach to Ca transport and distribution is attractive theoretically but needs to be applied cautiously. First, in order for Eqs. (5)–(13) to be valid, the platelets must be at steady state with respect to Ca. In other tissues, proof of steady-state conditions is obtained by demonstrating that total cellular Ca remains unchanged during the period of observation. We have performed such measurements in platelets under the conditions described in this chapter and found no changes in total platelet-associated Ca. There is, however, potentially a problem with such measurements. Unlike most cells, the majority of platelet Ca is located within storage granules. In the time frame of these studies, these granules exchange little, if at all, with extracellular Ca. Platelets contain approximately 20 nmol Ca/10^8 platelets. The exchangeable Ca pools represent only 1% of this amount. Therefore, even large changes in exchangeable Ca might be missed. Alternatively, proof of the existence of steady-state conditions can be obtained by comparing the results of ^{45}Ca influx and ^{45}Ca efflux studies performed under the same conditions. Good agreement in the results, especially if obtained over a range of extracellular free Ca concentrations, can be taken as support for the existence of steady state conditions.[4]

The second caveat lies in the fact that this type of analysis tends to lump together exchangeable Ca pools that have similar $t_{1/2}$. For example, the intracellular pool associated with the dense tubular system may not be homogeneous. Different parts of the dense tubular system may have different Ca transport properties and play different roles in Ca homeostasis in platelets. Such information cannot be obtained by this method, but inhomogeneities in the slowly exchangeable pool are suggested by the failure of digitonin to release all of the ^{45}Ca that has entered this pool (Fig. 3).

The third caveat concerns the length of time during which ^{45}Ca influx or efflux are observed. Ideally, the observations should be continued for at least as long as the $t_{1/2}$ of the slowest compartment, provided the cells maintain their integrity. In practice, we have found that extending the efflux studies to 4 or 5 hr does not alter the results shown in Fig. 1A. Therefore, any additional exchangeable Ca pools in unstimulated platelets must have $t_{1/2} >> 5$ hr.

The final caveat has already been alluded to. In essence, this method of analysis defines exchangeable Ca compartments mathematically, but does not provide direct evidence that the pools actually exist. Therefore, some additional evidence for distinct Ca pools, such as that gathered in the studies with digitonin (Fig. 3), must also be obtained.

Example: The Effect of the Na Gradient on Ca Transport

The mechanisms that control Ca transport across the platelet plasma membrane are unknown. It is known, however, that changes in the plasma membrane Na gradient affect platelet function.[20] One possible explanation for this phenomenon is that Ca efflux from platelets and, therefore, the cytosolic free Ca concentration, are controlled entirely or in part by Na/Ca exchange. This problem can be addressed using a combination of ^{45}Ca exchange techniques and cytosolic free Ca concentration measurements. Figure 4 shows the changes in steady-state ^{45}Ca influx and efflux that occur when the plasma membrane Na gradient is altered by replacing extracellular Na with *N*-methyl-D-glucamine. In the absence of Na, more ^{45}Ca accumulates within the platelets and the fractional rate of loss is less. Similar curves are obtained when extracellular Na is replaced with choline or when the plasma membrane Na gradient is reduced by increasing the intracellular Na concentration with ouabain.[5]

Although these observations are consistent with the presence of an Na/Ca exchange mechanism, further analysis using Eqs. (5)–(13) suggests that this is not the case. Removing the Na gradient increases the sizes of the intracellular Ca pools, but has no effect on k_{10}, the rate constant for Ca

[20] T. M. Connolly and L. E. Limbird, *J. Biol. Chem.* **258,** 3907 (1983).

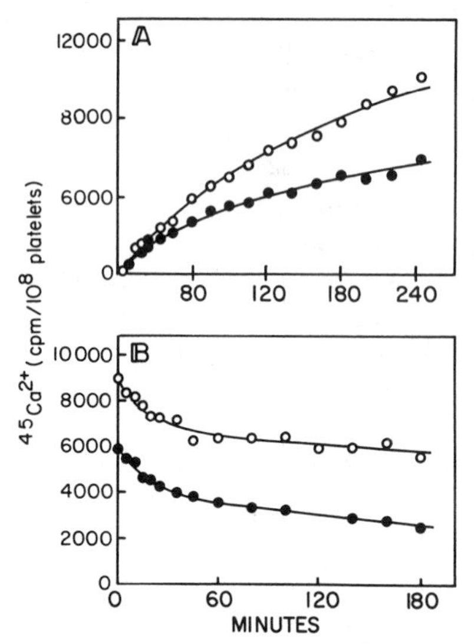

FIG. 4. The effect of removing extracellular Na on ^{45}Ca influx and efflux. Gel-filtered platelets were prepared in buffer containing either 137 mM NaCl (closed circles) or 137 mM N-methyl-D-glucamine (open circles) and incubated with 0.2 mM $CaCl_2$. After a 30-min equilibration period, ^{45}Ca was added. Calcium-45 influx (A) and efflux (B) were measured as described in Fig. 1. Reproduced with permission from Brass.[5]

efflux. At the same time, the rate of plasma membrane Ca flux (J_1) increases (Table III). These changes are inconsistent with an Na/Ca exchange mechanism. Equally important, the cytosolic free Ca concentration measured with Quin 2 and total platelet Ca are unaffected by decreasing $[Na]_o$ or increasing $[Na]_i$.[5] Therefore, although the plasma membrane Na gradient clearly affects Ca homeostasis in platelets, the effects produced by removing the gradient do not suggest a major contribution by a plasma membrane Na/Ca exchange system. The appearance of the ^{45}Ca influx and efflux curves is, therefore, somewhat misleading.[21] In contrast to these results in resting platelets, at least one recent report has suggested that Na/H exchange may affect agonist-induced changes in the cytosolic free Ca^{2+} concentration.[22] However, this conclusion has been challenged.[23]

Ca Transport in Agonist-Stimulated Platelets

Applying these methods to studies of activated platelets is somewhat problematic. Platelet activation by most agonists is associated with an

[21] A. Borle, *Cell Calcium* **2,** 187 (1981).
[22] W. Siffert and J. Akkerman, *Nature (London)* **325,** 456 (1987).
[23] A. W. Simpson and T. J. Rink, *FEBS Lett.* **222,** 144 (1987).

TABLE III
EFFECT OF REMOVING EXTRACELLULAR Na ON Ca EXCHANGE IN INTACT PLATELETS[a]

Parameter	Control	NMDG	p	Unit
S_1	54 ± 3	80 ± 12	<0.001	pmol/10^8 platelets
J_1	1.76 ± 0.16	2.43 ± 0.49	<0.005	pmol/10^8platelets/min
k_{10}	33.1 ± 3.5	31.1 ± 3.2	NS	min^{-1} (×1000)
S_2	153 ± 15	697 ± 74	<0.001	pmol/10^8 platelets
J_2	0.46 ± 0.02	1.15 ± 0.11	<0.01	pmol/10^8 platelets/min
k_{12}	8.7 ± 0.5	14.6 ± 1.5	<0.01	min^{-1} (×1000)
k_{21}	3.1 ± 0.3	1.8 ± 0.4	<0.005	min^{-1} 1×1000)

[a] Calcium-45 efflux from platelets in buffer containing NaCl or NMDG was measured as described in Fig. 4. The results shown are the mean ± SE ($n = 6$). Reproduced from Brass[5] with permission.

immediate rise in the cytosolic free Ca concentration into the micromolar range (see Chapter [33]). Studies with Quin 2 suggest that for thrombin the increase in cytosolic free Ca is due partly to increased Ca influx and partly to Ca release from intracellular sites.[1] Clearly, stimulated platelets are not at steady state. Borle *et al.*[24] have used the fractional ^{45}Ca efflux rate to examine the changes in Ca transport that occur immediately after addition of an agonist. This method has not yet been applied to platelets.

An alternative approach is to wait until the platelets enter a new steady state after the addition of the agonist before introducing ^{45}Ca into the system. Since the presence of the agonist precludes the second gel filtration required in the efflux studies, only ^{45}Ca influx studies can be performed. An example of the application of this approach is given elsewhere.[3] In those studies, gel-filtered platelets were equilibrated with 0.2 m*M* $CaCl_2$ for 15 min prior to the addition of 10 μM ADP or epinephrine. Fifteen minutes after addition of the agonist, $^{45}CaCl_2$ (4 μCi/ml) was added and tracer influx was observed for 3 hr. Since this was not sufficiently long for ^{45}Ca influx studies to delineate both intracellular Ca pools, the platelets were treated as a single compartment. The data suggest that the new steady state achieved after stimulation by ADP and epinephrine is characterized by an increase in the rate of Ca exchange across the platelet plasma membrane without a change in the amount of exchangeable intracellular Ca.[3]

[24] A. Borle, T. Uchikawa, and J. Anderson, *J. Membr. Biol.* **68,** 37 (1982).

[33] Measurement of Platelet Cytoplasmic Ionized Calcium Concentration with Aequorin and Fluorescent Indicators

By PETER C. JOHNSON, J. ANTHONY WARE, and EDWIN W. SALZMAN

Introduction

The importance of the cytoplasmic ionized calcium concentration ($[Ca^{2+}]_i$) to regulation of platelet function has been widely acknowledged.[1] Direct measurement of $[Ca^{2+}]_i$ in platelets has until recently been impossible, however, since the microinjection technique used to introduce Ca^{2+} indicators into large cells is not applicable to the 2-μm-diameter platelet. Two comparatively recent technical discoveries have enabled the direct measurement of $[Ca^{2+}]_i$ in intact platelets using the Ca^{2+}-sensitive fluorophore Quin 2 and the Ca^{2+}-sensitive photoprotein, aequorin. In addition, a second generation of Ca^{2+}-sensitive fluorophores (represented by Fura 2 and Indo 1) has been developed, in which signalling is much improved. This chapter will initially compare features of aequorin and Quin 2 as prototypes of their respective indicator classes. Then, the newer fluorophores will be described and their calibration and potential utility will be explored.

Tsien invented Quin 2 and described a technique for loading it into intact cells by reversibly masking its hydrophilic carboxylic acid functional groups with acetoxymethyl groups.[2,3] This lipophilic ester form can freely penetrate the plasma membrane, whereupon it is exposed to an intracellular esterase which cleaves the acetoxymethyl groups to yield free, trapped Quin 2. Intracellular accumulation of Quin 2 depends upon the extracellular concentration of the lipophilic form to which the cell is exposed.

Meanwhile, Sutherland *et al.* discovered that relatively large molecules (e.g., the photoprotein obelin) could enter intact cells when incubated together with them in solutions containing high concentrations of ethylene glycol bis(β-aminoethyl ether)-N,N,N',N'-tetraacetic acid (EGTA) and

[1] N. G. Ardlie, *Pharmacol. Ther.* **18,** 249 (1982).

[2] R. Y. Tsien, T. Pozzan, and T. J. Rink, *J. Cell Biol.* **94,** 325 (1982).

[3] R. Y. Tsien, *Biochemistry* **19,** 2396 (1980).

ATP.[4] Morgan and Morgan[5] loaded aequorin, a Ca^{2+}-sensitive luminescent protein (M_r 20,000), into vascular smooth muscle cells using this technique.

The Ca^{2+}-response characteristics of the fluorescent indicators and aequorin differ markedly, although fortunately in ways which complement one another. While each has disadvantages that necessitate cautious interpretation of results, the use of both groups of indicators in parallel experiments provides insight into platelet Ca^{2+} homeostasis not available from the use of either type of indicator alone.

Measurement of Platelet $[Ca^{2+}]_i$ Using Aequorin

As reviewed by Blinks *et al.*,[6] aequorin is synthesized by the photocytes of the jellyfish *Aequorea,* from which it can be extracted and purified. The techniques required to harvest and process *Aequorea aequorea* have been fully described,[6] as has the technique for purification of the protein in a Ca^{2+}-free environment.[7] A very pure preparation of the protein is available for purchase in lyophilized form from the Mayo Clinic.[7]

Aequorin as Indicator of $[Ca^{2+}]_i$: Theoretical Utility

The fact that aequorin has a biological source might suggest that its Ca^{2+}-response characteristics would be ideal for intracellular applications and, to some degree, this is true. For example, the range of $[Ca^{2+}]$ to which aequorin is sensitive ($10^{-8}-10^{-3}$ M) symmetrically encompasses the range of $[Ca^{2+}]_i$ ($10^{-7}-10^{-4}$ M) of greatest physiologic importance.[6] Further, the K_D of aequorin, which varies between 2 and 6 μM depending on the occupancy state of its three Ca^{2+}-binding sites, is similar to that of calmodulin ($K_D \sim 2.5$ μM)[8] (Fig. 1). Aequorin luminescence rises as the 2.5th power of the $[Ca^{2+}]$ within the physiologic range. This feature theoretically allows even small zones of elevated $[Ca^{2+}]_i$ to produce a bright aequorin signal. This feature also restricts the ability to determine mean $[Ca^{2+}]_i$ from data obtained using aequorin, a concept which will be addressed later in this chapter.

[4] P. J. Sutherland, D. G. Stephenson, and I. R. Wendt, *Abstr. Proc. Aust. Physiol. Pharmacol. Soc.* **11,** 160P (1980).

[5] J. P. Morgan and K. G. Morgan, *Pfluegers Arch.* **395,** 75 (1982).

[6] J. R. Blinks, P. H. Mattingly, B. R. Jewell, M. VanLeeuwen, G. Harker, and D. G. Allen, this series, Vol. 57, p. 292.

[7] J. R. Blinks, W. G. Wier, P. Hess, and F. G. Prendergast, *Prog. Biophys. Mol. Biol.* **40,** 1 (1982).

[8] P. C. Johnson, J. A. Ware, P. B. Cliveden, M. Smith, A. M. Dvorak, and E. W. Salzman, *J. Biol. Chem.* **260,** 2069 (1985).

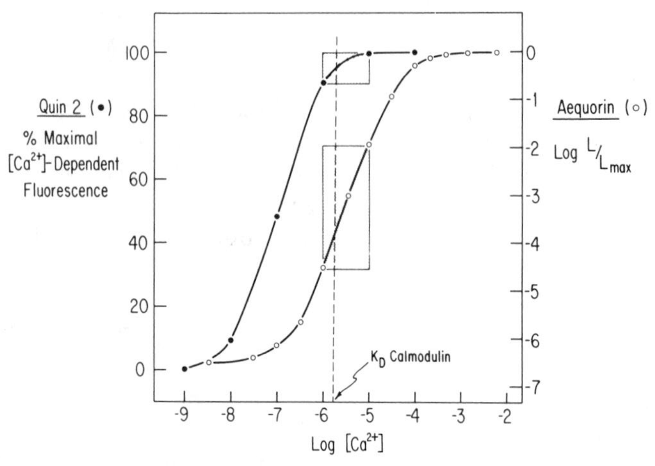

FIG. 1. Aequorin luminescence and Quin 2 fluorescence in response to various $[Ca^{2+}]$. Aequorin calibration curve (on log scale) prepared in a cell-free suspension containing 150 m*M* KCl and 1.25 m*M* Mg^{2+} at 22.5°. Quin 2 curve (on linear scale) was calculated by the authors, assuming a K_D of 115 n*M*. The shaded regions highlight the 1–10 μM $[Ca^{2+}]$ range, which is the region of greatest physiologic interest in stimulated platelets, and includes the K_D for calmodulin (2.5 μM). Aequorin curve from Ref. 9 by permission. It should be noted that adjustment of both Mg^{2+} and temperature to levels found physiologically in human platelets will significantly affect the left-hand portion of the aequorin curve.[24]

Aequorin is highly selective for Ca^{2+} over other physiologically important cations. Mg^{2+} inhibits the Ca^{2+}–aequorin-binding reaction by competition, but does not discharge the photoprotein. Therefore, a constant or near-constant $[Mg^{2+}]_i$ must be assumed when using aequorin in intracellular applications. This is not necessarily a safe assumption, considering our present ignorance regarding the fate of $[Mg^{2+}]_i$ in the stimulated cell. Silver and mercury ions cause rapid, irreversible loss of aequorin luminescence (without overt discharge of light) and must be prevented from having access to the photoprotein. While strontium ions can discharge aequorin, they are not present in cells in a concentration great enough to be of concern.

The quantum yield of aequorin wavers little within the ranges of pH and ionic strength found in the cell.[9] While the rate of aequorin discharge at a particular $[Ca^{2+}]$ is a function of temperature, this variable can be controlled through use of isothermic experimental conditions.

[9] J. R. Blinks, F. G. Prendergast, and D. G. Allen, *Pharmacol. Rev.* **28,** 1 (1976).

Preparation and Handling of Aequorin

The jellyfish *Aequorea aequorea* may be netted yearly in the fall at Friday Harbor, Washington. Their aequorin-bearing photocytes are dissected mechanically and lysed in hypotonic ethylenediaminetetraacetic acid (EDTA) solution (10–50 m*M*). Subsequent steps include ammonium sulfate precipitation of the crude extract, gel filtration on a Sephadex G-50 bed, and ion-exchange chromatography on QAE-Sephadex A-50 with pH step and salt gradient elution. The product, according to Blinks,[6,9] can be made 98% pure (as shown using SDS gel electrophoresis) with a yield of 10% relative to the amount of aequorin present in the undisturbed jellyfish. In the final separation step, the aequorin is eluted with a solution containing 150 m*M* KCl, 5 m*M* sodium acetate, and 5 m*M* HEPES. The aequorin concentration is then adjusted to approximately 1 mg/ml, and 1-ml aliquots are lyophilized and hermetically sealed in glass vials. Aequorin prepared in this way and sold by the Mayo Clinic is consistently of high quality.

The dried protein is reconstituted easily, using doubly distilled or Chelex-treated H_2O to avoid contamination with Ca^{2+}. The possibility of contamination of instruments and H_2O with Ca^{2+} must always be considered since aequorin luminescence is rapidly triggered upon contact with $[Ca^{2+}] > 10^{-8}$ *M* and, for practical purposes, the active photoprotein cannot be regenerated. This is especially important for large cell microinjection experiments, in which aequorin must be redissolved in medium containing no Ca^{2+} chelators (such as EDTA or EGTA) to protect against contamination.

Fortunately, when loading platelets with aequorin through a modification of Sutherland's technique,[8] EGTA is present with aequorin in the extracellular loading solution. Therefore, under these conditions, aequorin may be reconstituted from the lyophilized state using a solution containing enough EGTA (10^{-2} *M*) to ensure protection against stray Ca^{2+} contamination. Our standard practice is to dissolve 1 mg of lyophilized aequorin in 333 μl deionized H_2O containing 7–10 m*M* EGTA. The solution is divided into 20-μl aliquots which can then be stored at $-80°$ for at least 6 months without appreciable decay.[8]

Introduction of Aequorin into Cells

Since microinjection was required in the past to introduce aequorin into cells, a large body of data has accumulated in which this technique was used to measure $[Ca^{2+}]_i$ in relatively large cells from several invertebrate species.[10] Recently, several methods have been developed to load aequorin into cells too small to microinject. Two of these (exposure to reversible hypoosmotic conditions in the presence of the photoprotein, and scraping of cells against a Petri dish containing photoprotein-laden buffer) have been fully described elsewhere.[11,12] Solutions containing up to millimolar concentrations of EGTA and ATP have been used by several investigators to load small cells with molecules of increasing size, including nucleotides,[13] myosin light chain derivatives,[14] fluorescent dyes,[13] and the photoproteins aequorin[5] and obelin.[4]

Our method for loading aequorin into human platelets has been modified from the techniques of Sutherland *et al.*[4] and Morgan and Morgan,[5] for cardiac and smooth muscle cells, respectively. The latter investigators compared signals from cells loaded with aequorin via the EGTA/ATP technique to signals from cells microinjected with aequorin and found them to be similar in conformation and magnitude, after correction for the increased amount of aequorin loaded by the chemical method. Unfortunately, no such comparison is possible using platelets. Our procedure for loading aequorin into platelets is as follows:

Human blood is drawn by venepuncture into $\frac{1}{10}$ vol of 0.15 *M* trisodium citrate and gently mixed. Platelet-rich plasma (PRP) is prepared by centrifuging the whole blood sample at 2000 *g* at room temperature for 120 sec and aspiration of PRP to not less than 1 cm from the buffy coat.

Prostaglandin E_1 (PGE_1), 1 μM, is added to the PRP from a 1 m*M* stock solution in absolute ethanol. The PRP is then spun at 430 *g* for 15 min to form a soft platelet pellet. The supernatant plasma is discarded.

The pellet is then resuspended/washed in HEPES–Tyrode's buffer (NaCl, 129 m*M*; $NaHCO_3$, 8.9 m*M*; KCl, 2.8 m*M*; KH_2PO_4, 0.8 m*M*; $MgCl_2$, 0.8 m*M*; dextrose, 5.6 m*M*; HEPES, 10 m*M*; pH 7.4), also containing 10 m*M* EGTA and 1 μM PGE_1 at 0°. The sample is transferred to a microcentrifuge vial and spun at 11,700 *g* for 12 sec, and the supernatant is discarded.

The pellet is resuspended in 280 μl of ice-cold aequorin loading solution A, which contains 150 m*M* NaCl, 5 m*M* HEPES, 5 m*M* ATP, 2 m*M* $MgCl_2$, 10 m*M* EGTA, and 1 μM PGE_1, pH 7.0. Twenty microliters of the reconstituted aequorin solution (=3 mg/ml aequorin) is then added to provide a final extracellular aequorin concentration of 0.2 mg/ml (approx-

[10] C. C. Ashley and A. K. Campbell, "Detection and Measurement of Free Ca^{2+} in Cells." Elsevier/North-Holland, Amsterdam, The Netherlands, 1979.

[11] K. W. Snowdowne and A. B. Borle, *Am. J. Physiol.* **247,** C396 (1984).

[12] P. L. McNeil and D. L. Taylor, *Cell Calcium* **6,** 83 (1985).

[13] B. D. Gompertz, *Nature (London)* **306,** 64 (1983).

[14] W. G. L. Kerrick and L. Y. W. Bourguignon, *Proc. Natl. Acad. Sci. U.S.A.* **81,** 165 (1984).

imately 10^{-5} *M*). Platelets are incubated for 1 hr over melting ice, then spun at 11,700 *g* for 12 sec. The supernatant is aspirated, and the platelet pellet is resuspended in 1 ml solution B, which contains 150 m*M* NaCl, 5 m*M* HEPES, 5 m*M* ATP, 10 m*M* $MgCl_2$, 0.1 *M* EGTA, and 1 μ*M* PGE_1, pH 7.0, but no aequorin. The platelets are again incubated for 1 hr over melting ice.

At the end of the incubation in solution B, the cold platelets are recalcified with three 1-μl aliquots of 100 m*M* $CaCl_2$ added 5 min apart. The platelet suspension is then removed from the ice and allowed to equilibrate at room temperature for 10 min.

The aequorin-loaded platelet suspension is then layered onto a Sepharose 2B gel column (9 ml bed volume) preequilibrated with HEPES–Tyrode's buffer containing 1 m*M* [Ca^{2+}] and 0.1% bovine serum albumin or 0.1% gelatin, which may be omitted if desired. Elution and resuspension of platelets (to $2-6 \times 10^8$/ml) is carried out using the same buffer. Aequorin-loaded platelets are incubated at 37° for at least 10 min before use, but must be studied within 1 hr, since consumption of the aequorin occurs rapidly.

Once loaded into platelets, aequorin decays with a half-life of approximately 45 min at 22° and nearly twice as quickly at 37°. This is partially due to aequorin leakage from loaded platelets into the surrounding Ca^{2+}-rich extracellular fluid. However, the high basal $[Ca^{2+}]_i$ indicated by aequorin, even when EGTA is present to chelate extracellular Ca^{2+} (see below), suggests that consumption of aequorin also occurs by continuing exposure to intracellular Ca^{2+}.

Platelets can also be loaded with aequorin by incubation with DMSO,[15] and other cells have been loaded by brief exposures to hypoosmotic solutions (HOST).[16] We have been able to load platelets with aequorin using these methods, both of which load more aequorin than does the EGTA–ATP method; however, exposure of platelets to either the DMSO or HOST method diminishes the platelet response to exogenous agonists, which is not true of the EGTA–ATP method (unpublished observations).

Platelet Integrity during and after Loading with Aequorin

During the aequorin-loading procedure, platelets undergo shape change as a consequence of their exposure to cold. As previously reported,[17] cold platelets return to the discoid shape when rewarmed at 37°. Electron micrographs taken at each step of the aequorin-loading procedure demon-

[15] A. Yamaguchi, H. Suzuki, K. Tanoue, and H. Yamazaki, *Thromb. Res.* **44,** 165 (1986).
[16] A. B. Borle, C. C. Freundenrich, and K. W. Snowdowne, *Am. J. Physiol.* **251,** C323 (1986).
[17] J. G. White and W. Krivit, *Blood* **30,** 625 (1967).

strate the cold-induced shape change, but show no loss of membrane continuity or granule centralization (which, if present, would suggest activation of the contractile apparatus).[8] Separation of the glycocalyx from the plasma membrane does occur to some extent. The lack of plasma membrane damage is consistent with the findings of Kerrick and Bourguignon,[14] who found no loss of membrane continuity in similarly treated lymphocytes studied by electron microscopy.

As previously reported,[9] platelets preloaded with the cytoplasmic marker [^{3}H]adenine and the dense granule marker 5-hydroxy[^{14}C]tryptamine (serotonin) retain these markers throughout the aequorin-loading procedure. Both lactate dehydrogenase (another cytoplasmic marker) and β-thromboglobulin (a marker for platelet α granules) are also retained during this procedure. This indicates that treatment of platelets with EGTA and ATP under these conditions does not make the platelets freely permeable to constituents of the cytoplasm or induce secretion of granule contents.[8] The shape change induced by the cold solutions during the loading procedure resolves when platelets are rewarmed to 37° for 10 min.

Quantitation of Aequorin Concentration in Platelets

Signal Requirements. Even though intracytoplasmic and intragranular substances are retained during the loading procedure, which implies the presence of intact membranes, a small quantity of aequorin appears able to traverse the plasma membrane. This is evidenced by the failure to discharge all of the aequorin instantly when the platelets are suspended in a saturating concentration of Ca^{2+} (1 mM) at the end of the loading procedure. Further, agonist-induced increases in aequorin luminescence are obtained even in the presence of sufficient extracellular EGTA to reduce the external [Ca^{2+}] to $<10^{-7}$ M (Fig. 2).

Based upon the known number of photons emittable by 1 mg of pure aequorin (4.5×10^{15}),[18] the final intraplatelet aequorin concentration can be calculated to be approximately 1 nM (about 10 aequorin molecules per platelet). When ^{125}I-labeled aequorin[8] is loaded into platelets, a higher figure is obtained (10^{-8} M), which is explained by the fact that the former method measures only undischarged aequorin, while the isotopic method measures both intact and spent aequorin, including any aequorin which might be bound to the external surface of the platelet. It is possible to increase the aequorin content 10- to 100-fold by commensurate increases in the extracellular aequorin concentration in loading solution A. This is not necessary for routine experiments, however, since 10^{-9} M aequorin provides sufficient luminescence to allow detection in the instruments to

[18] F. H. Johnson and O. Shimomura, this series, Vol. 57, p. 271.

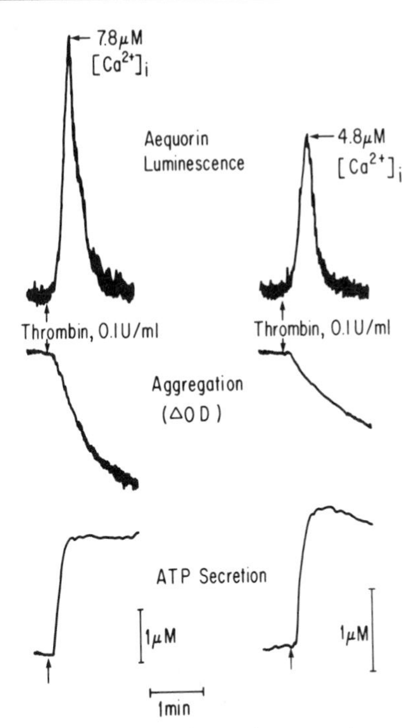

FIG. 2. Representative tracings of aequorin luminescence (top panel), aggregation (middle panel), and secretion of ATP (bottom panel) in response to thrombin (0.1 U/ml), in media containing Ca^{2+}, 1 m*M* with (right panel) or without (left panel) EGTA, 2 m*M*. ATP secretion was recorded using a Lumiaggregometer and the firefly luminescence assay.[26] Although external Ca^{2+} contributes to both the aequorin-indicated $[Ca^{2+}]_i$ and platelet functional response, thrombin can generate, and aequorin can detect, changes in $[Ca^{2+}]_i$ arising from apparent intracellular sources.

be described. The low concentration of intracellular aequorin provides assurance that the indicator will not significantly buffer $[Ca^{2+}]_i$ transients, as has been shown indirectly by comparison of aggregation and secretion of aequorin-loaded gel-filtered platelets (GFP) and GFP prepared by standard techniques without exposure to the aequorin-loading solutions.[8] These show similar concentration–response relationships for all common platelet agonists.

Despite the small number of indicator molecules/platelet, as long as these molecules are randomly distributed within the population of loaded platelets, the interpretation of signals from such cells, based upon the calibration procedure to be described, is valid. The calibration of aequorin depends on the probability of Ca^{2+} binding to individual aequorin molecules at various $[Ca^{2+}]$ compared to the probability of Ca^{2+} binding to aequorin when it is exposed to a saturating [Ca^{2+}]. Therefore, it makes no difference whether 1 molecule or 500 molecules of aequorin are contained in each platelet, as long as stimulated platelets mobilize [Ca^{2+}]$_i$ similarly, and aequorin does not by itself change the [Ca^{2+}]$_i$ to which it is exposed. Blinks has calculated that as much as 10^{-5} *M* intracellular aequorin may be used without the indicator itself providing significant buffering of [Ca^{2+}]$_i$.[9]

Location of Aequorin in the Platelet. The occurrence of an aequorin luminescent signal from platelets which are stimulated while suspended in an EGTA-containing (essentially Ca^{2+}-free) medium (Fig. 2) suggests an intracellular location of aequorin. More precise localization has not yet been possible, due to the small number of aequorin molecules per platelet. Since the method of loading apparently involves the transmembrane passage of aequorin, it is possible that the majority of aequorin molecules lie within or adjacent to the plasma membrane, rather than being uniformly distributed throughout the cytoplasm. Selective digitonin lysis suggests that undischarged aequorin is not in either the dense tubular system or secretory granules.[19] Aequorin signals cannot at present be used to infer the precise intraplatelet location of [Ca^{2+}]$_i$ transients.

Measurement of Aequorin Signals. Measurement of luminescence from aequorin-loaded platelets requires the apposition of a highly sensitive photomultiplier tube to a light-tight heated (37°) sample chamber. Our original apparatus, constructed by Dr. Robert Auty, combines customized circuitry with an EMI-Gemco refrigerated photomultiplier tube (#8401-A) and housing. The 55-mm photomultiplier photocathode is installed directly beneath a shuttered aluminum sample chamber containing a 50-mm disposable polystyrene Petri dish. A flanged aluminum lid prevents light leakage. A 300 rpm electric turntable motor (Teledyne Acoustic Research #881399, Norwood, MA), installed centrally in the lid, turns a disposable polystyrene T-bar stirrer immersed in the 5-ml sample. An 18-gauge spinal needle, whose shaft is turned through 360° to prevent light entry, is permanently inserted through the lid for agonist addition. Custom-built circuitry converts the photomultiplier current to voltage, and the analog signal is recorded on a Fisher Recordall strip chart recorder, having decade scales sensitive from 0.001 to 10 V. Motion (including aggregation) of platelets does not interfere with measurement of the aequorin signal.[7,9] The photomultiplier tube is calibrated using radioactive photon emission standards prepared according to a previously described method.[20]

While our initial studies used this custom-built apparatus to measure

[19] J. A. Ware, P. C. Johnson, M. Smith, and E. W. Salzman, *J. Clin. Invest.* **77**, 878 (1986).

[20] J. W. Hastings and G. Weber, *J. Opt. Soc. Am.* **53**, 1410 (1963).

luminescence, a modified commercially available instrument (Chronolog Lumiaggregometer—Chronolog Corp., Havertown, PA) is capable of measuring aequorin luminescence and platelet aggregation simultaneously.[21] A second generation device from the same manufacturer has been designed with these modifications as standard equipment, and is marketed as a "Platelet Ionized Calcium Lumiaggregometer (PICA)." This instrument is satisfactory for the study of aequorin-loaded platelets, provided that the factory-installed electronic delays and damping devices are removed.

Calibration of Aequorin Signals. Aequorin can be calibrated *in vitro* by rapid injection of a quantity of aequorin having a known total photon content into well-characterized buffers of known [Ca^{2+}] whose constituents and pH mimic cytosolic conditions.[6,7,9] The initial rate of photon emission (L) is then compared to the rate of photon emission (L_{max}) obtained when the same quantity of aequorin is injected into a buffer containing a saturating concentration of Ca^{2+} (e.g., $>10^{-3}$ M). The calibration apparatus should have a temperature-controlled chamber over which a rubber septum is situated. The latter is penetrated by a spring-loaded syringe (Hamilton Co., Reno, NV) which reproducibly ejects several microliters of the standard aequorin solution with a force that ensures rapid mixing with the test buffer.

Log L/L_{max} is then plotted versus log[Ca^{2+}] to provide the calibration curve (Fig. 1). Calibration of intracellular signals requires an analogous approach. The height of the aequorin luminescence tracing obtained from either resting or stimulated platelets can be read from an analog recorder and compared as a ratio (with normalized scales) to the height of the luminescence spike obtained when a detergent such as Triton X-100 lyses the cell membranes and exposes all of the aequorin in the preparation to the saturating [Ca^{2+}] in the medium. Immediate lysis is necessary to ensure that the maximal rate of photon emission (L_{max}) will occur. Alternatively, one can convert these signals to digital data for more accurate and efficient processing.

At least two problems complicate this type of calibration. First, since the ability of aequorin to emit photons is "consumed" upon its reaction with Ca^{2+}, the quantity of intracellular aequorin may differ between the time of agonist addition (when L is generated) and the time of cell lysis (to provide L_{max}). When aequorin is microinjected into large cells, an aequorin concentration as high as 10^{-5} M is achieved, and individual luminescent signals (L) consume an insignificant percentage of the intracellular indica-

[21] P. C. Johnson, J. A. Ware, and E. W. Salzman, *Thromb. Res.* **40,** 438 (1985).

tor. Multiple evoked signals can be obtained from these large cells without an appreciable alteration in the L_{max} obtained upon lysis of the cell.

In platelets, however, there is a significant basal rate of aequorin consumption, such that the L_{max} may change from the beginning of experimentation to the end. The components of this consumption are not established; they are not limited to leakage of aequorin into the Ca^{2+}-rich extracellular fluid (which does occur), since consumption also occurs when the platelets are suspended in buffers containing EGTA ($[Ca^{2+}] < 10^{-7}$ *M*). We determine L_{max} by lysis of a known number of aequorin-loaded platelets which have not been previously stimulated with an agonist to provide a luminescent signal, and which have been incubated at 37° for the same length of time to control for nonspecific consumption. This relationship between L and L_{max} remains constant from 10 to 60 min after incubation; however, wide variance is seen among samples incubated in Ca^{2+}-containing media either before or after this time frame.

Second, to determine L_{max} accurately, all of the intracellular aequorin must be exposed to a saturating $[Ca^{2+}]$ instantaneously, or at least as rapidly as occurs when aequorin-containing solutions are injected *in vitro* into calibration buffers. Intracellular aequorin is enclosed by the plasma membrane and perhaps also by other membranes. Lysis of these membranes may constitute a rate-limiting step in the exposure of intracellular aequorin to saturating $[Ca^{2+}]$. If the rate of egress of aequorin from a "lysed" cell is significantly delayed relative to the rate with which aequorin mixes with the buffer when injected alone *in vitro*, L_{max} will be artifactually diminished. Log L/L_{max} becomes larger under these circumstances, and declares a higher $[Ca^{2+}]_i$ than, in truth, develops. Blinks[6] has devised a method to correct for this artifact based on the following principle: the total number of photons emitted upon total lysis of an aequorin-loaded cell will be independent of the rate of aequorin discharge. The number of photons emitted is directly proportional to the area under the curve (i.e., the integral) of the complete luminescence tracing. When a known quantity of aequorin is injected into a standard buffer containing $>10^{-3}$ *M* $[Ca^{2+}]$, both L_{max} and the luminescence integral are found to be constants. Therefore, the ratio L_{max}/area is constant, under these conditions. When aequorin-loaded cells are lysed with a detergent into buffer containing $>10^{-3}$ *M* $[Ca^{2+}]$, any delay in membrane lysis is manifest as a decrease in this ratio, since L_{max} is dwarfed and the signal is protracted over time. Correction toward the true L_{max} requires multiplication of the area under the cell lysis curve by the *in vitro* peak-to-integral ratio. Thus,

$$L'_{max} = (L_{max}/\text{area})(\text{area}')$$

where L'_{max} and area′ apply to the parameters measured upon lysis of the cell preparation.

This correction becomes important only if the peak-to-integral ratio obtained upon aequorin discharge from a lysed cell differs significantly from the peak-to-integral ratio obtained upon discharge of aequorin into a solution containing a saturating $[Ca^{2+}]$ *in vitro.* Our studies with platelets, which lyse very rapidly when treated with 0.1–1.0% (final concentration) Triton X-100, show that the peak-to-integral ratio obtained upon lysis into saturating $[Ca^{2+}]$ is 0.45 times (in an appropriately modified PICA) that obtained when aequorin alone is discharged *in vitro.* Therefore, while the lysis of platelet membranes represents a rate-limiting step and requires that L_{max} be adjusted accordingly, the ultimate effect on the calculated $[Ca^{2+}]_i$ is relatively small.

In calibrating the response of aequorin to Ca^{2+} *in vitro,* it may be necessary to preequilibrate aequorin with a $[Mg^{2+}]$ identical to that found in the calibration buffer (and therefore to the expected $[Mg^{2+}]_i$). Earlier measurements made without this preequilibration yielded results which differ slightly from those obtained after preequilibration (Blinks, personal communication), but in practice we find that preequilibration has little discernible effect. However, variations in the $[Mg^{2+}]_i$ would produce a shift in the calibration curve for aequorin, such that at higher $[Mg^{2+}]_i$ a given L/L_{max} indicates a higher $[Ca^{2+}]_i$. The $[Mg^{2+}]_i$ has been estimated to be 1 mM in lymphocytes,[22] and therefore we have previously used an aequorin calibration curve carried out near this $[Mg^{2+}]$. However, if the $[Mg^{2+}]_i$ were much lower, as reported in erythrocytes,[23] both the resting and stimulated $[Ca^{2+}]_i$ calculated from a given L/L_{max} would be significantly reduced. More recent experiments have been performed using platelets, in which nuclear magnetic resonance techniques have been used to measure $[Mg^{2+}]_i$. The ^{31}P NMR spectra demonstrated that the $[Mg^{2+}]_i$, as calculated from the chemical shift values of ATP resonances, was 0.17 ± 0.06 mM in unstimulated platelets, which was similar to that obtained by null point titration.[24] When this $[Mg^{2+}]_i$ value was used to construct a Ca^{2+}-calibration curve for aequorin at 37°, a given L/L_{max} corresponded to a lower $[Ca^{2+}]_i$ in unstimulated platelets; however, these values are still significantly elevated compared to those obtained with fluorophores (see below).

Appearance of the Aequorin Signal in Response to Agonists

Figure 2 depicts a typical response of aequorin-loaded platelets to thrombin. When platelets are suspended in a medium containing 1 mM

[22] T. J. Rink, R. Y. Tsien, and T. Pozzan, *J. Cell Biol.* **95,** 198 (1982).

[23] R. K. Gupta, J. L. Benovic, and Z. B. Rose, *J. Biol. Chem.* **253,** 6172 (1978).

[24] M. Smith, J. A. Ware, E. T. Fossel, and E. W. Salzman, *Thromb. Haemostas.* **58** (Suppl. 1), 19 (Abstr.) (1987).

$[Ca^{2+}]$, 0.1 U/ml thrombin causes a rise in $[Ca^{2+}]_i$ from an apparent basal level of 1 – 2 μM to a peak level of nearly 10 μM. A similar experiment with platelets suspended in a medium containing 2 mM EGTA demonstrates that thrombin can initiate intracellular Ca^{2+} mobilization, that aequorin is able to detect this event, and that the peak $[Ca^{2+}]_i$ achieved in this setting is close to that obtained when aequorin-loaded platelets are stimulated by thrombin while suspended in a medium containing 1 mM Ca^{2+}.

Aequorin-loaded platelets indicate a rise in $[Ca^{2+}]_i$ in response to all common platelet agonists, including epinephrine, collagen, ADP, the prostaglandin H_2 analog U44619, the Ca^{2+}-ionophore A23187, arachidonic acid, thrombin, and phorbol ester.[19,25] The luminescent signal in response to these agonists is not secondary to platelet aggregation, since the addition of an agonist to aequorin-loaded platelets without stirring (thereby preventing aggregation) only minimally depresses the resulting aequorin signal, relative to a stirred control. Resumption of stirring, which then allows the development of full aggregation, stimulates no further rise in the aequorin signal.[21] Furthermore, omission of fibrinogen, which is essential for platelet aggregation induced by some agonists (e.g., ADP), only minimally affects the peak $[Ca^{2+}]_i$ indicated by aequorin (unpublished observations).

The aequorin-indicated rise in $[Ca^{2+}]_i$ precedes or coincides with the onset of the first visible platelet functional change (shape change or aggregation) in response to all of the agonists mentioned above.[19,21] Since secretion occurs after shape change and aggregation are underway,[26] the aequorin signal does not appear to require secreted platelet granule contents such as ADP or serotonin.

Upon addition of an agonist to platelets, the aequorin signal rises to a peak and then falls quickly to a lower level as shown in Fig. 2. The decline is more rapid in response to certain agonists (e.g., thrombin, A23187, arachidonic acid) than to others (phorbol ester, ADP, epinephrine), suggesting that the rapid decline in the signal, while perhaps in part due to the consumption of aequorin, may also be a reflection of other features: e.g., how effectively Ca^{2+} can be constrained within certain regions of the cytosol upon stimulation by various agonists.

The hypothesis that aequorin can detect $[Ca^{2+}]_i$ changes in isolated compartments of platelets is based on image intensification studies in other cells,[27,28] on comparing of aequorin responses with those of Quin 2 after

[25] J. A. Ware, P. C. Johnson, M. Smith, and E. W. Salzman, *Biochem. Biophys. Res. Commun.* **133,** 98 (1985).

[26] R. D. Feinman, J. Lubowsky, I. Charo, and M. P. Zabinski, *J. Lab. Clin. Med.* **90,** 125 (1977).

[27] B. Rose and W. R. Lowenstein, *Science* **190,** 1204 (1975).

[28] H. Harary and J. E. Brown, *Science* **224,** 292 (1984).

stimulation of platelets loaded with both indicators,[19] and on theoretical grounds.[8] In platelets loaded with Quin 2 and aequorin, it can be shown that aequorin will detect an elevated $[Ca^{2+}]_i$ under certain circumstances (e.g., after stimulation with epinephrine, phorbol ester, or diacylglycerol) when Quin 2 will not, suggesting that these indicators reflect different aspects of $[Ca^{2+}]_i$ homeostasis.[19,25] The response features of aequorin, described earlier, indicate that it is capable of detecting a localized elevation of $[Ca^{2+}]_i$, if one exists. Conversely, Quin 2 may not have the ability to detect a localized change, but is very unlikely to fail to indicate a generalized rise in $[Ca^{2+}]_i$. Combined, the two indicators may provide complementary information regarding the distribution of $[Ca^{2+}]_i$ within the platelet.

Interpretation of the Aequorin-Indicated Basal $[Ca^{2+}]_i$

An unexplained peculiarity of the aequorin-indicated basal $[Ca^{2+}]_i$ is its high apparent concentration relative to the levels of $[Ca^{2+}]_i$ thought necessary to activate cellular events. Using the calibration techniques described by Blinks,[6,9] aequorin indicates a basal $[Ca^{2+}]_i$ of about 1 μM when platelets are suspended in a buffer containing 1 mM $[Ca^{2+}]$. When the external $[Ca^{2+}]$ is abruptly dropped to $<10^{-8}$ M by addition of EGTA, however, the measured $[Ca^{2+}]_i$ falls only to 400–600 nM. We have hypothesized that aequorin is concentrated in a zone of high local $[Ca^{2+}]_i$ in the platelet. This zone might be submembranous, where the Ca^{2+} for aequorin discharge may be provided by the shuttling of Ca^{2+} ions between phospholipid head groups,[29,30] or via the pump–leak cycle in proximity to intracellular Ca^{2+}-bearing organelles.[31] It is also possible that aequorin contained in a few injured cells may be exposed to high $[Ca^{2+}]$. Since the log–response of aequorin luminescence to Ca^{2+} will preferentially expose the highest zone of $[Ca^{2+}]_i$ present in a cell, the high basal level prevents one from using aequorin to measure $[Ca^{2+}]_i$ lower than about 400 nM.

Measurement of Platelet $[Ca^{2+}]_i$ Using Fluorescent Indicators

The first direct measurement of $[Ca^{2+}]_i$ in platelets was made using the Ca^{2+}-sensitive fluorophore Quin 2. It represents the first generation prototype of a group of fluorescent indicators that have revolutionized the study of $[Ca^{2+}]_i$ in cells too small to permit microinjection. The compound binds Ca^{2+} via two of its four carboxylic acid groups, providing a one-to-one

[29] M. J. Broekman, *Biochem. Biophys. Res. Commun.* **120,** 226 (1984).
[30] M. J. Broekman, J. W. Ward, and A. J. Marcus, *J. Clin. Invest.* **66,** 275 (1980).
[31] N. L. Leung, R. L. Kinlough-Ruthbone, and J. F. Mustard, *Br. J. Haematol.* **36,** 417 (1977).

stoichiometry, a feature that is shared by its more recent relatives, to be described later.

Quin 2 as Indicator of $[Ca^{2+}]_i$*: Theoretical Utility*

Free Quin 2 is fluorescent, with a maximal excitation wavelength of 339 nm and a maximal emission wavelength of 492–500 nm. The intensity of the basal, or "Ca^{2+}-independent fluorescence," observed with free Quin 2 is enhanced up to 6-fold when the compound is exposed to a $[Ca^{2+}]$ that saturates the indicator (e.g., $>10^{-5}$ *M*). The range of sensitivity to Ca^{2+} of Quin 2 is 10^{-9} to 10^{-5} *M*, representing two log units of concentration on either side of its K_D for Ca^{2+}, 115 n*M* (Fig. 1). However, since the resting $[Ca^{2+}]_i$ in most cells is thought to be approximately 100 n*M*, Quin 2 is initially half-saturated in intracellular applications. Thus, the Quin 2 signal from stimulated cells can only double before saturation fluorescence intensity is reached. This feature of the compound makes it relatively insensitive to small and/or local changes in $[Ca^{2+}]_i$, in contrast to photoproteins. The importance of this point will be demonstrated later. This and other differences between Quin 2 and photoproteins are illustrated in Table I.

Quin 2 is highly selective for Ca^{2+} over Mg^{2+}. Its binding affinity for Mn^{2+} is so much higher than for Ca^{2+}, however, that micromolar concentrations of the former element are capable of completely quenching the Ca^{2+}-dependent fluorescence in the presence of up to millimolar $[Ca^{2+}]$. Therefore, reagents used in Quin 2 studies should be kept free of Mn^{2+}.

The Quin 2 signal at a fixed $[Ca^{2+}]$ varies slightly as a function of pH, but to a degree that is probably insignificant in intracellular applications, in which pH changes are minimized by endogenous buffers.[10] A more important variable in Quin 2 studies is temperature. In general, molecular fluorescence intensity is enhanced by a reduction in temperature. The slopes of temperature–fluorescence intensity curves can be steep, demanding accurate temperature control when using fluorescent indicators such as Quin 2.

The fluorescence responses of cells themselves must also be taken into account when using Quin 2 to indicate $[Ca^{2+}]_i$. The excitation wavelength employed (339 nm) is in the ultraviolet range, and is capable of stimulating the fluorescence of several cellular constituents, notably nucleotides, providing measurable signals in the visible spectrum (e.g., at 500 nm). Consequently, all studies performed using Quin 2-loaded platelets must be accompanied by parallel studies using an equal concentration of platelets without the fluorophore to enable correction of the calibration for $[Ca^{2+}]_i$ (Fig. 3).

Since the signal-to-noise ratio of Quin 2 in intracellular applications is small, a relatively high (e.g., >1 m*M*) intracellular concentration of the

TABLE I
COMPARISON OF FLUORESCENT INDICATORS AND AEQUORIN

Parameter	Aequorin	Quin 2	Fura 2	Indo 1
K_D (μM)	2–6	0.115	0.224	0.250
Molecular weight	20,000	800	1002	1010
Property measured	Luminescence	Fluorescence	Fluorescence	Fluorescence
Cytoplasmic concentration	1 nM	1 mM	10–30 μM	10–30 μM
Basal [Ca^{2+}] (μM)	0.5[a]	0.1	0.1	0.1
Peak [Ca^{2+}] (μM)	1–15	1	1–2	1–2
Consumed	Yes	No	No	No

[a] In EGTA-containing media.

indicator is necessary to furnish adequate signals. This fact, however, also introduces the possibility that mobilized Ca^{2+} is buffered by the indicator, providing an altered estimate of the $[Ca^{2+}]_i$ that would otherwise occur in cells not loaded with Quin 2. There is evidence that this indeed occurs[8,19] (Fig. 4).

An additional problem with Quin 2 and some other fluorescent indicators is that homogeneous suspensions of platelets are required to ensure that the real-time signal rises or falls solely as a function of $[Ca^{2+}]_i$. If platelets aggregate after stimulation, the peak Quin 2 signal will not be adequately represented, because, as in optical density measurements of platelet aggregation, the net concentration of single platelets in the measured column of fluid falls throughout the course of the experiment. There is no obvious way to correct for this artifact, and while nonstirred preparations have occasionally been employed to avoid this problem,[32] the question is left as to whether the close interplatelet contact achieved by stirring independently alters $[Ca^{2+}]_i$.

While many such factors must enter into the interpretation of data collected using Quin 2, it has nevertheless been shown to be a useful indicator of changes in $[Ca^{2+}]_i$, at least under circumstances in which such changes occur throughout much of the cytosol at once. One great advantage to the use of Quin 2 lies in the ease with which it can be loaded into platelets and other small cells.

Preparation and Handling of Quin 2. The acetoxymethyl ester of Quin 2 (Quin 2/AM) is available from several commercial suppliers, and it is easily stored as a 10 mM solution in dimethyl sulfoxide (DMSO) at −20°. It is very stable under these conditions. The free acid, Quin 2, may also be purchased and is useful in studies of agonist-induced fluorescence artifacts. It is stored in a similar fashion, and is also stable in DMSO.

Introduction of Quin 2 into Platelets. Since Quin 2 is a polar molecule

[32] T. J. Rink, S. W. Smith, and R. Y. Tsien, *FEBS Lett.* **148**, 21 (1982).

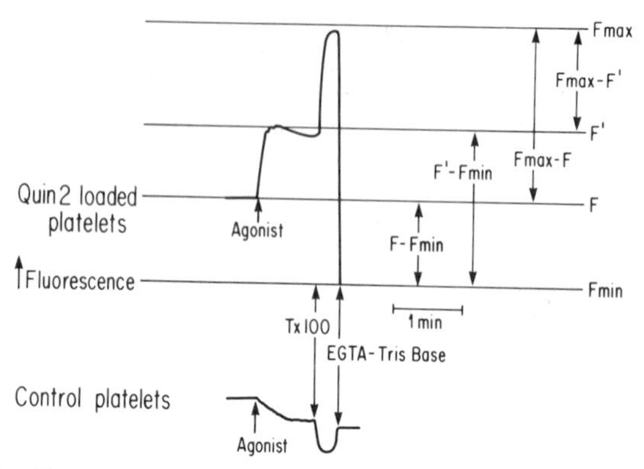

FIG. 3. Diagrammatic representation of a tracing obtained from Quin 2-loaded platelets (top) and control platelets (bottom) treated with an unspecified agonist, Triton X-100 to lyse platelets and expose intracellular Quin 2 to saturating $[Ca^{2+}]$ in the media, and EGTA with Tris–base to quench Ca^{2+}-dependent fluorescence. F, Unstimulated (baseline) fluorescence; F', stimulated (altered) fluorescence; F_{max}, maximum fluorescence, with all Quin 2 present in Ca^{2+}-bound form; F_{min}, minimum fluorescence, with all Quin 2 present in free (non-Ca^{2+}-bound) form. Deflections in the control tracing are added to or subtracted from the test tracing at corresponding points, prior to measurement of parameters for calculation of $[Ca^{2+}]_i$. See text for details.

which does not readily cross the platelet's lipid-rich plasma membrane, Tsien[3] blocked the carboxylic acid groups of Quin 2 by esterification with acetoxymethyl groups, thus creating the lipid-soluble parent molecule, Quin 2/AM. On exposure of intact platelets to Quin 2/AM in micromolar extracellular concentration, rapid transmembrane passage of the molecule is followed by intracellular hydrolysis of the blocking groups to yield free Quin 2, which accumulates at millimolar intracellular concentrations. The intracellular hydrolysis is apparently performed by an esterase whose characteristics have yet to be defined. It is also not known whether all of the blocking groups are removed from each molecule and, if not, whether variable cleavage alters the K_D of Quin 2 when it is used in intracellular applications, or produces fluorescent, non-Ca^{2+}-dependent intermediates as is the case with Fura 2 and Indo 1 (see below). Further, relative concentrations and activities of the "nonspecific esterase" between cell types or within a particular cell line, or the effects of such variables on Quin 2 signals, are not yet known.

Selective digitonin-induced membrane lysis has shown that Quin 2 is apparently confined to the platelet cytosol without significant quantities entering either granules or mitochondria.[3] It is well to note, however, that such entrance of a small proportion of the intracellular Quin 2 into Ca^{2+}-rich organelles would probably not have a large effect on the Quin 2 signal,

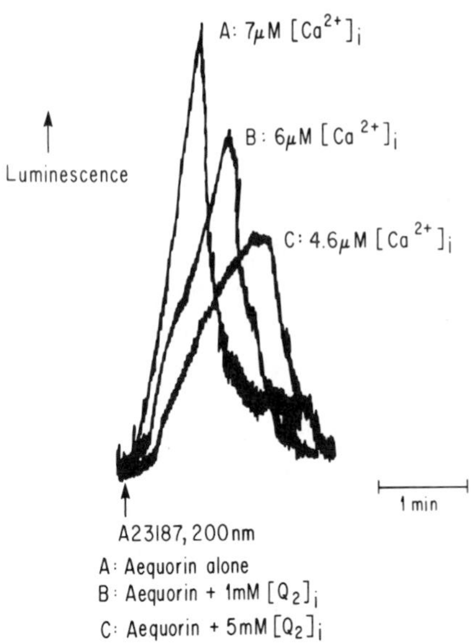

FIG. 4. Simultaneous loading of platelets with aequorin and Quin 2; effect of intracellular Quin 2 ($[Q_2]_i$) on aequorin luminescence in response to the Ca^{2+} ionophore A23187. All platelets are aequorin loaded, and suspended in media containing 2 m*M* EGTA. Increasing intracellular Quin 2 lowers and delays aequorin $[Ca^{2+}]_i$ signal, but does not eliminate it.

because of its relative insensitivity to local changes in $[Ca^{2+}]_i$. Our technique for loading Quin 2 into platelets is as follows.

To 10 ml of PRP, prepared as previously described, is added 5–10 μl of a 10 m*M* stock solution of Quin 2/AM in DMSO (final concentration range 5–20 μ*M*). A control aliquot of PRP is treated with 10 μl DMSO alone. PRP is incubated for 20 min at 22°, after which PGE_1 is added to a final concentration of 1 μ*M* from a 1 m*M* stock solution in absolute ethanol. PRP is then spun at 1100 *g* for 15 min, forming a soft pellet. The supernatant plasma is discarded.

The pellet is resuspended in 1 ml HEPES–Tyrode's buffer (previously described) containing 1 μ*M* PGE_1, 10 μ*M* EGTA, and no added Ca^{2+}. The resulting platelet suspension is layered onto a Sepharose 2B gel column (10 ml bed volume) preequilibrated with HEPES–Tyrode's buffer containing 10 μ*M* EGTA (to bind traces of Ca^{2+} on the column) and no added Ca^{2+}. The elution buffer is the same. After elution, the platelet count is adjusted to 1×10^8/ml using the same buffer, after which the extracellular $[Ca^{2+}]$ is restored to 1 m*M* from a 1 *M* $CaCl_2$ stock solution in distilled H_2O. Aliquots of 1.6 ml are placed in polystyrene spectrophotometer cuvettes, each containing a metal stir bar (which does not release ions that quench Quin 2 when free in solution), and incubated at 37° for 30–45 min. The latter incubation allows the $[Ca^{2+}]_i$, buffered downward during Quin 2 loading, to achieve a steady-state level of 50–100 n*M*. This long incubation cannot be used with Fura 2 or Indo 1 loaded platelets because of leakage of the fluorescent dye into the extracellular media.

Platelet Integrity during and after Loading with Quin 2

The addition of up to 50 μM Quin 2/AM to gel-filtered platelets does not induce shape change or stimulate aggregation or secretion of granule contents. There is no loss of LDH from the cytosol, and electron micrographs appear the same as with untreated control platelets (unpublished observations). Platelets loaded with Quin 2 and treated with common agonists change shape, aggregate, and, in some cases, secrete granule contents in delayed fashion, relative to untreated control platelets. The extent of this delay is directly proportional to the concentration of intracellular Quin 2, and most likely represents a Ca^{2+}-buffering artifact.[19]

Quantitation of Quin 2 Concentration in Platelets

Signal Requirements. In our experience, the minimum intracellular concentration of Quin 2 that allows detection of a rise in $[Ca^{2+}]_i$ above the basal fluorescence of the platelets themselves is 0.7 m*M*. Higher concentrations may be used and provide greater reproducibility in each measurement, but carry the increased risk of artifactual signals due to buffering. The average concentration of Quin 2 in a suspension of platelets can be determined by lysis of the platelets into an extracellular buffer containing 1 m*M* $[Ca^{2+}]$, followed by the addition of EGTA and Tris–base to reduce the extracellular $[Ca^{2+}]$ to $<10^{-9}$ *M*. The Ca^{2+}-dependent portion of the fluorescence is then measured and compared to the Ca^{2+}-dependent fluorescence of known standard concentrations of Quin 2 in solution, in order to determine the actual amount of Quin 2 released from the population of platelets. Then, based on the calculated combined volume of the platelets from which the Quin 2 was released, the intracellular Quin 2 concentration can be calculated. Reports of data regarding $[Ca^{2+}]_i$ in platelets obtained using Quin 2 should be accompanied by an estimate of its intracellular concentration, so that the degree of potential buffering in the preparation can be taken into account.

Location of Quin 2 in the Platelet. Quin 2 has been localized to the cytosol by selective lysis of the platelet membrane by digitonin, as previously described. When platelets are loaded with $[^3H]$Quin 2 and stimulated with agonists to secrete their granule contents, tritium is retained, indicating that Quin 2 does not partition into the granule compartment. Finally, direct visualization of fibroblasts and unstimulated neutrophils loaded with Quin 2 using fluorescence microscopy reveals Quin 2 to be distributed evenly throughout the cytosol.[33,34]

Measurement of Quin 2 Signals. Fluorescent signals from Quin 2-loaded platelets can be measured in a variety of conventional fluorescence detectors; thermostatting and stirring capabilities are essential. Experiments are conducted at 37° with constant stirring (unless otherwise indicated) at 600 rpm. An excitation wavelength of 339 nm and an emission wavelength of 500 nm are used. Agonists are added directly to samples through a porthole in the lid overlying the fluorimeter chamber. Aliquots of non-Quin 2-loaded platelets that are otherwise treated identically are also studied.

Calibration of Quin 2 Signals. The technique for calibration of Quin 2 signals is similar to that described by Tsien *et al.*,[2] with important modifications. Through reference to the upper tracing in Fig. 3, it is evident that four parameters can be measured directly from a Quin 2 tracing. $F - F_{min}$ is the magnitude of the Ca^{2+}-dependent fluorescence of Quin 2 at resting $[Ca^{2+}]_i$. $F' - F_{min}$ is the magnitude of the peak fluorescence achieved after stimulation of the suspension of Quin 2-loaded platelets by an agonist. $F_{max} - F$ is the magnitude of the Ca^{2+}-dependent fluorescence of Quin 2 when fully saturated with Ca^{2+} ($> 10^{-5}$ M) minus the magnitude of Quin 2 fluorescence at the $[Ca^{2+}]_i$ of the resting platelet. Similarly, $F_{max} - F'$ represents the difference between the fluorescence at saturation and that present at the peak stimulated $[Ca^{2+}]_i$. Based on a modification of the equilibrium equation, Tsien proposed the following as a theoretical calibration for $[Ca^{2+}]_i$:

For resting platelets: $$[Ca^{2+}]_i = \frac{(1.15 \times 10^{-7}\ M)(F_{max} - F)}{(F - F_{min})}$$

For stimulated platelets: $$[Ca^{2+}]_i = \frac{(1.15 \times 10^{-7}\ M)(F_{max} - F')}{(F' - F_{min})}$$

where 1.15×10^{-7} M is the K_D of Quin 2 for Ca^{2+} under conditions generally thought to mimic the cytosol.

F_{max} is obtained when platelets are lysed with Triton X-100 (0.1–1.0%), which exposes the intracellular Quin 2 to the extracellular $[Ca^{2+}]$ ($> 10^{-5}$ M). F_{min} is thereafter obtained when EGTA and Tris–base are added in concentrations adequate (10 and 20 mM, respectively) to provide Ca^{2+} chelation enhanced by pH adjustment to > 8.3. Optimal Ca^{2+} chelation by EGTA occurs at this pH, providing a final extracellular $[Ca^{2+}] < 10^{-9}$ M, below which Quin 2 is insensitive to Ca^{2+}. An alternative tech-

[33] F. S. Fay and R. W. Tucker, *J. Gen. Physiol.* **84,** 15a (Abstr.) (1984).
[34] D. W. Sawyer, J. A. Sullivan, and G. L. Mandell, *Science* **230,** 662 (1985).

nique for the measurement of F_{min} involves the addition of 100 μM Mn^{2+} (final concentration) to the Triton-lysed sample, thereby quenching the Ca^{2+}-dependent fluorescence of Quin 2. If it is not practical to lyse the cells, one should note that F_{max} has been obtained in other cells by addition of ionomycin[35]; however, this approach has not been validated in platelets.

Unfortunately, many other substances also fluoresce, including platelet agonists and plasma proteins, requiring further correction of the basic calibration. Furthermore, Triton X-100, added to lyse the platelets and release the trapped Quin 2, causes a fall in the autofluorescence of control platelets, and the magnitude of this fall must be used as a correction factor in the above calibration. Similarly, both EGTA and Tris–base, added late in the calibration, induce a rise in the fluorescence of lysed control platelets which must be subtracted at the appropriate point in the Quin 2 tracing before the aforementioned parameters are measured (see Fig. 3, lower half).

Since added substances fluoresce in an unpredictable fashion, side-by-side controls must be performed whenever the standard Quin 2 calibration technique is used to estimate $[Ca^{2+}]_i$. This procedure allows one to correct the Quin 2 calibration, but this correction is based on the assumption, still unproved, that the magnitude of the alteration in the Quin 2 signal upon addition of fluorescent substances will be identical to that observed in a control sample containing an equal concentration of cells, and that a linear correction of the Quin 2 signal will suffice. Should Quin 2 itself interact with such substances to either enhance or reduce the fluorescence artifact, this correction procedure might provide erroneous data. For this reason, it is wise to perform preliminary experiments in which Quin 2 in solution is treated, in sequence, with agonist, Triton X-100, EGTA, and Tris–base to evaluate the effect(s) of such added substances on the Quin 2 fluorescence itself, in the absence of intact or lysed cells.

Some evidence exists which suggests that this correction technique is valid. When platelets are loaded with increasing concentrations of Quin 2 and resting $[Ca^{2+}]_i$ is calibrated according to Tsien's technique without correction, the resting $[Ca^{2+}]_i$ appears to fall as a function of the intracellular Quin 2 concentration. Since the concentration of platelets is held constant in all samples, the magnitudes of the respective fluorescence artifacts observed upon addition of Triton X-100, EGTA, and Tris–base to control platelets remain constant between samples. The increasing quantities of Quin 2 in this series of samples, however, give rise to an increasing total fluorescence upon lysis of Quin 2-loaded platelets into the

[35] T. R. Hesketh, G. A. Smith, J. P. Moore, M. V. Taylor, and J. C. Metcalfe, *J. Biol. Chem.* **258**, 4876 (1983).

Ca^{2+}-rich extracellular fluid. The constant magnitude of the fluorescence artifacts due to added substances constitutes a progressively smaller fraction of the total fluorescence, and therefore has a declining influence on the calculated level of resting $[Ca^{2+}]_i$. When the correction for these fluorescence artifacts is performed as described above, the resting $[Ca^{2+}]_i$ is constant among samples, supporting the validity of the correction technique (unpublished observations).

Appearance of Quin 2 Signals in Response to Agonists: Interpretation of the Magnitude of the Peak $[Ca^{2+}]_i$

Typically, the baseline of the Quin 2 tracing indicates a stable $[Ca^{2+}]_i$. Upon addition of an agonist such as thrombin, a brisk rise in the signal to a peak is seen (Fig. 5). When platelets are prevented from aggregating (by discontinuation of stirring after agonist addition), the signal remains at the peak level, forming a plateau. If platelets are allowed to aggregate, the signal will artifactually fall to or even below the baseline value as the net concentration of unquenched Quin 2 in the light beam falls. It is therefore wise to trust as estimates of $[Ca^{2+}]_i$ only the basal and initial peak measurements from Quin 2-loaded platelets under stirred conditions. If aggregation is very rapid (e.g., when high doses of thrombin are used), then even the initial peak $[Ca^{2+}]_i$ may be artifactually reduced as a consequence of this phenomenon.

When aggregation and Quin 2 fluorescence studies are done in parallel, it can be shown clearly that the Quin 2 signal precedes or coincides with the earliest visible platelet functional response.[8,19] Elevation of the Quin 2 signal may be inhibited partially or fully by antagonists of platelet activation such as aspirin or agents that elevate cyclic AMP, depending on the agonist used.[36,37] The shape of the Quin 2 tracing is similar in response to most platelet agonists, its magnitude depending mainly on the concentration of agonist.

Interpretation of the Magnitude of the Basal $[Ca^{2+}]_i$

As previously stated, the basal $[Ca^{2+}]_i$ indicated by Quin 2-loaded platelets suspended in buffer containing 1 m*M* $[Ca^{2+}]$ is approximately 100 n*M*. If the basal $[Ca^{2+}]_i$ is measured immediately after the platelets have completed incubation with Quin 2/AM in a medium containing 1 m*M* $[Ca^{2+}]$ and at brief (5 min) intervals thereafter, the $[Ca^{2+}]_i$ apparently rises from a starting level of 50–60 n*M* to a plateau level of 100 n*M*.

[36] G. B. Zavoico and M. B. Feinstein, *Biochem. Biophys. Res. Commun.* **120,** 579 (1985).
[37] A. Pannocchia and R. M. Hardisty, *Biochem. Biophys. Res. Commun.* **127,** 339 (1985).

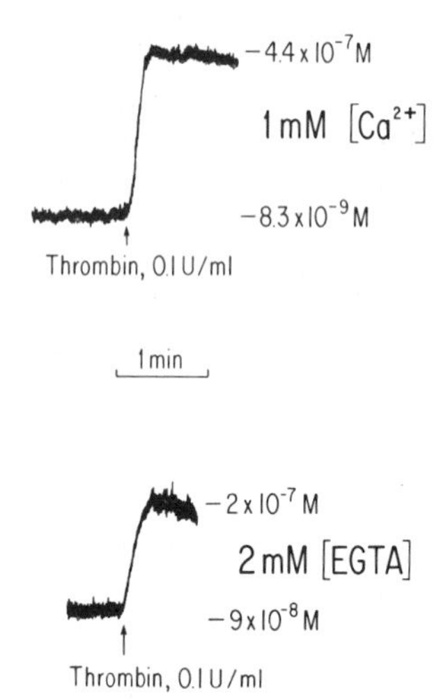

FIG. 5. Representative tracings of Quin 2 fluorescence in response to thrombin, 0.1 U/ml, in media containing 1 m*M* [Ca^{2+}] with (bottom) or without (top) 2 m*M* EGTA. Chelation of extracellular Ca^{2+} by EGTA does not lower the baseline [Ca^{2+}]$_i$ with Quin 2, but the peak [Ca^{2+}]$_i$ is partially inhibited.

The process requires 20–30 min of incubation at 37°, and may signify the return to a true basal [Ca^{2+}]$_i$, depressed during Quin 2 loading through rapid accumulation of a millimolar concentration of indicator/chelator. The Ca^{2+} supplied for this equilibration is thought to come mainly from the extracellular medium, though platelets loaded with Quin 2 and then incubated in medium devoid of Ca^{2+} also demonstrate a postloading baseline elevation, albeit slight. The final basal [Ca^{2+}]$_i$ in platelets suspended in medium containing no Ca^{2+} typically reaches only 60–80 n*M*.

An artifact that may alter the apparent basal [Ca^{2+}]$_i$ is leakage of Quin 2 from platelets, which occurs at a rate of about 10%/hr.[8] When Quin 2-loaded platelets are suspended in medium containing 1 m*M* [Ca^{2+}], this is a particular problem because the apparent baseline gradually rises as extracellular Quin 2 makes an ever-increasing contribution to the total signal. If 10 μM $MnCl_2$ or 2–10 m*M* EGTA is added, however, the baseline immediately returns to the plateau basal [Ca^{2+}]$_i$, or slightly below. Leaked Quin 2 interferes with the calibration of the basal and peak [Ca^{2+}]$_i$ because it contributes to an elevation of the basal signal without a concomitant increase in F_{max}. While linear corrections can be made for this elevation, their accuracy is not guaranteed. Thus, it seems best to simplify matters by experimenting only with freshly loaded preparations.

Measurement of $[Ca^{2+}]_i$ in Platelets Loaded with Both Aequorin and Quin 2

When platelets are loaded with both Quin 2 and aequorin, the basal $[Ca^{2+}]_i$ indicated by either indicator in the presence of 1 m*M* external $[Ca^{2+}]$ is unchanged. That is, Quin 2 still indicates a $[Ca^{2+}]_i$ of 50–100 n*M* at rest, while aequorin indicates a $[Ca^{2+}]_i$ of approximately 1 μM or 500 n*M*, depending on whether platelets are suspended in Ca^{2+} or EGTA. Aequorin exhibits no effect on the rate or magnitude of the Quin 2 signals, in comparison to platelets loaded with Quin 2 alone. However, the presence of Quin 2 in platelets also loaded with aequorin produces delays in the agonist-induced aequorin signal relative to those obtained from platelets loaded with aequorin alone (Fig. 4). The delay is in proportion to the intracellular concentration of Quin 2. The presence of Quin 2, however, only slightly reduces the peak $[Ca^{2+}]_i$ measured with aequorin.

Comparison of the Aequorin and Quin 2 Signals

The aequorin signal, like the Quin 2 signal, is manifested by a rapid upstroke to a peak, but unlike the latter, it falls back to its baseline rapidly (30–90 sec). Initially, the discrepancy between these configurations was not well understood. The rapid consumption of aequorin by a localized $[Ca^{2+}]_i$ transient probably does not account for the brevity of the signal, as subsequent agonist additions can provide similar signals, which suggests that aequorin is actually present in relative abundance. An alternative which might account for this could be the rapid removal of Ca^{2+} from a region of high concentration, either by diffusion or by an active sequestering mechanism. This latter possibility seems most likely, as tracings of the ratio between excitatory wavelengths of Fura 2 following stimulation closely resemble the aequorin signal, which suggests that the slower decay of the Quin 2 signal is an artifact of its lower K_D for Ca^{2+}.

There are circumstances in which an increased aequorin signal can be observed without a concomitant rise in the Quin 2 signal. In these cases, the increase in the aequorin signal occurs also in platelets loaded with both Quin 2 and aequorin. Phorbol ester, diacylglycerol, and collagen in the presence of aspirin have been postulated to stimulate platelets through a Ca^{2+}-independent pathway, because they have been shown to activate platelets without producing a Quin 2 signal.[38] However, these agonists always cause a rise in $[Ca^{2+}]_i$ indicated by aequorin which precedes or coincides with the onset of platelet activation, despite the absence of a Quin 2 signal.[19,25] This disparity in the relative signalling between aequorin

[38] T. J. Rink, A. Sanchez, and T. J. Hallam, *Nature (London)* **305**, 317 (1983).

and Quin 2-loaded platelets has sparked a lively controversy in recent years, which will likely abate only when the physical mechanisms underlying these differences in signalling become clearly understood.

Second Generation of Fluorescent Indicators of $[Ca^{2+}]_i$: Fura 2 and Indo 1

In 1985, Tsien and colleagues reported the development of several new Ca^{2+}-sensitive fluorophores.[39] Though similar in design and function to the molecule Quin 2, they have differences that improve upon several of the problems associated with the use of Quin 2.[8,40,41] Two members of the group described, Fura 2 and Indo 1, were proposed for use as Ca^{2+} indicators in biological systems. Since they are in principle quite similar, they will be considered together, and the differences cited where appropriate.

Like Quin 2, these indicators resemble the EGTA molecule, and bind Ca^{2+} with 1 : 1 stoichiometry. Also, they respond to a change in $[Ca^{2+}]$ with a change in fluorescence intensity at a measured emission wavelength. To some degree, like Quin 2, they may buffer $[Ca^{2+}]_i$ transients in the cytosol. Their differences from Quin 2, however, are significant.

Fura 2, for example, exhibits a quantum efficiency at one-half saturation with Ca^{2+}, which is five times that of Quin 2. This, combined with a 6-fold increase in the extinction coefficient of Fura 2 over that of Quin 2, results in a 30-fold relative increase in fluorescence intensity.[39] In and of itself, this feature enables one to use lower concentrations of intracellular fluorophore, thereby limiting the Ca^{2+}-buffering effect. In addition, Fura 2 has a lower affinity for Ca^{2+} than does Quin 2 ($K_D = 224$ and 115 nM, respectively), which not only further minimizes buffering, but also allows one to detect $[Ca^{2+}]_i$ levels greater than 1 μM with improved accuracy, a level at which the Quin 2 signal becomes nearly insensitive to changes in $[Ca^{2+}]_i$. There is some evidence that Fura 2 is able to detect subtle changes in $[Ca^{2+}]_i$ which are missed by Quin 2, and it is supposed that this is due to the increase in brightness of Fura 2, combined with decreased buffering.[39]

An alternative hypothesis that might account for the differences in signalling would be that Quin 2 and Fura 2 are located in different compartments in the cell. There is no good evidence to suggest this in platelets; but in other cells, sequestration of dye has been a major problem.[42-44]

[39] G. Grynkiewicz, M. Poenie, and R. Y. Tsien, *J. Biol. Chem.* **260,** 3440 (1985).

[40] R. Y. Tsien, T. Pozzan, and T. J. Rink, *Trends Biochem. Sci.* **9,** 263 (1984).

[41] R. Y. Tsien, T. Pozzan, and T. J. Rink, *J. Cell Biol.* **94,** 325 (1982).

[42] W. Almers and E. Neher, *FEBS Lett.* **192,** 12 (1985).

[43] M. Goligorsky, K. Hruska, G. Cofrus, and E. Elson, *J. Cell Physiol.* **128,** 466 (1986).

[44] D. A. Williams, K. E. Fogarty, R. Y. Tsien, and F. S. Fay, *Nature (London)* **318,** 558 (1985).

When the indicators are directly visualized through a fluorescence microscope, both Quin 2 and Fura 2 are found to be distributed throughout the cytosol, though microcompartmentation cannot be rigorously excluded on the basis of these studies.[33,34,45,46] Like Quin 2, Fura 2 binds little if at all to erythrocyte membranes.[39] Fura 2 may be altered by uncharacterized cytoplasmic enzymes (located in muscle on the sarcoplasmic reticulum) which give rise to fluorescent intermediates that may introduce error into the calculation of $[Ca^{2+}]_i$.[47]

Both Quin 2 and Fura 2 have been shown to leak from platelets at a significant rate (from 10 to 20%/hr), and Fura 2 has been shown to enter and become retained in granules of various types of cells.[48] Intracellular granules may contain esterases capable of converting Fura 2/AM to Fura 2, allowing the granules to compete with the cytosol for trapping of the latter. Whether this is a problem with platelet granules has not been conclusively shown.

The selectivity for Ca^{2+} over other metals is better for Fura 2 than Quin 2, by a factor of 4 for Mg^{2+}, 12 for Mn^{2+}, and 40 for Zn^{2+}.[39]

Fura 2 fluorescence is a function of temperature, pH, and ionic strength, and the K_D levels provided by Grynkiewicz *et al.*[39] are restricted to use under the conditions defined. Significant variance from these conditions should prompt the investigator to prepare Ca^{2+}–EGTA buffers and recalibrate the dye to determine the K_D under the appropriate unique experimental conditions.

Loading Platelets with Fura 2 and Indo 1

Fura 2 and Indo 1 are loaded into platelets in a fashion similar to that used to load Quin 2. The major difference lies in the concentration of extracellular indicator added as the acetoxymethyl ester form to achieve adequate cytosolic loading of the free dye. This is approximately 1–2 μM for Fura 2/AM versus 5–10 μM for Quin 2/AM. Final, usable intracellular indicator concentrations of Fura 2 can be as low as 10–30 μM, a significant improvement over the millimolar intracellular concentrations needed when Quin 2 is used.

Calibration of Fura 2 and Indo 1 Signals

Calibration of Fura 2 signals differs from the calibration technique for Quin 2 in that several methods may be applied. The fluorescence of Fura 2,

[45] D. A. Williams and F. S. Fay, *Am. J. Physiol.* **250,** C779 (1986).

[46] M. P. Timmerman and C. C. Ashley, *FEBS Lett.* **209,** 1 (1986).

[47] S. Highsmith, P. Bloebaum, K. W. Snowdowne, *Biochem. Biophys. Res. Commun.* **138,** 1153 (1986).

[48] W. Almers and E. Neher, *FEBS Lett.* **192,** 13 (1985).

like that of Quin 2, increases in intensity upon binding Ca^{2+}. Both indicators also undergo a spectral shift when Ca^{2+} is bound, but the Quin 2 spectral shift is so small that background fluorescence overwhelms it, and it is adversely influenced by smaller changes in $[Mg^{2+}]$.[2,39]

The spectral shifts demonstrated by Fura 2 and Indo 1 are relatively large, and signals are minimally perturbed by alterations in $[Mg^{2+}]$. The Fura 2 spectral shift is best seen as a change in the maximal excitation wavelength, while that of Indo 1 is best seen as a change in the maximal emission wavelength. Indo 1 is therefore thought to have better potential than Fura 2 in flow cytometry applications,[39] since in time-restricted experiments, simultaneous measurements of two emission wavelength intensities are easier than serial rapid measurements of an emission wavelength produced by changing excitation wavelengths, as is the case when Fura 2 is used. In other respects, however, the two indicators are quite similar.

When path length, effective dye concentration, ionic strength, and the physical components provoking excitation and measuring emission are held constant, the presence of a spectral shift allows one to measure the $[Ca^{2+}]_i$ on the basis of a fluorescence ratio alone. While Quin 2 must be calibrated using destructive techniques, $[Ca^{2+}]_i$ detected using Fura 2 and Indo 1 may be calibrated without cellular destruction. Grynkiewicz *et al.*[39] have demonstrated the mathematical steps leading to their calibration formula for fluorophores when using the ratio technique. This is

$$[Ca^{2+}]_i = K_D[(R - R_{min})/(R_{max} - R)](S_{f2}/S_{b2})$$

where K_D is the dissociation constant of the dye for Ca^{2+}, under the chosen experimental conditions. At 37° and at physiologic ionic strength, $K_D =$ 224 n*M* for Fura 2 and 250 n*M* for Indo 1.

R is the ratio of fluorescence intensities measured at two wavelengths under experimental conditions. For Fura 2, this ratio is obtained using fluorescence intensities measured at 510 nm (emission) based upon excitation at 340 nm (numerator) and 380 nm (denominator). Cell autofluorescence must be subtracted from the individual fluorescence intensity measurements prior to calculation of the ratio. The wavelengths used are held to be less likely to register changes in cell autofluorescence during metabolic changes induced by agonists, but each investigator should carry out the appropriate experiments to ensure that his or her own experimental system falls under this umbrella. Though not often discussed, changes in the milieu's fluorescence induced by the addition of agonists themselves should also be measured and subtracted from the measured intensities, as these have the same effect on the calculations as does native autofluorescence (see Quin 2 section). Accurate calculation of $[Ca^{2+}]_i$ by this method is dependent on the assumption that changes in the fluorescence of the free acid in cell-free solutions are the same as those that occur in the cytosol, which has been shown not to be true in other cells,[49,50] in which fluorescent non-Ca^{2+}-dependent intermediates arise from incomplete hydrolysis of the AM form.

As described by Kassotis,[51] R_{max} and R_{min} have been measured in intact cells using a calcium ionophore (100 n*M* ionomycin) in the presence of 1 m*M* extracellular Ca^{2+} (to obtain R_{max}), followed by 5 m*M* EGTA (to bind the external Ca^{2+} and evacuate the cell of Ca^{2+} to provide the R_{min}). When using this method, as opposed to a technique in which the dye is released by cell lysis, caution must be exercised in the interpretation of R_{max} and R_{min} signals, since this calculation assumes that all of the Ca^{2+} is being accessed by the ionophore and that the ionophore is equally capable of adding and depleting Ca^{2+} in a cell which may institute countermeasures that actually provide a steady-state compromise near the true R_{max} or R_{min}.

S is a fluorescence proportionality constant which takes into account the excitation intensity, extinction coefficient, path length, quantum efficiency of the dye, and the instrument's photon collection efficiency.[39] S_{f2} is a constant which applies under these conditions to the free form of the dye measured at the second wavelength (the denominator wavelength), while S_{b2} is its counterpart representing the bound form of the dye. The total fluorescence intensity at the denominator wavelength is related to concentration of Ca^{2+}-bound and free dye as:

$$F_2 = S_{f2}c_f + S_{b2}c_b$$

where c_f is the concentration of dye in the free form and c_b is the concentration of dye bound to Ca^{2+}. S_{f2} can be calculated from the fluorescence intensity of a solution of dye at zero [Ca^{2+}], and S_{b2} can be similarly calculated from the fluorescence intensity obtained when all of the dye present is bound to Ca^{2+}, i.e., at saturating [Ca^{2+}]. Based on these determinations, the ratio S_{f2}/S_{b2} can be calculated and entered as a constant under unchanging experimental conditions.

This complex series of measurements is made possible using specialized spectrofluorimetric hardware. Appropriate software with the memory capacity to allow registration of a large amount of data in real time is also necessary. In our studies, we have used a temperature-controlled, dual-excitation wavelength spectrofluorimeter with continuous stirring (SPEX Fluorolog-2, Edison, NJ). Using this device, the fluorescence signals at

[49] A. Malgaroli, D. Milani, J. Meldolesi, and T. Pozzan, *J. Cell Biol.* **105,** 2145 (1987).

[50] M. Scanlon, D. A. Williams, and F. S. Fay, *J. Biol. Chem.* **262,** 6308 (1987).

[51] J. Kassotis, S. F. Steinberg, S. Ross, J. P. Bilezikian, and R. B. Robinson, *Pfluegers Arch.* **409,** 47 (1987).

excitation wavelengths 340 and 380 nm are obtained once every 500 msec and stored in separate memories of a SPEX datamate microcomputer. Appropriate software is used to generate the calculations described above, and to allow generation of an analog plot depicting $[Ca^{2+}]_i$ versus time.

When such hardware and software are not available, Fura 2 and Indo 1 may be calibrated using the calibration formula for Quin 2, with replacement of the K_D by that of the dye in use, determined for the given experimental conditions. In this case, changes in cell autofluorescence and agonist fluorescence must be measured separately and added to or subtracted from the tracing values at the appropriate intervals, as described for Quin 2 experiments. Of benefit is the larger signal-to-autofluorescence ratios of the new dyes, which tend to reduce the significance of changes in autofluorescence or agonist fluorescence.

This technique requires a destructive step at the end of the experiment, to release the indicator into the extracellular medium where it is sequentially exposed to saturating $[Ca^{2+}]$, then to a $[Ca^{2+}] < 10^{-9}$ M, to measure F_{max} and F_{min}, respectively. Lysis may be achieved using >50 μM digitonin or $>.1\%$ Triton X-100. As with Quin 2, addition of 20 mM Tris–base followed by 10 mM EGTA allows determination of F_{min}.

Pollock *et al.* have described a third calibration technique.[52] Here, 2 mM Mn^{2+} is added to the platelet lysate in lieu of Tris and EGTA. Mn^{2+} quenches the released dye, allowing measurement of the autofluorescence (AF) of the cells. Using standard calibration curves, F_{min} is calculated on the basis of the relationship between F_{max}, F_{min}, and AF. The resulting equation includes a constant which must be calculated for each instrument under each set of experimental conditions. The general form of the equation is

$$F_{min} = AF + (F_{max} - AF)/C$$

where C is the constant. We have as yet made no comparisons between this and other calibration techniques.

In general, the levels of $[Ca^{2+}]_i$ calculated using calibration formulas must be viewed with the caution appropriate to any experiment whose results are calibrated using only one internal standard (the dye itself). The concentration–response curves on which the K_D and intensity ratios of these indicators depend are developed extracellularly using Ca^{2+}–EGTA buffer systems which are easily perturbed by changes in pH, temperature, and ionic strength, and are based on the assumption that fluorescence of Fura 2 acid form is similar to that of the hydrolyzed AM form. Since the $[Ca^{2+}]$ values of interest are too low to measure using ion-selective elec-

[52] W. K. Pollock, T. J. Rink, and R. F. Irvine, *Biochem. J.* **235,** 869 (1987).

trodes, there is presently no way to preclude an effect of the fluorophore on the K_D of EGTA itself, an effect that would provide a systematic error in all further calculations.

It is also clear that, when applied to cell suspensions, these calibration techniques do not allow one to determine if elevations of $[Ca^{2+}]_i$ occur inhomogeneously within single cells, a fact which in itself casts doubt upon strict interpretation of $[Ca^{2+}]_i$ derived from calibration formulas. Rather, calculated values based on measurements from cell suspensions may be better used at present as semiquantitative indices of degrees of change in $[Ca^{2+}]_i$.

To escape the problems and artifacts inherent in studying cells in suspension, many investigators are examining $[Ca^{2+}]_i$ changes as indicated by Fura 2 or Indo 1 with the techniques of quantitative epifluorescence microscopy and imaging, and flow cytometry. Although these techniques introduce their own artifacts and present several challenges for application in small cells, their use in platelets has been the subject of preliminary reports[53,54] and, with advances in technology, should add considerably to our understanding of $[Ca^{2+}]_i$ transients in platelet activation.

[53] T. J. Hallam, M. Poenie, and R. Y. Tsien, *J. Physiol. (London)* **377,** 123P (Abstr.) (1986).
[54] T. A. Davies, D. Drotts, G. Weil, and E. R. Simons, *Thromb. Haemostasis* **58** (Suppl. 1), 1672 (Abstr.) (1987).

[34] Endogenous Phosphatidylinositol 4,5-Bisphosphate, Phosphatidylinositol, and Phosphatidic Acid in Stimulated Human Platelets

By M. JOHAN BROEKMAN

Background

The lipids involved in the phosphatidylinositol (PI) cycle[1,2] have recently become recognized as important intermediaries in stimulus–response coupling, especially with regard to those cell responses where increases in intracellular free Ca^{2+} have been shown to play a role.[3,4] The

[1] M. R. Hokin and L. E. Hokin, *J. Biol. Chem.* **203,** 967(1953).
[2] D. J. Hanahan and D. R. Nelson, *J. Lipid Res.* **25,** 1528 (1984).

current working hypothesis is summarized in Fig. 1. Upon cell activation two different intracellular messengers are derived from a single molecule of phosphatidylinositol 4,5-bisphosphate (PIP_2) by phospholipase C-catalyzed hydrolysis. One messenger, diglyceride, is lipophilic, and is thought to activate protein kinase C[5], thereby transmitting receptor recognition to cellular responses involving protein phosphorylations.[4] The other messenger constitutes the remainder of the PIP_2 molecule, inositol 1,4,5-trisphosphate (IP_3; Fig. 2), and is thought to act by releasing sequestered calcium from storage sites associated with the endoplasmic reticulum of the cell.[6,7] Production of IP_3 in a variety of cell systems has been documented, including platelets.[8-11]

Changes in levels of PIP_2 in stimulated platelets yield insight in the ways PIP_2-derived second messengers might be generated. Measurements can rely on the use of radioactive phosphate, either in settings where the label in the 4 and 5 positions of inositol (Fig. 2) is in equilibrium with that of ATP,[7] or in nonequilibrium situations.[12] Radioactive inositol can also be used, but has the disadvantage of being incorporated very poorly into PI and PIP_2 of *human* platelets (better in rabbit platelets[10]), necessitating the use of large amounts of expensive radioactive inositol.[9] In contrast, as described in this chapter, total amounts of endogenous PIP_2 can easily be measured after isolation of PIP_2 by thin-layer chromatography (TLC), in either one of two ways: spectrophotometric phosphorus assay, or quantitative fatty acid analysis using gas–liquid chromatography (GLC).[13-16]

Platelet Collection and Processing

Reagents and Supplies

Whole blood, collected into ACD [trisodium citrate (85 m*M*), glucose (111 m*M*), citric acid (71 m*M*)] anticoagulant (6 vol blood/vol ACD), preferably in a triple pack, or

[3] R. H. Michell, *Biochim. Biophys. Acta* **415,** 81 (1975).
[4] R. H. Michell, *Life Sci.* **32,** 2083 (1983).
[5] Y. Nishizuka, *Science* **225,** 1365 (1984).
[6] H. Streb, R. F. Irvine, M. J. Berridge, and I. Schulz, *Nature (London)* **306,** 67 (1983).
[7] M. J. Berridge, *Biochem. J.* **220,** 345 (1984).
[8] B. W. Agranoff, P. Murthy, and E. B. Seguin, *J. Biol. Chem.* **258,** 2076 (1983).
[9] S. P. Watson, R. T. McConnell, and E. G. Lapetina, *J. Biol. Chem.* **259,** 13199 (1984).
[10] J. D. Vickers, R. L. Kinlough-Rathbone, and J. F. Mustard, *Biochem. J.* **224,** 399 (1984).
[11] S. E. Rittenhouse and J. P. Sasson, *J. Biol. Chem.* **260,** 8657 (1985).
[12] D. B. Wilson, E. J. Neufeld, and P. W. Majerus, *J. Biol. Chem.* **260,** 1046 (1985).
[13] M. J. Broekman, R. I. Handin, A. Derksen, and P. Cohen, *Blood* **47,** 963 (1976).
[14] M. J. Broekman, J. W. Ward, and A. J. Marcus, *J. Clin. Invest.* **66,** 275 (1980).
[15] M. J. Broekman, J. W. Ward, and A. J. Marcus, *J. Biol. Chem.* **256,** 8271 (1981).
[16] M. J. Broekman, *Biochem. Biophys. Res. Commun.* **120,** 226 (1984).

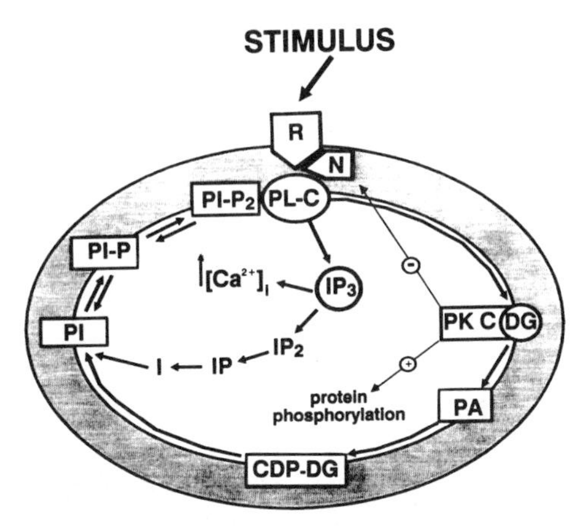

FIG. 1. The role of inositol phospholipid metabolism in stimulus–response coupling. In unstimulated cells, an equilibrium exists between phosphatidylinositol (PI), phosphatidylinositol 4-phosphate (PIP), and phosphatidylinositol 4,5-bisphosphate (PIP_2). Upon recognition of a stimulus, phospholipase C (PLC) is activated, probably via stimulation of a regulatory guanine nucleotide-binding protein (N), resulting in formation from PIP_2 of two intracellular messengers: Inositol 1,4,5-trisphosphate (IP_3) and diacylglycerol (DG). IP_3 is currently thought to release calcium intracellularly from endoplasmic reticulum-associated storage sites. Increases in intracellular free calcium can be measured, and lead to activation of a variety of calcium-dependent enzymes. DG promotes the association of protein kinase C (PKC) with the cell membrane, and dramatically reduces the PKC requirement for calcium. PKC in turn phosphorylates several proteins, among them myosin light chain (20 kDa) and 40- to 47-kDa proteins. Phosphorylated N-protein may inhibit PLC. Such feedback inhibition of PLC is observed when PKC is activated via pathways not involving receptor-mediated activation of PLC, such as by phorbol myristate acetate. IP_3 is rapidly degraded by phosphatases, resulting in free inositol (I), while DG is rapidly phosphorylated to form phosphatidic acid (PA), which can be further metabolized via a liponucleotide to reform PI.

Platelet-rich plasma (PRP, purchased from the local blood center, or prepared from whole blood by collecting plasma after centrifugation at 600 *g* for 10 min)
Citrate solution:
  Citric acid, 38 m*M*
  Trisodium citrate, 75 m*M*
Tris–citrate buffer:
  Tris, 63 m*M*
  NaCl, 95 m*M*
  KCl, 5 m*M*
  Citric acid, 12 m*M*
  Adjust pH to 6.5 with concentrated HCl
Incubation buffer:
  Tris, 75 m*M*

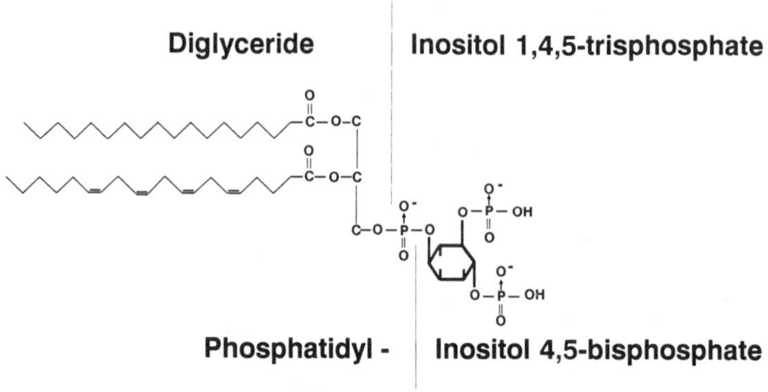

Diglyceride | Inositol 1,4,5-trisphosphate

Phosphatidyl - | Inositol 4,5-bisphosphate

FIG. 2. Structure of phosphatidylinositol 4,5-bisphosphate (PIP_2). Stearic acid is esterified to the *sn*-1 position, arachidonic acid to the *sn*-2 position of glycerol. A phosphodiester linkage, subject to phospholipase C action, links the *sn*-3 position of glycerol to the 1 position of myoinositol, which carries monoester phosphates at its 4 and 5 position. Phospholipase C-mediated hydrolysis results in formation of diglyceride and inositol 1,4,5-trisphosphate.

NaCl, 95 m*M*
Glucose, 7 m*M*
KH_2PO_4, 3 m*M*

Albumin, fatty acid free (e.g., Sigma A 6003), 0.35%
Adjust pH to 7.4 with concentrated HCl prior to adding albumin

Incubation tubes (125 × 20 mm round-bottom glass tubes with Teflon-lined screw caps. Tubes are reusable after cleaning with chromic acid solution, and scrupulously rinsing with deionized water; for studies using human platelets siliconizing is unnecessary)

Platelet Isolation. Washed platelet suspensions are prepared using the plastic bag system at 4°.[14,15] Initial centrifugation is at 1000 *g*, 10 min. PRP is expressed into the first satellite bag, and the bag containing erythrocytes and leukocytes is discarded. Platelet cyclooxygenase activity can be checked prior to further platelet processing by monitoring O_2 consumption following collagen stimulation of platelet-rich plasma.[17] Prior to the following centrifugation steps, the satellite bag is blown up with air until taut. This facilitates the separation of liquid phase from pellet. Contaminating erythrocytes and leukocytes remaining in the PRP are removed by a slow spin (225 *g*, 10 min). The PRP is expressed into the second satellite bag, and acidified with citrate solution (1 ml citrate solution/8 g PRP). The bag is filled with air again, and platelets are pelleted (2000 *g*, 10 min). The

[17] N. M. Bressler, M. J. Broekman, and A. J. Marcus, *Blood* **53,** 167 (1979).

platelet-poor plasma is expressed back into the bag containing erythrocytes and leukocytes pelleted in the 225 *g* spin and discarded. To wash, the pellet is resuspended in 3 ml Tris–citrate buffer by gentle massaging. Tris–citrate buffer (50 ml) is added, and the bag blown taut. The platelets are then pelleted by centrifugation (2000 *g*, 10 min), and finally resuspended (154 m*M* NaCl) and adjusted to a concentration of 10^{10} platelets/ml. Platelets so prepared are essentially devoid of polymorphonuclear leukocytes, and only rare erythrocytes or lymphocytes are encountered when examined by phase-contrast microscopy.

Incubations for Studies of Phospholipid Metabolism. Standard incubations are carried out at 37° in incubation tubes. Stirring, which enhances platelet aggregation, can be achieved by placing the incubation tubes (with stirrer magnets) in a plastic or stainless steel pan placed on two or more identical magnetic stirrers. Water in the pan is maintained at 37° by a constant temperature circulator which can pump into as well as out of the pan, such as a Lauda K-2/RD (Brinkmann, Westbury, NY). Standard incubations contain 0.5 ml of platelet suspension and 1.4 ml incubation buffer. After a 4-min preincubation period, stimuli or controls are added in 0.1 ml 0.154 *M* NaCl, resulting in a 2-ml total volume, containing 5×10^9 platelets. This is the optimal number of platelets for accurate phosphorus determination of minor platelet phospholipids such as PA and PIP_2.

Lipid Analyses

Reagents and Supplies

Chloroform
Methanol
C/M 2:5: Chloroform–methanol (2:5, v/v)
C/M 2:1: Chloroform–methanol (2:1, v/v)
EDTA–KCl solution: (0.1 *M* EDTA–2 *M* KCl, pH 6.5)
PIP_2 standard, e.g., Sigma #P 9763

All solvents are of reagent grade

Spotting tubes (120×17 mm conical glass centrifuge tubes with Teflon-lined screw caps. Cleaned in the same way as the incubation tubes)
Spotting syringes: 50-μl gas-tight syringe (Hamilton #1705), equipped with repeating dispenser (e.g., Hamilton #PB600-1)
Spotting boxes: The homemade variety (Fig. 3) is judged best and easiest to use

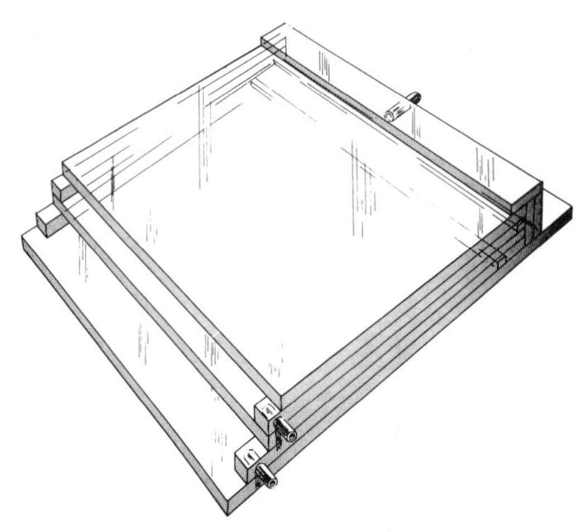

FIG. 3. Acrylic spotting box. The drawing illustrates the three inlets for nitrogen to superfuse two TLC plates in their entirety (through the inlet at the back of the box), as well as the extra inlets next to the lipid application spot, to enhance drying of the applied extract. The supports for the TLC plates are slightly staggered so as to make the lower 2 cm of each accessible for spotting of the lipid extract. Two thin strips of acrylic prevent the plates from sliding too far back in the box. The shelves upon which the plates rest should be scratched lengthwise to prevent the bottom of the flat glass plates from sticking to the equally flat shelves. The main parts are made from $\frac{3}{8}$-in. acrylic sheets.

TLC plates:

For two-dimensional TLC of phosphatidic acid (PA), PI, and all major phospholipids, silica gel H plates *with 7.5% magnesium acetate* (Cat. #88011, Analtech, Newark, DE) are utilized, without activation. In very dry climates, they probably should be equilibrated in 50% relative humidity prior to use. The Analtech plates are an improved version of Supelco's Ready-Coat 2D plates, which are no longer manufactured

For PIP_2 isolation, Whatman K5 plates (20 × 20 cm, 250-μm-thick layer; Thomas, Philadelphia, PA) are utilized. Plates are "washed" by running them with acetone as solvent. Lots are checked for occasional high phosphate background. Plates are activated at 110° for 1 hr

TLC tanks (4 × 9 × 11$\frac{3}{4}$ in.)

N-Evap model 111 (Organomation Associates, South Berlin, MA) to dry samples under nitrogen at 37°

Extraction Procedures

PI, PA, and Major Phospholipid Classes. Incubations are stopped and lipids extracted by a modified Bligh and Dyer[18] method: 7 ml of C/M 2:5 is added to each incubation tube containing 2 ml of aqueous cell suspension. Tubes are capped tightly, vortexed, and left at room temperature for 30 min or, after flushing with N_2, stored at −20° for up to 1 week. Two milliliters each of chloroform and water are then added to each incubation tube, which is vortexed and centrifuged (1000 *g*, 1 min). The lower, chloroform phase is transferred to a spotting tube with a Pasteur pipet, taking care not to carry over any of the aqueous phase (which also contains most of the methanol). The aqueous phase remaining in the incubation tube is reextracted with 3.5 ml chloroform and, after vortexing and centrifugation, the lower phase is added to the same spotting tube as the original chloroform extract. The combined organic extracts are dried at 37° under a gentle stream of N_2 using an N-Evap. The sides of the spotting tubes are washed down with 500 μl of C/M 2:1 to concentrate the lipid extract at the bottom of the tubes. This is again dried down under N_2. Upon addition of 200 μl C/M 2:1 the lipid extracts can be stored at −20° for up to 1 week, prior to spotting on TLC plates.

PIP and PIP_2. Extraction of PIP and PIP_2 is by a slightly different procedure, due to their physical properties: PIP_2 in particular exhibits rather high water solubility, and forms complexes with calcium and protein, which localize in the interfacial fluff between the aqueous and chloroform-rich phases of the extraction mixture.[17,19] This leads to large losses of PIP and PIP_2 in the normal extraction procedure. To interrupt these complexes, and to drive the inositides into the chloroform-rich phase, the Bligh and Dyer extraction is further modified as follows[20]:

Immediately after addition of 7 ml C/M 2:5 to the 2-ml aqueous cell suspension, 0.5 ml EDTA–KCl solution is added. Tubes are capped tightly, vortexed, and left at room temperature for 30 min, or stored under N_2 at −20° for up to 1 week. Two milliliters chloroform and *1.5 ml* EDTA–KCl are then added to each incubation tube, which is again vortexed and centrifuged (1000 *g*, 1 min). The lower, chloroform phase is transferred to a spotting tube with a Pasteur pipet, taking care not to carry over any of the aqueous phase. The remaining aqueous phase in the incubation tube is reextracted with 3.5 ml chloroform, and, after vortexing and centrifugation, the lower phase is added to the same spotting tube as

[18] E. G. Bligh and W. Dyer, *Can. J. Biochem. Physiol.* **37**, 911 (1959).

[19] R. M. C. Dawson, *Biochem. J.* **97**, 134 (1965).

[20] P. Cohen, M. J. Broekman, A. Verkley, J. W. W. Lisman, and A. Derksen, *J. Clin. Invest.* **50**, 762 (1971).

the original chloroform extract. The combined organic extracts are dried at 37° under a gentle stream of N_2. The sides of the spotting tubes are washed down with 500 μl of C/M 2:1 to concentrate the lipid extract at the bottom of the tubes. To free the extract from salts carried over from the aqueous phase, the spotting tubes are centrifuged (1000 *g*, 2 min), and the supernatant is carefully transferred to clean spotting tubes. This is repeated twice more to achieve quantitative transfer of the lipid extract, without carrying over salt, which can clog the needle of the spotting syringe. The combined C/M 2:1 extracts are dried again under a gentle stream of N_2 at 37°. Upon addition of 200 μl C/M 2:1 the lipid extracts can be stored at −20° for up to 1 week, prior to spotting on TLC plates.

Thin-Layer Chromatography

PI, PA, and Major Phospholipid Classes. The lipid extracts in spotting tubes are dried down under N_2, in order to concentrate them for spotting. They are redissolved in 35 μl C/M 2:1 for spotting. Two Analtech plates are conveniently spotted at the same time in a fume hood. At this point TLC tanks are prepared, to allow for equilibration of their solvent atmosphere (see below). Plates are placed in a spotting box (Fig. 3) under a gentle stream of N_2 through all three ports of the spotting box. The extract is applied as a streak of droplets at the lower right corner of the plate, 1.5 cm from bottom and right side of the plates (cf. Fig. 4). The extract is evenly divided over the streak in two or three series of droplets. The spotting tube is rinsed with 25 μl C/M 2:1, which is also applied to the plate. This is repeated once more. During spotting, care is taken that the droplets dry completely before lipid extract is reapplied to the same area. This serves to keep the streak narrow for optimal separations. Thus, two plates can be prepared at the same time by alternately applying a streak of lipid extract to each one. This procedure is used for the lipid extract from 5×10^9 platelets. If it is judged that less lipid is sufficient for the particular application, the streak can be made shorter.

After the extracts are applied, the plates are dried in the spotting boxes under N_2 for 10–15 min. They are then placed in TLC tanks equilibrated with solvent 1 (first dimension): chloroform/methanol/concentrated ammonia, 65:35:5 (v/v/v). Solvent equilibration is enhanced by a strip of Whatman 3MM filter paper (19.5 × 46 cm) placed inside the TLC tank. TLC tanks are kept inside a large plastic bag to minimize edge effects due to drafts in the laboratory. First dimension runs require approximately 65 min. Plates are removed from the tank and placed in a spotting box (in a fume hood) to dry under a gentle stream of N_2. After 15 min, plates are turned 90° and dried 45 min more. Second dimension TLC tanks are prepared 1 hr prior to use to allow for solvent equilibration similar to the

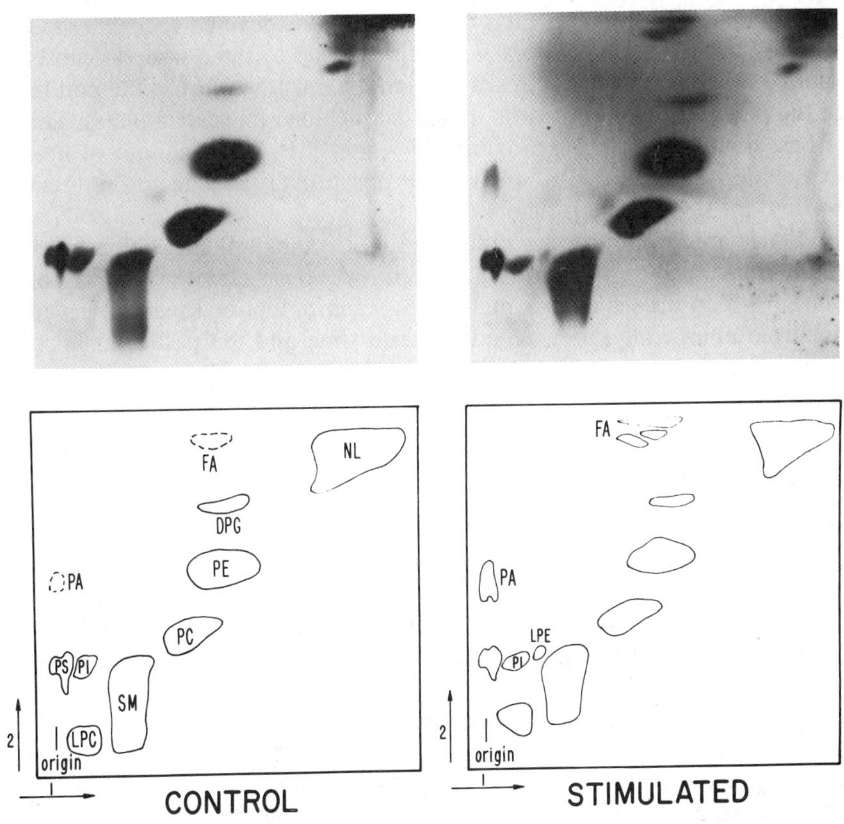

FIG. 4. Two-dimensional TLC of lipid extracts of control and thrombin-stimulated platelets. Photographs (top) and tracings (bottom) of iodine-stained TLC plates. Only trace amounts (broken lines) of PA and fatty acids (FA) are discernible in the control (left panels). After a 3-min incubation with 0.6 U/10^8 platelets (30 U/5 × 10^9 platelets, a saturating dose) the lipid pattern was changed: PI was clearly decreased, PA increased, and a new spot, LPE (absent in the control), appeared. Three distinct spots in the fatty acid area are tentatively identified as 12-hydroxyeicosatetraenoic acid (HETE), 12-hydroxyheptadectrienoic acid (HHT), and free fatty acids. Thromboxane is not extracted by the procedure used, unless acidified to pH <3.5, which leads to losses in phospholipid recovery. The line above the word "origin" indicates the position of application of the lipid extract. NL, neutral lipid. Reproduced from Broekman *et al.*,[14] by copyright permission of the American Society for Clinical Investigation.

first dimension tanks. Second dimension solvent is chloroform/acetone/methanol/acetic acid/water, 60:80:20:20:10 (v/v/v/v/v). The plates are placed in the second dimension tanks in such a way that the separation effected by the first dimension is at the bottom of the plate (see Fig. 4). Second dimension runs take approximately 75 min.

PIP and PIP$_2$. Lipid extracts are spotted onto activated K-5 plates in a similar manner as the 2-D plates, except that up to three samples can be spotted onto each plate in streaks of 4 cm length, 1.5 cm from the bottom of the plate. Initially, standards should be included on several plates. This is best done by adding approximately 7.5 μg PIP$_2$ to an aliquot of lipid extract derived from 5×10^9 platelets. Proportionally smaller amounts can be used for streaks of less than 4 cm in length.

Plates are developed in equilibrated TLC tanks utilizing chloroform/methanol/concentrated ammonia/water, 90:90:7:22 (v/v/v/v) as solvent.[16,21] Run time is approximately 55–65 min. Optimal separations are achieved empirically; they depend on run time and the particular lot of plates used.

Visualization

Reagents and Supplies

Rhodamine 6G, 0.002% in water/methanol, 1:1 (v/v)
Fluorescein, 0.04% in water/methanol, 1:4 (v/v). It has to be purified as follows[22]: Dissolve 500 mg 2′,7′-dichlorofluorescein in 250 ml methanol, dilute with 250 ml water, and wash three times with 500 ml hexane. Finally, dilute the aqueous phase to 1250 ml with methanol
HCl, 3 *N* in methanol (Supelco, Bellefonte, PA)
Pentane
Heptane
NaOH, 5 *N*
Internal standard (21:0, heneicosanoic acid; Supelco, Bellefonte, PA), 0.745 mg/ml in C/M 2:5
$HClO_4$, 70%
Phosphorus reagent[23]: Mix the following in sequence:

1 vol H_2SO_4, 6 *N*
1 vol ammonium molybdate, 2.5%
4 vol ascorbic acid, 2.5%
4 vol water

This solution should remain greenish-yellow. If it changes to bluish-green, contamination with phosphate is indicated, as from detergents. The ascorbic acid should be made up freshly before each use; other components can be stored almost indefinitely. The reagent is stable for several hours

[21] J. Schacht, *J. Lipid Res.* **19,** 1063 (1978).
[22] F. Parker and N. F. Peterson, *J. Lipid Res.* **6,** 455 (1965).
[23] P. S. Chen, T. Y. Toribara, and H. Warner, *Anal. Chem.* **28,** 1756 (1956).

Phosphate standard solution: 87.8 mg KH_2PO_4 is dissolved in 100 ml water (0.2 mg phosphate/ml). Five milliliters of this solution is diluted with 10 ml water to yield the standard solution (107.4 nmol phosphate/50 μl)

All solvents are pesticide grade.

Iodine tank (a TLC tank with some iodine crystals)
Phosphorus tubes (16×150 mm borosilicate disposable culture tubes)
Marbles to cap the phosphorus tubes. Wash with chromic acid solution
180° heating block
50° water bath
UV viewing apparatus, to provide 365-nm UV light, e.g., Chromato-Vue model CC-20 (Ultra-Violet Products, San Gabriel, CA)

After development, plates can be air dried overnight (in a fumehood) if phospholipid quantitation is by phosphorus assay. However, if quantitation will be by fatty acid analysis, plates have to be dried under N_2 in spotting boxes, and processed further as described below.

For lipid phosphorus determinations, plates are stained with iodine vapor. Spots are identified and marked, and most iodine is removed by sublimation in a fume hood. Marked areas of silica gel are scraped off the plate onto weighing paper, and placed into phosphorus tubes. In addition, a blank area, containing no phospholipid, and of approximately the same size as the phospholipid spots, is also scraped from each plate.

For fatty acid analyses, plates are dried for 1–2 hr under N_2 after chromatography. They are then sprayed with either rhodamine 6G or fluorescein. Rhodamine-sprayed plates are immediately viewed in 365-nm UV light. Identified spots are marked and plates are rigorously dried under N_2. Fluorescein-sprayed plates are first dried, then viewed in 365-nm UV light, and marked. Marked lipid areas, as well as blank areas of approximately the same size, are scraped into incubation tubes containing appropriate amounts of heneicosanoic acid (21:0) as internal standard (for PE and PC, 74.5 μg is dried down in the tubes in advance; for PI, 4.96 or 14.9 μg; for PA and PIP_2, 2.98 μg). Two milliliters methanolic HCl is then added to the tubes, which are flushed with N_2, at which point they can be stored at −20° for up to 1 week. In studying PIP_2, two samples, derived from 5×10^9 platelets each, are combined for fatty acid analyses of this minor phospholipid. All assays are carried out on triplicate samples of platelets for greatest accuracy.[24]

[24] M. J. Broekman, *J. Lipid Res.* **27**, 884 (1986).

Phosphorus Assay

It should be emphasized that it is important to ensure that *no* organic solvents are subjected to perchloric acid digestion, as this may lead to explosive breaking of the tube upon heating. Perchloric acid is added to the silica gel in the phosphorus tubes in 0.2-ml aliquots: six aliquots are used for PE and PC, two for PS, and one for PI, PA, PIP_2, other minor phospholipids, silica gel and reagent blanks, and standards. Since the absorbance response of the assay is linear with phosphate concentration to $A > 2.0000$, it is more convenient and accurate to prepare 107.4-nmol standards in quadruplicate than it is to perform routinely complete standard curves.

The marble-capped tubes are gently vortexed after addition of the perchloric acid, and then placed in the 180–190° heating block for 30 min. In view of concerns about the safety of using heated concentrated perchloric acid, it is recommended that this procedure be performed in an approved perchloric acid hood. If sphingomyelin is assayed (four aliquots), 45–60 min at 180–190° is necessary for complete destruction. After the tubes are cooled, 2.9-ml aliquots of phosphate reagent are added (six aliquots for PE and PC, two for PS, and one for PI, PA, PIP_2, other minor phospholipids, silica gel and reagent blanks, and standards). Tubes are vortexed and placed in a 50° water bath for 1 hr. Tubes are then vortexed again and, after cooling, centrifuged (5 min, 1000 *g*) to pellet the silica gel.

The absorbance of the samples is then read at 800–850 nm. The absorbance curve is sufficiently broad so that any wavelength between 800 and 850 nm is usable. However, the readout of the spectrophotometer should be reproducible and accurate to 0.0005 absorbance units or better. The spectrophotometer is zeroed with the reagent blanks, the standards are read, and then the silica gel blanks and samples are read. To prevent contamination with phosphates, glass spectrophotometer cells are stored in 0.01 *N* HCl between analyses.

The amount of phosphate in each sample is calculated as:

$$\text{Nanomoles phosphate} = \{[(A_{\text{sample}}) - (\text{avg } A_{\text{blanks}})] \times 107.4\}/(\text{avg } A_{\text{standards}})$$

Fatty Acid Quantitation

Fatty acid methyl esters (FAMEs) are prepared by heating the incubation tubes containing silica gel scrapings in 2 ml methanolic HCl at 100° for 40–50 min. FAMEs are isolated by adding 4 ml pentane and 2 ml water. After vortexing and short centrifugation to separate the phases, the upper, pentane phase is removed to a spotting tube. The lower aqueous phase is reextracted with 2 ml pentane, and the combined pentane layers are *completely* dried under N_2. The FAMEs are dissolved in $\pm$ 200 μl pentane. At this point they can be stored at $<-20°$ almost indefinitely without deterioration. Losses in arachidonate or other polyunsaturated fatty acids usually occur during extraction or chromatography, and are probably due to traces of iron in the lipid extract, or sometimes the silica gel of the plates.

When studying platelet PE, it is often desirable to also quantitate the dimethyl acetals, derived from PE plasmalogens (containing an alkenyl ether linkage at the 1-position, rather than a fatty acid). Since dimethyl acetals are acid labile, they are isolated *rapidly* as follows: PE samples are immediately cooled on ice, after heating in methanolic HCl as above. Subsequently, 4 ml ice-cold pentane and 2 ml ice-cold 5 *N* NaOH (not water) are added with stirring at 4°. The samples are then rapidly extracted identically to FAME samples from other phospholipids as described above.

Gas chromatography is performed utilizing a Hewlett-Packard 5880A gas chromatograph, equipped with two columns (1800×2 mm) packed with 10% SP2340 on 100/120 mesh Chromosorb WAW. This allows simultaneous analyses of two samples. Nitrogen carrier gas flow is 20 ml/min. Injection port and flame ionization detector are at 235 and 245°, respectively. Temperature programming is 5 min at 170°, followed by an increase of 5°/min to 225°, which is maintained for an additional 15 min. Fatty acids are quantitated relative to the amount of internal standard added to the sample tube prior to addition of methanolic HCl. For normal accuracy, the area under each peak is considered proportional to the weight of the FAME in that peak, although this is not strictly true, especially for short-chain fatty acids. From the weight of the FAME, division by its FW (FW of fatty acid + 14) yields the molar equivalent. In our hands this method is accurate (SD of triplicate samples <10% of their mean) for samples containing between 0.3 and 50 μg of individual FAMEs.

Applications

Our modification of Chen's phosphate assay[23] achieves a sensitivity of better than 0.3 nmol phosphate. This is accomplished by concentrating the original reagents. Chen's method is influenced by the pH of the assay mixture, and samples which are very acid or alkaline could introduce errors.[23] As described above, the assay of minor phospholipid components such as PIP_2, and low levels of PA (such as in resting platelets), necessitates use of relatively large numbers of platelets (up to 5×10^9 platelets per assay). Within these limitations, the method yields very sensitive indications of early (less than 5 sec) stimulation of platelets as demonstrated by our comparison of inositide metabolism and release of membrane-associated calcium (indicated by chlortetracycline).[16]

Figure 5 is an example of the fatty acid pattern of platelet PI and PIP_2 as determined in our recent work.[24] As shown, the original chromatograms show several impurities, especially in PIP_2 samples. However, palmitate and some stearate, as well as the material with a greater retention time than arachidonate are also present in blank areas of silica gel. Such blanks are easily subtracted from the raw data for accurate quantitation. For all fatty acid analyses, it is essential that the solvents utilized are of pesticide grade or, better, redistilled.

Several reagents can be utilized to transesterify fatty acids and aldehydes from phosphoglycerides: BCl_3 or BF_3 in methanol, or 3 *N* HCl in methanol, have similar efficiency. A choice between them is dictated by the purity of the reagents. This should be determined by heating fresh aliquots of each new batch alone, and with blank areas of a "washed" TLC plate, in the presence of small amounts (2.98 μg) of internal standard, and assay of these samples by GLC. Only very minor peaks (normally only a small peak of palmitate) are found if reagents and solvents are of high quality.

In addition, esterification efficiency can be determined, and absence of destruction of polyunsaturated fatty acids by the reagents can be verified as follows: Known quantities of arachidonic acid are transesterified in the presence of internal standard. Just prior to extraction with pentane, known quantities of a second standard, this time a fatty acid methyl ester such as 17:0-Me, are added. Comparison of different quantities of arachidonic acid and standards in a constant ratio will indicate whether the smallest quantities of arachidonate are efficiently esterified without destruction of double bonds. In our experience this will yield satisfactory results with most of the reagents mentioned, although an occasional bad batch unfortunately does come along.

Fatty acid methodology yielded sensitive measurements of net release of arachidonate from platelet phospholipids, including changes in such minor components as PIP_2 and PA.[15,24] This allowed us to conclude that phospholipase A_2 activation was responsible for more than 70% of the arachidonate released from platelets stimulated with high doses of thrombin.[24]

We also demonstrated that fMet-Leu-Phe stimulation of human neutrophils resulted in very early (5 sec) net synthesis of phospholipids, as determined by phosphate assay. In addition, fMet-Leu-Phe stimulation caused a redistribution of arachidonate among phosphoglycerides, and activation of the PI/PA cycle.[25]

Recently our methodology was applied to studies of medium condi-

[25] E. M. Wynkoop, M. J. Broekman, H. M. Korchak, A. J. Marcus, and G. Weissmann, *Biochem. J.* **236**, 829 (1986).

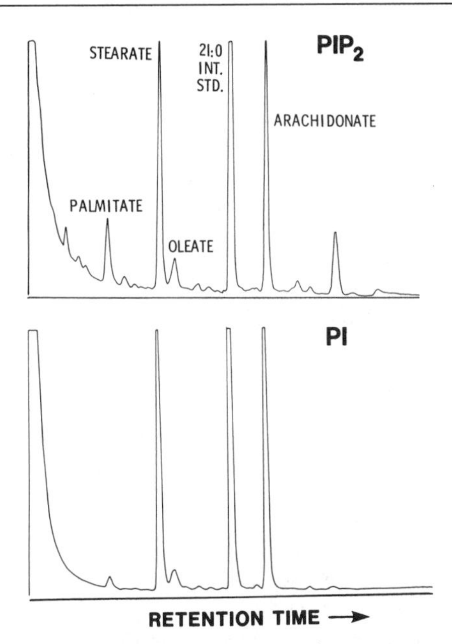

FIG. 5. Tracings of original gas chromatograms of PI and PIP_2 obtained from resting platelets. The fatty acid composition of the inositides is characterized by its simplicity: Virtually only stearate and arachidonate are present, as well as minor amounts of oleate. Palmitate is almost absent from these patterns, except for traces, which are also seen when appropriate blank areas of TLC plates are processed. Correction for these minute amounts is only necessary when processing very minor phospholipids such as PIP_2.

tions influencing cell–cell communications. Initial investigations indicated the importance of albumin in the medium both as a "sink" and as a "reservoir" for released arachidonate and its cyclooxygenase and lipoxygenase metabolites. Thus, depending on albumin concentration, time of exposure of platelets to thrombin, and concentration of thrombin, albumin will function both spatially and as a function of time in either one of two ways: When times of exposure to thrombin and/or concentration of thrombin are relatively low, albumin acts as a limiting factor in platelet stimulus–response coupling, reducing the number of affected platelets and the extent of their activation. In contrast, when thrombin concentration and time of exposure are increased, albumin acts as a vehicle for greater distribution of bioactive arachidonate metabolites resulting from platelet activation, increasing the effects of stimulation both spatially and as a function of time.[26]

[26] M. J. Broekman, *Clin. Res.* **36,** 406A (1988).

Acknowledgments

This research was supported by grants from the National Institutes of Health (HL 29034, HL 18828 SCOR), the New York Heart Association, and the Veterans Administration.

Dr. Broekman is an Established Investigator of the American Heart Association.

[35] Membrane-Permeable Diacylglycerol, Its Application to Platelet Secretion, and Regulation of Platelet Protein Kinase C

By NORIO KAJIKAWA, USHIO KIKKAWA, KIYOSHI ITOH, and YASUTOMI NISHIZUKA

Protein kinase C is abundant in platelets as an inactive enzyme under resting conditions. This enzyme absolutely requires Ca^{2+} and membrane phospholipid, particularly phosphatidylserine, for its activation.[1,2] Upon stimulation of platelets by thrombin, collagen, or platelet-activating factor, diacylglycerol is transiently produced in the plasma membrane from the receptor-mediated hydrolysis of inositol phospholipids.[3-5] The diacylglycerol markedly increases the affinity of this enzyme for Ca^{2+} as well as for phospholipid, and renders this enzyme fully active without increase in the Ca^{2+} concentration. In platelets, protein kinase C phosphorylates preferentially an endogenous protein having a molecular weight of approximately 40,000 (40K protein).[6] Some synthetic diacylglycerols such as 1-oleoyl-2-

[1] Y. Takai, A. Kishimoto, Y. Iwasa, Y. Kawahara, T. Mori, and Y. Nishizuka, *J. Biol. Chem.* **254,** 3692 (1979).

[2] A. Kishimoto, Y. Takai, T. Mori, U. Kikkawa, and Y. Nishizuka, *J. Biol. Chem.* **255,** 2273 (1980).

[3] Y. Kawahara, Y. Takai, R. Minakuchi, K. Sano, and Y. Nishizuka, *Biochem. Biophys. Res. Commun.* **97,** 309 (1980).

[4] K. Sano, Y. Takai, J. Yamanishi, and Y. Nishizuka, *J. Biol. Chem.* **258,** 2010 (1983).

[5] H. Ieyasu, Y. Takai, K. Kaibuchi, M. Sawamura, and Y. Nishizuka, *Biochem. Biophys. Res. Commun.* **108,** 1701 (1982).

[6] This protein has been highly purified and shown to be a 47K component which displays some microheterogeneity as judged from isoelectrofocusing analysis [T. Imaoka, J. A. Lynham, and R. J. Haslam, *J. Biol. Chem.* **258,** 11404 (1983)]. This protein has been identified as a substrate specific to protein kinase C both *in vivo* and *in vitro* by using the fingerprint procedure.[4] However, the function of this protein in platelet secretion remains unknown.

acetylglycerol are membrane permeable in intact platelets and directly activate protein kinase C without involvement of cell surface receptors for platelet agonists.[7] Thus, the synthetic diacylglycerol constitutes a tool for studies of the role of protein kinase C in platelet secretory reactions. It was originally found in the enzymatic assay that the active diacylglycerols appear to contain at least one unsaturated fatty acyl moiety. However, the diacylglycerols must not necessarily be unsaturated, since 1-palmitoyl-2-acetylglycerol was subsequently found to be fully active *in vitro* to support the activation of protein kinase C.[8] Recently, dihexanoylglycerol, dioctanoylglycerol, and didecanoylglycerol have been shown to be membrane permeable in intact platelets and capable of activating protein kinase C directly.[9,10] It is worth noting that tumor-promoting phorbol esters such as 12-*O*-tetradecanoylphorbol 13-acetate act as a substitute for diacylglycerol.[11] The platelet activation by this tumor promoter has been described elsewhere.[12] This chapter describes the chemical synthesis of a membrane-permeable diacylglycerol, 1-oleoyl-2-acetylglycerol, and its application to some studies on the secretory reaction involving platelet constituents, such as serotonin stored in dense bodies and acid hydrolases stored in lysosomes. Since protein kinase C appears to play a key role in platelet functions, an assay method, partial purification, and some properties of this enzyme in regard to its regulation will also be briefly presented.

Chemical Synthesis of 1-Oleoyl-2-acetylglycerol

Principle

The procedure is based on the method of Buchnea.[13] 1-Oleoylglycerol is protected first at the 3-position with a triphenylmethyl group, and then acetylated at the 2-position. The protecting group of the acetylated product is removed with trifluoroacetic acid. The product is purified with silicic acid column chromatography.

[7] K. Kaibuchi, Y. Takai, M. Sawamura, M. Hoshijima, T. Fujikura, and Y. Nishizuka, *J. Biol. Chem.* **258,** 6701 (1983).

[8] Unpublished observations (1984).

[9] E. G. Lapetina, B. Reep, B. R. Ganong, and R. M. Bell, *J. Biol. Chem.* **260,** 1358 (1985).

[10] R. J. Davis, B. R. Ganong, R. M. Bell, and M. P. Czeck, *J. Biol. Chem.* **260,** 1562 (1985).

[11] M. Castagna, Y. Takai, K. Kaibuchi, K. Sano, U. Kikkawa, and Y. Nishizuka, *J. Biol. Chem.* **257,** 7847 (1982).

[12] J. Yamanishi, Y. Takai, K. Kaibuchi, K. Sano, M. Castagna, and Y. Nishizuka, *Biochem. Biophys. Res. Commun.* **112,** 778 (1983).

[13] D. Buchnea, *Lipids* **6,** 734 (1971).

Step 1. Preparation of 1-Oleoyl-3-triphenylmethylglycerol. To 1-oleoylglycerol (91.4 g, 0.25 mol, Nakarai Chemicals, Japan) dissolved in 1.2 liters of a mixture of anhydrous pyridine and anhydrous benzene (1 : 2, v/v) a solution of pure triphenylmethyl bromide (80.8 g, 0.25 mol) in 200 ml of anhydrous benzene is added under anhydrous conditions. The reaction mixture is stirred for 24 hr at 30°, and then diluted with 1.2 liters of diethyl ether. The mixture is washed successively with two 1.0-liter portions of distilled water, two 1.0-liter portions of ice-cold 1 *N* hydrochloric acid, two 1.0-liter portions of distilled water, two 1.0-liter portions of a saturated potassium bicarbonate solution, and finally with two 1.0-liter portions of distilled water. The solution is dried with anhydrous sodium sulfate, and the solvents are removed under reduced pressure. The remaining materials are dissolved in dichloromethane, and applied to a silicic acid column, Wakogel C-200 (Wako Pure Chemical Industries, Japan, equivalent to Silica Gel 7734, Merck), 5 × 90 cm, previously washed with 2.0 liters of dichloromethane. The materials are eluted with dichloromethane until the eluate is free of solute. Fractions of 150 ml each are collected. An aliquot of each fraction is subjected to thin-layer chromatography on Silica Gel $60F_{254}$ (Merck) using dichloromethane as a solvent system. Organic materials on the plate are detected with iodine vapor followed by irradiation with ultraviolet light. Initially triphenylmethyl bromide appears. Subsequently the reaction product, 1-oleoyl-3-triphenylmethylglycerol, is recovered and finally the starting material appears. Triphenylmethyl bromide, 1-oleoyl-3-triphenylmethylglycerol, and the starting material exhibit R_f values of 0.74, 0.23, and 0.11, respectively. The fractions containing the product are pooled and concentrated under reduced pressure. The residue is dissolved in 200 ml of *n*-hexane and kept overnight at room temperature, then the white crystal, residual triphenylmethyl bromide, is removed by filtration. The filtrate is concentrated, and the remaining oily material that is 1-oleoyl-3-triphenylmethylglycerol is kept *in vacuo* until its weight is constant. The product weighs 52 g (yield 34%), and is almost homogeneous as judged by thin-layer chromatography.

Step 2. Preparation of 1-Oleoyl-2-acetyl-3-triphenylmethylglycerol. 1-Oleoyl-3-triphenylmethylglycerol (36.4 g, 0.06 mol), freshly prepared, is dissolved in a mixture of 10 ml of anhydrous pyridine and 120 ml of anhydrous benzene, and then acetyl chloride (4.8 g, 0.06 mol) in 45 ml of anhydrous benzene is added. The reaction mixture is stirred under anhydrous conditions for 2 hr at 30°, and then diluted with 600 ml of diethyl ether. The mixture is washed successively with 150 ml each of the same aqueous solutions as described in step 1, dried with anhydrous sodium sulfate, and then the solvents are removed completely by evaporation. The remaining oily material that is 1-oleoyl-2-acetyl-3-triphenylmethylglycerol weighs 32.5 g (yield 83%), and shows a single spot on thin-layer chromatography.

Step 3. Preparation of 1-Oleoyl-2-acetylglycerol. Ten milliliters of trifluoroacetic acid is added to a solution of freshly prepared 1-oleoyl-2-acetyl-3-triphenylmethylglycerol (32.5 g, 0.05 mol) in 500 ml of anhydrous ethanol. The mixture is stirred for 24 hr at room temperature. The reaction is terminated by adding 2.5 liters of diethyl ether and 1.5 liters of distilled water, and then the mixture is shaken. After removal of the aqueous phase the organic layer is washed with 800 ml of distilled water and dried. Then the solvents are evaporated. The remaining materials are dissolved in dichloromethane and applied to the Wakogel C-200 column described above (5 × 90 cm). A by-product, triphenylcarbinol, is eluted with 2.0 liters of dichloromethane and then 1-oleoyl-2-acetylglycerol and its 1,3-isomer spontaneously formed are eluted with dichloromethane–diethyl ether (80:20, v/v). The products in the effluent are monitored by thin-layer chromatography on Silica Gel $60F_{254}$ with a solvent system of dichloromethane–diethyl ether (70:30, v/v). 1-Oleoyl-2-acetylglycerol, its 1,3-isomer, and triphenylcarbinol exhibit R_f values of 0.55, 0.63, and 0.98, respectively. The fractions containing 1-oleoyl-2-acetylglycerol are pooled and the solvents are removed. The 1-oleoyl-2-acetylglycerol thus prepared weighs 4.7 g (yield 24%) and is shown to be almost homogeneous in thin-layer chromatography. The material sometimes contains a small amount of the 1,3-isomer. To remove the 1,3-isomer the method described by O'Flaherty *et al.*[14] is available. Silicic acid plates for preparative thin-layer chromatography are first washed with water–ethanol (1:1, v/v) containing 12.5% boric acid, and then activated at 180° for 3 hr. The diacylglycerol is chromatographed on this plate with diethyl ether–*n*-hexane (90:10, v/v) as a solvent. The main band is scratched off and the 1,2-isomer is extracted with diethyl ether through a sintered glass filter.

The product has been analyzed with a Hitachi M-52 mass spectrometer to confirm the product structure. In the mass spectrum, *m/e* 398 (molecular ion), *m/e* 399 (M + 1), *m/e* 380 ($M-H_2O$), *m/e* 338 ($M-CH_3COOH$), *m/e* 265 ($C_{17}H_{33}CO^+$), and *m/e* 43 (CH_3CO^+) are detected, and unambiguously confirm the product identity. The final product in the oily state or in *n*-hexane is stored at −70° until use.

This product seems to be a mixture of enantiomers. It has recently been reported that the diacylglycerol active in this role is highly specific 1,2-*sn*-diacylglycerol, while other enantiomers are inactive to support protein

[14] J. T. O'Flaherty, J. D. Schmitt, C. E. McCall, and R. L. Wykle, *Biochem. Biophys. Res. Commun.* **123,** 64 (1984).

kinase C activity.[15] The synthetic product described above can be used for the studies of platelet secretion. During storage, isomerization and migration of the fatty acyl moiety in diacylglycerols appear to take place very slowly.

Assay Method for Protein Phosphorylation in Rabbit Platelets

Blood (approximately 40 ml) is obtained from a healthy rabbit in a 50-ml polypropylene centrifuge tube containing acid–citrate–dextrose as anticoagulant. Washed platelets (4×10^9 cells) are prepared by the method of Baenziger and Majerus,[16] or alternative method (see Chapters 1 and 2 in this volume) and gently resuspended in a buffer solution consisting of 15 m*M* Tris–HCl at pH 7.5, 0.14 *M* NaCl, and 5.5 m*M* glucose (buffer A). The platelet suspension is centrifuged for 15 min at 2000 *g* at room temperature, and the platelet pellet is resuspended in buffer A containing 1.0 mCi of carrier-free $H_3{}^{32}PO_4$ (Japan Radioisotope Association) and then incubated for 1 hr at 37° as described by Lyons *et al.*[17] After centrifugation the labeled platelets are resuspended in buffer A at a cell count of 7.5×10^8 cells/ml.

1-Oleoyl-2-acetylglycerol is freed of *n*-hexane under a stream of nitrogen. The residue is suspended in buffer A containing 1% dimethyl sulfoxide by stirring for 1 min at room temperature with a Vortex mixer, followed by sonication for 1 min at 0° using a Kontes sonifier, model K881440. Buffer A (20 μl) containing 1% dimethyl sulfoxide and various concentrations of the synthetic diacylglycerol is added to the suspension of the radioactive platelets (160 μl) in a siliconized glass tube, and the mixture is incubated for 2 min at 37°. Subsequently, buffer A (20 μl) containing 10 m*M* calcium chloride and various concentrations of A23187 (Calbiochem) is added to the reaction mixture. After additional incubation for 1 min at 37° the protein phosphorylation is terminated by the addition of 100 μl of 186 m*M* Tris–HCl at pH 6.7, containing 9% sodium dodecyl sulfate (SDS), 15% glycerol, 6% 2-mercaptoethanol, and 0.003% bromphenol blue. The platelet proteins are dissolved by immersing the samples for 3 min in a boiling water bath.

A 60-μl aliquot of the samples is directly subjected to SDS–polyacrylamide slab gel electrophoresis prepared as described by Laemmli.[18] The separation gel of a length of 9 cm and an inside width of

[15] R. R. Rando and N. Young, *Biochem. Biophys. Res. Commun.* **122,** 818 (1984).
[16] N. L. Baenziger and P. W. Majerus, this series, Vol. 31, p. 149.
[17] R. M. Lyons, N. Stanford, and P. W. Majerus, *J. Clin. Invest.* **56,** 924 (1975).
[18] U. K. Laemmli, *Nature (London)* **227,** 680 (1970).

2 mm consists of a 5 to 18% linear acrylamide gradient, and the stacking gel contains 3% acrylamide. After electrophoresis the slab gel is stained with Coomassie Brilliant Blue followed by destaining. The destained gel is dried and an autoradiograph is obtained with a Kodak Royal X-Omat film. The autoradiograph is scanned in a Shimadzu dual-wavelength chromatogram scanner, model CS-910. The relative intensity of each band is quantitated by measuring the absorbance at 430 nm.

Assay Method for Platelet Secretion

Secretion of Serotonin

Platelet-rich plasma (about 20 ml) is prepared from rabbit blood, and is incubated for 30 min at 37° with 1 μCi of [2-^{14}C]serotonin (specific activity 58 mCi/mmol, Amersham) as described by Haslam and Lynham[19] (see also Chapter [17] in this volume). The labeled platelets are washed and suspended in buffer A as mentioned above. The platelet suspension is incubated with various concentrations of the synthetic diacylglycerol and then with A23187 under the conditions employed for the assay of protein phosphorylation. The secretion of [^{14}C]serotonin is terminated by the addition of 20 μl of 6% formaldehyde containing 50 m*M* ethylenediaminetetraacetic acid (EDTA) followed by immersion in ice using a modification of the method of Costa and Murphy.[20] After centrifugation for 40 sec at 10,000 *g*, the radioactivity in the supernatant is determined.

Secretion of β-N-Acetylglucosaminidase

For this assay no radioactive material is needed. The platelets are isolated and incubated with the synthetic diacylglycerol and A23187, and the secretion reaction is terminated in the same way as that employed in the serotonin secretion assay. The volume of each incubation mixture is twice that employed for the serotonin secretion assay. The samples are centrifuged for 15 min at 2000 *g* at 4°, and the supernatant is obtained as much as possible. The enzyme released is assayed with *p*-nitrophenyl-*N*-acetyl-β-glucosaminide (Sigma) as a substrate by a modification of the method of Dangelmaier and Holmsen[21] (see Chapter [29] in this volume) as follows. The supernatant (400 μl) is mixed with the assay medium (600 μl) that consists of 0.1 *M* acetate buffer at pH 4.5, 5 m*M* *p*-nitro-

[19] R. J. Haslam and J. A. Lynham, *Biochem. Biophys. Res. Commun.* **77,** 714 (1977).
[20] J. L. Costa and D. L. Murphy, *Nature (London)* **255,** 407 (1975).
[21] C. A. Dangelmaier and H. Holmsen, *Anal. Biochem.* **104,** 182 (1980).

phenyl-*N*-acetyl-β-glucosaminide, and 0.1% Triton X-100. The mixture is incubated for 45 min at 37°. The reaction is terminated by the addition of 1 *N* sodium hydroxide (50 μl). Under the alkali condition, *p*-nitrophenol develops color which is measured at 410 nm. The sedimented platelets are lysed in 500 μl of buffer A containing 0.1% Triton X-100, sonicated for 15 sec at 0° using a Kontes Sonifier, model K881440, and the sonicates are then centrifuged for 15 min at 2000 *g* at 4° to obtain the supernatant. Thus, the enzyme in the cells (400-μl aliquot of the supernatant) is similarly assayed. Assuming the total enzyme activity as the summation of that amount released into the medium and that amount remaining in the cells, the percentage of enzyme released is calculated.

Synergistic Action of Membrane-Permeable Diacylglycerol and Ca^{2+} Ionophore in Platelet Secretion

It has been repeatedly documented that the stimulation of receptors relating to inositol phospholipid turnover usually mobilizes Ca^{2+} at the same time.[22-24] When platelets are stimulated by either thrombin, collagen, or platelet-activating factor, another endogenous protein having a molecular weight of approximately 20,000 (20K protein) is heavily phosphorylated in addition to 40K protein. This protein has been identified as myosin light chain, and the reaction is catalyzed by a Ca^{2+}, calmodulin-dependent protein kinase that is specific for this protein.[25] Hence, the activation of protein kinase C and Ca^{2+} mobilization may be monitored by measuring the phosphorylation of 40K and 20K proteins, respectively.

When platelets are stimulated by 1-oleoyl-2-acetylglycerol alone (6–25 μg/ml), only 40K protein is phosphorylated to an extent that is very similar to that induced by extracellular messengers, and 20K protein is not phosphorylated to a measurable extent under the present conditions.[26] The reaction rate of 40K protein phosphorylation is not affected by the addition of Ca^{2+} ionophore, suggesting that the synthetic diacylglycerol greatly

[22] J. N. Hawthorne and D. A. White, *Vitam. Horm.* **33,** 529 (1975).

[23] R. H. Michell, *Biochim. Biophys. Acta* **415,** 81 (1975).

[24] R. H. Michell, *Trends Biochem. Sci.* **4,** 128 (1979).

[25] D. R. Hathaway and R. S. Adelstein, *Proc. Natl. Acad. Sci. U.S.A.* **76,** 1653 (1979).

[26] Myosin light chain is also phosphorylated by protein kinase C, but this reaction appears to proceed very slowly after the aggregation and secretion reactions are completed [M. Naka, M. Nishikawa, R. S. Adelstein, and H. Hidaka, *Nature (London)* **306,** 490 (1983)]. The site of phosphorylation by this enzyme differs from that phosphorylated by myosin light chain kinase [M. Nishikawa, J. R. Sellers, R. S. Adelstein, and H. Hidaka, *J. Biol. Chem.* **259,** 8808 (1984)]. The significance of this reaction is unknown. Perhaps it may constitute a feedback mechanism to prevent overresponse, because these two enzymes exert opposing effects on the activity of myosin light chain.

increases the affinity of protein kinase C for Ca^{2+} and induces full activation of this enzyme without a net increase in the intracellular concentration of Ca^{2+}. Inversely, the phosphorylation of 20K protein is induced by a low concentration (0.6 μM) of Ca^{2+} ionophore alone, and is not affected to a measurable extent by the synthetic diacylglycerol (6–25 μg/ml). Thus, the synthetic diacylglycerol and Ca^{2+} ionophore can selectively induce the activation of protein kinase C and mobilization of Ca^{2+}, respectively.[7,27]

Neither serotonin nor β-*N*-acetylglucosaminidase is secreted sufficiently by the stimulation with the synthetic diacylglycerol alone, although the phosphorylation reaction of 40K protein proceeds rapidly under the same conditions. Similarly, Ca^{2+} ionophore alone at lower concentrations (less than 0.6 μM) induces minimally the secretion of these platelet constituents. However, when the synthetic diacylglycerol and Ca^{2+} ionophore are added together, serotonin and lysosomal enzyme are released to the extent observed after stimulation with agonists such as thrombin (Fig. 1). In these experiments the concentration of A23187 (0.2–0.6 μM) is critical since at concentrations greater than 0.6 μM ionophore will itself cause the phosphorylation of 40K protein due to the nonspecific activation of phospholipases and protein kinase C, presumably by a large increase in Ca^{2+} concentration. Likewise, it is important to use no more than 25 μg/ml of the synthetic diacylglycerol because at higher concentrations (50 μg/ml) this compound induces a significant secretion of the platelet constituents. The exact reason for this enhanced release is unknown, but it is possible that diacylglycerol may act as a membrane fusigen or a weak Ca^{2+} ionophore at higher concentrations.[7] Nevertheless, the activation of protein kinase C and mobilization of Ca^{2+} are both essential, and act synergistically to evoke release reactions of various constituents from different platelet granules.

Partial Purification and Properties of Platelet Protein Kinase C

Among many tissues and cell types so far tested, platelets contain protein kinase C with highest specific activity. The enzyme does not appear to be tissue- or species-specific in its kinetic and catalytic properties.

Assay Procedure

Protein kinase C is routinely assayed by measuring the incorporation of radioactive phosphate from [γ-^{32}P]ATP into H1 histone in the presence of $CaCl_2$, phospholipid, and diolein. The reaction mixture (0.25 ml) contains 20 mM Tris–HCl at pH 7.5, 5 mM magnesium acetate, 10 μM [γ-^{32}P]ATP

[27] N. Kajikawa, K. Kaibuchi, T. Matsubara, U. Kikkawa, Y. Takai, Y. Nishizuka, K. Itoh, and C. Tomioka, *Biochem. Biophys. Res. Commun.* **116**, 743 (1983).

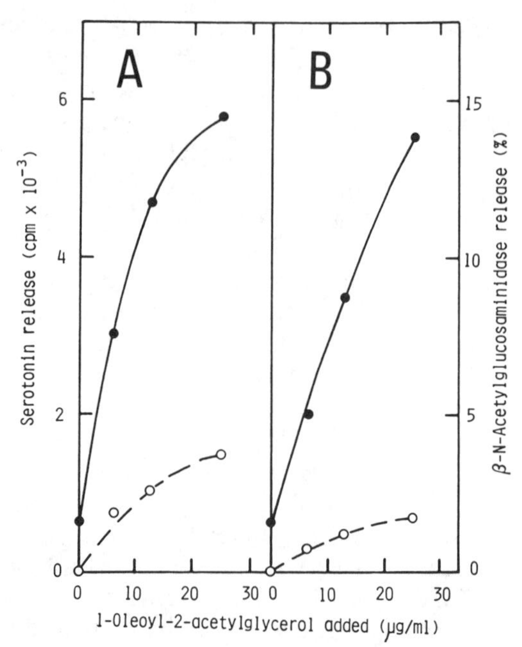

FIG. 1. Synergistic effects of 1-oleoyl-2-acetylglycerol and A23187 on platelet secretion. The platelets were stimulated with the synthetic diacylglycerol and A23187 under the conditions described in the section, Assay Method for Platelet Secretion. (A) Secretion of serotonin; (B) secretion of β-*N*-acetylglucosaminidase. (●—●), Stimulated by the synthetic diacylglycerol and A23187; (○- - -○), stimulated by the synthetic diacylglycerol alone. Adapted from Kajikawa *et al.*[27]

($5-10 \times 10^4$ cpm/nmol), 0.1 m*M* $CaCl_2$, 50 μg of H1 histone, 10 μg of phospholipid mixture, 0.2 μg of diolein, and enzyme preparation. Phospholipid and diolein are mixed first and then added to the reaction mixture as micelles.[2] Basal activity is measured in the presence of 0.5 m*M* ethylene glycol bis(β-aminoethyl ether)-*N,N,N′, N′*-tetraacetic acid (EGTA) instead of $CaCl_2$, phospholipid, and diolein. The reaction is started by the addition of a mixed solution of Tris–HCl, magnesium acetate, [γ-^{32}P]ATP, and H1 histone to the test tube that contains $CaCl_2$, lipid micelles, and enzyme. After 3 min at 30°, the reaction is terminated by the addition of 3 ml of 25% trichloroacetic acid. Acid-precipitable materials are collected on a nitrocellulose membrane filter (pore size, 0.45 μm) in a suction apparatus. The membrane filter is washed five times each time with 3 ml of 25% trichloroacetic acid. The radioactivity is quantitated by Cerenkov counting using a Packard Tri-Carb liquid scintillation spectrometer, model 3330.

Application to Crude Extract

It is often difficult to measure the amount of protein kinase C in crude extracts due to the presence of some undefined inhibitors and other protein kinases that utilize H1 histone as a phosphate acceptor. Protein kinase C may be detected as a sharp peak, requiring both phospholipid and Ca^{2+}, after chromatography on a Sephadex G-150 column (Fig. 2) or a DE-52 column (see below). The medium to homogenize platelets as well as the column elution buffer contains 2 m*M* EDTA and 5 m*M* EGTA to prevent the proteolysis of the enzyme, which is very susceptible to Ca^{2+}-dependent neutral protease.[28]

Partial Purification

This procedure is based on that described earlier for human platelet enzyme,[3] and may also be used for rabbit platelets.

Human blood is obtained from healthy volunteers, and washed platelets (5×10^{10} cells) are prepared by the method of Baenziger and Majerus[16] or by alternative methods as described in Chapters 1 and 2 in this volume. All subsequent operations are performed at 0–4°. The platelets (5×10^{10} cells) are suspended in 10 ml of a medium containing 20 m*M* Tris–HCl at pH 7.5, 0.25 *M* sucrose, 0.01% leupeptin, 2 m*M* phenylmethylsulfonyl fluoride, 10 m*M* 2-mercaptoethanol, 2 m*M* EDTA, and 5 m*M* EGTA. The suspension is subjected to sonication for 30 sec with a Kontes sonifier, model K881440, and the sonicate is centrifuged for 60 min at 100,000 *g*. Protein kinase C is recovered in the soluble fraction. The supernatant (10 ml, 33 mg of protein) is put onto a DE-52 column (1.4×5 cm, Whatman) equilibrated with 20 m*M* Tris–HCl at pH 7.5, containing 10 m*M* 2-mercaptoethanol, 2 m*M* EDTA, 2 m*M* EGTA, and 0.001% leupeptin (buffer B). Elution is carried out with a 96-ml linear concentration gradient of NaCl (0 to 0.4 *M*) in buffer B. The elution profile of the enzyme is shown in Fig. 3. The activity of protein kinase C is 20 times higher than that of cyclic AMP-dependent protein kinase which is measured with a common substrate, H1 histone, in the presence of 1 μM cyclic AMP. Cyclic GMP-dependent protein kinase is not detected under comparable conditions. The fractions of protein kinase C (fractions 12 through 22) are pooled and concentrated by an Amicon ultrafiltration cell equipped with a PM10 filter membrane. The enzyme solution is subjected to gel filtration on a Sephadex G-150 column (1.6×80 cm, Pharmacia Fine Chemicals) equilibrated with buffer B. The enzyme is eluted with the same buffer. The fraction of protein kinase C is pooled and concentrated by ultrafiltration.

[28] A. Kishimoto, N. Kajikawa, M. Shiota, and Y. Nishizuka, *J. Biol. Chem.* **258,** 1156 (1983).

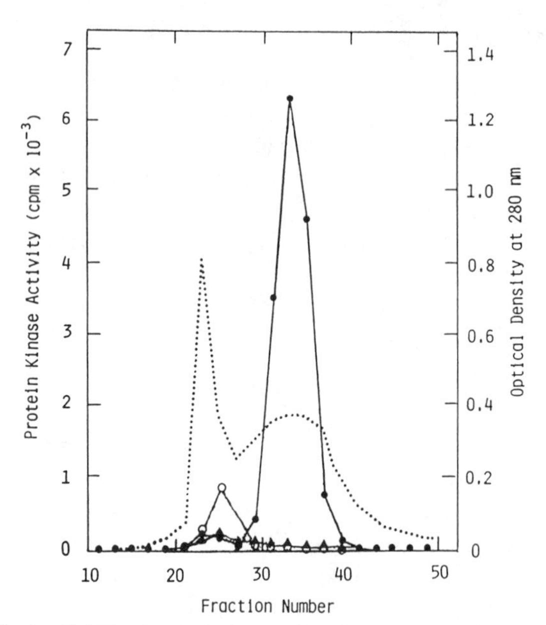

FIG. 2. Sephadex G-150 column elution profile of protein kinase C from human platelets. The soluble fraction of human platelets was subjected to a gel filtration column. Conditions and procedures are similar to those for the purification of the enzyme as described in the text. (●—●), With $CaCl_2$, phospholipid, and diolein; (○—○), with cyclic AMP; (▲—▲), with EGTA; (···), optical density at 280 nm. Adapted from Y. Nishizuka, Y. Takai, A. Kishimoto, U. Kikkawa, and K. Kaibuchi, *Recent Prog. Horm. Res.* **40,** 301 (1984).

The enzyme preparation thus obtained is free of other protein kinases, calmodulin, interfering enzymes, and endogenous phosphate acceptor proteins.

Mode of Activation and Some Properties

Protein kinase C apparently lacks tissue and species specificities in its kinetic and physical properties, and is essentially similar to the brain enzyme which has been extensively purified and characterized.[29,30] In short, the enzyme itself is totally inactive unless protamine is utilized as a phosphate acceptor, but is activated by diacylglycerol in the presence of Ca^{2+} and membrane phospholipid. The activation is very specific for Ca^{2+}. A small amount of diacylglycerol markedly increases the affinity of the enzyme for phosphatidylserine, and sharply decreases the Ca^{2+} concentra-

[29] U. Kikkawa, R. Minakuchi, Y. Takai, and Y. Nishizuka, this series, Vol. 99, p. 288.

[30] U. Kikkawa and Y. Nishizuka, *in* "The Enzymes" (P. D. Boyer and E. G. Krebs, eds.), Vol. 17, Part A, p. 167. Academic Press, Orlando, Florida, 1986.

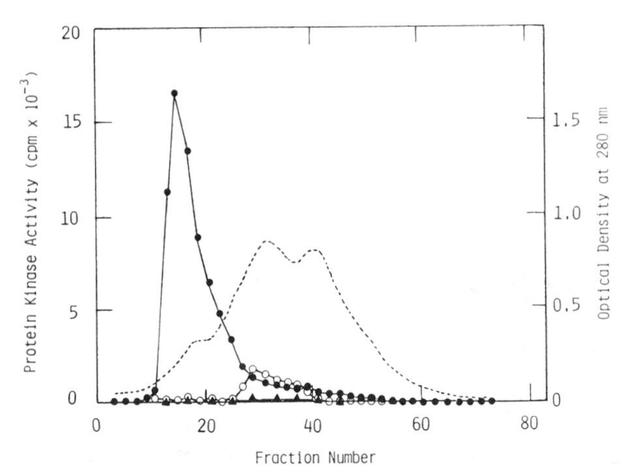

FIG. 3. DE-52 column elution profile of protein kinase C from human platelets. Conditions and procedures are described in the text. (●—●), With $CaCl_2$, phospholipid, and diolein; (○—○), with cyclic AMP; (▲—▲), with EGTA; (-----), optical density at 280 nm. Adapted from Kawahara *et al.*[3]

tion needed for activation of this enzyme to less than the 10^{-6} *M* range. Thus, protein kinase C may be synergistically activated by diacylglycerol and Ca^{2+} under physiological conditions. Among various phospholipids phosphatidylserine is indispensable. Phosphatidylethanolamine enhances further protein kinase activation, and renders the enzyme fully active at the 10^{-7} *M* range of Ca^{2+} concentration.[31] The precise biochemical mechanism of this unique lipid–protein interaction has not been elucidated. Protein kinase C is normally very unstable and may be stored at 0° for only a few days. However, in the presence of 10% (w/v) glycerol and 0.05% (w/v) Triton X-100 the enzyme is stable at least several months at −80°. The molecular weight of protein kinase C is estimated to be about 82,000 by SDS–polyacrylamide gel electrophoresis. The Stokes radius is estimated by a gel filtration analysis to be 42 Å, which corresponds to a molecular weight of 87,000. The enzyme is apparently composed of a single polypeptide chain with no subunit structure. The isoelectric point is approximately pH 5.6. ATP but not GTP serves as phosphate donor, and the K_m value of ATP with H1 histone as substrate is 6 μM. The optimum magnesium concentration is 5–10 m*M*. The optimum pH is 7.5–8.0 with 40 m*M* Tris–acetate as a test buffer.

[31] K. Kaibuchi, Y. Takai, and Y. Nishizuka, *J. Biol. Chem.* **256,** 7146 (1981).

Phosphate Acceptor Proteins

Protein kinase C shows broad substrate specificity when tested *in vitro*, and phosphorylates seryl and threonyl but not tyrosyl residues in many endogenous proteins.[29,30] In platelets, 40K protein described above appears to be a major phosphate acceptor protein.[3] Myosin light chain may serve as a substrate of this enzyme *in vitro* although the reaction rate is extremely slow,[26] and the physiological significance of this reaction remains unknown. Other substrate proteins in platelets have not yet been well defined.

Activator and Inhibitor

Tumor promoters such as 12-*O*-tetradecanoylphorbol 13-acetate,[11] mezerein,[32] teleocidin,[33] and debromoaplysiatoxin[33] are able to substitute for diacylglycerol, and thereby directly activate protein kinase C *in vitro* as well as *in vivo*. So far, no inhibitor has been found that is specific to protein kinase C. However, many phospholipid-interacting drugs, such as chlorpromazine, trifluoperazine, dibucaine, imipramine, phentolamine, verapamil, and tetracaine, inhibit this enzyme activation process.[30,34] H7 [1-(5-isoquinolinesulfonyl)-2-methylpiperazine], recently described, inhibits the enzyme by interaction with the catalytically active site of protein kinase C, but also inhibits cyclic AMP- dependent protein kinase by the same mechanism.[35]

[32] R. Miyake, Y. Tanaka, T. Tsuda, K. Kaibuchi, U. Kikkawa, and Y. Nishizuka, *Biochem. Biophys. Res. Commun.* **121**, 649 (1984).
[33] H. Fujiki, Y. Tanaka, R. Miyake, U. Kikkawa, Y. Nishizuka, and T. Sugimura, *Biochem. Biophys. Res. Commun.* **120**, 339 (1984).
[34] T. Mori, Y. Takai, R. Minakuchi, B. Yu, and Y. Nishizuka, *J. Biol. Chem.* **255**, 8378 (1980).
[35] H. Hidaka, M. Igarashi, S. Kawamoto, and Y. Sasaki, *Biochemistry* **23**, 5036 (1984).

[36] Calcium-Dependent Proteases and Their Inhibitors in Human Platelets

By JUN-ICHI KAMBAYASHI and MASATO SAKON

Calcium-dependent proteases in mammalian tissues (e.g., heart, lung, and skeletal muscle) and the role of brain-derived enzyme in the activation of protein kinase C, an intracellular signal transducer, were described in other volumes of this series.[1,2] The name "calpain" proposed by Murachi *et al.*[3] for calcium-dependent neutral proteases was recommended by the Nomenclature Committee of the International Union of Biochemistry.[4] Depending on the range of calcium concentration required for activation of calpain, two types are distinguished: calpain I, active at micromolar Ca^{2+} concentrations, and calpain II, active at millimolar Ca^{2+} concentrations. Both types share similar properties in terms of being a thiolprotease composed of two noncovalently bound subunits of M_r 75,000–80,000 and 25,000–30,000 based on mobility in polyacrylamide gel electrophoresis in the presence of sodium dodecyl sulfate.[5]

The existence of an intracellular protein which specifically inhibits calpain has been documented in various cells. A general name, calpastatin, was proposed for this protein inhibitor by Murachi *et al.*[3] The physiological role of this unique protease–inhibitor system has not been fully elucidated yet, though a wide variety of endogenous substrates such as myofibrillar proteins, cytoskeletal proteins, and enzymes has been reported.[6-10]

Since the original description of Ca^{2+}-dependent protease activity in human platelets by Phillips and colleagues in 1977,[11] a number of studies on the calpain–calpastatin system in human platelets has been reported, including those from our laboratory. This chapter is devoted to the description of the specific assay, purification, and properties of the platelet calpain–calpastatin system as developed in our laboratory. A brief discussion of potential endogenous substrates for platelet calpain is also included herein.

[1] L. Waxman, this series, Vol. 80, p. 664.

[2] N. Kajikawa, A. Kishimoto, M. Shiota, and Y. Nishizuka, this series, Vol. 102, p. 273.

[3] T. Murachi, K. Tanaka, M. Hatanaka, and T. Murakami, *Adv. Enzyme Regul.* **19,** 407 (1981).

[4] The IUPAC/IUB Committee on Enzyme Nomenclature, *in* "Enzyme Nomenclature Recommendations 1984." Academic Press, 1984.

[5] T. Murachi, *Biochem. Soc. Symp.* **49,** 149 (1984).

[6] P. F. Verhallen, E. M. Bevers, P. Comfurius, and R. F. Zwaal, *Biochim. Biophys. Acta* **903,** 206 (1987).

[7] Y. Ando, S. Imamura, Y. Yamagata, A. Kitahara, H. Saji, T. Murachi, and R. Kannagi, *Biochem. Biophys. Res. Commun.* **144,** 484 (1987).

[8] J. E. Fox, C. C. Reynolds, J. S. Morrow, and D. R. Phillips, *Blood* **69,** 537 (1987).

[9] M. C. Beckerle, T. O'Halloran, and K. Burridge, *J. Cell Biochem.* **30,** 259 (1986).

[10] T. Onji, M. Takagi, and N. Shibata, *Biochim. Biophys. Acta* **912,** 238 (1987).

[11] D. R. Phillips and M. Jakabova, *J. Biol. Chem.* **252,** 5602 (1977).

Assays for Calpains I, II, and Calpastatin

Principle

The activity of calpain is measured by caseinolytic assay. The caseinolytic activity of calpain is determined by measuring the amount of acid-soluble peptides in the supernatant after undigested substrate is precipitated by a solution containing trichloroacetic acid, sodium acetate, and acetic acid. This assay provides a good correlation between the amount of enzyme and the activity detected.[12] The acid-soluble peptides are detected by fluorometric assay with fluorescamine (4-phenylspiro[furan-2(3*H*),1′-phthalan]-3,3′-dione), which reacts with substrates containing primary amino groups to yield highly fluorescent products. The fluorometric assay with fluorescamine is approximately 10 times more sensitive than the colorimetric assay. An alternative method employing radiolabeled casein has been used in the assay of calpain as presented above.[13,14]

Reagents for Calpain Assay

4% Heat-denatured casein (Hammarsten grade, Merck), pH 7.0
Imidazole–HCl buffer, 0.2 *M*, pH 6.9
KCl, 1 *M*; dithiothreitol (DTT), 100 m*M*; $CaCl_2$, 40 m*M*; ethylenediaminetetraacetic acid (EDTA), 40 m*M*; TCA solution (0.11 *M* CCl_3COOH, 0.22 *M* CH_3COONa, 0.33 *M* CH_3COOH)[7]
Borate–NaOH buffer, 0.5 *M*, pH 8.5
Fluorescamine (Roche) in acetone (30 mg/100 ml)

Calpain Assay

The reaction mixture (200 μl) contains 0.5% heat-denatured casein (pH 7.0), 100 m*M* imidazole–HCl (pH 6.9), 50 m*M* KCl, 1 m*M* DTT, 2 m*M* $CaCl_2$, and 20 μl of enzyme sample. The reaction is initiated by the addition of the enzyme sample. Control reactions are carried out in the presence of 2 m*M* EDTA and the control values are subtracted from those obtained in the presence of calcium. After the incubation at 25° for 20 min, the reaction is terminated by adding 300 μl of TCA solution. The mixture is kept at room temperature for more than 30 min and then centrifuged at 10,000 *g* for 10 min. An aliquot (80 μl) of the supernatant is mixed with 0.8 ml of 0.5 *M* borate–NaOH (pH 8.5) and then 0.2 ml of fluorescamine in acetone is added to the mixture, while the test tube is being vigorously shaken. The fluorescence is measured with excitation at

[12] B. Hagiwara, *Annu. Rep. Sci. Works* **2,** 35 (1954).
[13] J. A. Truglia and A. Stracher, *Biochem. Biophys. Res. Commun.* **100,** 814 (1981).
[14] N. Yoshida, B. Weksler, and R. Nachman, *J. Biol. Chem.* **258,** 7168 (1983).

405 nm and emission at 475 nm, using leucine as a standard. When it is desired to assay calpain I activity, 100 μM $CaCl_2$ is added instead of 2 mM $CaCl_2$ in the reaction mixture because no appreciable activity of calpain II is detected in the presence of 100 μM $CaCl_2$ as shown in Fig. 2. One unit of activity is defined as the amount of enzyme which liberates 1 μmol of α-amino group as leucine equivalent per hour.

Assay for Calpain Inhibitor (Calpastatin)

Inhibitor activity (calpastatin) is determined by the same procedure as calpain assay except that the inhibitor sample (30 μl) is preincubated with a fixed amount (0.3 or 0.5 U) of partially purified platelet calpain at 25° for 10 min and then reaction mixture without DTT is added. One unit of calpastatin is defined as the amount of calpastatin that inhibits one unit of calpain. The values are calculated from the straight portion of the inhibitor concentration curve.

Purification Steps for Calpains and Calpastatin

Preparation of Platelet Lysate

All procedures are performed at 0–4°, unless otherwise stated. Human platelets are isolated from platelet-rich plasma anticoagulated with ACD solution. Contaminating red and white cells are removed by slow centrifugation at 225 *g* for 10 min followed by a microscopic examination of a platelet-enriched supernatant. Platelets are washed three times at 1000 *g* for 10 min in 10 mM Tris–HCl, pH 7.5, containing 154 mM NaCl, 5 mM EDTA. Platelets are finally suspended at a concentration of 5×10^9 cells/ml in 10 mM Tris–HCl, pH 7.5, containing 154 mM NaCl, 5 mM EDTA, and 1 mM DTT. Washed platelets are then disrupted in a sonifer (Bronson Sonic Power Company—model W-185E) with an index of 7 for 15 sec × 15 cycles. The homogenate is then centrifuged at 105,000 *g* for 60 min to obtain the supernatant fluid.

Separation of Calpain I, II, and Calpastatin

Before proceeding to any further purification, it is essential to separate calpain I, II, and calpastatin, coexisting in the soluble fraction of platelet lysate.[15] In certain tissues, such as liver[16] and cardiac muscles,[17] the activity

[15] M. Sakon, T. Tsujinaka, J. Kambayashi, and G. Kosaki, *Experientia* **38**, 1099 (1982).
[16] I. Nishiura, K. Tanaka, S. Yamato, and T. Murachi, *J. Biochem. (Tokyo)* **84**, 1657 (1978).
[17] L. Waxman and E. G. Krebs, *J. Biol. Chem.* **253**, 5888 (1978).

of calpain is suppressed until coexisting calpastatin is removed. The separation of three components can be achieved easily by a single run of platelet protein on an ion-exchange chromatography column, details of which is described herein.

The soluble fraction of platelet lysate is precipitated by adding solid ammonium sulfate to produce 70% saturation. The precipitate is dialyzed against 10 m*M* Tris–HCl, pH 7.5, 0.25 *M* sucrose, 1 m*M* ethylene glycol bis(β-aminoethyl ether)-*N,N,N′,N′*-tetraacetic acid (EGTA), 1 m*M* EDTA, and 1 m*M* DTT. The dialyzed supernatant is applied to a DEAE-Sepharose CL-6B column and eluted by a linear concentration gradient of NaCl as shown in Fig. 1. Since the assay mixture contains 0.5 U platelet calpain, positive peaks indicate the presence of calpain and negative deviation indicates the presence of calpastatin. Thereby, calpastatin is eluted at 0.15 *M* NaCl (arrow A) and calpains I and II are eluted at 0.2 and 0.3 *M* NaCl (arrows B and C), respectively. Much higher activity of calpain I is obtained in human platelets in contrast to bovine platelets in which calpain II is predominantly eluted.[15] The peak fractions of active calpain are precipitated by 70% ammonium sulfate saturation and applied to a Sephadex G-150 column (1.5 × 17 cm) equilibrated with 5 m*M* imidazole–HCl, pH 6.9, 20 m*M* KCl, 1 m*M* EGTA, and 1 m*M* DTT. Elution is carried out by the same buffer. Calcium dependency of calpains thus obtained is determined using a strictly adjusted Ca^{2+}–EGTA mixture.[18] As shown in Fig. 2, calpains I and II require 6 and 900 μ*M* free Ca^{2+}, respectively, for the half-maximal activity.

Purification of Calpain I from Human Platelets

As mentioned above, calpain I is the predominant form of calpain in human platelets, which is presumably activated first in stimulated platelets because of the lower Ca^{2+} requirement. Our procedure for purification of platelet calpain I is as follows:

Buffers

Buffer A: 10 m*M* Tris–HCl (pH 7.5), 0.25 *M* sucrose, 1 m*M* EGTA, 1 m*M* EDTA, 1 m*M* DTT

Buffer B: 20 m*M* Tris–HCl (pH 6.9), 0.25 *M* sucrose, 1 m*M* EDTA

Procedure

Step 1: Ammonium sulfate fractionation. The supernatant of lysed platelets (2.5 × 10^{11}), prepared as described above, is fractionated with ammonium sulfate. The precipitate obtained between 30 and 45% saturation is dissolved in buffer A and the solution is dialyzed against the same buffer for 15 hr, followed by centrifugation at 10,000 *g* for 10 min.

[18] Y. Ogawa, *J. Biochem. (Tokyo)* **64,** 255 (1968).

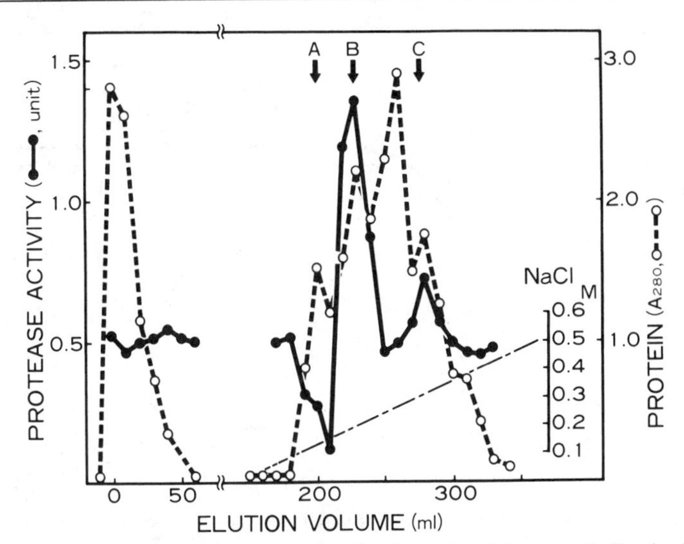

FIG. 1. Separation of calpains I and II and calpastatin of human platelet by DEAE-Sepharose CL-6B chromatography. Details are described in the text. (●—●), Ca^{2+}-Protease activity; (○- - -○), absorbance at 280 nm (A_{280}); (— — — — — —), NaCl concentration.

Step 2: Sephacryl S-300 gel filtration. The dialyzed supernatant (10 ml, 116 mg protein) is applied to a Sephacryl S-300 column (2.5 × 86 cm) equilibrated with buffer A and fractions (5 ml) are eluted with the same buffer at a flow rate of 16 ml/hr. The protein (A_{280}) and calpain activities are determined in each fraction. A sharp single peak of calpain activity is obtained in fractions 49–52. The activity of contaminated calpastatin eluted in fractions 34–37 is thus completely separated from calpain activity by this gel filtration.

Step 3: DEAE-Sepharose CL-6B ion-exchange chromatography. The active fractions of Sephacryl S-300 chromatography are pooled (20 ml, 17.9 mg protein) and applied to a DEAE-Sepharose CL-6B column equilibrated with buffer A. After washing out with the same buffer, elution is performed by a linear concentration gradient of NaCl (0 to 0.35 *M*) in 200 ml of the same buffer. The flow rate is 16 ml/hr and fractions of 2 ml are collected. Two peaks of calpain activity are eluted. The first prominent peak of calpain activity eluted at 0.175 *M* NaCl contains calpain I and the second small peak at 0.275 *M* NaCl contains calpain II. The active fractions of calpain I are combined (10 ml, 1.33 mg protein) and dialyzed against buffer B for 3 hr.

Step 4: DEAE-Sephadex A-25 ion-exchange chromatography. The dialyzed supernatant is applied to a DEAE-Sephadex A-25 column (1.5 × 17 cm) equilibrated with buffer B. After washing with buffer B, elution is

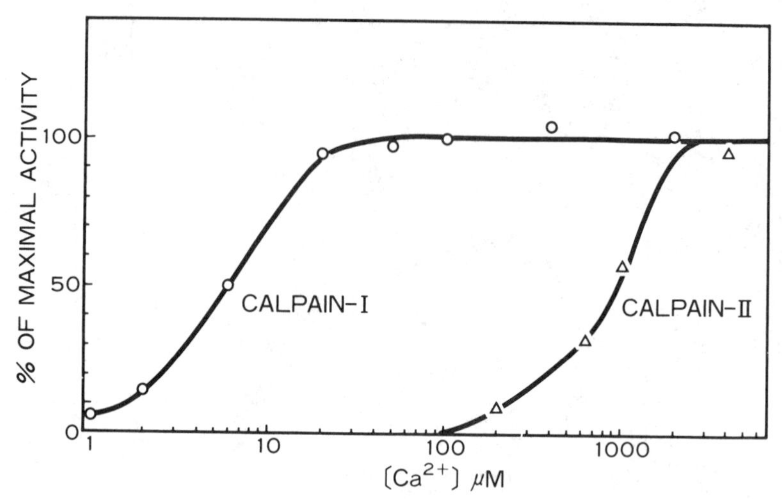

FIG. 2. Ca^{2+} Dependence of calpains I and II partially purified from human platelets. Details are described in the text.

carried out by a linear concentration gradient of NaCl (0 to 0.25 *M*) in 200 ml of buffer B. The flow rate is 18 ml/hr and fractions of 2 ml are collected. A single peak of calpain activity (fractions 47–50) is eluted at 0.12 *M* NaCl and stored at −80° until use.

The purified enzyme is homogeneous in polyacrylamide gel electrophoresis and the enzyme is composed of 80,000- and 25,000-Da subunits, determined by sodium dodecyl sulfate (SDS)–polyacrylamide gel electrophoresis. The purification summary is listed in Table I. The protease is purified approximately 475-fold with a recovery of 14%, although the true yield cannot be determined because of calpastatin, coexisting in crude enzyme preparations. The purified enzyme is activated half-maximally at 25 *μM* Ca^{2+}.

Several studies on purification of platelet calpain have been reported, as summarized in Table II. In comparison to our preparation,[19] enzymes obtained by Truglia and Stracher[13] and Yoshida *et al.*[14] require a much higher Ca^{2+} concentration for activity than calpain I, though the molecular weight and subunit structure are very similar in three independent reports. Data on purification of calpains I and II from rat kidney by Yoshimura *et al.*[20] indicate that the molecular weight and subunit structure are identical for two forms of calpain. Therefore, it is likely that some preparations of

[19] T. Tsujinaka, E. Shiba, J. Kambayashi, and G. Kosaki, *Biochem. Int.* **6,** 71 (1983).
[20] N. Yoshimura, T. Kikuchi, T. Sasaki, A. Kitahara, M. Hatanaka, and T. Murachi, *J. Biol. Chem.* **258,** 8883 (1983).

TABLE I
PURIFICATION OF HUMAN PLATELET CALPAIN I

Step	Total protein (mg)	Total activity (unit)	Specific activity (unit/mg)	Yield (%)	Purification (-fold)
1. Lysed platelets (2.5×10^{11})	1030	1500	1.46	—	—
2. 105,000 *g* supernatant	528	1500	2.84	100	1.95
3. $(NH_4)_2SO_4$ 30–45% precipitate	116	536	4.62	36	3.16
4. Sephacryl S-300	17.9	440	24.6	29	16.8
5. DEAE-Sepharose CL-6B	1.33	250	188	17	129
6. DEAE-Sephadex A-25	0.31	215	694	14	475

purified platelet calpain may represent a mixture of two forms of calpain.[13,14] In this respect, the use of inhibitor-conjugated affinity chromatography in purification of calpains should be judicious because two forms bind equally well to this ligand in the presence of Ca^{2+}.

Characteristics of Platelet Calpains I and II

Similar and dissimilar characteristics of platelet calpains I and II are summarized in Table III. It is now generally agreed that molecular weight and subunit structure of calpains I and II are identical (80,000 and

TABLE II
PURIFICATION OF CALPAIN FROM PLATELETS

Year	Source	Reported name	Molecular weights	Ca^{2+} requirement	Ref.[a]
1981[b]	Bovine	CANP-I	120 K	6 μM > for half-maximal activity	1
		CANP-II	135 K	600 μM	
1981	Human	CDSP	80K + 30K	1 mM for optimal activity	2
1982[b]	Human	μ-CANP		6 μM > for half-maximal activity	3
		m-CANP		900 μM	
1983	Human	μ-CANP	80K + 25K	25 μM for half-maximal activity	4
1983	Human	CAP	80K + 27K	0.52 mM for half-maximal activity	5

[a] Key to references: (1) M. Sakon, J. Kambayashi, H. Ohno, and G. Kosaki, *Thromb. Res.* **24,** 207 (1981); (2) J. A. Truglia and A. Stracher, *Biochem. Biophys. Res. Commun.* **100,** 814 (1981); (3) T. Tsujinaka, M. Sakon, J. Kambayashi, and G. Kosaki, *Thromb. Res.* **28,** 149 (1982); (4) T. Tsujinaka, E. Shiba, J. Kambayashi, and G. Kosaki, *Biochem. Int.* **6,** 71 (1983); (5) N. Yoshida, B. Weksler, and R. Nachman, *J. Biol. Chem.* **258,** 7168 (1983).

[b] Partial purification.

25,000–30,000) as determined in SDS–polyacrylamide gel electrophoresis, though there is a difference in amino acid composition between the two types.[20] Both types share similar pH optimum, activation by sulfhydryl protecting agents, and inhibition by sulfhydryl-reactive agents as shown in Table III. Two forms of platelet calpain differ in terms of Ca^{2+} requirement, activation by Mn^{2+}, Sr^{2+}, and autocatalytic degradation.

Purification and Properties of Platelet Calpastatin

Although the presence of inhibitor was mentioned by Truglia and Stracher during their purification of calpain,[13] our group provided the initial procedure for partial purification of a high-molecular-weight, heat-stable calpastatin from bovine platelets and demonstrated that it inhibits platelet calpains I and II specifically by forming a stoichiometric complex.[15] Subsequently, we have developed a purification method resulting in a homogeneous preparation of platelet calpastatin as described below.[21]

Buffers

Buffer C: 20 m*M* Tris–HCl, pH 7.5, 0.1 m*M* EGTA

Procedure

Step 1: First heat treatment. The supernatant of platelet lysate obtained as above is heated at 80° for 10 min and the formed precipitate is removed by centrifugation at 12,000 *g* for 15 min. The resultant supernatant is then fractionated with ammonium sulfate. The precipitate obtained with 0–70% saturation is suspended in buffer C and the solution is dialyzed overnight against the same buffer.

Step 2: DEAE-Sepharose CL-6B ion-exchange chromatography. The dialysate (11 ml, 31.5 mg protein) is applied to a DEAE-Sepharose CL-6B column (1.5 × 16 cm) equilibrated with buffer C. After the column is washed with buffer C, elution is carried out by a linear concentration gradient of NaCl (0 to 0.35 *M*) in 150 ml of the same buffer. Fractions of 3 ml are collected at a flow rate of 11 ml/hr. The calpain inhibitory activity is eluted at 0.2 *M* NaCl.

Step 3: Second heat treatment. The fractions containing calpastatin activity are pooled and subjected to the second heat treatment at 95° for 15 min. The precipitate is removed by centrifugation. The resultant supernatant is lyophilized and resuspended in buffer C.

Step 4: Ultrogel AcA 34 gel filtration. The solution is dialyzed for 3 hr against buffer C and then the dialysate (800 μl, 1.08 mg protein) is applied to an Ultrogel AcA 34 column (0.9 × 56 cm) equilibrated with buffer C and the elution performed with the same buffer at a flow rate of 6 ml/hr.

[21] E. Shiba, T. Tsujinaka, J. Kambayashi, and G. Kosaki, *Thromb. Res.* **32,** 207 (1983).

TABLE III
PROPERTIES OF PLATELET CALPAIN I AND II

Property	Calpain I[a]	Calpain II[b]
Molecular weight	80,000 + 25,000	110,000 − 120,000 (gel filtration)
Optimal pH	7.0	7.0
Ca^{2+} for 50% activation	25 μM	0.9 mM
Activation		
By cysteine, DTT[c]	(+)	(+)
By calmodulin	(−)	(−)
By other divalent cations	Ba^{2+}	Mn^{2+}, Ba^{2+}, Sr^{2+}
Autodigestion (0°, pH 7.5, 20 min)	18%	73%
Inhibition		
By calpastatin	(+)	(+)
By IAA,[c] pCMB,[c] leupeptin	(+)	(+)
By DFP, PMSF,[c] SBTI[c]	(−)	(−)

[a] Homogeneously purified enzyme.
[b] Partially purified enzyme.
[c] DTT, Dithiothreitol; IAA, iodoacetamide; pCMB, *p*-chloromercuribenzoate; PMSF, phenylmethylsulfonyl fluoride; SBTI, soybean trypsin inhibitor.

Fractions of 0.5 ml are collected. Two peaks of calpastatin activity are obtained. The results of the purification are summarized in Table IV. By running calibration proteins on the same column, the molecular weights of the first and second peaks are estimated to be 250,000 and 60,000, respectively. When an aliquot of the first and second peak is applied on SDS–polyacrylamide gel electrophoresis, a single band of 63,000 Da is obtained. These results indicate that platelet calpastatin exists as a monomer or a tetramer composed of subunits of apparent M_r 63,000.

TABLE IV
PURIFICATION OF HUMAN PLATELET CALPASTATIN

Step	Total protein (mg)	Total activity (unit)	Specific activity (unit/mg)	Yield (%)	Purification (-fold)
1. Lysed platelets (5×10^{11})	2030	—	—	—	
2. 105,000 *g* supernatant	1142	—	—	—	
3. Heat treatment	87.2	—	—	—	
4. 0–70% $(NH_4)_2SO_4$ precipitate	31.5	66.9	2.12	100	1
5. DEAE-Sepharose CL-6B	3.20	56.8	17.8	84.9	8.4
6. Heat treatment	1.08	22.8	21.1	34.1	10.0
7. Ultrogel AcA 34	0.18	6.0	33.3	9.0	15.7

TABLE V
NATURAL SUBSTRATES FOR PLATELET CALPAIN

Reported name	Enzyme source	Activation	Substrates	Ref.[a]
Platelet CAF	Lysed platelets	Ca^{2+}, 2 m*M*	225 kDA, 245 kDA	1
	intact platelets	Ca^{2+} + A23187	230 kDa, 220 kDa	
Ca^{2+}-Activated protease	Intact platelets	Ca^{2+} + A23187	240 kDa (ABP),[b] 220 kDa 77 kDa, 68 kDa (albumin?)	2
Ca^{2+}-Dependent sulfhydryl protease (CDSP)	Purified enzyme	Ca^{2+}, 4 m*M*	270 kDa (ABP)[c]	3
	Lysed platelets	No chelation	280 kDa (ABP), 240 kDa	
Ca^{2+}-Activated neutral protease (μ-CANP, m-CANP)	Partially purified, two forms	Ca^{2+}, 50 μ*M* (μ-CANP)	260 kDa (ABP),[d] 230 kDa	4
		Ca^{2+}, 2 m*M* (m-CANP)	280 kDa, 290 kDa (MAP I, II)[e]	
Ca^{2+}-Activated protease	Triton X-100 extract	Ca^{2+}, 4 m*M*	235 kDa (P_{235})[f]	5
Ca^{2+}-Activated neutral protease	Purified μ-CANP	Ca^{2+}, 0.2 m*M*	147 kDa + 150 kDa (caldesmon)[g]	6
	Lysed platelets	Ca^{2+}, 10 μ*M*	135 kDa (MLCK)[g]	
			60 kDa, 90 kDa (calmodulin-binding proteins)	
Ca^{2+}-Dependent protease	Lysed platelet	Ca^{2+}, 1.8 m*M*	270 kDa (ABP), 235 kDa (P_{235})	7
	Intact platelets	A23187, collagen, thrombin		
Ca^{2+}-Activated protease	Purified enzyme	Ca^{2+}, 3m*M*	150 kDa (GPIb)	8
Ca^{2+}-Dependent neutral protease (CNP)	Supernatant of lysed platelets		150 kDa (GPIb)	9

[a] Key to references: (1) D. R. Phillips and M. Jakabova, *J. Biol. Chem.* **252,** 5602 (1977); (2) G. C. White II, *Biochim. Biophys. Acta* **631,** 130 (1980); (3) J. A. Truglia and A. Stracher, *Biochem. Biophys. Res. Commun.* **100,** 814 (1981); (4) T. Tsujinaka, M. Sakon, J. Kambayashi, and G. Kosaki, *Thromb. Res.* **28,** 149 (1982); (5) N. C. Collier and K. Wang, *J. Biol. Chem.* **257,** 6937 (1982); (6) T. Tsujinaka, E. Shiba, J. Kambayashi, and G. Kosaki, *Biochem. Int.* **6,** 71, (1983). (7) J. E. B. Fox, C. C. Reynolds, and D. R. Phillips, *J. Biol. Chem.* **258,** 9973 (1983); (8) N. Yoshida, B. Weksler, and R. Nachman, *J. Biol. Chem.* **258,** 7168 (1983); (9) M. O. Spycher, U. E. Nydegger, and E. F. Luescher, *Adv. Exp. Med. Biol.* **167,** 241 (1984).

[b] Abbreviations: ABP, actin-binding protein; MAP, microtubules-associated proteins; MLCK, myosin light chain kinase; GPIb, glycoprotein Ib.

[c] Partially purified, origin not stated.

[d] Triton X-100 supernatant.

[e] Partially purified from bovine brain.

[f] Purified from human platelets.

[g] Purified from chicken gizzard.

The purified calpastatin is heat stable, withstanding 100° for 20 min without a loss of inhibitory activity. Platelet calpastatin inhibits both platelet calpains I and II in the presence of respective Ca^{2+} concentration. The activities of other proteases tested, such as α-chymotrypsin, trypsin, plasmin, papain, and ficin, are not affected by platelet calpastatin. Kinetic studies using the purified platelet calpain I and calpastatin revealed that the inhibition is noncompetitive with a low K_i value ($3.2 \times 10^{-8} M$).[21]

Natural Substrates for Platelet Calpain

In order to search for the physiological role of the calpain–calpastatin system in platelets, it is mandatory to identify specific endogenous substrates for calpains. Several potential endogenous substrates of platelet calpains have been studied by activating intact platelets, adding Ca^{2+} to lysed platelets, and using the purified enzyme and/or substrates. The list of endogenous platelet substrates for calpains is summarized in Table V. Cleavage of these proteins involved in actin polymerization was seen not only in lysed platelets but also in intact platelets stimulated by physiological agonists. We have recently identified several calmodulin-binding proteins, of which M_r 90,000 and 60,000 polypeptides are specifically cleaved by calpain.[22] Myosin light chain kinase and caldesmon, typical calmodulin-binding proteins, purified from chicken gizzard are also good substrates for the purified platelet calpain.[22] These substrates are related to cytoskeletal or contractile systems of platelets but detailed analysis of calpain's role therein is not completed yet. In addition, platelet membrane glycoprotein Ib (GPIb) has been reported as the substrate of platelet calpain.[14,23] This cleavage involves an extramembrane segment of GPIb, while calpain is located in the cytoplasm and not secreted during platelet stimulation.[24-26] In this regard, platelet factor XIII, localized in the cytoplasm, can be activated by calpain.[27] Its role in enzymatic crosslinking of intraplatelet proteins awaits elucidation as do other reactions mediated by calpain.[28]

[22] G. Kosaki, T. Tsujinaka, J. Kambayashi, K. Morimoto, K. Yamamoto, K. Yamagami, K. Sobue, and S. Kakiuchi, *Biochem. Int.* **6,** 767 (1983).

[23] M. O. Spycher, U. E. Nydegger, and E. F. Luescher, *Adv. Exp. Med. Biol.* **167,** 241 (1984).

[24] E. B. McGowan, K.-T. Yeo, and T. C. Detwiler, *Arch. Biochem. Biophys.* **227,** 287 (1983).

[25] Unpublished observations (1984).

[26] J. A. Samis, G. Zboril, and J. S. Elce, *Biochem. J.* **246,** 481 (1987).

[27] Y. Ando, S. Imamura, Y. Yamagata, A. Kitahara, H. Saji, and T. Murachi, *Biochem. Biophys. Res. Commun.* **144,** 484 (1987).

[28] A. H. Schmaier, P. M. Smith, A. D. Purdon, J. G. White, and R. W. Colman, *Blood* **67,** 119 (1986).

[37] Measurement of Changes in Platelet Cyclic AMP *in Vitro* and *in Vivo* by Prelabeling Techniques: Application to the Detection and Assay of Circulating PGI_2

By RICHARD J. HASLAM and MARIANNE VANDERWEL

Many of the physiological and pharmacological agents that inhibit platelet aggregation and degranulation in response to all stimuli are believed to exert their effects by activation of adenylate cyclase (EC 4.6.1.1) or inhibition of cyclic AMP phosphodiesterase and thus by increasing the intracellular concentration of cyclic AMP.[1,2] Measurements of platelet cyclic AMP have therefore assumed considerable importance in analysis of the mechanisms of action of both old and new antiplatelet agents. Although use of a sensitive radioimmunoassay for cyclic AMP, such as that of Harper and Brooker,[3] is appropriate in experiments with suspensions of washed platelets, this approach cannot detect small but nevertheless physiologically significant increases in platelet cyclic AMP, when the platelets are suspended in plasma or blood. Thus, under basal conditions, about 75% of cyclic AMP has been shown to be extracellular in human platelet-rich plasma containing $3-4 \times 10^8$ platelets/ml.[1] Both for this reason and to save time and expense, many workers have preferred to measure cyclic [^{14}C]AMP or cyclic [^{3}H]AMP in platelets in which the metabolic adenine nucleotide pool has been labeled by preincubation of the platelets with [^{14}C]adenine or [^{3}H]adenine. Thus, under basal conditions, little or no labeled cyclic AMP is found extracellularly in suspensions of platelets that have been incubated with labeled adenine.[1] This approach was first applied to platelets by Ball *et al.*[4] and was subsequently adopted by other groups, who have used several different methods for isolation of the labeled cyclic AMP from the platelets.[5-8] The precise methodology chosen is crucial to

[1] R. J. Haslam, *Ciba Found. Symp.* **35,** 121 (1975).

[2] R. J. Haslam, M. M. L. Davidson, T. Davies, J. A. Lynham, and M. D. McClenaghan, *Adv. Cyclic Nucleotide Res.* **9,** 533 (1978).

[3] J. F. Harper and G. Brooker, *J. Cyclic Nucleotide Res.* **1,** 207 (1975).

[4] G. Ball, G. G. Brereton, M. Fulwood, D. M. Ireland, and P. Yates, *Biochem. J.* **120,** 709 (1970).

[5] D. C. B. Mills and J. B. Smith, *Biochem. J.* **121,** 185 (1971).

[6] R. J. Haslam and A. Taylor, *Biochem. J.* **125,** 377 (1971).

the success of this technique, as some of the more complicated earlier methods[4,6] greatly reduce the number of samples that can be processed conveniently, whereas failure to remove all the labeled impurities can obscure significant increases in the labeled cyclic AMP present. Under basal conditions, labeled cyclic AMP should amount to no more than 0.04% of the total radioactivity in human platelets[1,9] or 0.03% of that in rabbit platelets.[8] Use of prelabeling methods has been validated by demonstration of linear relationships between increases in cyclic AMP radioactivity and mass under a variety of conditions in both human and rabbit platelets.[7,9,10] The method described here,[8] which is that adopted most recently in this laboratory, permits convenient processing of up to 200 samples in a single experiment and detection of increases in platelet cyclic AMP in the presence of all concentrations of activators of adenylate cyclase and inhibitors of cyclic AMP phosphodiesterase that are capable of inhibiting platelet responses.

It is possible to use the formation of labeled cyclic AMP in platelets as a method of detecting and assaying compounds that activate platelet adenylate cyclase, such as prostacyclin (PGI_2). In principle, this is no different from methods[11,12] that use the associated inhibition of platelet function as the basis of an assay. However, by including an inhibitor of cyclic AMP phosphodiesterase, such as 3-isobutyl-1-methylxanthine (IBMX), the accumulation of labeled cyclic AMP is greatly enhanced and assay sensitivity is increased about 10-fold. This is not so easily achieved in assays based on platelet function, because high concentrations of IBMX alone block platelet responses. An essential feature of assays of this type must be the inclusion of some means of identifying the compound responsible for increases in platelet cyclic AMP. In the case of PGI_2, this has been achieved by use of an antibody that binds PGI_2 and closely related compounds and by comparison of the half-life of the compound assayed with that of authentic PGI_2.[8]

More recently, we have extended the prelabeling technique to studies of the relationship between changes in platelet cyclic AMP and platelet re-

[7] J. P. Harwood, J. Moskowitz, and G. Krishna, *Biochim. Biophys. Acta* **261,** 444 (1972).

[8] R. J. Haslam and M. D. McClenaghan, *Nature (London)* **292,** 364 (1981).

[9] R. J. Haslam, *in* "Platelet Function Testing" (H. J. Day, H. Holmsen, and M. B. Zucker, eds.), NIH Publ. No. 78-1087, p. 487. U.S. Department of Health, Education and Welfare, 1978.

[10] R. J. Haslam and M. Vanderwel, *J. Biol. Chem.* **257,** 6879 (1982).

[11] N. L. Baenziger, M. J. Dillender, and P. W. Majerus, *Biochem. Biophys. Res. Commun.* **78,** 294 (1977).

[12] E. Dejana, J.-P. Cazenave, H. M. Groves, R. L. Kinlough-Rathbone, M. Richardson, M. A. Packham, and J. F. Mustard, *Thromb. Res.* **17,** 453 (1980).

sponsiveness in whole blood from rabbits that have been transfused with labeled platelets.[13] As platelet cyclic [^{3}H]AMP can be extracted from whole-blood samples without isolation of the platelets, this provides the only means of directly correlating inhibition of platelet function with increases in platelet cyclic AMP in a physiological milieu.

Labeling of Platelets in Platelet-Rich Plasma

Human Platelets. Many studies of the relationships between changes in labeled cyclic AMP and platelet function have been carried out using human citrated[4,5,14] or heparinized[1,6,15] platelet-rich plasma containing 3–5 $\times$ 10^8 platelets/ml. In these studies, 9 vol of blood is collected into 1 vol of 3.8% (w/v) trisodium citrate dihydrate or 99 vol of blood is mixed with 1 vol of 0.154 *M* NaCl containing 1000 units of heparin/ml. The blood is centrifuged at 160 g_{av} for 15 min and the supernatant platelet-rich plasma is stored in a sealed siliconized tube at 37°. Loss of platelets may occur if heparinized blood is allowed to cool below 30°. The platelets are labeled by incubation of the platelet-rich plasma with 2 μM [^{3}H]adenine for 90–120 min at 37°. A maximum of 80–90% of the ^{3}H is taken up by the platelets. For accurate measurement of increases in platelet cyclic [^{3}H]AMP in 1-ml samples of platelet-rich plasma, it is sufficient to use [^{3}H]adenine at a specific activity of 1–2 Ci/mmol. As it is not possible to centrifuge platelets from citrated or heparinized platelet-rich plasma prepared as described above without the platelets undergoing activation, experiments with platelet-rich plasma must be performed without further manipulation of the platelets. For the same reason, cyclic [^{3}H]AMP is always extracted from platelets in experimental samples by addition of trichloroacetic or perchloric acid to whole platelet-rich plasma (or washed platelet suspension), rather than to a platelet pellet obtained by centrifugation. [^{14}C]Adenine (~250 mCi/mmol) has often been used to label human platelets[4–6,14] but is more expensive than [^{3}H]adenine. Moreover, its use leads to recovery of fewer cpm in cyclic AMP.

Rabbit Platelets. Studies on cyclic [^{3}H]AMP in rabbit platelets have mostly been carried out using washed preparations labeled by incubation of platelet concentrates with [^{3}H]adenine during the washing procedure.[8,10] This may reflect the slower rate of uptake of [^{3}H]adenine by rabbit platelets (about 60% of that seen with human platelets at the same temperature and platelet concentration). However, it is certainly possible to label rabbit

[13] M. Vanderwel and R. J. Haslam, *J. Clin. Invest.* **76,** 233 (1985).
[14] D. E. Macfarlane and D. C. B. Mills, *J. Cyclic Nucleotide Res.* **7,** 1 (1981).
[15] R. J. Haslam, M. M. L. Davidson, and J. V. Desjardins, *Biochem. J.* **176,** 83 (1978).

platelets with [^{3}H]adenine in a plasma medium. Incubation of citrated rabbit platelet-rich plasma ($4-6 \times 10^8$ platelets/ml) with 0.5 μM [^{3}H]adenine for 60 min at 37° results in maximal (80–90%) uptake of ^{3}H and, provided [^{3}H]adenine with a specific activity of at least 8 Ci/mmol is used, leads to adequate labeling of the metabolic adenine nucleotide pool for experiments with 1-ml samples of platelet-rich plasma.

Preparation and Labeling of Suspensions of Washed Platelets

Rabbit Platelets. These are washed essentially according to the method of Ardlie *et al.*,[16] which can readily be modified to include incubation of the platelets with labeled adenine.[8] Blood is most easily collected by carotid cannulation of animals anesthetized with about 30 mg of sodium pentobarbital/kg injected intravenously as a 6.5% (w/v) solution in 0.154 M NaCl. This avoids the hemolysis observed with some commercial pentobarbital formulations. The blood is immediately mixed with 0.175 vol of ACD anticoagulant, containing 85 mM sodium citrate, 65 mM citric acid, and 111 mM dextrose,[17] which gives a final blood pH of 6.5 and thus suppresses platelet reactivity. Platelet-rich plasma is then obtained by centrifuging the blood at 160 g_{av} for 15 min; this is repeated, if necessary. These steps and all washes are carried out at room temperature. Platelets are isolated from the platelet-rich plasma by centrifugation at 2000 g_{av} for 10 min and resuspended, avoiding any residual red cells, in a modified Tyrode's solution at pH 6.5 (solution I) containing 137 mM NaCl, 2.7 mM KCl, 11.9 mM $NaHCO_3$, 0.42 mM NaH_2PO_4, 2 mM $MgCl_2$, 5.6 mM dextrose, 5 mM PIPES, and 0.35% bovine serum albumin (Sigma, fraction V). For this step only, solution I is supplemented with 0.25 mM EGTA. Solution I is prepared by first mixing 75 vol of water with 5 vol of concentrated stock containing 2.74 M NaCl, 54 mM KCl, 238 mM $NaHCO_3$, and 8.4 mM NaH_2PO_4, and 2 vol of 0.1 M $MgCl_2$. Solid dextrose and bovine serum albumin are then dissolved in this mixture and the pH is adjusted to 6.5 before addition of PIPES (buffered to pH 6.5 with NaOH) and water to give a final 100 vol of this solution.

After resuspension of the platelets in solution I (plus EGTA), the platelet count is determined and the platelets are isolated by centrifugation at 1400 g_{av} for 10 min. They are then resuspended in solution I at 2.5×10^9 platelets/ml and incubated with 2 μM [^{3}H]adenine for 90 min at room temperature. The optimal uptake of ^{3}H ($1-2$ μCi/10^8 platelets) is obtained by using [2,8-^{3}H]adenine (ICN Radiochemicals, Irvine, CA) at its original

[16] N. G. Ardlie, M. A. Packham, and J. F. Mustard, *Br. J. Haematol.* **19,** 7 (1970).
[17] R. H. Aster and J. H. Jandl, *J. Clin. Invest.* **43,** 843 (1964).

specific activity of 25–50 Ci/mmol. With this level of labeling, small increases in cyclic [^{3}H]AMP can readily be measured using only 10^8 platelets. For experiments in which samples containing severalfold more platelets are required, as in aggregation studies, it is more economical to decrease the specific activity of the [^{3}H]adenine to a proportional extent.

After they have been labeled, the platelets are again isolated by centrifugation at 1400 g_{av} for 10 min and are finally resuspended at the desired concentration in a modified Tyrode's solution (pH 7.4) at 37°. This medium (solution II) contains 137 mM NaCl, 2.7 mM KCl, 11.9 mM $NaHCO_3$, 0.42 mM NaH_2PO_4, 1 mM $MgCl_2$, 2 mM $CaCl_2$, 5.6 mM dextrose, 5 mM HEPES, and 0.35% bovine serum albumin and is supplemented with apyrase in most experiments. Solution II is prepared in the same manner as solution I, but with the addition of 1 vol of 0.1 M $MgCl_2$ and 2 vol of 0.1 M $CaCl_2$ in place of 2 vol of 0.1 M $MgCl_2$, and of HEPES buffer, pH 7.4, in place of PIPES buffer. When high concentrations of platelets are required (e.g., 2.5×10^9/ml), 30 μg of apyrase/ml is added to solution II to prevent stimulation of the platelets or inhibition of platelet adenylate cyclase by traces of ADP released from the platelets. Suspensions stored at 37° and containing high platelet concentrations should, however, be used within 3 hr. For these studies, apyrase is prepared by adsorption on calcium phosphate and fractionation with ammonium sulfate, as described by Molnar and Lorand[18] (see also Chapters [1] and [2] in this volume). This material is dialyzed against 0.154 M NaCl, diluted to 3 mg of protein/ml, and stored at −20° until used. Apyrase at 30 μg/ml is usually sufficient to reverse the platelet aggregation caused by 10 μM ADP within 30 sec and must be omitted or used at a much lower concentration in experiments with ADP as the aggregating agent.

Human Platelets. The similar method of Mustard *et al.*[19] for washing human platelets (see Chapter [1] in this volume) can also be modified for labeling the platelets with [^{3}H]adenine. This technique differs from that described for rabbit platelets in the following respects: (1) g_{av} values 15% lower are sufficient for centrifugation of human platelets; (2) all steps are carried out at 37°; and (3) solution II, containing 30 μg of apyrase/ml, is used to wash the platelets, but with the addition of heparin (50 units/ml) when the platelets are first resuspended from plasma (to prevent thrombin formation). Labeling of the platelets with [^{3}H]adenine is also carried out in solution II containing apyrase. To measure cyclic [^{3}H]AMP in experimental samples containing 4×10^8 platelets, suspension containing $1-2 \times 10^9$

[18] J. Molnar and L. Lorand, *Arch. Biochem. Biophys.* **93,** 353 (1961).

[19] J. F. Mustard, D. W. Perry, N. G. Ardlie, and M. A. Packham, *Br. J. Haematol.* **22,** 193 (1972).

platelets/ml is incubated at 37° for 60 min with 2 μM [^{3}H]adenine (5 Ci/mmol). However, similar results are obtained if suspension containing 2.5×10^9 platelets/ml is incubated for 30 min with 2 μM [^{3}H]adenine (10 Ci/mmol). The platelet concentration in the final suspension (usually containing 3 μg of apyrase/ml) should not exceed 10^9/ml.

Labeling of Platelets with [^{14}C]Serotonin. In experiments in which release of dense granule contents by aggregating agents is measured at the same time as changes in platelet cyclic [^{3}H]AMP, rabbit or human washed platelets are preincubated with [^{14}C]serotonin (~50 mCi/mmol) at the same time as [^{3}H]adenine. No ^{14}C derived from [^{14}C]serotonin copurifies with cyclic [^{3}H]AMP. As [^{14}C]serotonin is taken up almost quantitatively, it is adequate to incubate platelets with 0.05–0.1 μCi of [^{14}C]serotonin/10^9 platelets (see also Chapter [17] in this volume). When high platelet concentrations are used, [^{14}C]serotonin should be added at 15-min intervals during the incubation with [^{3}H]adenine, so that the initial concentration of the former does not exceed 1 μM.

Measurement of Cyclic [^{3}H]AMP in Labeled Platelets

Extraction of Labeled Cyclic AMP. Incubations of platelets containing ^{3}H-labeled adenine nucleotides with agents that affect cyclic [^{3}H]AMP levels are carried out at 37° and are terminated by addition of 0.5–2 vol of ice-cold trichloroacetic acid to give a final acid concentration of 10% (w/v). Addition of perchloric acid to give a final concentration of 0.5 M is equally effective. A precisely known amount of radiochemically pure cyclic [^{14}C]AMP (600–1200 dpm) is then added to each experimental sample and all samples are vortexed thoroughly, allowed to stand at 0° for 30 min, and centrifuged. Extraction of platelet cyclic [^{3}H]AMP into the acid supernatant does not require homogenization or sonication of the samples.

Isolation of Labeled Cyclic AMP. This compound is separated from all other labeled nucleotides in the acid supernatants by a modification[8] of a method initially described by Jakobs *et al.*,[20] that involves sequential chromatography on alumina and Dowex 50 resin. Acid extracts are applied to columns (15 cm $\times$ 0.4 cm^2), containing 1.5 g of alumina (Sigma, WN-3) that has been washed with 10% trichloroacetic acid. The columns are then washed with a further 9 ml of 10% trichloroacetic acid, followed by 9 ml of water and 2 ml of 0.2 M ammonium formate (pH 6.0). Labeled cyclic AMP is eluted with a further 3 ml of 0.2 M ammonium formate and the eluates are applied to columns containing 1.5 ml (packed volume) of

[20] K. H. Jakobs, E. Böhme, and G. Schultz, *in* "Eukaryotic Cell Function and Growth" (J. E. Dumont, B. L. Brown, and N. J. Marshall, eds.), p. 295. Plenum, New York, 1976.

Dowex 50 resin (Bio-Rad AG 50W-X8, 100–200 mesh, H^+ form). These columns are then washed with 6 ml of 1 m*M* KH_2PO_4 (adjusted to pH 7.3 with KOH) and labeled cyclic AMP is finally eluted with a further 9 ml of the same solution. Resolution of cyclic AMP from other nucleotides is improved by use of dilute potassium phosphate buffer as the eluant rather than water. The eluates are lyophilized and counted for ^3H and ^{14}C by liquid scintillation.

Purity of the Cyclic [^3H]AMP Isolated. If required, confirmation that the ^3H-labeled compound isolated consists solely of cyclic [^3H]AMP can be obtained by supplementary methods demonstrating that further purification or enzymatic conversion to AMP does not affect the ratio of ^3H to ^{14}C in the product. In the latter approach,[13] some of the material isolated is acidified and rechromatographed on Dowex 50 from which it is eluted with water to remove potassium phosphate. The eluate is then lyophilized and the residue redissolved in 0.1 ml of 1 m*M* cyclic AMP, as carrier. Samples (45 μl) are then incubated for 30 min at 30° with and without 2 units of Sigma cyclic nucleotide phosphodiesterase (in 5 μl of solution containing 100 m*M* TES buffer, pH 7.5, 40 m*M* $MgSO_4$, and 2.5 m*M* EDTA). These incubations are terminated by heating at 100° and cyclic AMP and AMP are then separated by thin-layer chromatography on cellulose, using *n*-butanol/acetone/acetic acid/14.8 *M* NH_3/H_2O (90:30:20:1:60, by volume) as the solvent. Cyclic AMP and AMP are eluted from the cellulose with water and counted for ^3H and ^{14}C.

Typical Results. Values for platelet cyclic [^3H]AMP are corrected for the recovery of cyclic [^{14}C]AMP (50–70%) and are expressed as percentages of the platelet ^3H, which is determined by counting small samples of platelet suspension and of platelet-free supernatant immediately before and after the experiment. Platelet ^3H varies from 92 to 97% of the total ^3H present in washed platelet suspensions, depending on the period for which they are stored before use. The precision of the method is such that when incubations are performed in triplicate, using the same platelet suspension, changes in platelet cyclic [^3H]AMP greater than 10% are usually statistically significant ($p < 0.05$ in a two-sided *t* test). Variations in platelet cyclic [^3H]AMP between experiments are somewhat greater. The basal level of cyclic [^3H]AMP found in suspensions of washed rabbit platelets amounts to $0.024 \pm 0.002\%$ of platelet ^3H and this value is increased to $0.086 \pm 0.006\%$ after incubation of the suspensions for 0.5 min with 1 m*M* IBMX (mean values ± SEM from 30 preparations of labeled platelets). Addition of activators of adenylate cyclase, such as PGI_2, can cause much larger increases in cyclic [^3H]AMP (Fig. 1A) and these increases are enhanced severalfold more in the presence of IBMX (Fig. 1B). The lowest concentration of PGI_2 causing significant increases in cyclic [^3H]AMP in washed

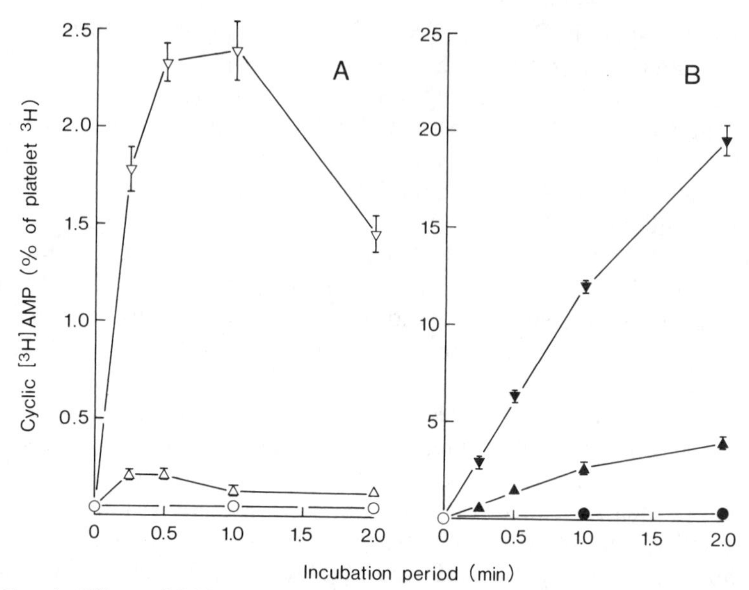

FIG. 1. Effects of PGI_2 on the accumulation of cyclic [^{3}H]AMP in rabbit platelets incubated for varying periods without IBMX (A) and with IBMX (B). Incubation mixtures (0.5 ml) contained 10^8 platelets (labeled with [^{3}H]adenine and suspended in modified Tyrode's solution) and the following additions (final concentrations): ○, none; △, 4 n*M* PGI_2; ▽, 100 n*M* PGI_2; ●, 1 m*M* IBMX; ▲, 1 m*M* IBMX + 4 n*M* PGI_2; ▼, 1 m*M* IBMX + 100 n*M* PGI_2. Values are means ± SEM from three identical incubation mixtures containing the same preparation of labeled platelets.

rabbit platelets is 0.4 n*M*, which is also the threshold concentration at which inhibition of platelet aggregation can be detected under optimal conditions.[8] In the presence of 1 m*M* IBMX, significant increases in platelet cyclic [^{3}H]AMP are observed with PGI_2 concentrations of 0.04 n*M* and above.

Measurement of Circulating PGI_2 in Rabbits

The high concentrations of PGI_2 initially found in circulating rabbit blood by Moncada *et al.*[21] were almost certainly an artifact of the vascular trauma associated with the tendon superfusion assay used by these workers. More recently, bioassay of PGI_2 by the method described here has given much lower control values of about 0.05 pmol/ml, which are well below the levels that affect platelet function.[8] However, the ability of high concentrations of physiological agonists, such as angiotensin II, to release

[21] S. Moncada, R. Korbut, S. Bunting, and J. R. Vane, *Nature (London)* **273**, 768 (1978).

large amounts of PGI_2 into the circulation of rabbits[8] and other species,[22,23] and the possibility that platelet function is modulated locally by PGI_2 released at sites of vascular injury,[24] justify further studies with methods that measure PGI_2, rather than its biologically inactive hydrolysis product, 6-keto-$PGF_{1\alpha}$. Values for the latter, which can be determined by radioimmunoassay, include PGI_2 that has broken down *in vivo* as well as *ex vivo* and, though useful to give limiting values for PGI_2 when levels are low, may exaggerate the inhibitory concentration of PGI_2 in the blood after it has been released by pharmacological means. One group[25] has reported a 6-keto-$PGF_{1\alpha}$ level of 0.75 ng/ml (2 pmol/ml) in rabbit plasma, though another[26] has obtained values of less than 0.1 ng/ml (0.3 pmol/ml). Early values for 6-keto-$PGF_{1\alpha}$ in human plasma were also high and the more recent results indicate that, as in the rabbit, circulating levels of PGI_2 are negligible under normal physiological conditions.[27,28] Thus, the principal value of the method described here is in the measurement of functional PGI_2 (or related compounds) that have been released into the circulation by pharmacological agents.

Cyclic [^{3}H]AMP Formation in Blood in Response to Standard Amounts of PGI_2. Rabbit platelets, for use in the measurement of PGI_2 in blood samples from other rabbits, are washed, labeled with [^{3}H]adenine, and suspended at 2.5×10^9 platelets/ml in a modified Tyrode's solution, as described earlier. The responses of these platelets to standard amounts of PGI_2 are measured after their addition to blood taken from each experimental animal. Usually, 13.5 ml of blood is withdrawn from the central artery of the ear into a syringe containing 1.5 ml of 3.8% trisodium citrate dihydrate; this blood is incubated for 1 hr at 37° in a sealed tube to allow time for complete breakdown of any endogenous PGI_2. Incubation mixtures (0.5 ml) are then constituted, as follows. Samples of citrated blood (0.4 ml) are first mixed with 0.5 μmol of IBMX in 0.04 ml of 0.154 *M* NaCl. The stock solution of IBMX (12.5 m*M*) must be heated immediately before use to dissolve the compound fully. Standard amounts of PGI_2 (0–2 pmol) are then added in 0.02 ml of 0.139 *M* NaCl containing 9.4 m*M* Na_2CO_3 and, exactly 0.1 min later, 0.04 ml of the suspension of labeled platelets (10^8 platelets containing 1–1.5 μCi of ^{3}H) is introduced.

[22] J. Swies, M. Radomski, and R. J. Gryglewski, *Pharmacol. Res. Commun.* **11,** 649 (1979).
[23] K. M. Mullane and S. Moncada, *Prostaglandins* **20,** 25 (1980).
[24] J. G. Kelton, J. Hirsh, C. J. Carter, and M. R. Buchanan, *J. Clin. Invest.* **62,** 892 (1978).
[25] A. L. Cerskus, M. Ali, and J. W. D. McDonald, *Thromb. Res.* **18,** 693 (1980).
[26] H. Bult, J. Beetens, and A. G. Herman, *Eur. J. Pharmacol.* **63,** 47 (1980).
[27] I. A. Blair, S. E. Barrow, K. A. Waddell, P. J. Lewis, and C. T. Dollery, *Prostaglandins* **23,** 579 (1982).
[28] R. A. Forder and F. Carey, *Prostaglandins Leukotrienes Med.* **12,** 323 (1983).

Incubations are continued for 0.5 min at 37°, at which time the reactions are stopped by addition of 1.0 ml of 15% trichloroacetic acid and 1000 dpm of cyclic [^{14}C]AMP. All incubations are performed in triplicate. After the amounts of cyclic [^{3}H]AMP formed are determined, the mean value with IBMX alone (no PGI_2) is subtracted from the mean values with PGI_2 and the linear regression of log (Δ cyclic [^{3}H]AMP) on log[PGI_2] is determined (Fig. 2A). As the K_a for activation of adenylate cyclase by PGI_2 in this system is 30 pmol/ml,[8] this log/log plot is linear over the limited concentration range of PGI_2 used in the assay. The principal source of variation in the effects of standard amounts of PGI_2 derives from slow changes in the responsiveness of the washed labeled platelets. It is, therefore, important that incubations containing PGI_2 standards are carried out not more than 40 min before or after assay of PGI_2 in the experimental samples, as described below.

Inactivation of PGI_2 by Antibody to 6-keto-$PGF_{1\alpha}$. Antibodies directed against 6-keto-$PGF_{1\alpha}$ conjugated to bovine serum albumin have been prepared and shown to block the inhibition of platelet aggregation by PGI_2.[29] These antibodies also prevent the increases in platelet cyclic [^{3}H]AMP caused by PGI_2 (Fig. 2B) and can therefore be used to enhance the specificity of the PGI_2 assay.[8] It is usually satisfactory to use a crude γ-globulin preparation precipitated from antiserum by addition of an equal volume of saturated ammonium sulfate. The precipitate is washed with 50% saturated ammonium sulfate, dialyzed against 0.154 *M* NaCl, and used at a concentration that blocks at least 90% of the increase in platelet cyclic [^{3}H]AMP caused by addition of 2 pmol of PGI_2 to the assay mixture. It is important to demonstrate that, at the concentration used, the γ-globulin does not aggregate preparations of washed rabbit platelets. The specificity of the antibody should be established. In our experience,[8] these antibodies may also bind 6-keto-PGE_1 which, like PGI_2, has been postulated to be a circulating hormone capable of inhibiting platelet function.[30] It may therefore be necessary to use additional methods (see below) to establish that any circulating PGI_2-like material is, in fact, PGI_2 itself.

Assay of PGI_2 in Experimental Blood Samples. Blood must be obtained with minimal trauma to avoid artifactually high PGI_2 values. We have therefore chosen to take blood from progressively more proximal segments of the central arteries of the ears of unanesthetized rabbits, rather than by cannulation of a major artery. Blood samples (2.7 ml) are withdrawn into syringes containing 0.3 ml of 3.8% trisodium citrate dihydrate; only sam-

[29] J. B. Smith, M. L. Ogletree, A. M. Lefer, and K. C. Nicolaou, *Nature (London)* **274**, 64 (1978).

[30] C. P. Quilley, P. Y.-K. Wong, and J. C. McGiff, *Eur. J. Pharmacol.* **57**, 273 (1979).

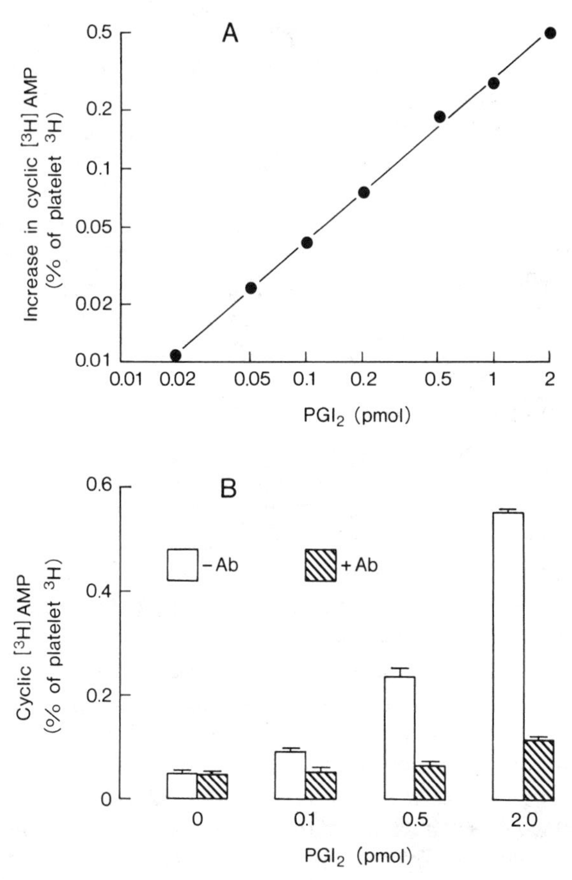

FIG. 2. Effects of standard amounts of PGI_2 on the formation of cyclic [^{3}H]AMP in rabbit platelets in citrated blood; inhibition by antibody to 6-keto-$PGF_{1\alpha}$. Incubation mixtures (0.5 ml) contained 10^8 platelets labeled with [^{3}H]adenine, preincubated citrated blood (0.4 ml), 0.5 μmol IBMX, and the indicated amounts of PGI_2 and were incubated for 0.5 min at 37° without or with 60 μg of γ-globulin from rabbits immunized with 6-keto-$PFG_{1\alpha}$. Values are means $\pm$ SEM from three identical incubation mixtures containing the same preparation of labeled platelets. (A) Increases in platelet cyclic [^{3}H]AMP caused by different amounts of PGI_2 present in blood samples. (B) Cyclic [^{3}H]AMP content of labeled platelets after incubation of blood with PGI_2 alone (open columns) or with PGI_2 and antibody (hatched columns).

ples taken cleanly and quickly are used. Citrated blood (0.4 ml) is pipetted rapidly into six tubes containing 0.5 μmol of IBMX (in 0.02 ml of 0.154 *M* NaCl), three without the above antibody and three with antibody (usually 60 μg in a further 0.02 ml of 0.154 *M* NaCl). Labeled platelet suspension (0.04 ml, 10^8 platelets) is added immediately and the mixtures (0.5 ml) are incubated for 0.5 min at 37° before extraction of cyclic [^{3}H]AMP. All six incubations should be completed within 2.5 min of arterial puncture, to avoid *ex vivo* breakdown of PGI_2 (half-life 9 min in citrated blood[8]). Accurate timing usually requires the coordinated efforts of three individuals. The PGI_2 present in the blood sample is determined from the difference between the mean values for the cyclic [^{3}H]AMP formed in the presence and absence of antibody. This is compared with the effects of standard amounts of PGI_2, determined as above. PGI_2 concentrations in blood are best expressed as picomoles/milliliter (rather than nanomolar) to avoid any implication that the compound is distributed homogeneously.

Measurement of the Half-Life of Presumptive PGI_2 in Blood. As PGI_2 is unstable at pH values below 8.4,[31] measurement of the rate of breakdown of the compound in citrated blood (pH 7.6–7.8) provides a method of confirming the identity of any presumptive PGI_2 found in the above assay. Blood (e.g., 8.1 ml) taken into citrate (e.g., 0.9 ml) is incubated at 37° and assayed for PGI_2 after 0, 10, and 20 min. The breakdown of PGI_2-like material in experimental samples can be compared with that of authentic PGI_2 incubated with citrated blood that has been preincubated at 37° for >1 hr (Fig. 3). The half-lives of the compounds are derived from semilog plots of the results. Whereas PGI_2 has a half-life of 9.3 ± 0.6 min (mean ± SEM, $n = 6$) in citrated rabbit blood, 6-keto-PGE_1 has a half-life of >40 min.[8] The half-lives of the PGI_2-like compounds found in rabbit blood 2 min after intravenous injection of either PGI_2 itself or angiotensin II are consistent with the presence of PGI_2 alone (Fig. 3).[8] Any experimental evidence that the breakdown of PGI_2-like material in blood is slower than that of PGI_2 itself or diverges from an exponential pattern would require a more detailed analysis, in view of the possibility that more than one compound could be present.

Application of Prelabeling Techniques to Studies on Platelets *in Vivo*

Measurement of cyclic [^{3}H]AMP in blood from rabbits transfused with labeled platelets permits the investigator to correlate changes in platelet function and cyclic AMP under conditions that are as nearly physiological as possible. This may be particularly valuable in evaluation of the mechanisms of action of pharmacological agents that inhibit platelet function indirectly, for example by increasing the plasma concentrations of activators of platelet adenylate cyclase, such as PGI_2[13] or adenosine.[32] Measure-

[31] R. A. Johnson, D. R. Morton, J. H. Kinner, R. R. Gorman, J. C. McGuire, F. F. Sun, N. Whittaker, S. Bunting, J. Salmon, S. Moncada, and J. R. Vane, *Prostaglandins* **12,** 915 (1976).

[32] M. Vanderwel, J. K. Hrbolich, and R. J. Haslam, *Thromb. Haemostasis* **54,** 183 (1985).

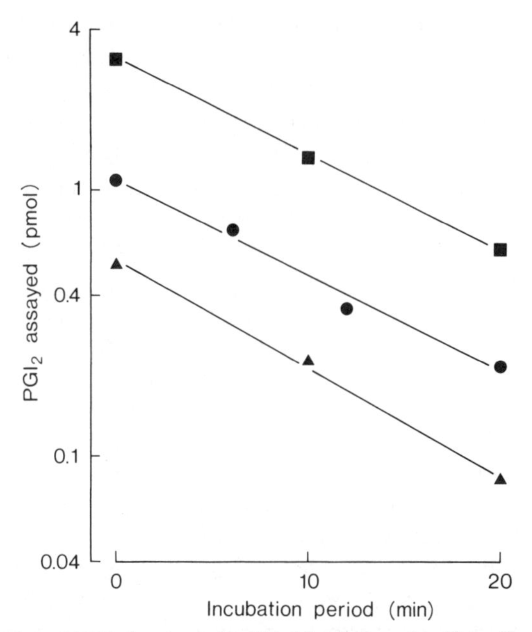

FIG. 3. Stability of PGI_2 in citrated rabbit blood. Samples (0.4 ml) of citrated blood collected over 1 hr beforehand were each incubated with 1 pmol of added PGI_2 and assayed for PGI_2 at the times shown, as described in the text (●). In separate experiments, blood was also taken into citrate 2 min after injection of rabbits with 5 nmol of angiotensin II/kg (■) or 1 nmol of PGI_2/kg (▲) and was incubated at 37° for the indicated times. Samples of citrated blood (0.4 ml) from these incubations were also assayed for PGI_2.

ment of changes induced *in vivo* in the responsiveness of platelets to aggregating agents presents greater experimental difficulties than detection of increases in platelet cyclic [^{3}H]AMP. Platelet function can be evaluated in whole blood *ex vivo* by the electrical impedance method[33] or by measurement of the release of platelet ATP in a lumiaggregometer.[34] However, in our hands, the former method proved unsatisfactory for detection of low levels of PGI_2 in blood,[13] as has also been found by others.[35] The latter technique gave acceptable results in rabbit blood, when platelet-activating factor was used to induce the release of ATP, but lower experimental errors were obtained when the transfused platelets were labeled with [^{14}C]serotonin, as well as [^{3}H]adenine, and platelet degranulation was monitored by the release of ^{14}C into the plasma.[13] In methods based upon the transfusion of labeled platelets, it is important to establish that the labeled platelets are

[33] D. C. Cardinal and R. J. Flower, *J. Pharmacol. Methods* **3,** 135 (1980).
[34] C. M. Ingerman, J. B. Smith, and M. J. Silver, *Thromb. Res.* **16,** 335 (1979).
[35] C. M. Ingerman-Wojenski, J. B. Smith, and M. J. Silver, *Thromb. Res.* **28,** 427 (1982).

functionally identical to the unlabeled platelets in the blood and to demonstrate that all the ^3H and ^{14}C are retained within platelets during the time course of the experiment; this was established for the method we describe.[13]

Preparation of Labeled Rabbit Platelets for Transfusion. Rabbits are anesthetized and cannulated and platelet-rich plasma containing ACD anticoagulant is prepared, as described above. After centrifugation of this material at 2000 g_{av} for 10 min, the platelets from each rabbit are resuspended in 10 ml of the supernatant ACD plasma. This procedure is repeated to remove residual red cells and the suspension, adjusted to contain 2.5×10^9 platelets/ml, is then incubated for 90–120 min at room temperature with 4 μM [2,8-^3H]adenine (25–50 Ci/mmol). When studies on the platelet release reaction, as well as measurements of cyclic [^3H]AMP, are undertaken, 1 μM [^{14}C]serotonin (~50 mCi/mmol) is added every 15 min during this incubation. About 65% of the [^3H]adenine and 85% of the [^{14}C]serotonin are taken up. This procedure is designed to incorporate as much ^3H into the platelet adenine nucleotides as possible, even at the cost of inefficient utilization of the [^3H]adenine. After further centrifugation, the platelets are finally resuspended at 2.5×10^9/ml in fresh ACD plasma containing no pentobarbital (prepared from blood taken from the central artery of the ear of an unanesthetized rabbit). This suspension, containing ~75 μCi of ^3H and ~0.2 μCi of ^{14}C/ml, can be stored at room temperature in a sealed tube for up to 3 hr before injection into other rabbits via the marginal ear vein. These experimental animals (usually unanesthetized) receive 2 ml of platelet suspension/kg, which increases the circulating platelet count by less than 10^8/ml (~20%).

Measurement of Cyclic [^3H]AMP in Blood from Transfused Rabbits. Rabbits are usually transfused with labeled platelets 30–40 min before dosing with the agent under study, so that the cyclic [^3H]AMP in the circulating platelets can be measured in at least two separate blood samples taken beforehand, as well as in several samples taken afterward (e.g., Fig. 4). Small blood samples (2.7–4.5 ml) are most easily taken into 3.8% trisodium citrate dihydrate (0.3–0.5 ml) from successively more proximal segments of the central arteries of the ears of conscious animals, as described for the assay of blood PGI_2. Three 0.8-ml portions of each sample of citrated blood are immediately mixed with 0.8-ml amounts of 20% trichloroacetic acid and 600 dpm of cyclic [^{14}C]AMP. After standing in ice for at least 30 min, these samples are centrifuged. Small fractions of the acid supernatants (e.g., 20 μl) are counted for ^3H to determine the total platelet ^3H in the blood samples and the remainder of each acid supernatant is used for determination of platelet cyclic [^3H]AMP, as described earlier. As the ^3H recovered in cyclic [^3H]AMP under basal conditions is low (200–400 dpm), the latter samples should be counted for at least 30

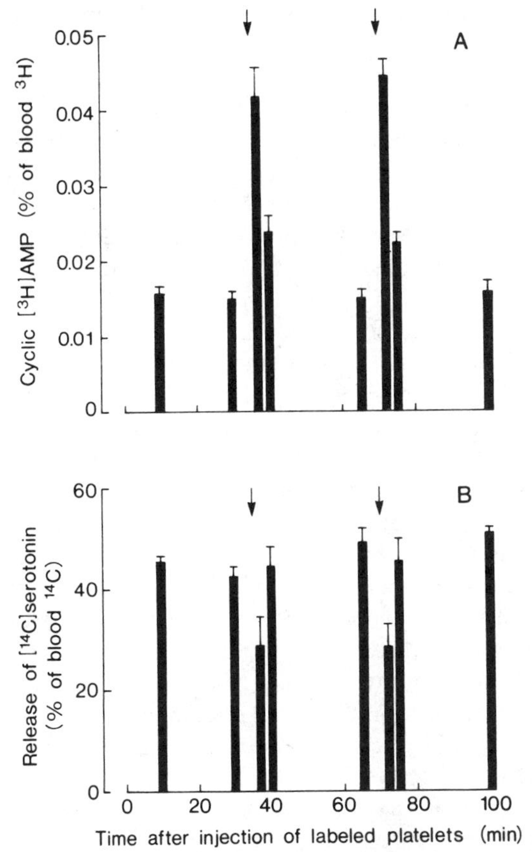

FIG. 4. Effects of two successive injections of PGI_2 into rabbits on platelet cyclic [^{3}H]AMP and on the release of platelet [^{14}C]serotonin by platelet-activating factor. Rabbits that had been transfused with labeled platelets received iv injections of 10 nmol of PGI_2/kg at the times shown by the arrows. Arterial blood samples were taken as indicated, and both blood cyclic [^{3}H]AMP (A) and the release of [^{14}C]serotonin by 40 pmol of platelet-activating factor/ml (B) were measured immediately. Results are shown as means ± SEM from determinations on four to six rabbits.

min. Values for cyclic [^{3}H]AMP (means ± SEM, $n = 3$) are expressed as percentages of the total blood (i.e., platelet) ^{3}H, as in Fig. 4A.

The validity of the method has been established for the 2-hr period following transfusion of the labeled platelets into rabbits,[13] but there appears to be no reason why changes in platelet cyclic [^{3}H]AMP should not be followed for considerably longer. Thus, assuming a blood volume of 65 ml/kg, all the injected platelets are found in the circulation 10 min after their injection and no significant decrease in blood radioactivity occurs for at least 2 hr. After this period, only about 2% of blood ^3H is found in the plasma and no transfer of ^3H to other blood cells can be detected.[13] These results indicate that changes in blood cyclic [^3H]AMP reflect exclusively changes that occur in the transfused platelets. Platelet cyclic [^3H]AMP decreases by about 30% immediately after injection of labeled platelets into rabbits and, in control animals, then remains unchanged for at least 2 hr. This decrease is largely attributable to dilution and warming of the platelets.[13] Injection of PGI_2 (e.g., 10 nmol/kg) after steady-state platelet cyclic [^3H]AMP levels are established leads to a rapid increase in platelet cyclic [^3H]AMP, which returns to control levels within 30 min (Fig. 4A). Identical results are obtained after a second injection of PGI_2 (Fig. 4A), indicating both the reproducibility of the method and the failure of the platelets to desensitize to PGI_2 at this dose level. Injection of angiotensin II causes increases in cyclic [^3H]AMP in circulating platelets[13] that parallel the release of PGI_2 into the circulation.[8] Increased platelet cyclic [^3H]AMP is also detected after injection of dipyridamole (10 mg/kg).[32] These observations provide examples of the sensitivity and application of this method.

Measurement of the Release of Platelet [^{14}C]Serotonin in Rabbit Blood. When citrated blood from rabbits transfused with labeled platelets contains a labile inhibitor of the platelet release reaction, such as PGI_2, measurements of the release of platelet [^{14}C]serotonin must be carried out within 2–3 min of arterial puncture. Samples of citrated blood (0.4 ml) are added to siliconized glass tubes containing 20 μl of 0.84 μM synthetic platelet-activating factor (in 0.154 M NaCl containing 3.5 mg of rabbit albumin/ml). These mixtures are stirred with magnetic stir bars for 0.5 min at 37° in an aggregation module (Payton Associates Ltd., Scarborough, Ontario, Canada). The incubations are then terminated by addition of 0.5 ml of ice-cold 0.154 M NaCl containing paraformaldehyde and EDTA [final concentrations 1.5% (w/v) and 5 mM, respectively], a modification of the method of Costa and Murphy[36] that prevents hemolysis. After centrifugation (0.5 min at 12,000 g), samples of supernatant are counted for ^{14}C to determine the release of [^{14}C]serotonin. For each blood sample, incubations with platelet-activating factor are carried out in triplicate, and a single incubation is carried out with vehicle, to determine the blank ^{14}C that should be subtracted. This is very low in cleanly taken blood samples. To normalize the results, the [^{14}C]serotonin released by platelet-activating factor is expressed as a percentage of the total blood [^{14}C]serotonin, which is extracted by a modification of the method of Anderson *et al.*,[37] as trichloroacetic acid

[36] J. L. Costa and D. L. Murphy, *Nature (London)* **255,** 407 (1975).

[37] G. M. Anderson, J. G. Young, D. J. Cohen, K. R. Schlicht, and N. Patel, *Clin. Chem.* **27,** 775 (1981).

extracts contain little ^{14}C. For each arterial blood sample, three 0.1-ml portions are lysed with 0.6 ml of water and mixed with 0.2 ml of 1 *M* ascorbic acid, followed by 0.2 ml of 5 *M* perchloric acid. After 10 min at 0°, these mixtures are centrifuged. The supernatants are neutralized with 0.08 ml of 10 *M* KOH and, after removal of $KClO_4$, the ^{14}C present is determined by liquid scintillation.

The synthetic platelet-activating factor used by us (1-*O*-octadecyl-2-*O*-acetyl-*sn*-glyceryl-3-phosphorylcholine) causes release of about 50% of the platelet [^{14}C]serotonin in blood at a final concentration of 40 pmol/ml.[13] This permits detection of the inhibitory effects of PGI_2 concentrations in blood in the range of 1–20 pmol/ml.[13] Measurements of the release of platelet [^{14}C]serotonin, carried out in parallel with determinations of cyclic [^{3}H]AMP in blood samples obtained before and after injections of 10 nmol of PGI_2/kg into rabbits, are shown in Fig. 4B. As occasional blood samples contain small clots and must be discarded, we have found it desirable to pool results from experiments with four to six animals, as in this study. The results show that the increases in platelet cyclic [^{3}H]AMP caused by injections of this dose of PGI_2 are associated with brief decreases in platelet responsiveness. Studies using this technique have also shown that injection or release of small amounts of PGI_2 into the circulation can cause a later enhancement of platelet sensitivity that is observed while the platelet cyclic AMP levels are decreasing toward control values.[13]

Acknowledgment

This work was supported by a grant from the Heart and Stroke Foundation of Ontario, Canada (T443).

Author Index

Numbers in parentheses are footnote reference numbers and indicate that an author's work is referred to although the name is not cited in the text.

A

B

D

E

F

G

H

L

M

N

X Y Z

Subject Index

D

G

H

M

N

O

P

Q

R

S

T